Advances in Intelligent Autonomous Systems

International Series on

MICROPROCESSOR-BASED AND INTELLIGENT SYSTEMS ENGINEERING

VOLUME 18



*The titles published in this series are listed at the end of this volume.*

# Advances in Intelligent Autonomous Systems

*Edited by*

SPYROS G. TZAFESTAS

*Department of Electrical and Computer Engineering, National Technical University of Athens, Greece*

KLUWER ACADEMIC PUBLISHERS

DORDRECHT / BOSTON / LONDON

Library of Congress Cataloging-in-Publication Data

ISBN 0-7923-5580-6

---

Published by Kluwer Academic Publishers,
P.O. Box 17, 3300 AA Dordrecht, The Netherlands.

Sold and distributed in North, Central and South America
by Kluwer Academic Publishers,
101 Philip Drive, Norwell, MA 02061, U.S.A.

In all other countries, sold and distributed
by Kluwer Academic Publishers,
P.O. Box 322, 3300 AH Dordrecht, The Netherlands.

*Printed on acid-free paper*



Printed in the Netherlands.

# Contents

## PART 2

*MOBILE, WALKING AND SNAKE-LIKE ROBOTS*

## PART 3

*APPLICATIONS*

**Chapter 19**

AUTONOMOUS AGENTS IN CELLULAR MANUFACTURING

*E. S. Tzafestas*



**Chapter 20**

MAN-MACHINE COOPERATION FOR THE CONTROL OF AN INTELLIGENT POWERED WHEELCHAIR

*G. Bourhis and Y. Agostini*



**Chapter 21**

THE DESIGN, DEVELOPMENT AND OPERATION OF AN AUTONOMOUS ROBOTIC EXCAVATOR

*D. A. Bradley and D. W. Seward*

# Preface

The field of Intelligent Autonomous Systems (IAS) has attracted over the years the attention of numerous research and industrial groups and has by now arrived at an advanced level of development. The results have been achieved through the synergetic use of concepts, techniques and technologies drawn from electrical and mechanical engineering, control engineering, systems science, computer science and management science. Currently, the majority of working systems in practice are of the semi-autonomous type needing some level of human intervention. Therefore much effort is presently devoted in academic, research and industrial environments towards further increasing the level of autonomy.

This book provides a collection of essays which cover the latest research in the IAS field and present a rich set of results accompanied by detailed descriptions of the relevant concepts, tools, techniques and hardware/software designs.

The book contains twenty three chapters grouped in the following parts:

**Part 1**: *General concepts, architectures and technologies*

**Part 2**: *Mobile, walking and snake-like robots*

**Part 3**: *Applications*

**PART 1** involves the first seven chapters which deal with generic issues.

*Chapter 1* (by S. G. Tzafestas) provides some background material, accompanied by a description of two research IAS prototypes, namely a car-disassembly robotic system and a semi-autonomous/autonomous robotic wheelchair.

*Chapter 2* (by G. Bolmsjö, M. Olsson and K. Brink) presents a generic event-based control system structure for the control of a robotic workcell, including its implementation, where the autonomous operation is achieved via reactive replanning and configurable corrections.

*Chapter 3* (by M. Seitz, A. Weigl, K. Hohm and K. Kleinmann) gives an overview of visual, tactile and range sensing techniques for perceiving the required information of the robot work space, and discusses some experiences on sensor-integrated robotic manipulators.

*Chapter 4* (by R. Sanz, F. Matia and E. A. Puente) outlines the ongoing developments on an intelligent-control architecture (ICa) which is a metarchitecture designed to support the design and implementation of IAS.

*Chapter 5* (by W. Jacak and S. Dreiseitl) develops and implements the concept of an autonomous robotic agent which is capable of showing machine (life long) learning and reactive behavior.

*Chapter 6* (by V. Klingspor) describes a distributed planning and plan execution system (SHARC) which applies to a real world robot operational concepts learned from training sequences.

*Chapter 7* (by E. M. Nebot) describes a set of internal and external sensors commonly used for robot navigation with emphasis on land vehicle applications.

**PART 2** starts with *Chapter 8* (by D. Floreano J. Godjevac, A. Martinoli, F. Mondada and J. -D. Nicoud) which gives an overview of the theory, the practice and the applications of autonomous mobile robots (AMRs) with details on the work carried out at the LAMI laboratory of EPFL (Lausanne).

*Chapter 9* (by E. Stella and A. Distante) analyzes key problems related to the use of visual techniques for AMR navigation, considering two approaches: viz. absolute and relative techniques.

*Chapter 10* (by R. Jarvis) is concerned with the interplay between localization, environmental mapping and path planning/execution in support of indoor and outdoor autonomous robotic navigation in time-varying obstacle fields.

*Chapter 11* (by E. M. Nebot) addresses the problem of developing ultra-high integrity navigation systems for large outdoor autonomous guided vehicles.

*Chapter 12* (by S. G. Tzafestas, M. P. Tzamtzi and G. G. Rigatos) presents a brief overview of the existing results on path/motion planning of mobile robots, manipulators and mobile manipulators. Then it develops a solution to the AMR motion planning and control problem in uncertain terrains via a combination of sliding-mode control with the minimum interference strategy.

*Chapter 13* (by K. Watanabe, Y. Shiraishi, J. Tang, T. Fukuda and S. G. Tzafestas) presents a dynamic model for an holonomic omnidirectional mobile robot with orthogonal-wheel assemblies, and develops a resolved acceleration control system based on this model. Practical experimental results are included.

*Chapter 14* (by E. Pereira da Silva, F. Lobo Pereira and J. Borges Sousa) proposes a hierarchical control architecture for AMRs that considers structural, functional and behavioral (operational) directions, and satisfies autonomy, mission reliability, modularity and fault tolerance requirements.

*Chapter 15* (by R. E. Reeve) summarizes the history of walking research from the study of body parts and the construction of clockwork automata to the understanding of neural rhythm generators and the control of robotic somersaulting. The interdisciplinary issues of the field are highlighted and the commercial use of walking robots are discussed.

*Chapter 16* (by M. A. Jiménez, P. Gonzalez de Santos and J. Tabera) presents a technique for changing the crab angle for a quadruped robot based on the wave-crab gait. The algorithm allows a numerical stability analysis to be carried out which is a key requirement in quadruped robot design .

*Chapter 17* (by T. E. Krikochoritis, C. S. Tzafestas and S. G. Tzafestas) deals with the kinematic and dynamic modeling of a 9-link biped robot, and presents two controllers based on the model derived, namely a sliding-mode robust controller and an adaptive controller which are shown via simulation to exhibit similar performances.

Part 2 ends with *Chapter 18* (by K. J. Kyriakopoulos, K. Sarrigeorgides and G. Migadis) which presents the design issues, the components' specifications and a motion planning algorithm for a novel snake-like robot which is kind of multiarticulated mobile robot. Test case results are also included.

**PART 3** presents five applications of IAS. *Chapter 19* (by E. S. Tzafestas) reviews and critically discusses the fundamental issues of the autonomous agent approach to modern cellular manufacturing and describes a case study concerning the tool management giving a comparative study of several motivational schemes for the tool manager robot.

*Chapter 20* (by G. Bourhis and Y. Agostini) discusses the class of intelligent (smart, sensor-based) wheelchairs with particular reference to the compromise between small cost and high robustness, and then focuses on the human-machine interface embodied in the wheelchair developed in the research project VAHM. Experimental results obtained in a test environment are included.

*Chapter 21* (by D. A. Bradley , D. W. Seward) presents the developments made in the Lancaster University Computerised Intelligent Excavator (LUCIE) project giving details of the machine kinematics, the real time control and the design implications.

*Chapter 22* (by P. A. M. Renton and H. A. Elmaraghy, M. Greenspan and H. Zghal) proposes a sensor-based (Plan-N-Scan) technique for collision-free autonomous exploration, gaze planning and workspace mapping using a wrist-mounted laser range camera. It demonstrates that the Plan-N-Scan system is capable of incrementally acquiring range information and successfully scanning both targets and workspace volumes.

Finally, *Chapter 23* (by S. -N. Huang, S. C. Chan and W. Ren) presents a mixed traffic control strategy that allows automatically controlled and manually controlled vehicles to co-exist in a highway. An intelligent vehicle driving system (IVDS) and a manual driver model that mimics the human driver are developed. Results from large scale simulation with many vehicles running on a highway simulator are included.

In overall, the book examines many important elements of modern IAS and provides techniques and tools that are readily applicable in real practice. The book is suitable for the researcher, practitioner and senior student in the field. The editor expresses his thanks to all contributors for their acceptance to offer some of their experience and for their work in preparing comprehensive essays. Special thanks are also due to Kluwer's (Dordrecht) editorial members, especially to Dr. Karel Nederveen and Ms. Catherine Murphy for their assistance throughout the lengthy editorial process.

*Spyros G. Tzafestas*
August 1998

## CONTRIBUTORS


| | |
|---|---|
| **AGOSTINI Y.** | *LASC, Metz Univ. , Iledu Saulcy, France* |
| **BOLMSJO G.** | *PME Dept., Lund Univ., Lund , Sweden* |
| **BORGES SOUSA J.** | *Eng. Fac., Porto Univ., Porto Portugal* |
| **BOURHIS G.** | *LASC, Metz Univ., Iledu Saulcy, France* |
| **BRADLEY D.A.** | *SEECS, Wales Univ., Bangor, U.K.* |
| **BRINK K.** | *PME Dept., Lund Univ., Lund, Sweden* |
| **CHAN S.C.** | *California Univ., Berkley , U.S.A* |
| **DISTANTE A.** | *IESI, Natl. Res. Council , Bari, Italy* |
| **DREISEITL S.** | *Johannes Kepler Univ., Linz, Austria* |
| **ELMARAGHY H.A.** | *Windsor Univ., Windsor, Canada* |
| **FLOREANO D.** | *EPFL., Lausanne , Switzerland* |
| **FUKUDA T.** | *Nagoya Univ. , Nagoya, Japan* |
| **GODJEVAC J.** | *EPFL., Lausanne , Switzerland* |
| **GONZALEZ DE SANTOS P.** | *Inst. Autom. Ind. (CSIC) , Madrid, Spain* |
| **GREENSPAN M.** | *NRC of Canada, Ottawa, Canada* |
| **HOHM K.** | *Darmstadt Univ. Tech., Germany* |
| **HUANG S.-N.** | *California Univ. , Berkley, U.SA.* |
| **JACAK W.** | *Johannes Kepler Univ., Linz, Austria* |
| **JARVIS R.** | *IRRC, Monash Univ., Clayton, Australia* |
| **JIMENEZ M.A.** | *Inst. Autom. Ind. (CSIC) , Madrid, Spain* |
| **KLEINMANN K.** | *Darmstadt Univ. Tech., Germany* |
| **KLINGSPOR V.** | *Informatik, Dortmund Univ., Germany* |
| **KRIKOHORITIS T.E.** | *IRAL, Natl. Tech. Univ. of Athens, Greece* |
| **KYRIAKOPOULOS K.J.** | *Natl. Tech. Univ. of Athens, Greece* |
| **LOBO PEREIRA F.** | *Engrg. Fac., Porto Univ., Portugal* |
| **MARTINOLI A.** | *EPFL., Lausanne , Switzerland* |
| **MATIA F.** | *DISAM, Univ. Polit. Madrid, , Spain* |
| **MIGADIS G.** | *Natl. Tech. Univ. of Athens, Greece* |
| **MONDADA F.** | *EPFL, Lausanne , Switzerland* |
| **NEBOT E.M.** | *Sydney Univ., Sydney, Australia* |
| **NICOUD J. -D.** | *EPFL, Lausanne , Switzerland* |
| **OLSSON M.** | *PME Dept., Lund Univ., Lund , Sweden* |
| **PEREIRA DA SILVA E.** | *Inst. Sup. Eng. do Porto, Porto , Portugal* |
| **PUENTE E.A.** | *DISAM, Univ. Polit. Madrid , Spain* |
| **REEVE R.E.** | *AI Dept., Edinburgh Univ., U.K.* |
| **REN W.** | *California Univ., Berkeley, U.S.A* |
| **RENTON P.A.M.** | *McMaster Univ., Hamilton, Canada* |
| **RIGATOS G.G.** | *IRAL, Natl. Tech. Univ. of Athens, Greece* |
| **SANZ R.** | *DISAM, Univ. Polit. Madrid ,, Spain* |

| | |
|---|---|
| **SARRIGEORGIDES K.** | *Natl. Tech. Univ. of Athens, Greece* |
| **SEITZ M.** | *Darmstadt Univ. Tech.,Germany* |
| **SEWARD D.W.** | *Eng. Dept. , Lancaster Univ., U.K.* |
| **SHIRAISHI Y.** | *ME Dept., Saga Univ., Saga , Japan* |
| **STELLA E.** | *IESI, Natl. Res. Council, Bari, Italy* |
| **TABERA J.** | *Inst. Autom. Ind. (CSIC), Madrid , Spain* |
| **TANG J.** | *Yamaguchi Univ., Tokiwadai, Japan* |
| **TZAFESTAS C.S.** | *IRAL, Natl. Tech. Univ. of Athens, Greece* |
| **TZAFESTAS E.S.** | *IRAL, Natl. Tech. Univ. of Athens, Greece* |
| **TZAFESTAS S.G.** | *IRAL, Natl. Tech. Univ. of Athens, Greece* |
| **TZAMTZI M.P.** | *IRAL, Natl. Tech. Univ. of Athens, Greece* |
| **WATANABE K.** | *ME Dept., Saga Univ., Saga , Japan* |
| **WEIGL A.** | *Darmstadt Univ. Tech., Germany* |
| **ZGHAL H.** | *IMS Centre, Windsor Univ., Canada* |

# PART 1

# GENERAL CONCEPTS , ARCHITECTURES AND TECHNOLOGIES

# 1
# INTELLIGENT AUTONOMOUS ROBOTIC SYSTEMS: SOME GENERAL ISSUES AND TWO REPRESENTATIVE RESEARCH PROTOTYPES

S. G. TZAFESTAS
*Intelligent Robotics and Automation Laboratory*
*National Technical University of Athens*
*Zographou 15773, Athens, GREECE*
*tzafesta@softlab.ece.ntua.gr*

## 1. Introduction

Autonomous robotic systems (manipulators, mobile robots, mobile manipulators) are used in factories to perform difficult tasks such as assembly, welding, painting, material handling etc. and in other real-life environments for various service tasks (domestic, health care etc.) [1,2]. Two basic problems in autonomous robotic systems are : (i) the path and motion planning problem, and (ii) the motion control problem. Given the path provided by the path planner, the robot motion motors/wheels are driven by appropriate controllers, so that the motion is performed smoothly and accurately along the planned path to arrive at the destination points. If a robotic manipulator is equipped with sensors, its trajectory and control can be modified by contact forces or tactile signals occurring during motion (compliant motion control).

The current trend is not to use special-purpose robots that characterized automation in the 1960s, but general-purpose robots that are able to perform a wide repertory of industrial or non-industrial tasks with no change in hardware and only minor changes in programming. To this end, not only must the hardware be general purpose, but also the software must be general enough to simplify the task of reprogramming the robot to perform a new job. In practice, it is very difficult, if not impossible, to anticipate and list all situations which the robot may face in its future work. Therefore, the designer of the robot software program has to specify general classes of situations and equip the robot with sufficient intelligence and problem solving abilities so that it is able to cope with any situation belonging to one of the taught classes. Sometimes, the situation faced by the robot is ambiguous and uncertain, and so the robot must be able to evaluate different possible courses of action. If the robot's environment is not changing, then the robot is given a model of its environment that enables it to predict the results of its actions. If the environment is varying, then the robot must be able to learn it. These and other requirements call for the development and embedding in the robot system of artificial intelligence (AI) capabilities (such as learning, reasoning, and problem solving) [3-5]. These robots are called "Intelligent Robots", and rely heavily, besides AI, on appropriate modern sensors (such as force, tactile, distance, visual, etc.).

In this introductory chapter, we discuss the basic architectural and path planning issues of intelligent autonomous robotic systems (IARS), and we outline two working prototypes, a car disassembly system and an autonomous/semiautonomous powered wheelchair.

## 2. Basic Architecture of Intelligent Autonomous Robotic Systems

*Intelligent Robotic Systems* (IRS) belong to the general class of intelligent machines which are designed on the basis of the principle of *increasing precision with decreasing intelligence* (IPDI) [6], a fundamental feature of hierarchical intelligent controls. The majority of modern-life autonomous systems, including manufacturing systems and process control, are designed and controlled by employing hierarchical control architectures [1,7]. IRSs possess decision-making capabilities, and use multi-sensory information about their internal state and workspace environment to generate and execute plans for carrying-out complex tasks. The general architecture of an intelligent controller involves the following key elements as shown in Fig. 1 [8]:

- Cognition
- Perception
- Planning
- Control
- Sensing
- Actuation

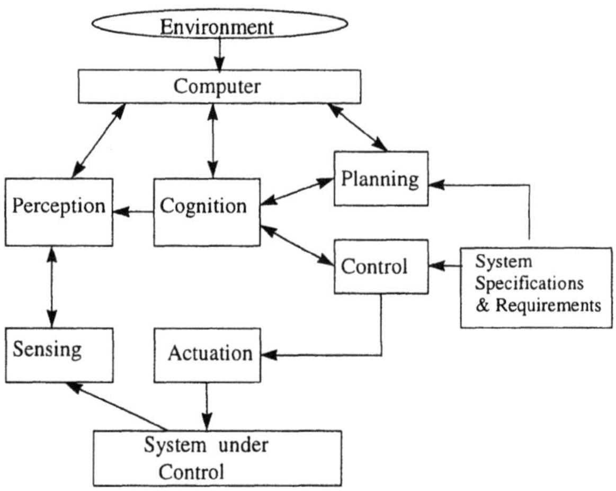

Fig.1: *General Architecture of an intelligent control system.*

The computer of an IRS communicates with the system's environment and performs the tasks of cognition, perception, planning, and control. It also sends information to, and receives information via the sensing elements from, the system under control. The control is executed via the actuation elements. A useful problem solving model used in

IRS is the blackboard model, which is of the distributed type and can integrate several reasoning techniques and several representation schemes, not only complementary, but also as different approaches of the same data. The operation of an intelligent robotic system (like any intelligent control system) is usually structured according to the IPDI principle in at least three main hierarchical levels (from top to down) [7,8].

- Organization level
- Coordination level
- Execution level,

as shown in Fig. 2.

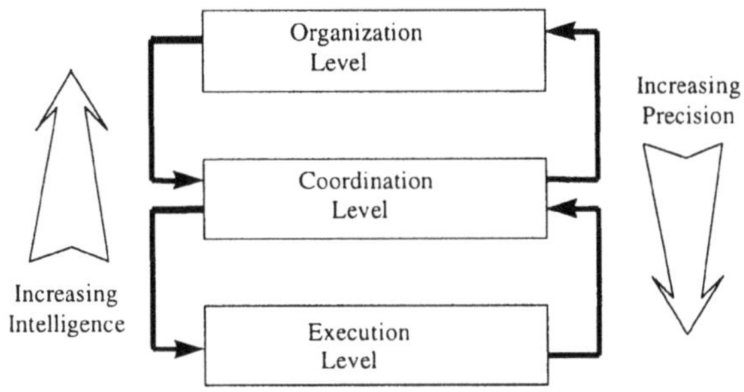


Fig. 2: *Tree-like structure of IRS according to IPDI*

Each level may contain several layers. The organization level implements the higher-level functions (e.g. learning, planning, decision making) which imitate functions of human performance, and can normally be represented and treated by artificial intelligence/knowledge-based techniques as discussed in the next section. The execution level involves the actuators, the hardware controllers and the sensing elements (visual, tactile, etc.). The coordination level consists of several coordinators (each implemented by a piece of S/W or a dedicated microprocessor) which receive the task(s) from the organization level. All necessary details must be provided so that the chosen task's-plan is successfully executed. To execute tasks simultaneously, some kind of parallel-processing task-scheduler is usually employed, which synchronizes the required operations in case they can be executed in parallel. This is found by consultation of a respective parallel execution table.

### 3. Path Planning of Intelligent Autonomous Robotic Systems

Robot path planning techniques are distinguished in *model-based* and *non-model-based* techniques. The first category assumes that all the knowledge about the robot's workspace is available (learned) *a priori* and used to construct geometric models of parts and their relations (known navigation map), whereas the second class assumes that some or all of the knowledge is obtained through sensors. According to the type of obstacles (static or moving) we have *path planning among stationary obstacles* and *path planning among moving obstacles*. To solve the path planning problem among moving obstacles the problem is usually decomposed in two subproblems : (i) plan a path to avoid collision with stationary obstacles, and (ii) plan the velocity along the path to avoid collision with moving obstacles. Robot path planning supports robot task planning by determining and planning collision-free paths for the completion of the task. A solution path in a known environment must be collision free, must be the shortest one, and must have the greatest minimum clearance among the paths along the collision free path. Three kinds of robots have been considered in the literature : robot manipulators, mobile robots, and mobile robots with manipulators (shortly: mobile manipulators). Path planning for mobile robots is distinguished in *global* and *local* path planning. In global path planning, we fix the robot (e.g. we consider it to be a circle or a polygon) and we assume that the environment is arbitrary, but in local path planning the problem is to move an arbitrary shaped robot to pass a low clearance locality such as a door, a corner or a corridor. The effectiveness of both global and local path-planning is measured on the basis of their complexity and their convergence rate. A robot manipulator is usually represented in its joint space, where each manipulator configuration is represented by a point. The sets of points in the joint space that correspond to the configurations of the robotic manipulator that collide with obstacles are called *joint space obstacles.* Once the joint space obstacles are determined, the path planning problem can be transformed to the problem of finding a path for a point from the initial to the goal configuration in the joint space that does not intersect with the joint space obstacles. Using this formulation one can apply for the manipulators' path planning, the mobile robots' path planning techniques. The problem is much more difficult for mobile manipulators since they have more degrees of freedom, which of course give greater maneuverability and higher flexibility in avoiding collisions.

The first accurate (exact, non-heuristic and non-approximate) solution to the global path planning problem of mobile robots was given for the so-called "piano movers" problem by [9-13]. This problem was shown to have a solution in polynomial time with respect to the smooth surfaces of the walls (boundaries and surfaces), the bodies, and the maximal number of equations describing them, but the complexity is exponential with respect to the degrees of freedom (DOF) of the system [9]. The problem was then reduced from three to two dimensions, and the solution was derived by studying the space of all collision-free positions of the robot (with n DOFs). This algorithm has complexity $O(m^5)$ where m is the number of distinct convex walls involved in the original problem [10]. Another version of the piano mover's problem was treated via the *connectivity graph* approach [11]. This problem was to find a continuous motion connecting two given configurations of m 2-dimensional circular bodies in a region bounded by a collection of walls, or else to establish that no such motion exists. The complexity of the solution with respect to the number of walls for a fixed number of moving circles is polynomial. The motion planning of 2-D robotic systems involving multiple arms, all joined at a common end point and free to rotate for passing each other, was solved in [12] with an algorithm of complexity $O(n^{k+4})$ where k is the number of arms. Finally the same authors have solved the problem for the case of moving in 3-dimensional space amidst polyhedral obstacles [13].

Lozano-Perez presents a method for finding the shortest path via the *configuration space* (C-space) concept [14,15]. Conceptually, the C-space shrinks a moving object (robot) to a point and expands the obstacles to the space of the moving object. The crucial step is to compute the C-space obstacles for each obstacle in the workspace. This technique was applied to polygonal (2-D) and polyhedral (3-D) objects. The shortest path from the initial to the final position among the obstacles is the *visibility graph* (V-graph). The C-space approach works well for translational object motions but needs special care for rotational motions [16-18].

Other approaches for global path planning include the following :

- Free space decomposition approach [19]
- Octree representation approach [20]
- Potential field approach [21]

The *free space* is described by generalized cones and the search attempts to find a collision free path of the moving robot inside the generalized cones. The algorithm starts by computing the generalized cones that describe the space. Then the cones are pairwise examined to see if and where their spines intersect. Finally, a path is found by the $A^*$ search algorithm, from the initial to the goal position following the spine intersection. Bourbakis [22] defines the navigation path using the shapes of the current free spaces selecting low-cost corridors in an unknown environment.

An *octree* is a hierarchical decomposition of a cubic space into subcubes, where initially the root node is used to represent the workspace. If the whole workspace is completely filled by objects or completely empty, it is decomposed in eight equal subcubes (octants) which become the children of the root. The procedure is repeated until all nodes are completely filled or until some resolution limit is reached. The search through an octree can be carried out by a combination of various search methods such as hypothesize-and-test, hill climbing, $A^*$, etc.

*Potential field* methods are not based on graph search but they use some artificial potential field functions for the obstacles and the goal positions to control the robot path. A first way to determine the free-to-move space is through the use of repulsive and attractive poles [21]. The repulsive poles cause a robot to move away from an object, whereas attractive poles work in the opposite direction. The big problem in potential field methods is the *trapping at local minima*. Naturally, the planner prefers the low-potential areas, and so the robot may reach a *potential basin* (which is an equilibrium state) and become trapped. An early attempt to overcome the trapping at a potential basin was made by Okutomi and Mori [23]. A new potential-field method based on the vector-field histogram of Borenstein and Koren [24,25] was proposed by Katevas and Tzafestas [26,27]. This method relies on 1-dimensional environment representations named *active kinematic histograms* that are derived by the available information for the obstacle space and the robot. A histogram shifting procedure is applied to determine the kinematic feasibility and result on the robot's steering limits. This active kinematic histogram method is suitable for fast, collision free and kinematically feasible path planning for non-point, non-holonomically constrained mobile robots. The method was applied successfully on a wheel chair (the SENARIO chair [28]) where, for overcoming the local minima traps, a local minima trap detector was developed and used. Newer solutions to the global path planning problem involve the topology-based discretization method [29] and the state-space augmentation method [30].

As already mentioned, in local path planning the problem is to move a robot which is arbitrarily specified through a fixed environment. Thus a local path planner specialized for a fixed environment is a "local expert". A solution to the local path planning problem for a rectangular robot that has to pass a corner in a corridor was given by Lee [31]. If the robot is longer than say "l" then it would not be able to pass the corner. The value of "l" is determined by the so-called "critical configuration algorithm". The problem of moving a chair through a door was solved by Yap [32], by representing the chair by a polygon. The smallest door allowing the chair to pass through it was rigorously determined by an algorithm of complexity $O(n^2)$.

The path planning problem for the robot manipulators can be solved by the C-space technique. The basic issues here are how to map obstacles from the Cartesian space to the joint space, and how to plan a collision-free path for manipulators in a high dimensional joint space. The first issue was treated by Lumelsky [33] and Hwang [34], whereas the second issue can be dealt with via the potential field approach [35] by splitting the potential field into two paths ; one for points inside a C-space obstacle and one for points outside.

The path planning for mobile manipulators and shape-changeable robots needs more sophisticated algorithms. As we saw, a mobile robot is usually approximated by a fixed-shape object (a circle or a polygon), whereas a robot manipulator is represented by points in the joint space. To solve the path planning problem for mobile manipulators and other shape-changeable robots (e.g. multilinked walking robots) a good model for robot and environment representation should be found. The collision-free path planning problem in a two-dimensional environment for a mobile robot possessing two arms was considered and solved by Xue and Sheu [36]. The planner derived involves an algorithm which finds the *collision-free feasible configurations* for the reconfigurable object once the position and orientation of the carried object are specified and fixed (CFFC algorithm), and an algorithm which selects the best candidate path (on the basis of an objective criterion). Then, all the collision-free feasible configurations along the best candidate path are found with the CFFC algorithm, and a collision free path is searched in these collision-free configurations. Other path planning solutions in this area were proposed by Mayorga and Wong [37], Xue et al. [38], and Lee and Cho [39].

We close our discussion on the robot path planning with a look at the case where the robot environment is unknown. In this case no attempt is made to minimize the robot path as in the known environment case. Here, the robot must be equipped with suitable sensors and may or may not have memory. When no memory is used, the path planning is exclusively based on proximity information provided by the sensors, and use is made of algorithmic strategies for moving the robot through the unknown environment. When the system has memory, some kind of learning is incorporated for incrementally constructing a visibility graph of the environment. Learning is accomplished as the robot visits a number of goal points. In both cases (memory, no-memory) the solution is of the heuristic type [40,41,33]. The critical issues in multi-robot navigation (in the same dynamic environment) are : (i) detection of other moving robots, (ii) traffic priority rules, and (iii) communication among robots. For issue (i) the basic questions are how to detect other moving robots in the same space, what is the size of the moving object, what is the motion direction of the moving object, and how many objects exist in the robot area. The first and second questions are answered using visual and sonar sensors, the third question is dealt with by registering the direction of the changes that occur in the shape of the navigation space, and for the final question each robot acquires and defines the changes in different locations of the shape of the navigation space. For issue

(ii) we distinguish two cases : objects of the same velocity and objects moving in different nonparallel directions. In the first case, if the objects move in the same parallel direction they have equal traffic priority. If the objects possess opposite parallel motion, they are given the same priority if the moving lanes are not identical, and if they move on the same lane they are shifted to the right a proper distance if space is available, otherwise one of the objects has to retract for a while until the open navigation lane allows a parallel-shifting function. In the second case (nonparallel directions) the robots estimate the distance which they have to travel until they reach the "anticipated" intersection point. If this distance is large enough to secure no-collision, they continue with the same velocity, but if this is not so, then the robot with the free space has the higher priority, while the other robot has to wait or reduce velocity or change direction. For issue (iii), the robots communicate to exchange information/knowledge about the structure of the common navigation environment in order to be able to improve their future navigation paths. In many cases there exists man-robot interaction, i.e. the human makes decisions that affect the operation of the robot, and the robot provides information to the human which affects these decisions. A complete navigation system for a mobile robot that involves a local navigator, a global navigator, and a map recalling subsystem was presented by Tzafestas and Zikidis [42]. The local navigation (obstacle avoidance and goal seeking) is performed using ultrasonic sensor data fed to an actor-critic type reinforcement learning neurofuzzy controller. The global navigation is accomplished using a topologically orderer Hopfield neural network which works with an ultrasonic sensor-built environment map represented by the so called *eligibility factor matrix* [43].

## 4. An Autonomous Robotic System for Car Disassembly (Neurobot)

This system was designed and built (in prototype) by a European Union Consortium* within the framework of the ESPRIT CIM project NEUROBOT (Neural Network-Based Robot for Car Disassembly) [44]. The system is capable of autonomously dissasembling the components of a car presented to it. It is based on neural network concepts and guided by multiple sensors. The overall architecture of NEUROBOT is shown in Fig. 3 and involves the following basic subsystems:

- Database (DB)
- Path planner (PP)
- Error recovery subsystem (ERS)
- Executor of dissasembly sequences (EDS)
- Sensory subsystem (SS)
- Controller (C)

* TUV, BIBA, ZENON, NTUA, FFT, HUT, KUUSAKOSKI, NedCar.

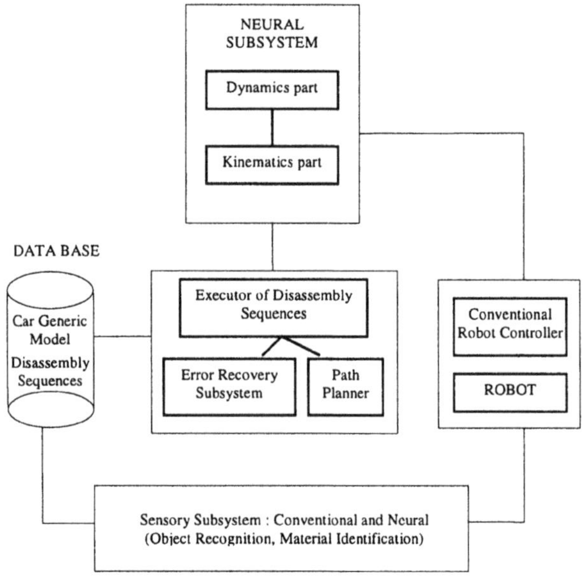


Fig. 3: *NEUROBOT system architecture*

- The DB stores in a unified way the car generic model and the disassembly sequences.
- The PP determines the optimal path on the basis of the desired initial and final positions of the end effector.
- The ERS takes care of the restoration of the operation from possible errors (static and dynamic error restoration).
- The EDS unit acquires the disassembly operations from the DB and supervises their full execution.
- The SS consists of two subunits, namely the object recognition unit and the material identification unit.
- Control: The motion control is performed by a conventional controller enhanced with a neural part based on the adaptive learning architecture.

The heart of the NEUROBOT system is the *Executor of Disassembling Sequences* which is linked with all subsystems and coordinates their operation. It retrieves a disassembly sequence from DB, initiates the disassembly procedure, and then supervises it until it is finished.

The *Object Recognition* unit recognizes and localizes objects randomly placed in the workspace so that the robot may reach them. Parts should not be in prespecified positions or precisely fed to the robot by special mechanisms. This way the resulting work-cell is flexible and easier to set up. Inspection of objects, like measurement of component dimensions and positions or verification of the presence of subassemblies and label, can also be achievable. The object recognition unit functionality has as follows: A vision system is used for data acquisition. The vision data processing module performs image processing and identifies the object. Complementary, the image contour can be extracted, compressed and invariantly transformed and subsequently fed into a Hopfield neural network. A high resolution, grayscale vision system supporting advanced imaging, processing and graphics tools, is used. The system's architecture includes separate processors for image analysis, program execution, and graphics. It is based on V+ multitasking language and operating system, which provides extensive vision tools for image capture, enhancement, and analysis ([45]). Multiple cameras are supported: up to eight physical cameras, with which one or more virtual cameras can be associated.

*Rough Scanning Module:* The object recognition unit is initially activated to perform the rough scanning procedure, a generalized location procedure of the car contour and of approximate positions of components of interest. Its necessity can be depicted from the fact, that when the car is initially placed in the disassembly cell, it is not put to an accurate predefined position, therefore its location relative to the robot is not known. The rough scanning module is designated to provide the system with an estimation of the car's coordinate system origin and locations for further more precise operations. So suggested positions for robot mounted sensor equipment (e.g. cameras, multisensor head) to accuretaly acquire components' info, can be retrieved. The basic system employed is the vision system. Visual sensors positioned in predefined fixed positions in the cell are used: three cameras situated at the top, side, and front of the disassembly cell, provide correspondingly a plan, side, and front view of the vehicle.

*Component Recognition:* The main task of the object recognition unit is to identify the components under disassembly and provide information for their removal. The components may not have their original position or shape. When displacement and deformation of a component is of such degree that it does not allow its identification by the system, it would possibly be unfeasible altogether to remove it with an automatic method. The automotive components have a wide range of positions, dimensions and shapes. Similarly, a large number of statically positioned cameras would be necessary to provide images of the disassembly items. Still in this case the system would not be able to deal with unexpected requirements, that might come-up from conditions conflicting to the manufacturing specifications or from the need for a "closer look". A robot mounted camera was considered as the solution to this problem, as it provides the system with the necessary flexibility in the image acquisition.

*Image Processing:* The image processing is performed in two alternative ways: *Conventional* and *Neural Network based*. In both approaches a feature-based object recognition is accomplished, where the features are pieces of image boundaries, like lines, arcs, corners, and holes.

*Field of View Size and Image Resolution:* One of the difficulties that had to be overcome in the disassembly cell, was that of the recognition of objects with a wide range, e.g. components from the size of a bolt up to that of a windscreen. The vision system, and more specifically the visual sensors, should be able to deal with this variety of dimensions. When choosing the size of the field of view there is a trade-off between image size and image resolution, and a compromise should be made in order to preserve the whole object in the scene, with an acceptable resolution. For the robot mounted camera an autofocus technique was considered the best solution, since images might be captured from various distances from the objects.

*Neural Network-Based Object Recognition:* The use of neural networks for invariant recognition is a topic of extensive research. There are basically two solutions to the problem [46]: (i) a two step procedure, during which at the first step the image is processsed in order to become invariant under geometrical transformation and/or to extract a set of features; at the second step a neural network classifies the image based on these features, (ii) there are neural networks specialized in spatial image recognition, implementing in a great extend biological concepts derived from the human retina system. Due to this, the ability of invariant recognition is inherent to them, but on the other hand they are extremely complicated. Thus these systems are used only in extreme cases.

In NEUROBOT the second approach was followed, with suitable image transformation/feature extraction techniques, and the Hopfield neural network was used as the classifier [47].

*Material Identification Unit:* Knowledge of object materials is important for the car recycling procedure and the material identification task is headed towards this goal. Since a universally applied method is not known yet, the approach followed takes advantage of the fact that differences in the properties characterize materials. A combination of sensors, each of which examines a different property, integrated in a robot compatible head, was the solution adopted. A brief description of the module's functionality has to follows: A multisensor head is used for acquisition of sensor's measurement sets. A preprocessor converts voltage indications to digitized signals. The material data processing module applies a set of rules to the data and estimates the material of the object under examination. Complementary, sensor data are fed into a neural network (Multi Layer Perceptron) for identification of the material.
The sensors employed for the detection of properties characterizing the above materials are the following:

- A *Standard Inductive Proximity Sensor*, which detects metals positioned in the area of its active face. The nominal switching distance refers to targets of mild steel, but it is reduced for non-ferrous metals.
- An *Inductive Proximity Sensor* with Selective Features, which gives the highest sensing distance for non-ferrous metals, while it is lowly or not at all influenced from ferrous metals.
- A *Capacitive Proximity Sensor*, which senses a change in the dielectric medium surrounding its active face, so when an object, with different dielectric constant that the air's, approaches its surface it switches its output. The switching distance varies according to the material.
- An *Ultrasonic Sensor*, which returns an analog signal proportional to the distance from the measuring surface.

After the selection of the sensors, the main challenge was to find a method to integrate them, and an operating scheme that would take advantage of their capabilities in a robot compatible way.

*Neural Network-Based Material Identification:* The output of the sensors that are going to be used in the material identification procedure (inductive, capacitive, etc.) depends on their distance from the pattern under examination. Hence, in order to recognize it we have to take several sets of measurements at different distances. During the measuring procedure it came out that different sensors gave the same output, at the same distance for different materials. As a consequence, for the material classification task the following architecture was designed. The Multi-Layer Perceptron (MLP) FUSER was fed with the outputs of the three sensors (inductive, standard inductive, capacitive) acquired at 20 different distances (Fig. 4). To train and test the network, 40 measurements were taken using the sensor head corresponding to 10 tests per material (steel, aluminum, plastic, textile). For each material, 5 patterns were used for training, and 5 for testing. In every case, the classification accuracy, for both training and patterns was 100%.

Fig. 4: *The neural network fuser of the material-sensor signals.*

The system was proved to work successfully (under carefully selected lightening conditions). It was able to disassemble car tyres, batteries, screens and seats.
A view of the NEUROBOT system at work is shown in Fig. 5.

Fig. 5: *The NEUROBOT system at work.*

## 5. An Autonomous Intelligent Wheelchair System (SENARIO)

The SENARIO (Sensor Aided Intelligent Wheelchair Navigation) system was designed and implemented by a TIDE consortium (1994-97)* and can operate in two modes [28,48,49] :

- Semi-autonomous mode
- Fully-autonomous mode

In *Semi-Autonomous* mode, the system accepts speech commands from the user to move on a direction, or to take an action (e.g. go ahead, turn left, stop, etc.). The system realizes the instructed action, while preventing risk conditions and avoiding obstacles during execution. Each time a risk is detected, SENARIO informs the user and takes appropriate corrective measures to preserve safety and continue with execution of the instruction. In Semi-Autonomous mode, the user can override system actions (e.g. approach closer to a wall than the system's alarm distance). In those cases the system applies minimum speed limit in all instructed commands. In any case, if SENARIO detects an emergency situation, it stops moving and asks for user instruction to recover. So, the responsibility of actions is shared between the system and the user. This mode requires *Risk Detection* and *Risk Avoidance* functionality.

*Fully-Autonomous* mode is a superset of Semi-Autonomous mode. The system accepts all the commands of Semi-Autonomous mode, along with "go to goal" commands. For example, the user can issue commands such as "go to living room". In this case, the system locates itself and then the target position in the environment map. It then plans and executes a path to the specified destination, avoiding all obstacles and risks on the way. During goal execution the user can interfere with the system, as in Semi-Autonomous mode, or he/she can specify a new destination. In this mode, the system takes full responsibility for execution. Full autonomy requires *Path Planning*, along with *Risk Detection* and *Risk Avoidance* functionality. The general architecture of the SENARIO system is shown in Fig.6.

According to this architecture the system performs the following four principal functions :

i. *Task planning* which schedules the execution of all the other processes and is responsible for overall system control. The planning, path following, and goal monitoring procedures reside in the task planner. Task planning is on the top of the control hierarchy of the robot.
ii. *Environment Perception*, where a set of sensors is used to track the interactions for the robot with its environment. Sensor data processing, environment features extraction, positioning, as well as user input, are grouped together as environment perception processes.
iii. *Risk Management*, which is responsible for both risk detection and risk avoidance. This process uses the output of environment perception to detect potential risks and determine their importance. It then computes the necessary sequence of actions for risk avoidance.
iv. *Actuation*, which handles the interface with the robot actuators.

---

* SENARIO Consortium : Zenon S.A.(GR), Microsonic GmbH (DE), NTUA (GR), Reading Univ. (UK), and Montpellier Univ. (FR)

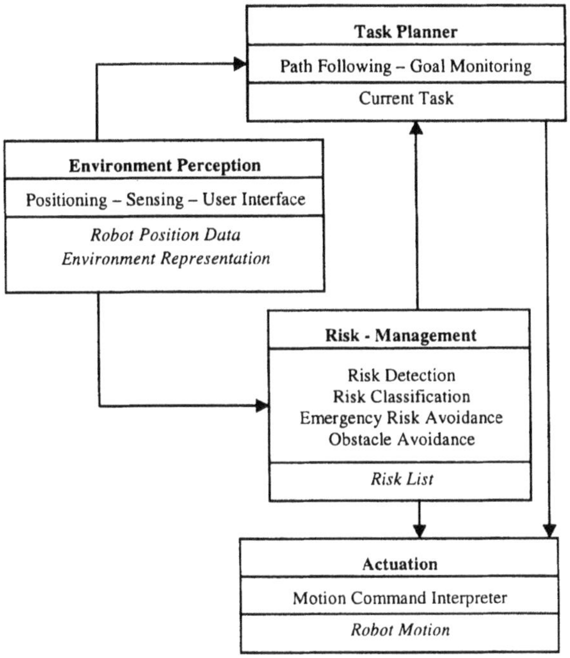


Fig.6: *SENARIO architecture*

This architecture (Fig.6) is an instance of a centralized/hierarchical intelligent scheme, because Task Planning assumes the responsibility for the co-ordination of all the other processes in the system. However, some of the processes can override Task Planning and can communicate directly between them, in cases of emergency. This option offers a distributed alternative to this centralized organization. This hybrid solution was named *virtually centralized control* [50]. Virtually centralized control combines the reactivity of the distributed approach, with the high-level control of the centralized one.

Each process (level) of the hierarchy consists of a series of executive tasks. These are specialized procedures computing the parameters that characterize each process. In particular, Environment Perception consists of the *Positioning* and *Sensing* and *User Interfacing* executive tasks. Similarly, Risk Management is split into *Risk Detection*, *Risk Classification*, *Emergency Risk Avoidance,* and *Obstacle Avoidance* tasks. The Task Planner monitors the overall system functionality through the *Path Following* and *Goal Monitoring* tasks. The Task Planner computes the *current task* of the robot using the Risk Management and Environment Perception outputs.

*Positioning* is the task responsible for reporting the robot's position, whenever it is asked to report, while running individually in its loop. As a task belonging to Environment Perception, it employs sensors and processes their information. The output of this task is referred to the *position estimation* of the robot or, equivalently, to the robot *position data*.

*Sensing* is supplementary to the Positioning task. Both tasks employ sensors, and, occasionally, share the same environmental information. The output of the sensing task is called the *environment representation* for this task. SENARIO supports multiple environment representations ranging from simple combinations of sensor data to occupancy grid representations for the environment.

*Risk Detection* has the dual responsibility to detect: i) external risks threatening the robot and originating from the environment, and ii) risks internal to the control system, such as malfunctions. The detection and reaction methods to these risks are different. The former, allows the subsequent use of Risk Classification, while the later is implemented by low-level, robust, reliable and fast components, that do not require additional processing and that react directly with the actuators through emergency lines.

*Risk Classification* uses a default set of criteria for classifying risks according to their emergency. All the risks identified during the detection process are classified according to the criteria used in the risk classification task. The outcome of these tasks is a *risk list* sorted in decreasing order of emergency. This risk list is further processed, either by Task Planning, or by Risk Avoidance (emergency risk avoidance task), or directly by the Actuation processes, thus supporting the virtually centralized control scheme specified above.

*Obstacle Avoidance* accepts input from three processes : i) Environment Perception (positioning-position estimation, and sensing-environment representation), ii) Risk Detection (risk classification-risk list) and iii) Task Planning (current target position or direction). The task needs only one of the three input sources in order to maintain reliable operation : the environment representation. Any additional information is affecting the robot route, according to the needs of either task planning or risk detection with varying priority. This scheme is another instance of virtually centralized control, due to the fact that either Risk Avoidance or Task Planning have the absolute control of the system, but there is a *supervised distribution of control* among the subsystems based on a dynamic priority ordering.

Summarizing the interactions between Risk Detection and Risk Avoidance, the difference between emergency risk avoidance and obstacle avoidance tasks, should be noticed. Emergency risk avoidance is triggered by an emergency risk situation (i.e. a risk with a highest emergency classification in the risk list), while obstacle avoidance covers the rest of the cases. Obstacle Avoidance uses a local path planning module.

*Actuation* realizes the commands of the rest of the supervised distributed control scheme described above. It consists of a *Motion Command Interpreter* task, which receives commands by the Risk Management and Task Planner tasks in a common format, and translates these instructions into motion commands for the actuators. The output of the Actuation task is identical to the output of the overall system and is referred to as the *Robot Motion.*

The SENARIO testing prototype is shown in Fig.7. It is equipped with proximity (ultrasonic) and orientation (encoder) sensors. The fifteen ultrasonic sensors are connected to two sonar control units which communicate with the control computer via two serial ports. Two sonars are located at the front of the chair to check for obstacles. The others are clustered in eight groups associated to eight proximity sensors located in a ring around the chair (a sensor every $45^0$). A pair of encoders attached to the two back wheels provides a rough estimate of the absolute orientation of the wheelchair.

Fig.7: *The SENARIO testing prototype*

The *user interface* supports speech recognition using an automatic speech recognition module. The module can record and interpret user commands, providing the appropriate control signals to the Task Planner. The emphasis is on user dependent speech recognition, so as to avoid accidental command initiation, due to other people talking nearby. The speech interface is trained during the SENARIO installation to recognize user commands. During operation, voiceprints already stored in the vocabulary lists are matched to the voiceprints of the verbal input. A voiceprint file can contain up to 160 commands. The number of commands can be increased if more RAM becomes available.

For the development of the integrated SENARIO system several new methodologies and technologies were used. These include :

(i) *A novel global path planning method* based on qualitative representations of the robot's interaction with the environment to structure planning and reaction. This method was proved to be better than traditional or reactive path planning approaches in the sense that the planning comes out to be tolerable to sensor/actuator noise and sufficiently reactive to navigate in indoor environments.

(ii) The off-line planning algorithm accepts as inputs a topological diagram of an indoor environment, along with the qualitative map and the robot destination. Based on these points, it computes a set of intermediate map points at doorways. During execution, some of these points form intermediate goal points, called *Attractors*, for reaching the destinations specified by the user.

On-line, the planner accepts as input the current position estimate, consisting of a set of points in the qualitative map. For each point in this estimate, the planner retrieves the active *Preferred* directions and chooses to move the robot in the direction suggested by the majority of points. If there are more than one such direction, the planner chooses the one closest to the current orientation of the robot.

*A separate risk management subsystem* was designed for SENARIO in order to represent and treat explicitly all types of possible risks (internal and external). This subsystem facilitates the evaluation of the capabilities of the system and enables more efficient debugging procedures. Furthermore, risk management uses a local obstacle avoidance method based on the Vector Field Histogram technique suitably enhanced to take into account the kinematic peculiarities of a typical wheelchair (especially the two front castor wheels). This method is used in conjunction with the qualitative map to detect and avoid dynamic obstacles (e.g. people moving) not included during qualitative map construction. When an obstacle (stationary or moving) blocks the current wheelchair path, SENARIO employs a corrective action sequence to avoid it, and keeps cruising on the instructed direction. This sequence satisfies the following goals given in a decreasing priority order : a) protect the wheelchair from all the emergency risks detected, b) avoid the obstacle, and c) return to the direction instructed by the user. Risk Management maintains a risk list that is updated by the risk Classification task. This task accepts the output of risk Detection and labels and sorts all the current risks. A demon in the Risk Avoidance task reads the contents of this list, and issues appropriate corrective actions in a highest-emergency-first order.

Two *novel positioning technologies* were explored and used in SENARIO towards obtaining increased accuracy and reliability. The first is based on the qualitative environment map, while the second uses a stochastic search approach implemented on a connectionist architecture. The *Qualitative Position Estimation* subsystem computes the current position of the wheelchair as a set of points in the qualitative map. Initially, the positioning algorithm receives a set of map points, corresponding to the starting position of the robot. Whenever the wheelchair moves, the algorithm computes a template describing the variation in the behaviour of all proximity sensors. Then, the algorithm computes another template, indicating which of these sensors are currently active and which are not. Both templates are compared with the corresponding templates stored for all points in the previous position estimate and their neighbors. Those that match either one or more closely (i.e. their respective positions hold equal values) comprise the new set of possible locations. The *stochastic search positioning* subsystem is based around two $180^0$ infrared diode range scanners. The scanners produce a range at 0.5 intervals, and so provide a 720-dimensional range vector to the positioning subsystem. The task of this subsystem is to convert the range vector into an environmental position and orientation, *$(x,y,\theta)$ - robot position*, that can be utilized by the risk avoidance subsystem. In addition to the two $180^0$ range finders, passive radio beacons are used to distinguish between rooms that produce the same range profile. Finally, odometery is also used to allow the wheelchair to interpolate between positions determined by the range finders.

The SENARIO wheelchair system was implemented on the commercial wheelchair of Fig. 7 and tested in both semi-autonomous and fully autonomous operation. During the semi-autonomous tests, the user guided the chair manually in several crowded office environments. The chair was able to avoid all possible collisions in these environments. During the fully-autonomous tests, the wheelchair was instructed to move to a specific qualitative map point from a known starting position and orientation. The case in which the wheelchair reached a point in space that was topologically equivalent to its destination was counted as a success. The wheelchair was tested in spaces with average complexity. In all of these cases the wheelchair was successfully navigated between points occluded by various obstacles.

Finally, the SENARIO system was tested in environments with above average complexity, involving points in adjacent rooms connected with narrow corridors (Fig. 8). A 70% success was achieved. The main reason for the 30% failure was caused by areas A and B in this figure. The width of the door and corridor openings in these areas was 1.2 times the width of the wheelchair. This forced the wheelchair to perform a series of manoeuvres in order to fit through the openings. This behaviour combined with the inaccuracies introduced by the front castor wheels of the chair, introduced a great amount of noise in the encoder readings for the robot heading, which can be faced by more accurate sensors (such as laser range finders).

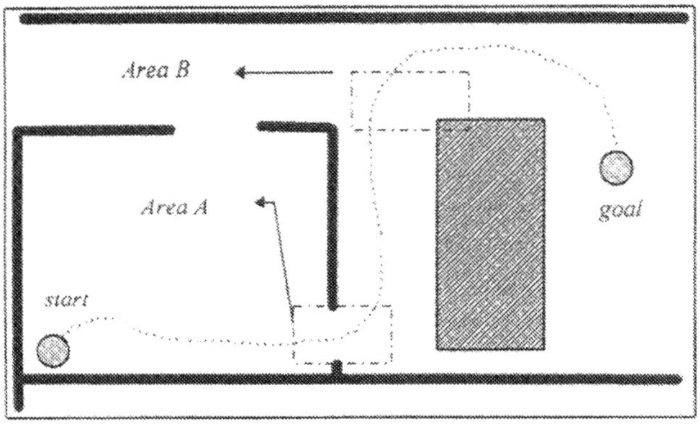


Fig.8: *Trial arena with above average complexity and trajectories*

## 6. Conclusions

This chapter has provided a discussion of some generic architectural issues of IARS and a review of the principal approaches to the solution of the global and local path planning problems. Then, two representative IARS examples were outlined, i.e., the NEUROBOT intelligent car disassembly robotic system, and the SENARIO autonomous wheelchair system.

Planning is a kind of human problem solving, an old well-established area of AI research. In contrast to expert systems, designed as aids to human problem solving in specialized domains using a knowledge-rich approach, planning systems are used most commonly to problems having well-defined logical structures, such as path planning and task planning in robotics. Three basic requirements in any robot path planner/navigator are : (i) detection of other moving objects in the same navigation space, (ii) selection of traffic priority rules, and (iii) intercommunication among cooperating robots. Path planning and navigation involves several subproblems such as : determination of the shortest path, generation of the free space, determination of the critical curves, and determination of the location of predefined obstacles and targets.

The NEUROBOT prototype was able to disassemble several car components (batteries, screens, tyres, seats) with a relatively low speed, due to the delays of the current task planning module. Special care is needed to ensure the required lightening conditions in real industrial environments.

The M3S (Multiple Master Multiple Slave) bus used in the SENARIO wheelchair system increases the accessibility of disabled persons to technical aids through a variety of input devices. Further work is planned for developing flexible "uncertainty management" components based on hybrid fuzzy-conventional technologies and for combining the SENARIO experience with the experiences gained from other intelligent wheelchair prototypes, The ultimate goal is to produce a user/market oriented product as a commercialized wheelchair accessory for the full range of indoor and outdoor powered autonomous wheelchairs.

**References**

[1] Tzafestas, S.G., Ed.: *Intelligent Robotic Systems*, Marcel Dekker, New York, 1991.

[2] Tzafestas, S.G., Koutsouris, D.G. and Katevas, N.I., Eds.: *Mobile Robotics Technology for Health Care Services (Proc. 1st MobiNet Sump.)*, Athens, May, 1997.

[3] Tzafestas, S.G. and Verbruggen, H.B., Eds.: *Artificial Intelligence in Industrial Decision Making, Control and Automation*, Kluwer, Boston/Dordrecht, 1995.

[4] Tzafestas, S.G., Ed.: *Knowledge-Based Systems: Advanced Concepts Techniques and Applications*, World Scientific, Singapore - New Jersey, 1997.

[5] Tzafestas, S.G. and Tzafestas, C.S.: Fuzzy and Neural Intelligent Control: Basic Principles and Architectures, In: S. G. Tzafestas (ed.), *Methods and Applications of Intelligent Control*, Kluwer, Boston/Dordrecht, 1997.

[6] Valavanis, K.P. and Saridis, G.N.: *Intelligent Robotic Systems: Theory, Design, and Applications*, Kluwer, Boston/Dordrecht, 1992.

[7] Tzafestas, S.G., Ed.: *Methods and Applications of Intelligent Control*, Kluwer, Boston/Dordrecht, 1997.

[8] Meystel, A.: Intelligent Control: A Sketch of the Theory, *J. Intelligent and Robotic Systems*, Vol. 2, 1989, pp. 97-107.

[9] Schwartz, J.T. and Sharir, M.: On the Piano Movers Problem: II — General Techniques Topological of Real Algebraic Manifolds, *Advances in Appl. Maths*, Vol. 4, 1983, pp. 298-351.

[10] Schwartz, J. T. and Sharir, M.: On the Piano Movers Problem: I — The Case of 2-Dimensional Rigid Polygonal Body Moving Amidst Polygonal Barriers, *Comm. Pure and Appl. Maths*, Vol. XXXVI, 1983, pp. 345-398.

[11] Schwartz, J. T. and Sharir, M.: On the Piano Movers Problem: III — Coordinating the Motion of Several Independent Bodies: The Special Case of Circular Bodies Moving Amidst Polygonal Barriers, *The Intl. J. of Robotics Research*, Vol. 2, No. 3, 1983, pp. 46-75.

[12] Schwartz, J. T. and Ariel-Sheffi, E.: On the Piano Mover's Problem: IV — Various Decomposable 2-Dimensional Motion Planning Problems, *Comm. Pure and Appl. Maths*, Vol. XXXVII, 1984, pp. 479-493.
[13] Schwartz, J. T. and Sharir, M.: On the Piano Movers Problem: V — The Case of Rod Moving in 3-Dimensional Space Amidst Polyhedral Obstacles, *Comm. Pure and Appl. Maths*, Vol. XXXVII, 1984, pp. 815-848.
[14] Lozano-Perez, T.: Spatial Planning: A Configuration Space Approach, *IEEE Trans. Computers*, Vol. 32, No. 2, 1983, pp. 108-120.
[15] Lozano-Perez, T.: A Simple Motion-Planning Algorithm for General Robot Manipulators, *IEEE J. Robotics and Automation*, Vol. 3, No. 3, 1987, pp. 224-238.
[16] Moravec, H. P.: Obstacle Avoidance and Navigation in the Real World by a Seeing Robot, *Tech Report AIM-340*, Stanford Univ., 1980.
[17] Gouzenez, L.: Strategies for Solving Collision-Free Trajectories Problems for Mobile and Manipulator Robots, *The Intl. J. Robotics Research*, Vol. 3, No. 4, 1984, pp. 51-65.
[18] Hatzivassiliou, F.V. and Tzafestas, S.G.: A Path Planning Method for Mobile Robots in a Structured Environment, In: S. G. Tzafestas (ed.), *Robotic Systems: Advanced Techniques and Applications*, Kluwer, Dordrecht-Boston, 1992, pp. 261-270.
[19] Brooks, R.: Solving the Find-Path Problem by Good Representation of the Free Space, *IEEE Trans. Systems, Man, and Cybernetics*, Vol. 13, No. 3, 1983, pp. 190-197.
[20] Herman, M. Fast 3-Dimensional Collision-Free Motion Planning, *Proc. IEEE Intl. Conf. Robotics and Automation*, San Francisco, Calif., 1986, pp. 1056-1063.
[21] Andrews, J.R. and Hogan, N.: Impedance Control as a Framework for Implementing Obstacle Avoidance in a Manipulator, *Proc. ASME Winter Annual Meeting on Control of Manufact. Processes and Robotic Systems*, Boston, 1983, pp. 243-251.
[22] Bourbakis, N.A.: Heuristic Collision Free Path Planning for an Autonomous Navigation Platform, *J. Intell. & Robotic Systems*, Vol. 1, No. 4, 1989, pp. 375-387.
[23] Okutomi, M. and Mori, M. Decision and Robot Movement by Means of a Potential Field, *Advanced Robotics*, Vol. 1, No. 2, 1986, pp. 131-141.
[24] Borenstein, J. and Koren, Y.: The Vector Field Histogram — Fast Obstacle Avoidance for Mobile Robots, *IEEE Trans. Robotics and Automation*, Vol. 7, No. 3, 1991, pp. 278-288.
[25] Borenstein, J., and Koren, Y.: Histogramic In-Motion Mapping for Mobile Robot Obstacle Avoidance, *IEEE Trans. Robotics and Automation*, Vol. 7, No. 4, pp. 535-539, 1991.
[26] Katevas, N.I., and Tzafestas, S.G.: Local Mobile Robot Path Planning: A New Technique Based on Active Kinematic Histogram, *Proc. 1st Mobile Robotics Technology for Health Care Services Symp. (MOBINET '97)*, Athens, Greece, May 1997, pp. 195-204.
[27] Katevas, N.I., and Tzafestas, S.G.: The Active Kinematic Histogram Method for Path Planning of Non-Point, Non-Holonomically Constrained Mobile Robots, *J. Advanced Robotics*, 1998 (In Press).
[28] Katevas, N.I., Tzafestas, S.G., Koutsouris, D.G., and Pneumatikakos, C.G.: The SENARIO Autonomous Navigation System, In: S.G. Tzafestas, D.G. Koutsouris

and N.I. Katevas (Eds.), *Mobile Robotics Technology for Health Care Services (Proc. 1st MobiNet Symp.)*, Athens, 1997, pp. 87-99.

[29] Lück, C.L.: Self-Motion Representation and Global Path Planning Optimization for Redundant Manipulators Through Topology-Based Discretization, *J. Intell. & Robotic Systems*, Vol. 19, No. 1, 1997, pp. 23-38.

[30] Zhou, Z.-L., and Nguyen, C.C.: Globally Optimal Trajectory Planning for Redundant Manipulators Using State Space Augmentation Method, *J. Intell. & Robotic Systems*, Vol. 19, No. 1, 1997, pp. 105-117.

[31] Lee, C.-T.: Critical Configuration Path Planning and Knowledge-Based Task Planning for Robot Systems, *Ph.D. Thesis*, School of Electr. Engrg., Purdue Univ., 1990.

[32] Yap, C.-K.: How to Move a Chair Through a Door, *IEEE J. Robotics and Automation*, Vol. 3, No. 3, 1987.

[33] Lumelsky, V.J.: Dynamic Path Planning for a Planar Articulated Robot Arm Moving Amidst Unknown Obstacles, *Automatica*, Vol. 23, No. 5, 1987, pp. 551-570.

[34] Hwang, Y.K.: Boundary Equation of Configuration Obstacles for Manipulators, *Proc. 1990 IEEE Intl. Conf. on Robotics and Automation*, 1990, pp. 298-303.

[35] Warren, C.W.: Global Planning Using Artificial Potential Field, *Proc. 1989 IEEE Intl. Conf. on Robotics and Automation*, 1989, pp. 316-321.

[36] Xue, Q., and Sheu, P.: Collision-Free Path Planning for Reconfigurable Robot, *Proc. 4th Intl. Conf. on Advanced Robotics*, Columbus, Ohio, June, 1989.

[37] Mayorga, R.V., and Wong, A.K.C.: A Robust Method for the Concurrent Motion Planning of Multi-Manipulator Systems, *J. Intell. & Robotic Systems*, Vol. 19, No. 1, 1997, pp. 73-88.

[38] Xue, Q., Sheu, P., Maciejewski, A., and Chen, S.Y.P.: Planning Collision-Free Paths for Reconfigurable Dual Manipulator Equipped Mobile Robot, *J. Intell. & Robotic Systems*, Vol. 17, No. 1, 1996, pp. 223-242.

[39] Lee, J.K., and Cho, H.S.: Mobile Manipulator Motion Planning for Multiple Tasks Using Global Optimization Approach, *J. Intell. & Robotic Systems*, Vol. 18, No. 2, 1997, pp. 169-190.

[40] Tzafestas, S.G., and Stamou, G.B.: A Fuzzy Path Planning Algorithm for Autonomous Robots Moving in an Unknown and Uncertain Environment, *Proc. EURISCON '94 : The 2nd Europ. Robotics and Intelligent Systems Conf.*, Malaga, Spain, Vol.1, pp.140-149, August, 1994.

[41] Lumelsky, V.J., and Stepanov, A.A.: Dynamic Path Planning for a Mobile Automaton with Limited Information on the Environment, *IEEE Trans. Autom. Control*, Vol. 31, No. 11, 1986, pp. 1058-1063.

[42] Tzafestas, S.G., and Zikidis, K.C.: Complete Mobile Robot Navigation via Neural and Neurofuzzy Control, *Proc. First Mobile Robotics Technology for Health Care Services Symp. (MOBINET '97)*, Athens, Greece, May, 1997, pp. 205-217.

[43] Elfes, A.: Sonar-Based Real-World Mapping and Navigation, *IEEE J. Robotics and Automation*, Vol. 3, No. 3, 1987, pp. 249-265.

[44] Tzafestas, S.G., Anthopoulos, Y., Katevas, N., and Spyropoulou, E.: Architecture and Implementation of an Autonomous Car-Disassembly System, *Systems Analysis, Modeling, Simulation (SAMS)*, Vol. 29, 1997, pp. 129-149.

[45] ADEPT Vision AGS-GV, Adept Technology Inc., San Jose, CA, USA, User's Guide and Reference Guide.

[46] Delopoulos, A., Tirakis, A., and Kollias, S.: Invariant Image Classification Using Triple-Correlation-Based Neural Networks, *IEEE Trans. Neural Networks*, Vol. 5, No. 3, May 1994.

[47] Haykin, S.: *Neural Networks: A Comprehensive Foundation*, Mcmillan, New York, 1994.

[48] Van Woerden, J.A.: A Safe and Easy to Use Integrated Control and Communications Method: M3S, In: E. Ballabio, I. Placencia-Porrero, R. Puig de la Bellasca (eds.), *Rehabilitation Technology*, IOS Press, Amsterdam, 1993, pp. 75-80.

[49] Katevas, N.I., Sgouros, N.M., Tzafestas, S.G., Papakonstantinou, G., Beattie, P., Bishop, J.M., and Tsanakas, P.: SENARIO: A Sensor-Aided Intelligent Navigation System for Powered Wheelchairs, *IEEE Robotics and Automation Magazine*, 1997.

[50] Katevas, N.I., and Tzafestas, S.G.: Analysis of the Risk Detection and Avoidance Architecture for Indoor Autonomous Mobile Robots, *Proc. First ECPD Intl. Conf. on Advanced Robotics and Intelligent Automation*, Athens, Greece, pp. 464-470, 1995.

# 2

# INCREASED AUTONOMY IN INDUSTRIAL ROBOTIC SYSTEMS: A FRAMEWORK

GUNAR BOLMSJÖ, MAGNUS OLSSON
and KRISTER BRINK
*Dept. of Production and Materials Engineering*
*Lund University*
*P.O. Box 118, SE-221 00 Lund, SWEDEN*

## 1. Introduction

Programming of industrial robots by traditional methods results in sequential programs with predefined actions. The programs are in general interpreted by the control system and the actions described in the programs are carried out in the work-cell [12]. It is in most cases possible to attach sensors to the control system for receiving work-cell information. However, in today's industrial robotic systems sensors have limited influence on the present realization of a primitive operation. Moreover, present sensor integration does not support future use of the program in similar tasks. In practice, most problems related to sensors are connected to limited robustness in both the sensor itself and the controlling actions initiated by the controller due to sensor signals. An example of such problems is a laser triangulation sensor used for seam-tracking in arc welding operation where every weld joint with its different geometrical and optical properties need adaptations in the set-up of the sensor.

The adaptivity of a robot control system is often described as control parameters depending on load, configuration etc. In the context of interaction with the environment, adaptivity also include the capability to take advantage of real time information from the work-cell, such as the work process the robot is involved in, information that typically comes from sensors. Most of the existing robot control systems are rather static, meaning that the adaptivity with respect to the environment is very low. Even if sensors are used to monitor process features, it is in the general case difficult to take advantage of this information to take proper reactive action due to the complexity of processes in the real world. For example, in GTAW[1]-orbital tube welding, a list of all possible practicable parameters totals

[1] Gas Tungsten Arc Welding

more than 80[5].

Several solutions for the improvement of robot programming and control are suggested in the literature. In [7] a hierarchical approach to the control system with a division of functionality at different levels is described. A control structure consisting of two parts, a scheduler and a supervisor, is presented in[11]. In this approach the scheduler utilize fuzzy logic and the supervisor consist of several levels using Petri net representation. Methods for solving reactive related planning problems are discussed and exemplified in [16]. For instance, methods to determine the execution order to achieve a complete task. In [8] a task directed sensor planning system is presented which have sensor processing strategies based on the desired robot task to be executed. Further approaches to robotic programming and control systems are discussed in [4, 15]. An interesting architecture for open robot control supporting different programming levels is presented in [13].

The presented control system structure is applicable on almost all types of robots. Most examples in this chapter concern industrial robots, especially arc welding application processes and associated sensors. Our experimental equipment consist of: VME based computer system and workstations, simulation software from DENEB Robotics inc.[2], ABB industrial robot, Servo Robot laser sensor and Migatronic arc welding power source.

Sensors in robot systems collect information from a part of the working area and share this information with the control system. Several types of sensors can be used in robot work-cells, for instance ultrasonic sensors, optical sensors and force sensors. A robot also provides sensor information about its state such as position and joint velocities.

When sensors are used together with a traditional robot control system the sensor information is usually used for either motion corrections or control of application processes. In reality some information is valuable for both motion and process control, for instance weld joint geometry information [10]. The presented event based robot control structure makes the sensor information available for all modules in the control system. The system is exemplified by an arc welding application in section five.

## 2. Task oriented programming

The event based robot control system structure presented in this chapter has requirements on the format of a task description applied within the framework of autonomous robots. This section is a brief description of the task oriented robot programming method which combines a geometrical model approach with process models for specific tasks to performed [3, 15]. The method focuses on the objects in a robot work-cell, their states and possibilities for state transitions or events

[2]In the work described in this chapter the work was mainly based on IGRIP and Envision TR simulation software

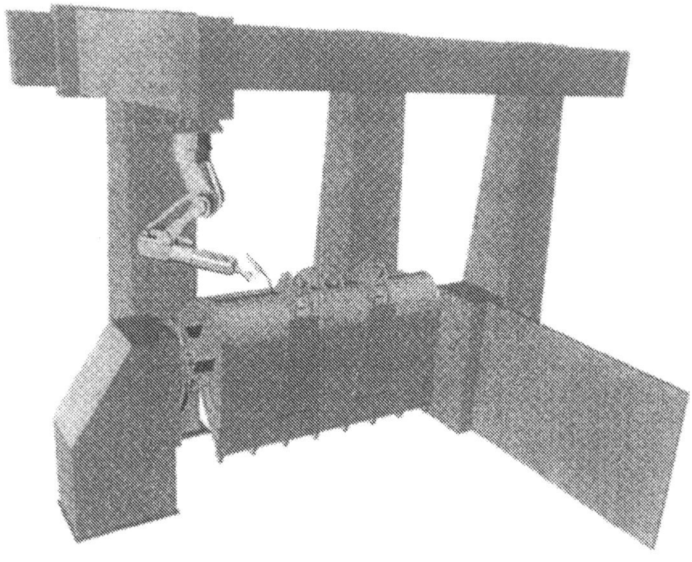

**Fig. 1.** *The world model of the case project, displayed in IGRIP.*

and it is assumed that a task list and description or models related to subtasks exist including planning mechanisms.

A programming session in the presented task oriented programming system includes creation of a *World model* through identification of all *Manipulated Objects in the Robot Work-cell* (MORW's) and their states (it is worth noting that MORW's can be created both through a traditional modeling approach similar to CAD work as well as autonomous creation through sensor data which detects and identifies specific geometrical parts and their location in the work-cell). Furthermore, the task is described through specification of sub-tasks. A sub-task is an aggregation of executable events possible for the control system to realize. Both MORW's and sub-tasks may be inherited from libraries with predefined MORW's and sub-tasks.

Sub-tasks are normally considered to be defined and interactively attached to MORW's in a graphical interactive way of the *World model* in IGRIP on a workstation, see figure 1. The purpose of creating a new programming environment is to provide re-usability, maintainability and reliability to robot program code, all key factors for efficient programming. An important aspect on re-usability is that existing world models, MORW's and sub-tasks can be re-used. The main part of the *World model* is identical for all tasks in the same work-cell. Sub-tasks have an even wider area of re-use because they consist of logical methods applicable in any work-cell which include the same kind of sub-task. The hierarchical structure of the model simplifies the maintenance of programs and submits modifications to the appropriate layer. The reliability of the programs is increased through simulation where logics in the task-plan and motion related characteristics (work space, collisions, singularities, etc.) can be checked.

The result of the programming session is a *World model* and a ***Task-plan*** where the *World model* is a description of all significant physical objects in the environment and their possible states. The ***Task-plan*** is a specification of the work to be done, expressed as state transitions for involved MORW's.

A general problem related to task programming is the "loose" coupling between the task model and the real robot system. To validate a task plan it must be possible to create a fairly true simulation of the system which represent the actual activities. There are several problems involved in this including mechanical systems with sensors, programming languages and translators, emulation of controller characteristics, system behaviors close to singularities and redundant systems, etc. Such problems must be handled separate as they all represent important but complex issues towards the creation of autonomy in robotic systems.

### 3. Event based robot control

Traditional industrial robot control systems interpret sequential manipulator level robot programs and control in general one single robot (in some cases two robots with possibilities for synchronization of processes), perhaps including some of its application process equipment. Robot control systems used in industry are constructed for high repeatability of motions and production reliability, not for communication with other control systems nor for handling of major changes in the environment.

An integrated control of all equipment within a robot work-cell and an increased autonomy with respect to changes in the environment would give rise to several advantages:

- An improved adaptability to changes in the environment, variations in the process equipment, deviations in workpieces, etc., maintaining the desired quality.
- Extended application process control through robot related process parameters such as travel speed in gluing applications and stand off in arc welding.

In industrial robotic systems these advantages of a high level task control system with extensive autonomy are very important. As an approach on such a system the event based robot control structure is presented in this chapter.

The world model and task-plan, created in a task oriented robot programming session described in the previous section, constitute the input to the event based control structure. These inputs are model descriptions of the work-cell, its conditions and the task to be performed respectively. To reach a high adaptation to the work-cell environment the MORW's in the world model must be up to date, since the task performance depends on the model. The world model is updated by information from sensors and notified when an executable event is realized. The functionality in the presented control system most closely related to sensors is described and exemplified in [4].

## 3.1 CONTROL STRUCTURE

Execution of a task-plan mentioned in section two needs a supporting control system. The control system is event driven and operates with MORW's as a base for the realization of events. The control system search for executable events which fulfill preconditions in the task-plan. Several executable events may be in progress at the same time. The *World model* is, when needed, consulted for information along the realization of an event in the work-cell.

The event based control system operates down to joint level. Everything below joint level, for instance motion control and dynamic issues, are solved by traditional methods. The layout of the control structure proposed in this chapter is shown in figure 2 where the frames symbolize processes and the connecting arrows represent the flow of information. It also in part reflect the work presented in the model based event control presented.

The control system interprets a *World model* and a *Task-plan* to execute a task defined by a user. The world model describes the objects in the *Robot work-cell* as parameters related to geometry, material and process. A task is expressed as executable events in the task-plan, as described in previous section, but includes no exact work-cell data like positions nor the time for firing of specific executable events. The positions are requested by the concerned module from the MORW model during task realization and the control system decide when an executable event will be fired. The context within the task-plan forms a number of preconditions for every executable event. An executable event is fired when its preconditions are fulfilled. Typical preconditions are object in position and robot available with suitable end-effector (tool/gripper). All executable events, also those concerning control of application processes, can request *World model* information.

The capacity to solve unexpected problems autonomously depends on the amount of appropriate knowledge in the task-plan. The knowledge in the presented system is increased for every new task described as in case-based reasoning, where new problems are approached by remembering old similar ones and building on those solutions [9].

## 3.2 CENTRAL MODULES OF THE CONTROL SYSTEM

The event based control system has several central modules, *Event dispatcher*, *Task scheduler* and global *Path planner*. In general, an *Event dispatcher* is a module that handles messages in an event based system. The *Event dispatcher* is in this

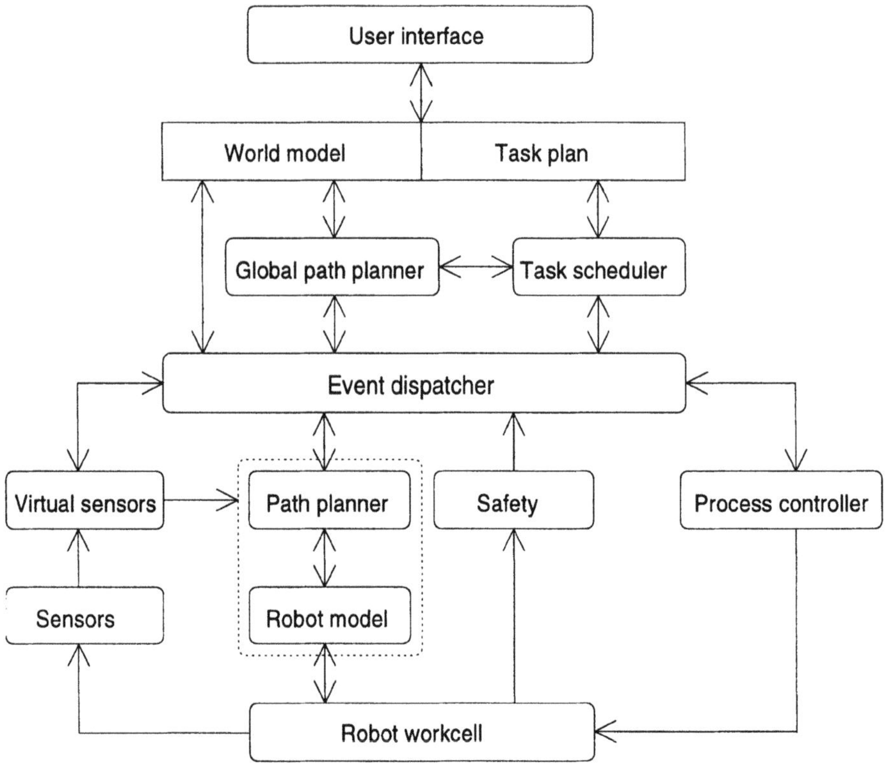


**Fig. 2.** *Layout of the event based robot control system including connections to process controller and sensors.*

event based robot control system also a supervisor for the execution of tasks.

The *Event dispatcher* directs messages to appropriate processes in the control system. All events passing are represented as messages and sent to the *Event dispatcher.* Every message has a priority and is perhaps requested, which the dispatcher must take into consideration. A request is created when several executable events are waiting on the same event to fulfill their preconditions.

The information in the *Task-plan* is a non-sequential description of the task to be realized. The information consists basically of stand alone sub-tasks and executable events expressed in state transitions of MORW's. Preconditions for the sub-tasks and the executable events are requirements on the state of one or several MORW's. These preconditions represent a net of rules for the execution of the sub-tasks. To realize the task a solution to the net is required [3, 15]. The *Task scheduler* interprets the net and schedules a solution for the task. Already during the task oriented programming phase when sub-tasks and executable events are stated the *Task scheduler* is involved scheduling for logical solutions on the net of the task.

The *Task scheduler* collaborates closely with the *Global path planner* during the net solving process. Gross motion planning based on the information in the *World model* is performed in advance by the *Global path planner* for estimation of motions connected to executable events and sub-tasks. However the whole task is not scheduled in advance by the task scheduler; it is only investigated for avoiding deadlocks in the solution.

Executable events which are waiting to be fired are sent by the task scheduler from the task-plan to the *Event dispatcher*. The final firing of executable events with fulfilled preconditions are managed by the *Event dispatcher*. The event concept with preconditions enables advanced features of the system; if the system has a lack of work to carry out within a sub-task while waiting for preconditions to be fulfilled, executable events of other sub-tasks can be fired.

The autonomous functionality of the system is divided into two parts, reactive re-planning and configurable correction. The reactive re-planning take care of major deviations in the work-cell providing sensors are capable to detect these deviations. The event dispatcher report to the *Task scheduler* when an executable event fails to reach its goal or when sensor information is outside the acceptable limits. Such a report initiates the *Task scheduler* in collaboration with the *Global path planner* to find a new solution for the remaining part of the task. The control system has ability to autonomously search for and create new solutions of problems appearing during task execution and reactive re-planning.

Second main part of the autonomous functionality of the system is the configurable correction. Its main purpose is to correct and compensate for small deviations in the work-cell. Configurable correction consist mainly of two ways of interaction within the system; sensor supporting the application *Process controller* and sensors supporting the robot motion controller. Some basic and limited interaction between sensors and application process controllers exists today, but this system provides a configurable connection between information measured in the work-cell and controllable process parameters.

The configurable interaction between sensors and robot motion controller enables unique possibilities for improved performance of application processes and less sensitivity to disturbances through a more open control system. This motion controller interaction is further discussed and exemplified in section four and five.

A concrete example on reactive re-planning in a work-cell with equipment for arc welding and one single industrial robot including collision avoidance sensors on a wrist will be given. The robot carry out arc welding of fillet joints on ship sections. In many cases one of the welding plates require a strut to keep the plates in position during welding. The strut is placed in advance by manually weld it to the plates so it closes a triangle together with the fillet joint. Thus, it constitute a serious obstacle for the robot which must be taken into account during operation. When the collision avoidance sensors detect a strut the arc welding process is interrupted and a reactive re-planning is carried out. A new trajectory around the strut is required and if possible reach the point where the welding were interrupted and then resume the arc welding. The re-planning functionality is reactive with aspect to process sensor data. In this case some restart attempts will probably be required before a possible trajectory around the strut is found. When the robot moves the weld-gun around the obstacle and a proper position to resume welding is reached the application process is restarted and the disturbance is solved by the system without any direct actions from the operator.

### 3.3 REAL-TIME ASPECTS

The *Event dispatcher* and all process modules below it in figure 2 have high real-time requirements. They are running in "hard" real-time, which imply from a few up to approximately 100 milliseconds in time perspective, depending on the process the robot is involved in.

The hardware used in our implementation is a VME computer system with processors from the Motorola 68k family. To achieve the time requirements *Process controllers* and *Virtual sensor* processes will in the future run on separate processors within the computer system.

*Task scheduler*, *Global path planner* and the functions carried out in IGRIP have lower real-time requirements than the rest of the event based control system. The most time consuming parts of these modules are only activated between sub-tasks or in case of major re-planning when the time is not critical. These modules are therefore running in "soft" real-time, meaning approximately response in the order of one second in time perspective. The "soft" real-time processing is carried out on a workstation with support for fast communication with the VME computer system. This communication is very important because it is the interface between hard and soft real-time, specially for the timing between *Task scheduler* and *Event Dispatcher*. An interesting modular software approach to sensor integration for welding process and robot control is presented in [1].

### 3.4 INTERACTION WITH A ROBOT WORK-CELL

The robot work-cell is the real world environment where tasks are realized. It consists of one or several robots, external devices, process equipment, positioners, sensors and safety devices. The modules between the *Event dispatcher* and the *Robot work-cell* in figure 2 are interacting with the work-cell. The rest of this section describes the four types of modules interacting with the work-cell; *Virtual sensors*, *Robot control*, *Process controller* and *Safety*. The modules are discussed as agents of a control system [17].

Sensors produce feedback data from the work-cell to the *World model*. The information is processed in the *Virtual sensor* module, described in more detail in [4]. The virtual sensor can be seen as an observer interfacing directly to different physical sensors. Sensors may have local data extraction but the purpose of the virtual sensors is to provide a common platform for interfacing sensors to the controller. In addition, models can be developed for sensors which map certain features measured by the sensors in relation to the task to be performed. Example of such models include, for a sensor in welding applications, extraction of specific data for gap, target position, root face, angles related to the weld joint, and data related to collision warnings, see figure 3.

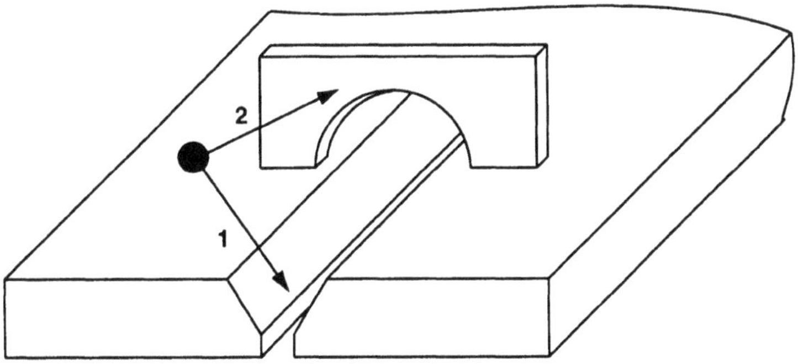


1: Weld joint data related to seam tracking (position) and welding procedure specification (orientation, joint data such as groove angle, gap and root face), all parameters important to control the welding process.

2: Detection of distance to object for collission avoidance and recognition of such objects (to invoke a procedure for welding to the object, stop and move around, and start welding from the other side).

**Fig. 3.** *Functionality of the virtual sensor in arc welding application.*

Every robot in the work-cell has its own robot motion control, consisting of a *Path planner* and a *Robot model.* The *Path planner* receives all requests on motions for the specific robot from the *Task-plan.* The *Path planner* calculates a trajectory based on information requested from the *World model.* The trajectory is confirmed or rejected by the manipulator model, calculating the trajectory concerning kinematics for the specific robot. If the trajectory is rejected the *Path planner* must create a proposal for a new trajectory. When the trajectory is finally accepted a path is created in the *Path planner.* The low level software of the robot is then fed with joint values.

Path planning problems caused by multiple robots working together in a common workspace are discussed in [16]. It is important to note that a *Global path planner* supervises the involved path planners and the timing of the realizations along the path to avoid collisions.

The control of an application process is initiated by a message sent from a sub-task in the *Task-plan.* The *Process controller* receives hardware specific configuration information and initial conditions for the process. The *Process controller* transforms a vector with general process control information to an equipment specific vector. The transformed vector is then applied to the process equipment.

Emergency stop initiated in the *Robot work-cell* is handled by the *Safety* feedback system. An emergency stop results in an immediate standby mode and the *Event dispatcher* performs a total stop of all controllable actions in the *Robot work-cell.* All events concerning the safety of the work-cell has the highest priority in the control system.

The difficulty to restart after an emergency stop depends on the complexity of the work-cell and the type of application process. If changes in the work-cell occurred when the emergency stop was pressed and after it was detected by sensors the *World model* is updated as usual. To restart the task realization valid trajectories and process related data for the specific operations are needed. Therefore, the reactive re-planning, described in section , is invoked to generate proper solutions and the activity in the work-cell may resume. Special functionality for restart of advanced application process may be needed. For instance, an interrupted arc-welding process require at least to be resumed from the correct position in the weld joint while other processes such as gluing and spray painting may not be restarted on the same work-piece.

## 3.5 OFF-LINE VALIDATION OF A TASK-PLAN

When a new task is implemented much of the work to define the task-plan can be made off-line. In this way many situations (normal and abnormal) can be investigated and solved in an early state of the implementation phase. This strategy usually preferable in economical terms, since late changes in an implementation is rather expensive. This strategy is the motive of robotics simulation and off-line programming and is the main reason of the growth of the virtual reality industry.

One important aspects is often forgotten in the existing robotics simulation systems; the dynamic environment where the process takes place. The most common strategy to bypass this problem is to avoid the use of sensors. If we treat the environment to be static the reliability of the simulation is increased. To be able to treat the environment as static the fixtures holding the workpiece and the workpiece have to be manufactured with very close tolerances. Furthermore, the model of the robot has to be very accurate and calibration is often necessary. This strategy is in most cases very expensive and sometimes impossible to use due to small-batch production or work-piece with inadequate geometry information.

The presented system makes it possible to simulate the behavior of the robot in a dynamic environment including the feedback from sensors. The behavior of the system both with a nominal model and in extreme situations can in this way be tested. The entire system is used with only one addition, a new layer to the virtual sensor process. This layer is copying the behavior of the sensors low-level interface and the virtual sensor and the rest of the system can be used without any modification.

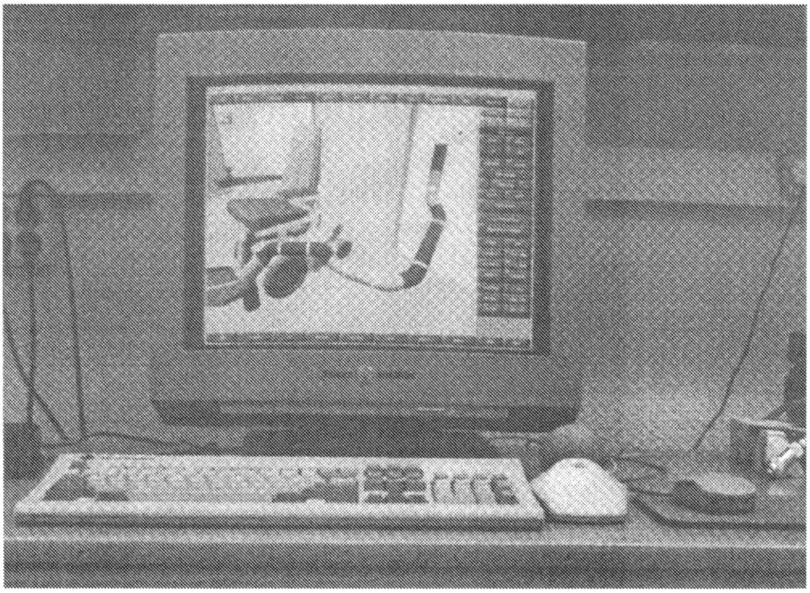

**Fig. 4.** *Computer hardware used for system implementation of "soft" real time.*

## 4. Implementation

A subset of the event based control system described in this chapter is implemented in an experimental platform. The main features updating the *World model* by information from sensors, obstacle avoidance, path planning, reactive re-planning and configurable correction are implemented. The *Task scheduler* and the application process control is not enabled at this stage. Initially hard real-time functionality is developed for execution in "soft" real-time. All source code is written in the C programming language. The computer hardware used is a SGI Impact workstation, figure 4, connected to a VME system with Motorola 68040 processors.

The case work-cell and the control system implementation deals with one robot. A multi robot work-cell will mainly increase the task planning and path planning demands on the system. The *User interface* and *World model* is implemented in the graphical environment of IGRIP, see figure 1. Every single graphical object in the *World model* is connected to a data structure. It is possible to extend the structure to represent all of the occuring features of objects residing in the work-cell and their relations. Examples of such features include motions, weld joints (objects to be welded) and grab positions (grasping operations).

## 4.1 UPDATE THE WORLD MODEL

The real *Robot work-cell* corresponds to the *World model* in IGRIP. Through a sequence of conventional calibrations of tool, cell, workpiece and signature a good static accordance between the real robot work-cell and the model can be reached.

Also a high dynamic accordance is desirable, specially for control of application processes sensible for geometrical and peripheral deviations. Perceptional equipment serves significant data to the system through socket connections. Specific information for instance a length value is connected to the corresponding distance in the world model. The data is read from the socket and code developed for LLTI[3] functionality within the simulation system and is used to update the world model. Positions and orientations of central objects in the *World model* is updated with a frequency of 30 Hz.

In the implementation a seam-tracking laser sensor from Servo Robot is used for measuring the distance between two weld plates. The sensor data is connected to the distance between the corresponding objects in the *World model.*

By using several sensors it is possible to identify objects from a group of known objects. New objects of the *World model* may be created based on information from sensors. Furthermore, a complete work-cell can theoretically be created from sensor data, provided access to manoeuvrable sensors or a large number of sensors.

An example where a parametric *World model* is created by sensor data in shipyard welding is presented in [6].

## 4.2 PATH PLANNING

The path planning functionality is developed as modules in a shared library to which IGRIP has access to. The behavior of this functionality is possible to adapt to current circumstances. It is, for instance, possible to choose the direction or sequence of directions used in automatic path planning applied to the creation of collision free paths.

Collision detection is activated during path planning operations to avoid impacts between robot, tool, workpiece and other objects in the robot work-cell environment. Objects significant for collision checking are specified and arranged in several groups for increased checking speed. The allowed clearance distance for the collision detection system used during path planning can be configured.

A specific problem associated with path planning is related to the motion planning of the robot which include handling joint limits (this problem is further extended if external track motions are integrated to the robot), robot configuration selection (change of configuration may result in critical turns of robot axis to obtain equivalent locations of the TCP), singularities, etc.

---

[3] Low Level Telerobotics Interface, Deneb Robotics inc.

## 4.3 REACTIVE RE-PLANNING

When the current state of the work-cell differ too much from the initial state known at planning time, paths and sub-tasks may need to be re-planned. Combinations of deviations and their allowed amount which does not cause a re-planning are stated by the user of the system. The deviation checks are performed by the source code for *World model* update. If the allowed deviations stated are exceeded the motion in progress is interrupted and the task is re-planned with current conditions, causing a reactive re-planning. The task realization continue automatically when a suitable solution is found.

Information from perceptional equipment can, besides *World model* update and reactive re-planning, be used for real-time corrections of the task in progress. The implemented event based control system enables connections between specific information flows and correctable operations. These connections are specified and configured by the system user for proper corrections in the current work-cell or configurable corrections.

Application process parameters like, for instance, power source parameters in arc welding applications, are corrected indirectly by information from perceptional equipment. The information is used by the application *Process controller* which decide suitable changes to the process parameters. Robot motions can be corrected directly by a connected configurable flow of information from *Virtual sensors*.

## 4.4 STRUCTURE OF THE SIMULATION SYSTEM

A central part of the system is the simulation module. In this module the world model is defined and visualized, and the sub-tasks are developed and validated. IGRIP is the base for this module and are extended to enable communication with external processes.

IGRIP provides an interactive, graphic simulation tool for design, evaluation, and analysis. A great variety of manufacturing process may be constructed, programmed, and analyzed. IGRIP has an open structure and can be extended in many ways. All major modules can be replaced with user-defined code. The most important technique to invoke or modify functionality is through shared library. This enable dynamically linked procedure calls to user code. Another way is to use the variety of different communication tools to access other applications during execution time.

In figure 5 the structure of the IGRIP modules is presented. The heart is the simulation engine including the motion pipeline. This module take information from the programming interface and execute the motion of the programmable device. The program is written in GSL (Graphic Simulation Language), a high-level programming language. A GSL process can be connected to each object in the work-cell providing an object oriented way to handle objects, processes and operations. Information to the GSL-program can come from external applications through the communication interface. This feature enable the connection to an external *Event Dispatcher*, compare with figure 2. Instead of executing a predefined static program, sub-tasks can be dynamically executed.

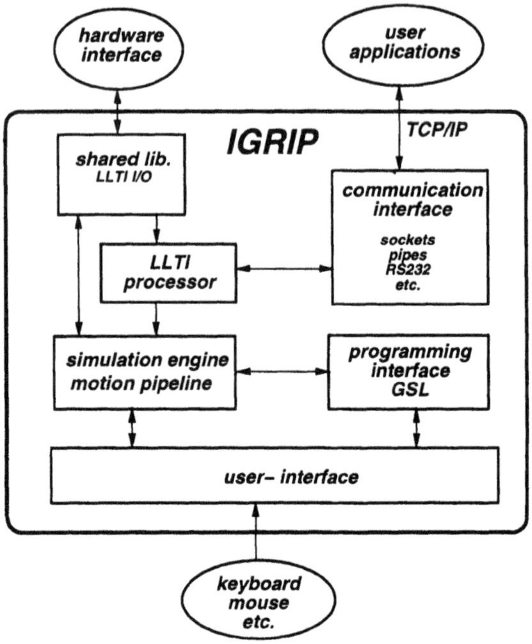


**Fig. 5.** *The major modules in IGRIP, simulation system.*

The motion pipeline is a number of modules responsible making a multiple-degree-of-freedom device to move, typically a manipulator. Users are able to write their own functions to work as components in the motion pipeline, either partially or completely replacing the existing components. In figure 6 the structure of the motion pipeline is presented.

The LLTI (Low Level Tele-robotics Interface) gives a possibility to access external hardware and software in real-time. A LLTI-process is called about 30 times per second, and enable control of external hardware and update of the simulated work-cell through sensory feedback. The motion-pipeline can be accessed to get current information about the manipulator state and use it to control the physical manipulator. In contrast, information from the real world, perhaps a seam-tracker, can also be used to modify or control the current motion of the virtual manipulator in the simulation system.

## 4.5 EXPERIMENTAL PLATFORMS

The described system implementation is evaluated in two different platform work-cells. The first one is an experimental service robot. The second case is an industrial robot system. The ABB robot used for the industrial robotic case has

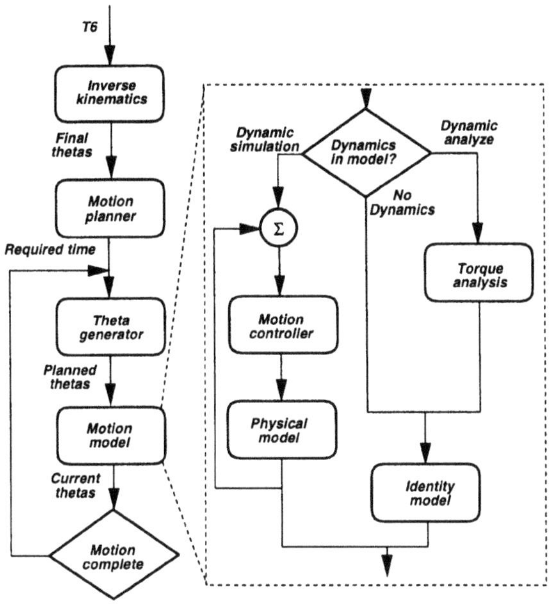


**Fig. 6.** *The motion pipeline, responsible of making a reliable simulation of the manipulator movement.*

been reconfigured [2] and extended with the required interface and supports input at several levels in the controller hierarchy. The event based control system can control any robot accepting joint values and joint velocities as input to its servo controller.

### 4.5.1 Service robot

In service robotics work-cells, sensing capability is of great importance. The robot is acting in an unstructured environment and the possibility to adapt to different situations is crucial. The design of the robot is specifically made to be used for physically disabled people.

The robot hardware in this work-cell is a redundant experimental service robot with eight axes developed at DPME[4], see figure 7. The robot has oblique angled oriented joints for a high flexibility within a large working area. The manipulator have some design features that are not so commonly used on industrial robots: carbon fiber composite links; self contained joints that include motor, reducer, bearings and control circuits; bus communication that handles positioning and feedback information.

The *Robot model* in the event based control system must include kinematics for the

[4]Department of Production and Materials Engineering, Lund University

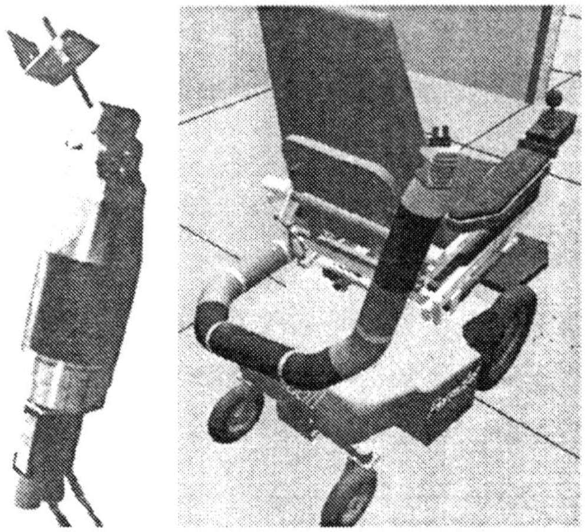

**Fig. 7.** *A picture of the wrist with a dummy gripper and a snapshot from the simulation software IGRIP.*

current robot. A fast numerical solution on the kinematics problem of the eight axis robot is developed. The part of the system implemented on the workstation produce joint values and joint speeds at a frequency of 30 Hz. Simultaneously the corresponding simulation is running in IGRIP. By serial communication the joint information is transferred at 38400 baud to the VME computer system where it is transformed to pulse information suitable for the servo controller. This information is distributed through the arm via a CAN-bus. In the developed servo controller module, a PID-regulator controls speed and position of the individual link in joint level. Via the CAN-bus is it also possible to update system parameters of the servo controller during execution. The modular approach of the design support robots configured for individual needs.

#### 4.5.2 Industrial robot

The system is tested together with an ABB IRb-6 industrial robot at the department of Automatic Control, Lund University. The low level software in the ABB system is rewritten [14] to accept joint information as input to the servo control system. In this case the joint information is transferred by TCP/IP[5] from the workstation running the event based control system to the robot controller host.

Both control of the robot and feedback from the real work-cell execution is implemented. In figure 8 the appearance of the computer screen is presented during the execution of the system consisting of one window with the interface in IGRIP

[5] Transmission Control Protocol / Internet Protocol

where the robot sub-tasks are executed, a second window with visual feedback of the execution via a vision system located by the robot, and a third window with a IGRIP model updated by the low-level robot controller.

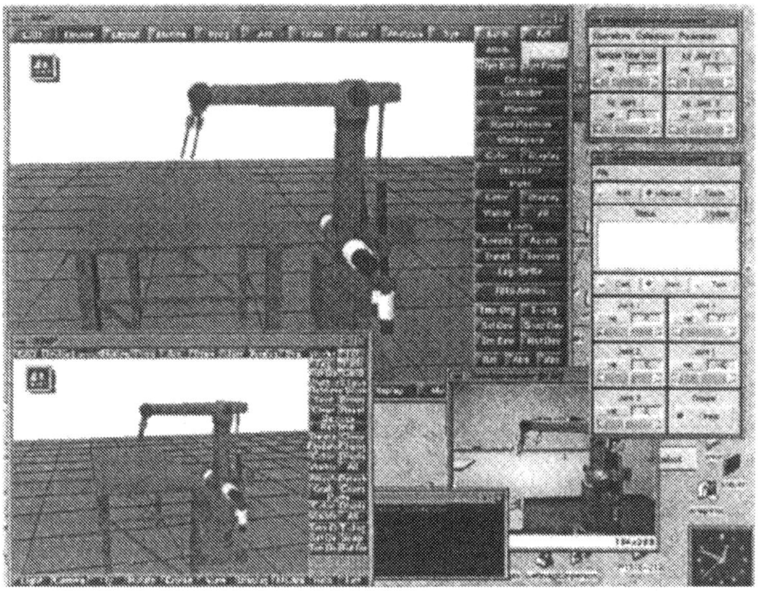

**Fig. 8.** *A snap shot from the screen during execution of the implemented software.*

4.6 SUMMARY

These implementations show that the presented event based control system is suitable as a framework for control of robot work-cells based on a task model approach. Actually, any robot accepting joint information as input for instant control is possible to control by this system, if computer power and communication with enough performance is available. Furthermore, the functionality of *World model* update, reactive re-planning and configurable correction are verified.

## 5. Robotic arc welding case project, programming and control

A case project focusing on programming and control of an industrial robot work-cell has been performed. The work-cell consisted of a six axis robot with two translation gantry axes and one revolute axis in the positioner, figure 9, with corresponding world model in figure 1. Arc welding of buckets for road machines are performed in the work-cell. The robotic system has support for through-the-arc seam tracking. This section focuses on one of the fillet joints with uncertainty in gap (distance between welding plates), an example is shown in figure 10.

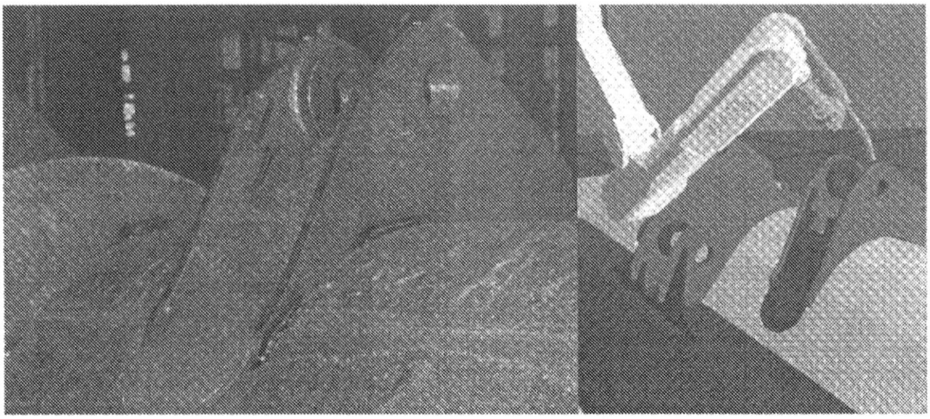

**Fig. 9.** *The case work-cell, in real life and as a virtual model in IGRIP.*

The robot systems of today requires a major effort to manage. Just local programming of the fillet joint of figure 10 require six positions with orientation; approach, weld start, weld stop, retract and two reference positions for weaving. Furthermore, information about stand off, travel speed, weaving direction, amplitude and frequency must be specified. Finally, power source parameters must be specified such as current, voltage, etc. The robotic arc welding system follows the seam and applies this static information independent of the geometry of the seam. Robotic control systems can be extended with functionality for switching between predefined sets of power source parameters, initiated by changes in seam geometry measured by a sensor during the arc welding process. This feature is however not sufficient for compensating for fluctuations in, for instance, gap, root-face or seam orientation.

The control system presented in this chapter enables several new possibilities for process control in the current case project. The global task to weld a complete bucket construction involve a large number of sub-tasks. During a weld operation several sub-tasks are running at the same time, for instance; process supervision, measuring plate positions, process control and the weld sub-task. The relationship between different sub-tasks in the case task is presented in figure 11. Only one instance of a weld sub-task can be executed at a time because there is just one robot and one installation of welding equipment in the work-cell.

Besides the two links to the involved weld plates the weld sub-task consist of arc welding process model information. Furthermore, the sub-task can be completed with a specification of configurable corrections through a fast low level interface between the virtual sensors and the robot motion controller. In the case of fluctuations in gap, figure 10, the specification contains a function for weld-gun position and motion control, e.g. offset along the vertical plate and weaving data express in the gap value measured by a laser scanner sensor. These configurable correc-

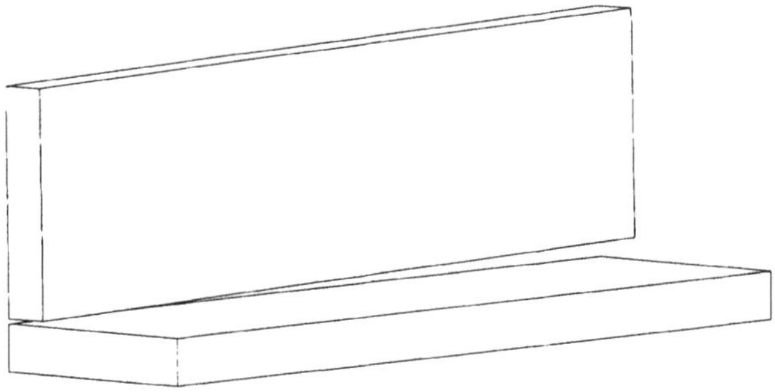

**Fig. 10.** *A fillet joint with increasing gap, from one to approximately three millimeters.*

tions influencing the motion control and enables the weld-gun to retract or even start weaving if necessary. An extended process control is offered through motion control which is aware of the circumstances. This indicates the importance of interacting process models which produce control action to both the motion control and the welding power source. This is in fact a common way to compensate for gap fluctuations in manual arc welding (lowering wire feed rate or current, decrease travel speed, start weaving to avoid burn through), but impossible to establish in industrial arc welding robot systems with traditional controllers.

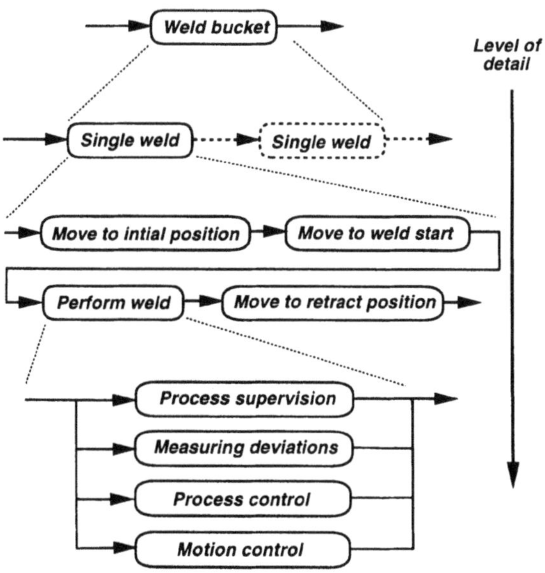


**Fig. 11.** *The task-plan can be decomposed to sub-tasks in several levels.*

The *Task scheduler* determines when it is time for execution of sub-tasks and events considering the states of MORW's in the world model. First, configuration information for the arc welding process controller and configuration of the interface between *Virtual sensors* and the robot motion controller is down-loaded to respective module. Then executable events get the weld-gun in position, start the arc welding process and move the weld-gun along the seam. The deviations in gap of the seam geometry is dynamically compensated for by the position of the weld-gun with respect to the seam, through the reconfigurable interface to the robot motion controller.

The *Virtual sensor* compare the measured gap with the gap limit of the process model originally specified in the sub-task. If the gap exceeds the limit an alarm message is sent to the *Event dispatcher* causing the arc-welding process to terminate. Other executable events can be fired by the *Event dispatcher* and the task realization proceeds.

Besides the exemplified position compensation, low level robot motion control can be used to control of stand off, orientation angle, travel speed and weaving motion. All these robot related parameters have important influence on the results of the arc welding process. The event based robot control system enable robot motion control for application process controlling purposes in; arc welding, spot welding, grinding, gluing, assembly, etc.

## 6. Conclusion

The presented task oriented robot programming method allows improved programming through re-use of MORW's and tasks. Simulation takes place directly in the programming environment, using a copy of the event based control system. This provides high reliability to the control system and the task to be realized.

The presented control system uses one single model of the world as base for all controllable events and targets for feedback. The system is effectively modularized and divided into proper abstraction levels. Furthermore, it is simple to extend the structure when a new robot or process is introduced in the robot work-cell.

The autonomous functionality of the presented system is established through reactive re-planning of the task and configurable corrections for local deviations in the work-cell. The control system has the ability to autonomously search for and create new solutions to problems appearing during task execution and reactive re-planning. Configurable connections between information measured in the work-cell and controllable process parameters are enabled. Furthermore, the open and more accessible control system allows configurable interaction between sensors and robot motion controller which offers unique possibilities for improved control of application processes.

The presented event based robot control system structure increases the efficiency in the robot work-cell. A robot does not have to wait for a device if it is possible to carry out other executable events, perhaps from other sub-tasks.

Sensor feedback is influencing the control system at a high level, directly on the

*World model.* Major deviations in the *World model* results in re-planning of sub-tasks or tasks. The presented event based control structure can take care of the dynamic environment in the work-cell by adaptation of the task realization. Task oriented programming and event based control of robots increase the possibilities for the system to autonomously solve unexpected problems appearing during task realization.

# References


[1] B. Ågren. *Sensor Integration for Robotic Arc Welding.* PhD thesis, Department of Production and Materials Engineering, Lund University, Lund, Sweden, 1995.

[2] R. Braun, L. Nielsen, and K. Nilsson. Reconfiguring an asea irb-6 robot system for control experiments. Technical Report TFRT-7465, Dept. of Automatic Control, Lund Institute of Technology, 1990.

[3] K. Brink, M. Olsson, and G. Bolmsjö. Task oriented robot programming. In *Proceedings of the Scandinavian Symposium on Robotics*, pages 103–109, Robotics Society in Finland, October 1994.

[4] K. Brink, M. Olsson, and G. Bolmsjö. Event based robot control, focusing on sensors. In *Proceedings of the International Symposium on Measurement and Control in Robotics*, Bratislava, Slovakia, June 1995.

[5] G. Engelhard, D. Pellkofer, and K. Schuchardt. WIG-OrbitalschweiSSen. In *Proceedings of DVS/SLV-München, WIG-Orbitaltechnik Conf.*, pages 32–40, Erding, Germany, May 1995.

[6] Esprit Project 6042 'Hephaestos 2'. Intelligent robotic welding systems for unique fabrications. Final report, wp1, European Commission, 1994.

[7] E. Freund and J. Rossmann. Autonomous operation of multi-robot-systems in factory of the future scenarios. *6th International Conference on CAD/CAM Robotics and Factories of the Future*, 1, 1991.

[8] G. Grunwald. *Modelling and Planning for Sensor Based Intelligent Robot Systems (Task-directed planning of cooperating sensors: A framework)*, volume 21 of *Machine Perception and Artificial Intelligence*, pages 143–159. Worl Scientific, 1995.

[9] J. L. Kolodner. *Case-Based Reasoning.* Morgan Kaufmann Publishers, Inc., 1993.

[10] L-O. Larsson. *The Modelling of Geometrical Parameters in Robotic GMA Welding.* Department of Production and Materials Engineering, Lund University, Lund, Sweden, 1995. Licentiate thesis.

[11] A. Naumann and P. Gu. Real time control of configurable manufacturing cells. In *Proc. 10th ISPE/IFAC International Conference on CAD/CAM, Robotics, and Factories of the future*, 1994.

[12] G. Nikoleris. *On the Programming of Welding Robots.* PhD thesis, Department of Production and Materials Engineering, Lund University, Lund, Sweden, 1988.

[13] K. Nilsson. *Application Oriented Programming and Control of Industrial Robots.* Department of Automatic Control, Lund Institute of Technology, Lund, Sweden, 2 edition, 1992. Licentiate thesis.

[14] K. Nilsson. *Industrial Robot Programming.* PhD thesis, Dept. of Automatic Control, Lund Institute of Technology, 1996.

[15] M. Olsson, K. Brink, and G. Bolmsjö. A visual task oriented robot programming system. In *Proceedings of the International Symposium on Measurement and Control in Robotics*, Bratislava, Slovakia, 1995.

[16] P. C.-Y. Sheu and Q. Xue. *Intelligent Robotic Planning.* World Scientific Publishing, 1993.

[17] A. Steiger-Garcao and L. M. Camarinha-Matos. A conceptual structure for a robot programming system. *Robotics International Journal*, 3, 1987.

# 3

# COMBINING VISUAL, TACTILE AND RANGE SENSOR INFORMATION TO INCREASE THE AUTONOMY OF ROBOTS

M. SEITZ, A. WEIGL, K.HOHM
and K. KLEINMANN
*Control Systems Theory and Robotics Laboratory*
*Darmstadt University of Technology*
*Landgraf-Georg-Str. 4, D-64283 Darmstadt, Germany*

## 1. Introduction

Up to now, most industrial robots employ sensors just to measure their internal state and to keep their end-effector on a desired trajectory. However, sensors monitoring contact situations and/or giving information about the environment can enhance the capabilities of robot manipulators. Reasons, why they are not used in general are additional costs, sceptism about their reliability, limitations in the robot controllers and computational burden. But cost reductions in sensor hardware as well as the development of industrial interfaces to Personal Computers with their rich environment of easily handable software has changed this picture. The exploitation of the abilities of further sensors has opened a large field of new robot applications in industry – e.g. robot assisted disassembly – as well as in direct human support, the broad field of so-called service robotics.

Research and first applications in the extension of robot manipulators to robot systems by adding actuation degrees of freedom and environmental sensor information are mainly restricted to simple extensions, e.g. by using just one additional sensor. Let us therefore examine at first shortly the most important sensors and their areas of applications.
Vision based on CCD-cameras is a powerful tool for robot work space inspection and object recognition and thus useful for motion planning and grasping. Surveys of basic image processing methods for robotics are given in [1], [2], [3], however, applications to sensor-integrated robotic systems and experimental trade-offs are missing therein.
Range sensing is normally based not on CCD-cameras but on laser and/or infra-red sensors. Besides triangulation methods also image processing methods are applied to the data acquired [4]. In [5] an approach is presented using feature extracting methods to analyze a range image of an object to be grasped.
Since range sensors have the advantage of more exact range information, which is in addition robust against lighting conditions, video cameras can scan large areas faster. So instead of using one of these sensors, a combined use of both sensor types as often used in mobile robot navigation would be reasonable for complex applications as object grasping.
Force/torque information is up to now mostly gained by force/torque sensors in

the manipulator wrist. Since the real contact in manipulation tasks is established by the gripper such sensors provide just indirect tactile information. But detailed direct tactile information at the gripper fingertips can improve the manipulation of an object [6]. Therefore, touch sensors with a spatial matrix arrangement of simple touch sensitive elements (taxels) [7], [8] are under development since several years. They provide three types of information: Firstly, they give an idea of the quantity of an established grasp by the number of activated taxels; furthermore, during manipulation they can detect slipping of an object and, last but not least, they measure the resulting forces applied to the object for grasp force control.
The presence of an object between the gripper fingers can be detected by light barriers before touching the object. Using infra-red (IR) diodes with IR-detectors in the opposite fingers has the advantage of spectral independence against normal day light and neon light, the latter widely used in industrial environments.

Work on the use of multiple sensors in cooperation with robot manipulators is relatively rare. One of the first who combined different sensor types for robotic object recognition was Allen [9]. Also robot manipulators guided by vision were introduced, e.g. for bin picking [2]. An early driving force for sensor integration was the development of dexterous, multi-fingered grippers for object manipulation: Most of these multi-fingered grippers are provided with joint position and joint torque sensors, which are sufficient for stable grasping and object manipulation by the fingertips if some geometric parameters as well as a reasonable estimation of friction coefficients are known. However, the latest systems also include touch sensors for slip detection [10]. A real robot system including vision and force sensors together with a multi-fingered gripper used for sorting postal parcels and clearing dishes from a tray is described in [11]. Also other interesting approaches to robot systems exist. But research is mainly concentrated on subareas with the problem that the system elements examined are difficult to integrate into a more complex robot system and allow therefore only limited experimental verification of the efficiency of the experiments [12].

In the following, experimental examinations concerning the possibilities and necessities to combine different kinds of sensors will be discussed. Background is the aim to increase the autonomy of robots. Therefore, the respective requirements will be put forward in section 2. Section 3 treats the collection of information about unmodelled and/or unknown objects by vision as a prerequisite for grasping such objects. Strategies to grasp those objects on the basis of their visual inspection and the transition from visual to tactile information in this connection forms the contents of section 4. Actually a combination of visual, tactile and range measuring techniques is shown to be appropriate in this context. Finally, the experimental examples of section 5 demonstrate the necessity to employ multisensory robot systems for complex tasks as highly autonomous assembly and disassembly. Conclusions are drawn from our experiments in section 6.

## 2. Requirements to Autonomous Robot Systems

System autonomy means in the context of robotics the ability to support human

| capability | sensor/actuator type | submodule | |
|---|---|---|---|
| segmentation of global scene, object isolation & recognition | color vision | | |
| 3D shape reconstruction, accessibility analysis | range sensors | | |
| estimation of position & size, object isolation | motion stereo | eye | |
| shape analysis, selection of suitable contact regions for grasping | contour vision | | |
| contact with the environment | force/torque sensors | | sensors |
| detecting and optimizing contact points | touch | | |
| slip detection | IR-light barriers, touch | | |
| contact force optimization | force/torque sensors | | |
| object manipulation and regrasping | force/torque sensors & touch | hand | |
| object manipulation and disassembly | force/torque sensors | | |
| dexterous object manipulation in hand | multi-fingered gripper | | |
| high grasp forces | parallel-jaw gripper | | |
| accessibility, graspability | slim gripper jaws | | actuators |
| object reachability | gripper on arm | | |
| object observability | tiltable camera on arm | arm | |
| maneuverability, reachability & observability | mobile basis | | |

Table 1: Capabilities necessary for autonomous manipulation and the required sensors and actuators.

beings in their work like a human assistant. That requires besides an information processing system, which is able to detail general task descriptions into a sequence of subtasks of primitive robot movements and to handle even major disturbances and/or failures in this context, also a robot system with fairly similar sensoric and handling abilities as human beings.

So the basic requirement for autonomous robot systems is the integration of various sensors, especially vision and tactile sensors, as well as adequate actuators, like arms with dexterous grippers, in order to interact humanlike with the environment.

As documented in table 1 visual sensors in form of color and gray scale cameras but also of range sensors are required for segmentation of the global scene, for shape analysis and 3D shape reconstruction and for determination of position and size of a-priori unknown objects. Besides these capabilities of the electronic eye, especially tactile sensors are used for detecting and reacting on contacts and collisions with the environment. In order to optimize contact points and prevent slipping IR-light barriers and touch sensors can be employed.

Dexterous object manipulation requires numerous degrees of freedom (at least six) for moving the end-effector and/or the grasped object to any arbitrary position in the robot work space. Figure 1 shows two examples of multisensory grippers employed for our experiments.

The jaws of the gripper in fig. 1a can integrate different types of sensors. First, an optical arrray of pulsed IR-diodes enables the detection of object presence between

a) b)

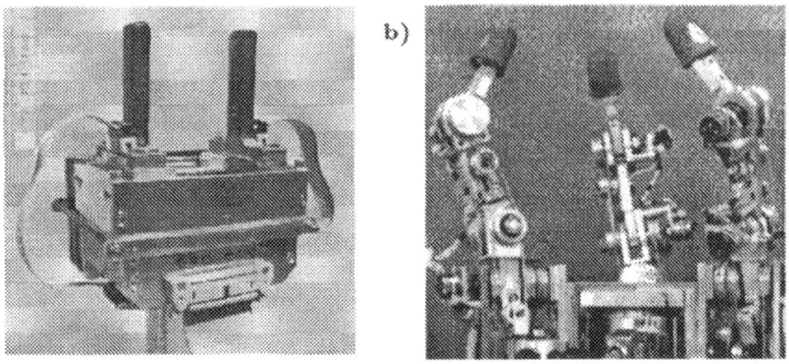

Figure 1: Two multisensory grippers: A parallel-jaw gripper with slim rigid fingers integrating IR-light barriers (a) and a three-fingered dexterous gripper for object manipulation in hand using both joint position and joint torques sensors (b).

the gripper jaws. Experiments have shown that an array of only six elements is sufficient in most cases. Depending on the application the relatively thin gripper jaws can be exchanged by thicker jaws integrating a touch sensor. It consists of a $16\times16$–matrix of force sensor elements on an area of $20\times20$ $mm^2$, measuring the contact force distribution over the surface using conductive rubber. It provides information about the quality of an established grasp, the resulting normal grasp force and it can be used for slip detection during manipulation.
The three-fingered Darmstadt hand (see fig. 1b) is a 9 degrees of freedom (DOF) system which consists of three identically designed fingers mounted symmetrically around a palm. Each finger has three tendon driven joints, each provided with a joint position and a joint torque sensor which is based on the measurement of the tendon tension [13]. On the surface of the finger limbs and on the palm tactile sensors can directly detect object contact.

The key for solving complex manipulation tasks autonomously lies in the ability of coordination. For this, incremental hand, arm and eye motions are guided and supervised by the employed sensor systems. Due to the different ranges of hand and arm, large distances to a desired goal position are covered by the arm, whereas small deviations between a grasped object and the goal can be compensated by motions of the fingers. Since large distances are measured by the eye small deviations between gripper and goal are sensed by internal gripper sensors. So, the combined use of global and local sensors allows a hand-arm-eye system, e.g. as sketched in fig. 9a, to grasp and manipulate a-priori unknown objects.

## 3. Visual Inspection of Unknown Objects

Since position, size and shape of the objects to be handled are mostly unknown in unstructured environments, gray level and color vision have to be employed for scene segmentation, object localization and shape analysis. The main difficulty of image processing is to achieve a symbolic representation of the acquired image of a real world scene, which means that the system somehow must understand the scene and distinguish between the background and the different objects within the scene.

## 3.1 IMAGE ACQUISITION AND ENHANCEMENT

The first step of image analysis is the acquisition of images which is of evident importance for the results of all succeeding steps. The goal is to get an image with as less noise and as much contrast as possible. For image acquisition especially the careful illumination is very important if there is the possibility to have an influence as e.g. in industrial environment.
After the acquisition image enhancement can be achieved by operations in spatial as well as in frequency domain, using e.g. morphological or algebraic operations, linear or non-linear filters and transformations, which either can be adapted to the signal or are signal independent [1], [3].

## 3.2 IMAGE SEGMENTATION

Considering an application, e.g. the disassembly of electronic devices as discussed in section 5.2, image segmentation is needed first to isolate the device to be disassembled in the scene. If a single module inside the overall device should be grasped and disassembled this module has to be isolated by segmentation.
Basically pixel-, edge- and region-oriented segmentation methods can be distinguished. Most pixel-based approaches applied are classification methods, e.g. thresholding in gray level images or clustering of color-based feature vectors in color images, where the clusters could be generated e.g. by histogram analysis. Then all connected pixels belonging to the same class or cluster are regarded to belong to the same object.
Tracking the boundary of an image segmented by a pixel-oriented approach already leads to edge detection methods. The main improvement is the reduction of data necessary for describing the content of an image. Alternatively, gradient-based methods can be employed for edge detection by filtering operators like e.g. the Sobel or the Kirsch operator. While contours of relatively large volumetric objects (e.g. the bottle in fig. 2a+b) are extracted by sobel-filtering and by contour thinning and following, smaller and planar objects can be segmented by thresholding methods generating binary images. Tracking the boundary of the silhouette results in the contours shown in fig. 3.
Both, pixel- and edge-oriented approaches work well in case of unicolored objects, but if there are surfaces with differently colored texture they run into troubles. With region-based methods it is possible to generate features of a local region and this also enables modelling of texture. The main approaches of this kind are region growing and split and merge algorithms. As a mixture of different methods one should also consider pyramidal approaches, which can combine the advantages of different algorithms.

Besides the problem of image enhancement and segmentation another difficulty is the recovery of the object's shape in order to guide the fingers of the gripper to those object parts which seem to be most favorable for establishing a good contact for further manipulation.

## 3.3 SHAPE AND MOTION FROM MOTION

Basically the 3D position of an image feature point representing e.g. a favorable contact point for one finger of a gripper, can be determined by a laser range finder as shown in section 4. Though its precision is excellent the size of laser range scanners with the required high resolution is up to now too big for being carried by the robot. That's why interesting parts of the workspace are coarsely localized by stereo vision.

**Motion Stereo:** Humans get a 3D impression of their environment mainly by stereo vision, but also by analysis of reflections and shades or by recognizing perspective distortions of textures and known shapes. Analogously to these strategies different methods for 3D shape recovery from video images have been developed like shape from stereo, shape from shading, shape from texture, shape from shape, etc. [3]. Since in unstructured environments location and power of the effecting light sources vary and texture and shape of objects are unknown, the only possible vision-based method is to recover the object's shape from stereo.
The ideal model of imaging sensors is the pinhole camera [1]. The mapping of a 3D point $P(x|y|z)$ inside the robot's work space to the image plane set up by the image coordinates $(u, v)$ can be expressed by the *Direct Perspective Transformation* based on the given image center $(u_0, v_0)$ and scaling factors $\lambda_u, \lambda_v$ computed by camera calibration:

$$u = \lambda_u \frac{y}{z} + u_0; \qquad v = \lambda_v \frac{x}{z} + v_0 \tag{1}$$

For stereo analysis two images from different positions have to be acquired. Instead of simultaneous acquisition by two static cameras one mobile camera can acquire one stereo image from position $C$ and the second from position $C'$ by moving its lens center from $C$ to $C'$. Based on this well-known transformation ${}^{C'}T_C$ the position of $P$ can be expressed as $(x', y', z')^T = {}^{C'}T_C(x, y, z)^T$ with respect to the new camera position $C'$. Again the *Direct Perspective Transformation* can be performed for the second stereo image mapping $P$ to the new image coordinates:

$$u' = \lambda_u \frac{y'}{z'} + u_0; \qquad v' = \lambda_v \frac{x'}{z'} + v_0 \tag{2}$$

Since equations (1) and (2) provide four equations and only depend on the three unknown coordinates of $P(x|y|z)$ the *Inverse Perspective Transformation* can be easily solved on the basis of just one feature point.
Though the employment of two static scene cameras for stereo is more precise, there are several advantages of using a flexible mobile camera and employing motion stereo for recovering *shape from motion.* Apart from the small size of mono-cameras the dexterity also allows to inspect objects from far and near.

**Automatic Object Zooming:** Based on the determined 3D position and orientation of the object's center and principal axis, the camera can be moved closer to the object. As discussed in [14], large distances are covered by camera motions which are planned and controlled by the 3D position and orientation determined by motion stereo. When the camera has been moved closely to the object fine alignments

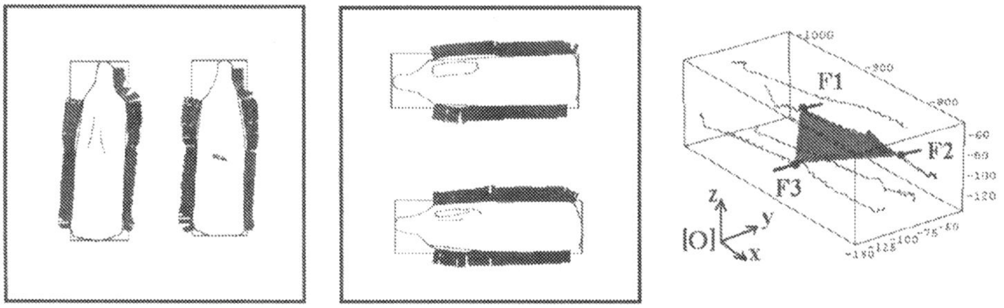


a) b) c)

Figure 2: Inspecting an object to be grasped from the top (a) and an orthogonal side view (b) allows to recover the 3D shape of four contours bounding the object as well as to plan favorable contact points $F_1, F_2, F_3$ for the three fingers of the dexterous gripper in fig.1b (c).

are performed by feature-based control meaning that the distances $\Delta u = u_{des} - u$ and $\Delta v = v_{des} - v$ of a feature to a goal position in the image are tracked in order to plan suitable camera motions. Assuming a constant depth $z = z'$ the substraction of equations (1) and (2) results in suitable motion commands for a robot carrying the camera:

$$\Delta x = \frac{z}{\lambda_v}\Delta v; \qquad \Delta y = \frac{z}{\lambda_u}\Delta u \tag{3}$$

This feature-based control is also called *arm-eye coordination* because it allows a direct coordination between image features and robot motions without the necessity of an expensive exact camera calibration [15]. Another advantage of the mobile camera is that the variable field of vision avoids occlusions and allows to recover the entire 3D object shape from multiple views.

**Orthogonal Inspection:** Grasping an unknown object requires the analysis of its entire 3D shape. For this, the 3D positions of multiple contour points extracted by edge segmentation (sec. 3.2) have to be determined as in case of the bottle sketched in fig. 2. Since multiple contour points are existing in the two stereo images acquired from the top view, the correspondence problem arises. In order to detect corresponding feature points in both images which describe the same 3D point in the robot's work space three aspects have to be taken into account. At first the contours bounding the bottle can be approximated by a pair of parallel straight lines of similar length. Secondly, the distance of corresponding feature points in the images is equivalent to the well-known distance covered by the camera motion for acquisition of both images. And finally, the sequence of successive contour points allows to find correspondences inside the contour. For each pair of corresponding contour points the 3D positions are determined by solving the *Inverse Perspective Transformation* of eq. (1) and (2). The length of the gray colored contour normals in fig. 2a+b symbolizes the distance between the lens center of the camera and the respective contour point. Based on the average position and orientation of the 3D points recovered from the top view (fig. 2c) the camera can be moved to a position from which the lateral object areas can be inspected as visible in fig. 2b. The two stereo images analyzed from this side view allow to recover the 3D points of the contour bounding the lateral areas of the bottle. Since for manipulation objects can be approximated by genetic primitives, e.g. cylinders, the four recovered contours bounding the bottle (fig. 2c) are sufficient for characterizing the entire object's shape, including the information about its length, width and height.

This coarse analysis of the 3D shape of the object to be handled allows the planning of favorable contact points and forces for the three fingers of the employed dexterous gripper shown in fig.1b [16].

## 4. Multisensor-Guided Grasping and Manipulation

In contrast to most model-based approaches found in literature, we present grasp planning algorithms based on local grasp features independent of the specific object shape and therefore applicable to even irregularly shaped unknown objects. While vision is used for non overlapping and freely accessible objects, 3D range data acquired by a laser sensor are analyzed for determination of grasp regions on hardly accessible objects as necessary for disassembling very compact electronic products.

### 4.1 VISION-BASED GRASP PLANNING

After the object of interest has been isolated by segmentation (sec. 3.2) and inspected (sec. 3.3) in the scene, the search for suitable grasping points requires a fine analysis of the object contour where the fingers can contact the object. Therefore, the contour bounding the object is segmented into favorable contour parts dependent on the given context.
If the goal is to perform a stable fingertip grasp primarily concave contour parts are selected because a finger placement in concave contour parts (see fig. 3a) results in a maximum contact area between fingers and object in order to avoid yawing torques. Additionally, the friction forces prevent that the fingers can easily slip away. But some applications require regrasping of the object during manipulation. Such a grasp goal leads to a selection of straight or convex contour parts as favorable contact areas within which the fingers can easily be moved on the surface of the object without collision as illustrated in fig. 3b.
The work space of the gripper is modelled as the gray area in fig. 3 indicating that part of the image plane which is accessible for the fingers. Since the three fingers of the gripper used in this work are arranged in $120^o$ angles apart on the palm, the system prefers those three favorable contour parts which also enclose approximately an angle of $120^o$ in order to provide maximum joint range availability for further manipulations (cf. fig. 3). In case that the object contour may only provide parallel contour parts a parallel finger arrangement is planned as shown in fig. 2c.

**Search for Favorable Grasp Points:** For the determination of force closure fingertip or precision grasps it is assumed that the directions of the finger forces are perpendicular to the object contour thus establishing maximum friction. Furthermore, it is necessary for a stable grasp, that the internal finger forces ${}^o f_{int} =$

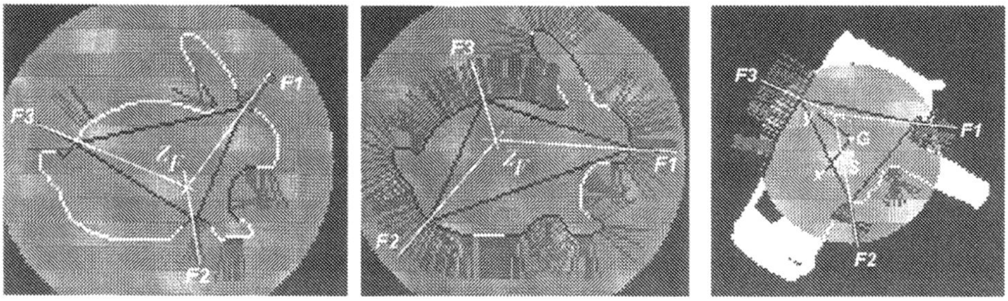


a) b) c)

Figure 3: Selection of favorable gripping positions $F_1, F_2, F_3$ for the three fingers in concave contour parts for the sake of grasp stability (a) and in convex contour parts for more grasp flexibility (b). In case of a camcorder a grasp is selected characterized by the favored gripping positions and the position [G] where the palm center should be located when contact is established (c).

$(f_x, f_y, f_z)^T$ as well as the resulting torques in the grasp plane are zero:

$$\sum_{i=1}^{3} f_{ix} = 0, \quad \sum_{i=1}^{3} f_{iy} = 0, \quad \sum_{i=1}^{3} {}^{o}m_{iz} = 0 \tag{4}$$

In order to prevent yawing torques ${}^{o}m_{iz}$ it is a basic requirement for a stable grasp that the lines of force (indicated by the white lines in fig. 3) through the three contact points $F_1$, $F_2$ and $F_3$ intersect at one single point, the grasp focus $Z_F$:

$$\{Z_F\} = \bigcap_{i=1}^{3} \{F | \vec{OF} = \vec{OF_i} + \mu \vec{f_i} \quad \forall \quad \mu \in \mathcal{R}\} \tag{5}$$

With respect to this background the contour normals extracted from the top view object silhouette are analyzed in order to find a set of three normal straight lines intersecting at one point. If the fingers contact the object on the respective contour points no yawing torques will result and equilibrium is accomplished in case that the internal finger forces are compensated.

Normally, various grasp foci and respective gripping positions are detected. By use of a quality measure assessing the distances of the determined contact points according to an equilateral grasp triangle an optimal grasp can be found ensuring a configuration within which the fingers are approximately placed in $120^o$-angles apart (cf. the black triangle in fig. 3). If a grasp with a satisfactory quality measure has been found the search can be stopped.

This vision-based method allows to determine a stable fingertip configuration for grasping e.g. the casing of a camcorder represented by the contour bounding its silhouette in fig. 3c. In spite of the three concave contour areas only two fingers are planned to contact concave areas whereas one finger is to be placed at the opposite straight contour area in order to form an almost equilateral grasp triangle meeting the requirements of the specific gripper geometry.

Estimating the 3D position of the center of area $S$ by monocular motion stereo allows to determine the 3D position of the three selected, most favorable contact points $F_1, F_2, F_3$ with respect to the center of area. After hand and arm have approached the object with respect to the planned reference frame [G] (see fig. 3c) the gripper is preshaped according to the determined gripping points. The desired gripper position $G$ is the center of the inscribed circle of the grasp triangle. The grasp forces to be executed by the fingertips of the gripper can at least be estimated by the constraints of eq. (4) and the geometric relations between $S$ and $F_1, F_2, F_3$ provided that the object's weight is approximately known by the visible size of the object in the image [17].

The presented approach plans suitable contact points for grasping freely accessible objects like the top of the camcorder casing (fig. 3c). After removing the top some of the modules inside the casing have to be disassembled to enable reuse.

### 4.2 GRASP PLANNING BASED ON 3D RANGE IMAGES

One of the main problems in grasping disassembly objects is the difficult accessibility, if the object is located inside a complex arrangement of several objects. Particularly, electronic products with their compact construction offer only very small clearance for approaching and placing the gripper fingers on objects inside the casing because access is restricted by neighboring objects as well as the casing. For this reason, grasp planning requires not only an analysis of the object to be grasped but of its neighborhood as well. While some clearance is necessary for approaching the gripper fingers, a sufficient grasp depth is essential to provide a stable grasp which is indispensable in order to disassemble the respective object afterwards. Since stereo image processing provides no exact depth information for narrow and deep object areas, which in general are insufficiently illuminated, a laser range finder has been used to acquire the object height data to obtain a 3D sensor image. In the following, an approach for processing such 3D range images to determine suitable gripping points (for details see [18]) will be illustrated by the example of the objective module located inside a camcorder, as shown in fig. 4a.

**3D Range Image Processing:** Based on 3D range images (see e.g. fig. 4b) local grasp features can be extracted for objects of a-priori unknown shape. In our context, we search for suitable gripping points providing enough clearance and a sufficient grasp depth in the area around each contact point for immersing the parallel-jaw gripper shown in fig. 1a into the camcorder. Because the grasp is to be established with a parallel-jaw gripper, the detection of antipodal contact faces on the object is required in order to achieve a stable grasp. The following approach proposed for analyzing the acquired object data is partly based on image processing methods for 2D gray level images. The idea to interpret range data as gray tones and apply the same methods as for vision is proposed e.g. in [4]. Height differences in the range data are represented by gray level edges in the resulting gray level image. For the example of the objective module of the camcorder, interpreting the object height differences (fig. 4b) as gray tones results in a gray level image as shown fig. 5a. To find contact faces arranged in parallel to each other the generated

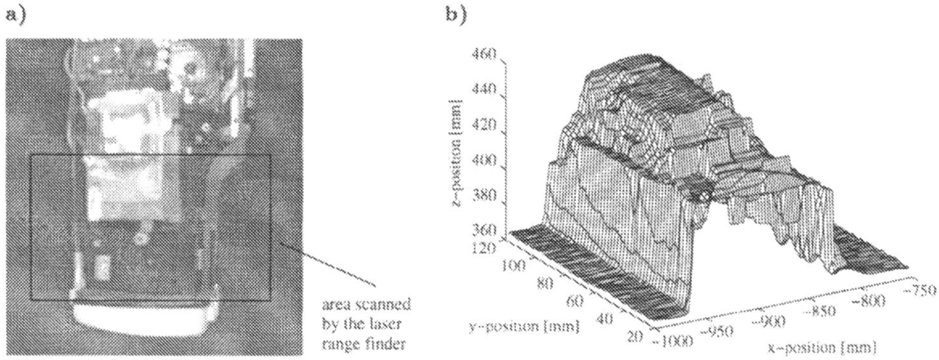


**Figure 4: By moving the robot over the interesting part of the electronic device (a) relevant range data are scanned. After interpolation, median filtering and transformation a 3D sensor image (b) is obtained representing the object heights of the objective module.**

gray level image is transformed into a contour image (cf. fig. 5b) by edge extraction and thinning operations as well as by contour following (sec. 3.2). After filtering and further data compression pairs of parallel contour segments of sufficient length for the gripper jaws and enough distance to the next neighboring edge are extracted from the resulting contour image (cf. fig. 5c). For detecting those parallel segments which allow an optimum contact by the gripper jaws, their distance and length as well as the available clearance and grasp depth have to be analyzed.
So for selecting optimum grasp regions the entire 3D information about the camcorder has to be analyzed again. By inspecting the respective part of the gray level image the area around every pair of parallel straight lines extracted from the contour image is examined, how far the following requirements are fulfilled:

1. To ensure that the detected parallel straight lines belong to the same object, the range data between the assumed antipodal grasp points should be homogeneous within certain limits.

2. The overlapping part of the pair of parallel straight lines must not be smaller than a minimum length required for contact with the gripper jaws.

3. The area behind each selected pair of straight lines must be free of other objects to provide sufficient clearance for approaching the gripper jaws.

4. The depth in the area behind each selected pair of straight lines must be deep enough for immersing the gripper jaws into the device.

In case of our example – the camcorder objective module – only one pair of parallel straight lines meets these requirements and the optimum gripping positions are in the center of each contour segment (indicated by the arrows in fig. 5c) in the desired depth. If more than one pair fulfills the requirements, the best grasp quality is given in object regions providing maximum grasp space, which means that the clearance and the length of the selected contact regions have to be a maximum.

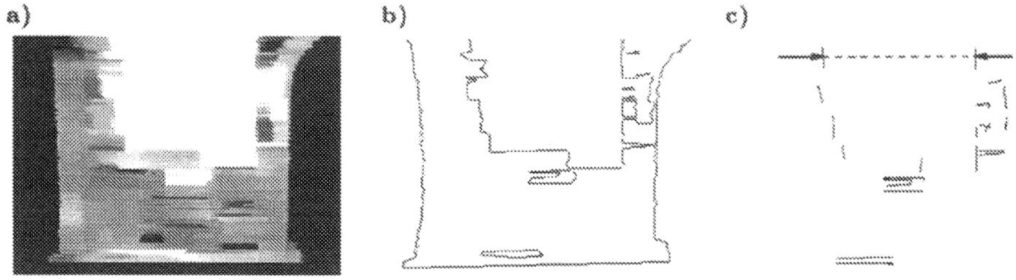


Figure 5: Interpreting the relevant object height differences as gray tones results in the gray level image encoding ranges of an interval length of $40mm$ (a). The contour image (b) aquired from this gray level image is further processed, before pairs of parallel straight lines can be extracted as favorable grasp features. Among all these pairs of parallel straight lines only one pair (indicated by the two arrows) was selected after further examination as suitable gripping points providing enough clearance and a sufficient grasp depth for the parallel-jaw gripper (c).

The grasp planning approach proposed could further be improved by applying a combination of video (color) image processing and range image processing. Especially if the scene to be analyzed is very large, a kind of pre-segmentation by video image processing is reasonable in order to determine a region of interest which, subsequently, will be scanned by the laser range finder, possibly with maximum resolution.

## 4.3 TACTILE GRASPING STRATEGY

After the grasp planning has been performed, in the next step, the robot gripper has to approach and subsequently grasp the object at the planned gripping points. Due to inaccuracies of the external sensor system and the robot control it may happen that gripper and fingertips are not exactly placed at the desired positions. In order to cope with the difficult accessibility as well as with inaccuracies for example resulting from the sensor-based planning approach, a fault-tolerant grasping strategy based on force/torque sensor information is necessary.
The tactile strategy described in this section (for details see [19]) requires only a coarse determination of suitable gripping points planned. According to these gripping points a coarse approaching direction is specified as a global motion for the robot system. The fine motions to be superposed to the global motion during the approaching and grasping process – in the following also called immersing operation – are then planned online based on the current force/torque sensor information supplied by a wrist force/torque sensor. In this way, no exact model of the environment is necessary and the system is robust against uncertainties.
The tactile grasping strategy allows the robot system to act in a way similar to the human behavior solving the problem of immersing, e.g. into an electronic product to be disassembled, in order to grasp hardly accessible objects located inside. For that, it is necessary to specify typical situations occurring during the immersing operation (see fig. 6) and to determine the corresponding way of acting, i.e. motion

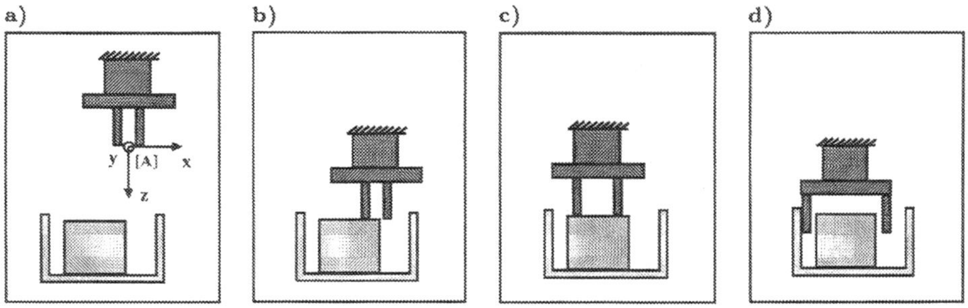


Figure 6: Typical situation clusters during the immersing motion of the parallel-jaw gripper in order to grasp a disassembly object which is hardly accessible: (a) `<free space>` cluster, (b) `<striking on the object with one gripper jaw>` cluster, (c) `<striking on the object with both gripper jaws>` cluster, (d) `<any other situation>` cluster comprising all situations not contained in one of the other clusters.

rules for the robot system:

$R_1$: IF the system is in the `<free space>` cluster, THEN continue movement in the global immersing direction.

$R_2$: IF the system is in the `<striking on the object with one gripper jaw>` cluster, THEN lift the gripper and correct the lateral displacement according to the experienced torque.

$R_3$: IF the system is in the `<striking on the object with both gripper jaws>` cluster, THEN lift the gripper and increase the gripper opening width.

$R_4$: IF the system is in the `<any other situation>` cluster, THEN execute evading movements according to the experienced forces and torques.

**Realization by a Fuzzy Approach:** For implementing the motion rules of the grasping strategy a fuzzy approach was used leading to smooth transitions between the different situation clusters as well as to an easy man-machine interface. A superposition of the situation clusters and consequently of the corresponding motion rules can easily be realized.

Based on the external forces and torques experienced by the robot end-effector, which is the parallel-jaw gripper in our case, the different situations can be classified into the typical situation clusters with their corresponding generic motion rules presented above. Therefore, the resulting rule base consists of various rules of the following form, e.g.

IF (*force error in $x$-direction is about zero*)
AND (*torque error in $y$-direction is negative*)
AND (*global velocity in $x$-direction is about zero*),
THEN (*nominal velocity in $x$-direction is negative*).

In addition to commanding the robot, the adaptation of the gripper opening width is determined and performed in parallel to the robot motion.

The tactile approaching strategy described above is stopped if a sufficiently large part of the object is located between the gripper jaws, which means that a sufficiently large contact area for grasping is accessible.

**Optimizing the Grasp Depth by Infra-Red (IR) Light Barriers:** To detect whether the object is within the gripper jaws, e.g. of a parallel jaw gripper, an optical array of pulsed IR sensors, consisting of IR diodes and IR detectors on the opposite jaw, can be used. The fact that they operate in the IR spectrum and are pulsed makes the system robust against normal daylight as well as neon light as it must be used in industrial environments if video image processing is employed for object and scene analysis. In this way, the approaching motion to an object during the grasping process can be stopped automatically as soon as the object is detected with a certain number of optical elements so that a sufficient grasp depth has been reached.

This fault-tolerant tactile motion strategy enables successfully approaching the object and reaching the planned contact points - even in case of uncertainties and difficult accessibility.

## 4.4 TOUCH SENSORS FOR ESTABLISHING CONTACT

As soon as the desired grasp positions are reached and the object is detected within the gripper jaws the grasp is to be established. Besides the control of the desired grasp force, especially when establishing the grasp, also its "quality" has to be evaluated in order to meet the stability requirements. A very simple but effective method is counting the number of activated taxels of the tactile matrix sensor integrated in the gripper fingers, because the more taxels detect a contact force the more stable is the grasp. If furthermore the force at each taxel can be measured, a tactile sensor image is generated which can be analyzed by image processing methods (cf. sec. 3). The force distribution over the sensor taxels characterizes the size and intensity of the contact surfaces and allows to optimize the grasp force and contact surfaces before lifting and disassembling the object.

**Slip Avoidance:** During object manipulation direct tactile sensors are needed for slip detection. Slipping occurs if e.g. the grasp force is insufficient to lift a heavy object. During manipulation collisions of the grasped object with the environment and/or jamming and wedging can be reasons for the object slipping out of the gripper. Whenever a slipping occurs the grasp force has to be increased or other measures have to be taken in order to keep contact to the object.
To detect slip e.g. frequency analysis of the force signals acquired by a tactile matrix sensor is employed [20], [21]. The start of slipping can be detected by a short appearance of a dominating 10 $Hz$ peek. Slipping itself leads to a 65 $Hz$ peak which is characteristic for the stick-slip-effect. The stick-slip-effect means that the object periodically moves and deforms the elastic sensor surface tangentially until the sensor surface comes off the object and snaps back. During supervising the center point of the contact only noise can be found in the spectrum as long as no slip occurs but immediately before slip starts the low frequency components become

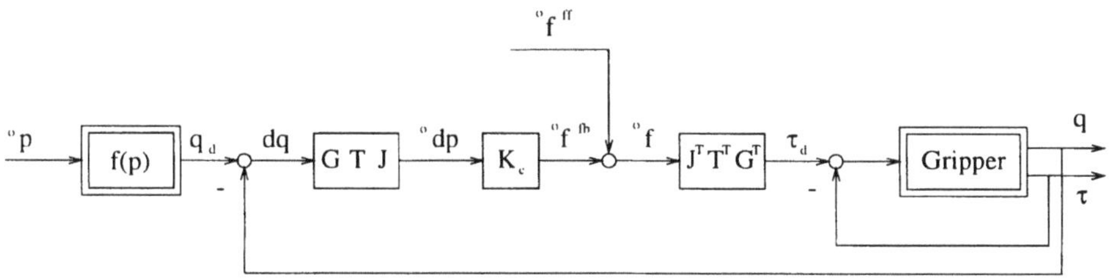


Figure 7: Blockdiagram of the stiffness control approach.

dominant. In other approaches dynamic sensors measure micro-vibrations caused by the stick-slip-effect also in combination with force analysis [22], [23], [24]. If a very fast slipping has to be expected due to a heavy object, optical arrays can be used for slip detection.

The described sensor-based strategies allow to establish a secure contact to the object. For further manipulation the grasped object has to be moved without loosing contact which can be achieved by stiffness control as explained in the following.

## 4.5 CONTACT FORCE OPTIMIZATION FOR MOVING THE OBJECT

While and after grasping an object by multi-fingered dexterous grippers the finger links have to be moved. Furthermore, after a suitable contact has been established the grasped object has to be lifted. Unlike parallel-jaw grippers multi-fingered grippers provide numerous degrees of freedom for grasping and moving an object. While moving the object by finger motions the grasp forces must ensure a secure contact. Therefore, object motions and grasp forces are controlled by stiffness control as proposed in [25] which allows to specify a cartesian behavior of the object while implicitly enabling an active finger coordination mechanism, i.e. the fingers move the object by so-called external forces as well as they keep a stable contact with it by applying so-called internal forces (the grasp forces). A practicable implementation is shown in fig. 7 which was successfully used for assembly tasks discussed later in section 5. As sample experiment the peg-in-hole insertion (see fig.9) is examined in the following. The finger forces are implied to the object for moving the object as well as for keeping in stable contact with it. Stiffness control (fig. 7) provides both control of object motions and grasp forces for establishing contact.

The joint reference value $q_d$ is computed in dependence on the desired cartesian object trajectory ${}^o p$. The joint torques $\tau_d$ required to compensate the joint position deflection $dq$ are obtained by:

$$\tau_d = J^T \, T^T \, G^T \, K_c \, G \, T \, J \, dq \tag{6}$$

where $J$ is the hand jacobian, $T$ the transformation from the center of the fingertips to the contact points and the so-called grasp matrix $G$ the transformation from the contact points to an object coordinate frame.

The desired stiffness behavior is defined in cartesian object coordinates by the matrix $K_c$. ${}^o dp$ denotes the cartesian position error of the object.

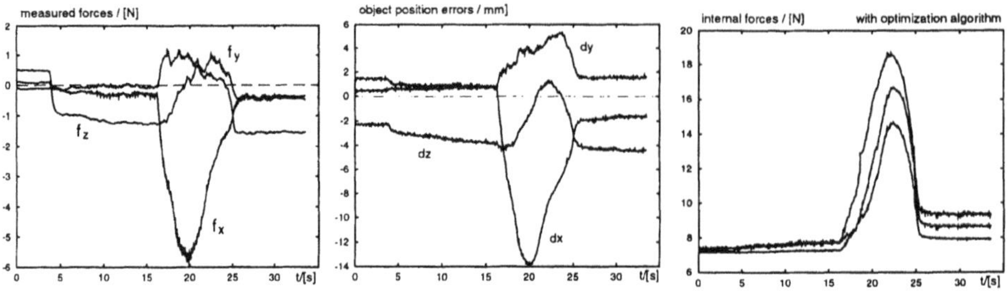


a) b) c)

Figure 8: Asymmetric online adaptation of the grasping forces during external disturbances.

Additionally, a bias force is introduced specifying a-priori known feedforward forces to be exerted by the gripper, e.g. for compensation of gravitational acceleration when the object orientation changes. Thus the desired force in object coordinates ${}^o f$ contains a feedback part of the stiffness loop ${}^o f^{fb}$ and a feedforward part of the bias ${}^o f^{ff}$ as

$$ {}^o f = \begin{bmatrix} {}^o f_{ext} \\ {}^o f_{int} \end{bmatrix} = {}^o f^{ff} + {}^o f^{fb} \,. \tag{7} $$

The desired values ${}^o p$ and ${}^o f_{ext}^{ff}$ can at least coarsely be adjusted by the contact points and forces planned in sections 4.1+2. But the main problem using the stiffness control for grasping a-priori unknown objects is their unknown mass. To compensate this unknown force, several mechanisms are used in practice:

- Heuristical compensation by applying high internal forces ${}^o f_{int}^{ff} = const.$, but this destabilizes the grasp due to modelling errors.
- Empirical compensation by measuring and positive feedback of the external force ( also destabilizing due to the feedback).
- Analytical compensation by online optimization of the internal forces [26], which requires high speed algorithms and computers.

The practicability of the analytical approach is described by the following experiment (fig. 8), which is documented by the optimized internal force ${}^o f_{int}^{opt}$, the cartesian position error of the object ${}^o dp$ and the measured external forces ${}^o f_{ext}^{m}$ ($= J^{-T} T^{-T} G^{-T} \tau$ obtained from the joint torque sensors):

After grasping a cylindrical peg (radius 25mm, weight 200g) the hand was lifted (t=5sec.) and translationally moved by the robot ($t = 5...35sec.$) despite of a collision between the peg and a table ($t = 18...25sec.$).

As shown in the right picture of fig. 8 there is an online adaptation of the internal forces in order to strengthen the grasp in presence of disturbances. The fingers are arranged symmetrically with an angle of 120° between them suggesting equal internal forces. When the disturbance occurs internal forces become non-symmetric. After the disturbance a small asymmetry is observed which is due to a change in the grip transform $G$ caused by the disturbance. This collision experiment could

a) b)

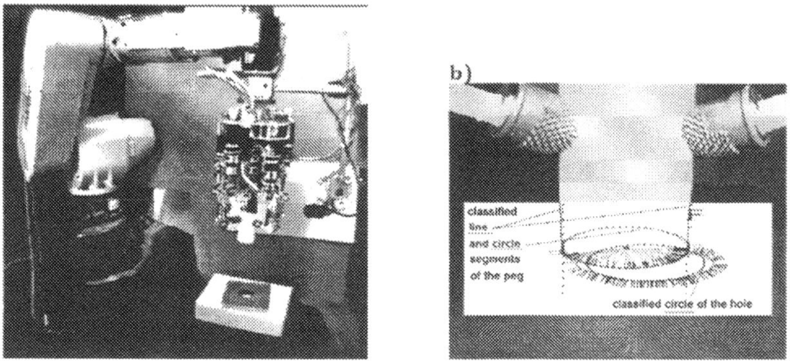


Figure 9: For assembling a cylindrical peg into a hole (a) the position and orientation of peg and hole are online determined by vision classifying the parallel straight lines as well as the visible circle segments of the cylinder and the hole (b).

not successfully be accomplished with the stiffness control approach alone, even if choosing very high constant internal forces heuristically! Simply setting high internal forces $f_{int,i} = 18$ caused stability problems.

## 5. Manipulation Experiments Using Visual and Tactile Skills

Since the previous section presented strategies for establishing contact, avoiding slipping and keeping in contact with the object during manipulation the basis for autonomous manipulation procedures is given. Besides applications to disassembly vision and force/torque sensor information can also be used for guiding and supervising assembly motions which will be exemplified by the benchmark problem of peg-in-hole insertion.

### 5.1 SENSOR-BASED ASSEMBLY

We considered a 3D peg-in-hole problem representing a complex manipulation task. In our experimental setup, a cylindrical peg (radius $50\,mm$) should be inserted into a nearly chamferless hole. As a consequence of several errors/disturbances exact modelling is not possible and task oriented strategies have to be formulated. We performed the task with a hand-arm-eye system (fig. 9a) consisting of a three-fingered dexterous gripper (cf. fig. 1b), a six-axes puma-type robot and a tiltable camera, which supplies seven degrees of freedom for observation and fifteen for manipulation. In the experiment, we divided the assembly of the peg into three stages:

**Phase I - Pick and Place:** In the first stage the camera looks into the global scene within which peg and hole are visible. A coarse 3D localization of both object centers is performed by stereo analysis of two images acquired during a short robot motion (sec. 3.3). For fine inspection of the peg to be grasped the robot moves the camera almost perpendicular above the peg. In order to get an optimum representation of the object in the image, the camera is moved as long as the object is zoomed to a desired size in the image center. Using this precise 3D relation between object and camera, the preshaped gripper approaches the object due to the known camera-gripper transformation and the coarsely determined distance between camera and object. After the camera has been tilted, it supervises the object from a side view while the robot moves down, until the fingers contact the peg in the desired grasp height[1]. Alternatively, this could be assured by light barriers which are more robust against a light background. But the arrangement of IR-diodes and IR-detectors inside the fingertips of the dexterous gripper (fig. 1b) is rather difficult due to the fact that the positions of the fingertips are not constant. According to the determined coarse position of the hole, the grasped object can be placed to the hole within a radius of $25\,mm$, which may be decreased by a fine localization of the hole as well. In this stage, robot and gripper motions are guided by vision alone while the internal joint torque sensors of the gripper supervise the motions and the fingers perform compliant evading motions in case of sensed collisions of the object with the environment (e.g. during placing the peg onto the plane ground surface).

**Phase II - Approaching:** For getting the peg closer to the hole, the second stage is activated. At first, the peg is lifted a little so that the camera can look beneath the peg in order to detect and localize the hole. In the contour image of the scene, a circle segment representing the hole is searched for (fig. 9b). A region of interest is selected due to the detected position of the hole in the image where the peg is also supposed to appear. In this picture of interest, which is indicated by the white box in fig. 9b, the contours of peg and hole are approximated by line and circle segments. If the peg is seated on a plane surface, a 2D measurement of the relative position of the object with respect to the position of the hole allows a coarse determination of the distance and direction the object has to be moved. Since collisions between peg and hole are necessary for assembly the danger that the peg slips out of the hand can be lowered by employing touch sensors for online slip supervision. By online visual analysis of the scene after each step of motion, the peg is positioned very close to the hole, but due to disturbances and certain ambiguities a misalignment can occur up to $3\,mm$ and even if this error would disappear, the small clearance leads to jamming and wedging which can not be detected by vision.

**Phase III - Insertion:** Therefore, in the third stage the task has to be completed by employment of the internal force/torque sensing of the gripper and vision is only used to supervise the upper edge of the peg in order to detect when this edge moves down and the peg is inserted into the hole. During this stage dexterous insertion capabilities are required. The basis of the approach [27] is to perform manipulation primitives in dependence on discrete situations, which can be considered as an abstraction of the insertion states obtained by classification of the measured forces. Actually, the success rate for the peg-in-hole problem in phase III ("insertion") is strongly depending on the initial state of it, which results from the previous phases "pick-and-place" and "approaching". To maximize the success rate, two goals have to be met by them: Firstly, minimization of grasping errors, which

[1] The fine localization of the object is necessary for providing an optimum joint range availability of the fingers due to the controlled grasp height and for keeping the initial state error described below as small as visually sensible.

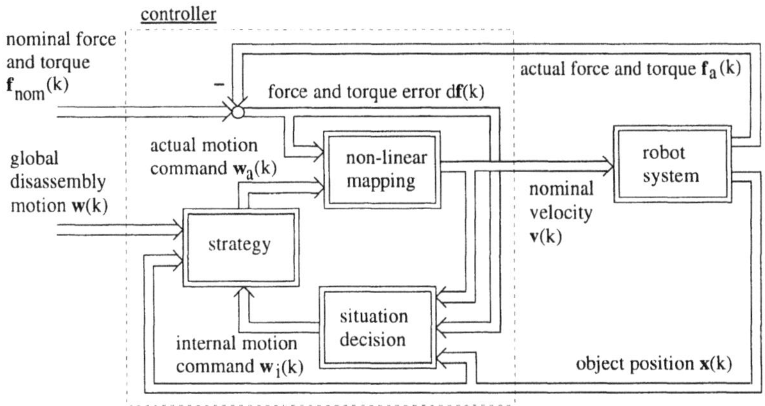


Figure 10: Control structure: a strategy level consisting of the function blocks "*situation decision*" and "*strategy*" for realizing global motions (coarse motion planning) and a non-linear mapping for generating local evading motions (fine motion planning).

guarantees an optimal versatility of the peg in hand and secondly, minimization of the distance of the center of the peg from the center of the hole before starting the "instertion"-phase. Therefore the three phases have been implemented as separate modules allowing separate optimization of the established manipulation primitives.

## 5.2 SENSOR-BASED DISASSEMBLY

Disassembly cannot be considered generally as the reversal of assembly because of a raised uncertainty: The condition of the product to be disassembled may change during its life-cycle and disassembly difficulties resulting from aging, use or product repair can occur. Therefore, disassembly strategies to be developed should not depend on an exact modelling of the environment. By integrating internal and external sensor information in a control loop structure, however, it is possible to react to these model errors in an efficient way, as it will be shown in the following (for details see [28]). In order to reduce the expenditure of system design, only the coarse disassembly direction is specified in advance while fine motions are planned online based on sensor information. In this way, no exact model of the environment is necessary and the system is robust against uncertainties. The fine motions to be applied are mostly strategies based on force/torque sensor information to overcome jamming and wedging of the object to be manipulated.

Following the behavior of human beings who perform manipulation tasks by executing global motions and superposed local evading motions a corresponding control structure for the robot system has been developed (see fig. 10). It consists of a strategy level for realizing global motions (coarse motion planning) as well as a non-linear mapping for generating local movements (fine motion planning).

**Global motions:** During a disassembly operation there are very often situations in which several reasonable motion possibilities exist. Since, however, because of lacking knowledge about the environment no favorable motion can be specified, different possibilities have to be tested by executing global search motions. These superior nominal global motions are generated in the strategy level consisting of the function blocks "*situation decision*" and "*strategy*" (cf. fig. 10). The situation decision determines a new internal nominal motion according to the force/torque sensor information, the current object position and the commanded nominal velocities. Then the strategy decides whether the superior disassembly motion is continued or the disassembly motion proposed by the situation decision is executed. A combination of both as the resulting new nominal global motion is possible as well.

**Local motions:** While the strategy level provides coarse motion specifications only, the non-linear mapping-based controller determines local motions, i.e. nominal velocities $v$ to command robot motion are generated depending on the current force/torque error $df$ and the nominal global motion $w_a$ given by the strategy level. Thus, the non-linear mapping-based controller is responsible for the fine motion and a compliant behavior. In this way, the coarse disassembly direction is followed and obstacles are avoided by evading movements depending on the forces and torques resulting from contact with the environment. In this work, the non-linear mapping for generating local evading movements has been realized by a fuzzy controller to facilitate the integration of heuristic knowledge. For the disassembly, following the human behavior simple heuristic IF-THEN rules can be stated:

- IF (*forces/torques are low*), THEN (*move object in disassembly direction*).
- IF (*forces/torques are high*), THEN (*execute evading movement*).

Resulting from these abstract rules, identical IF-THEN rules are obtained for each degree of freedom, e.g. for the nominal translational velocity $v_{t,x}$ in $x$-direction:

- IF $df_x$ = *negative*, THEN $v_{t,x}$ = *negative*.
- IF $df_x$ = *about zero* AND $w_{t,x}$ = *negative*, THEN $v_{t,x}$ = *negative*.
- IF $df_x$ = *about zero* AND $w_{t,x}$ = *about zero*, THEN $v_{t,x}$ = *about zero*.
- IF $df_x$ = *about zero* AND $w_{t,x}$ = *positive*, THEN $v_{t,x}$ = *positive*.
- IF $df_x$ = *positive*, THEN $v_{t,x}$ = *positive*.

Using normalized input and output variables enables an easy adaptation of the controller to different tasks or situations, e.g. to the sensitiveness of object and/or environment, without changing the principal behavior of the controller. In case the local evading movements should have a fundamentally different behavior depending on the respective task (e.g. disassembly or immersing motions in order to approach an object to be grasped), this can be realized by switching between different non-linear mapping-based controllers. The strategy level selects one of them according to the given task and in this way, a switching between different motion behaviors is possible.

**Experiments:** Applying the control loop structure proposed, various disassembly tasks could successfully be performed, e.g. the disassembly of the objective module of a camcorder. Fig. 11 shows a schematic representation of a typical sequence of

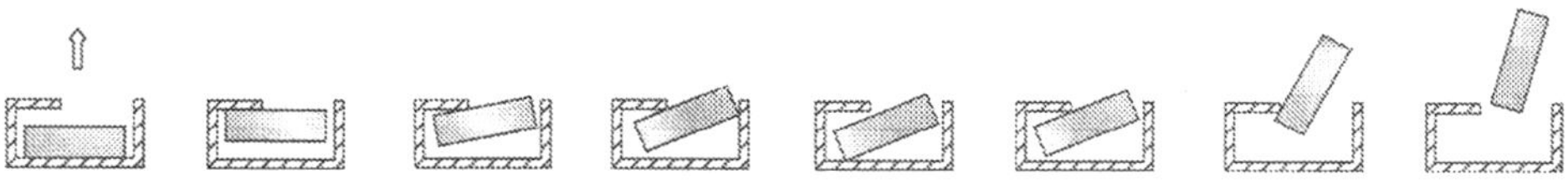

Figure 11: Typical sequence of situations during disassembling the objective module of a camcorder.

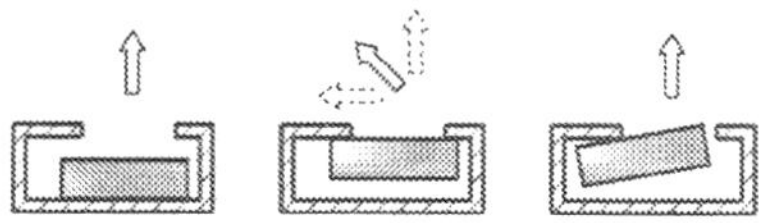

Figure 12: Disassembly problem in which the global motion direction (characterized by the arrow) has to be changed during the disassembly operation.

situations during the disassembly operation. The objective module is located inside the camcorder casing, which allows a disassembly motion upwards only. However, this motion is constrained by the undercut of the box, so that during the disassembly operation the strategy has to avoid the obstacle. The information given to the system is only the coarse disassembly motion "upwards", the fine motion – in this case, the necessary object rotation – is automatically generated by the system.
In this example, even the application of the non-linear mapping-based controller only without the global strategy is sufficient, while for more complex problems, e.g. in dead end situations, the global strategy is required in order to adapt the superior disassembly direction, if necessary (see e.g. fig. 12).

## 6. Conclusions

This chapter based mainly on [29], [30], [31] presented an approach for the integration of various sensors into a dexterous robotic system in order to solve autonomous object manipulation. Basic requirements have been discussed for:

- object inspection by a mobile camera and a laser range finder in order to acquire information necessary for approaching and grasping objects of a-priori unknown position, shape and accessibility,
- fault-tolerant approaching of planned contact points by force/torque sensor-based motions and in addition by touch sensor-based optimization of contact surfaces and forces,
- manipulation of the grasped object for assembly/disassembly using vision and tactile sensors for motion guiding and supervision.

The experimental validation showed that vision and range as well as internal force/torque and touch sensor information have to be combined in order to maximize the autonomy of the system. As documented by fig. 13 the answer to the question of when to use which kind of sensor depends on the distance of the end-effector to the goal which is assumed to be the object to be grasped at first. Before a contact can be established color vision is employed for segmentation. By arm-eye

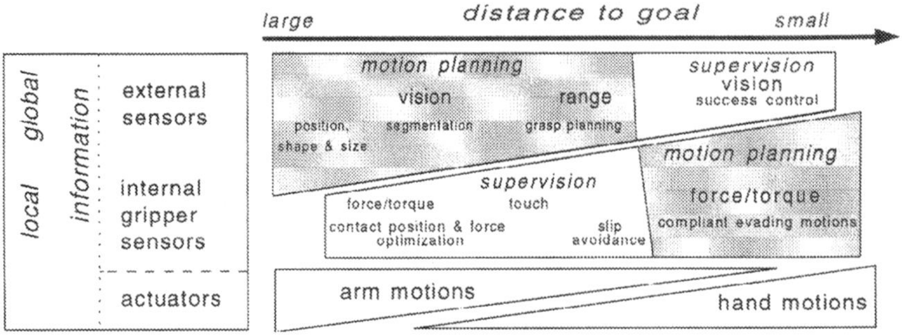


**Figure 13: Effective range of the presented methods and limits of vision and internal gripper sensors dependent on the distance of the hand-arm-eye system to the goal position.**

coordination the object is zoomed to an optimum representation. This allows the inspection of interesting object regions from a close distance and subsequently a detailed planning of contact points and forces by using vision and/or laser range sensors. Since range sensoring is relatively slow as long as no 3D-scanner is employed only a segmentation of relatively small object parts can be performed. So, range sensor-based grasp planning is a more detailed analysis of the object regions of interest and the sensor must be much closer to the object than in case of vision-based grasp planning.
The closer the distance to the object, the more local tactile sensors of the gripper are used for avoiding collisions, for optimizing contact surfaces and forces and for establishing contact at the most suitable positions. After grasping the object, the further manipulation task, e.g. for assembling a peg into a hole, is to move the grasped object to a goal position which is now e.g. the center of the hole where the peg is to be inserted in. For guiding the system to this position, again vision is used as long as the distance to be covered is relatively large. But when the resolution of the camera becomes insufficient to detect the remaining deviations to the goal, fine manipulation is guided by force/torque sensor information, alone, and vision is used for supervision, only (e.g. success of the insertion). The experiments in sec. 5 demonstrated that a combined use of different sensor types increases the security and the success rate of manipulation on the one hand because some information is analyzed redundantly. On the other hand, every sensor type presented is to be employed for guiding the robot in a specific section on its way to reach the goal. So, the limited capabilities of visual and tactile sensors complement each other.

Whereas future development od vision, range and force-torque sensoring techniques will focus on more powerful software algorithms, the trend in touch sensor-based slip detection is to create more complex sensors which gives more detailed information. But for most applications, a reduced number of taxels for slip detection in combination with traditional vision, range and force/torque sensors is a good compromise between costs and benefits.
Besides the problem of applicability of the different sensor types during the phases of the manipulation process, a major difficulty is the efficient design of interfaces for a combined use of sensors. Since we employed the different sensor types in most

cases in parallel every sensor was expected to provide a certain kind of information which is used to generate a specific kind of motion behavior. Those sensor-based motion commands were superposed on an upper scheduling level of our hierarchical information processing structure. Basically it is also possible to map different kinds of sensor signals to the favored motion strategies by training for example neural nets or non-linear mappings as explained in section 5.2. The advantage of this form of sensor fusion in comparison to the hierarchical structure is that the security and efficiency of the planned motions is increased. For example the classification during the approach for grasping is improved if infra-red light barriers are employed in addition to force/torque sensors. Also the peg-in-hole insertion could be achieved quicker by classifying visual and tactile sensor information simultaneously, e.g. superposing the approaching and insertion stage. However, this being a line of research in the future requires an extended sensor signal preprocessing and the structure of information processing and motion generation remains more unobvious which is for some problems hard to control.

## References

1. Haralick, R., and Shapiro, L.: *Computer and Robot Vision*, Vol. II, Addison Wesley Pub., 1993.
2. Horn, B.: *Robot Vision*, The MIT Press, Cambridge, USA, 1986.
3. Shirai, Y.: *3D Computer Vision*, MIT Press, Cambridge, 1987.
4. Joseph, R. and Rowland, J.: *Fusing Diverse Sensor Data By Processing Abstract Images*, Proc. Int. Conf. Intelligent Autonomous Systems, Karlsruhe, Germany, 1995.
5. Trobina, M., Leonardis, A. and Ade, F.: *Grasping Arbitrarily Shaped Objects*, DAGM Symp. Mustererkennung, Vienna, Austria, 1994.
6. Howe, H.: *Tactile Sensing and Control of Robotic Manipulation*, Advanced Robotics, Vol. 8, No. 3, 1994.
7. Seekircher, J. and Hoffmann, B.: *Improved Tactile Sensors*, Int. Symp. Robot Control, Karlsruhe, Germany, 1988.
8. Holweg, E.: *Autonomous Control in Dexterous Gripping*, Phd Thesis, Delft University of Technology, 1996.
9. Allen, P.: *Robotic Object Recognition Using Vision and Touch*, Kluwer Academic, Norwell, USA, 1987.
10. Maekawa, H., Tanie, K., and Komoriya, K.: *Tactile Sensor-Based Manipulation of an Unknown Object by a Multifingered Hand with Rolling Contact*, Proc. Int. Conf. Robotics and Automation, Nagoya, Japan, 1995.
11. Vischer, D.: *Cooperating Robot with Visual and Taktile Skills*, Proc. Int. Conf. Robotics and Automation, Nice, France, 1992.
12. Bard, C., Laugier, C. and Milesi-Bellier, C.: *An Integrated Approach to Achieve Dexterous Grasping from Task-Level Specification*, Proc. Int. Conf. Intelligent Robots and Systems, Munich, Germany, 1994.
13. Paetsch, W. and Kaneko, M.: *A Three Fingered, Multijointed Gripper for Experimental Use*, Proc. Int. Workshop Intelligent Robots and Systems, Tsuchiura, Japan, 1990.
14. Corke, P.: *Visual Control of Robot Manipulators - A Review*, Visual Servoing, K. Hashimoto (Ed.), World Scientific Series in Robotics and Automation - Vol. 7, 1993.

15. Yoshimi, B. and Allen, P.: *Active, Uncalibrated Visual Servoing*, Proc. Int. Conf. Robotics and Automation, San Diego, USA, 1994.
16. Seitz, M., Holeschak, U. and Kleinmann, K.: *Active Inspection and Handling of Unknown Objects Using an Autonomous Hand-Arm-Eye System*, Proc. Int. Conf. Intelligent Autonomous Systems, Karlsruhe, 1995.
17. Weigl, A. and Seitz, M.: *Vision Assisted Disassembly Using a Dexterous Hand-Arm System: An Example and Experimental Results*, Proc. Int. Symp. Robot Control, Capri, Italy, 1994.
18. Weigl, A., Hohm, K. and Seitz, M.: *Processing Sensor Images for Grasping Disassembly Objects with a Parallel-Jaw Gripper*, Proc. TELEMAN Telerobotics Research Conference & ERNET Workshop, Noordwijkerhout, The Netherlands, 1995.
19. Weigl, A., Hohm, K. and Tolle, H.: *A Flexible Tactile Grasping Strategy for Autonomous Robotic Disassembly*, Proc. Int. Symp. on Industrial Robots, Milano, Italy, 1996.
20. Holweg, E., Hoeve, H., Jongkind, W., Marconi, L., Melchiorri, C. and Bonivento, C.: *Slip Detection by Tactile Sensors: Algorithms and Experimental Results*, Proc. Int. Conf. Robotics and Automation, Minneapolis, USA, 1996.
21. Marconi, L. and Melchiorri, C.: *Incipient Slip Detection and Control Using a Rubber-Based Tactile Sensor*, IFAC World Congress, San Francisco, USA, 1996.
22. Howe, H. and Cutkosky, M.: *Sensing Skin Acceleration for Slip and Texture Perception*, Proc. Int. Conf. Robotics and Automation, Scottsdale, Arizona, USA, 1989.
23. Tremblay, M. and Cutkosky, M.: *Estimating Friction Using Incipient Slip Sensing During a Manipulation Task*, Proc. Int. Conf. Robotics and Automation, Atlanta, Georgia, USA, 1993.
24. Mingrino, A., Bucci, A., Magni, R. and Dario, P.: *Slippage Control in Hand Prostheses by Sensing Grasping Forces and Sliding Motion*, Proc. Int. Conf. Intelligent Robots and Systems, Munich, Germany, 1994.
25. Mason, M. and Salisbury, J.: *Robot Hands and the Mechanics of Manipulation*, MIT-Press, Camebridge, USA, 1985.
26. Buss, M. and Kleinmann, K.: *Multi-Fingered Grasping Experiments Using Real-Time Grasping Force Optimization*, Int. Conf. Robotics & Automation, ICRA'96, Minneapolis, USA, 1996.
27. Kleinmann, K., Bettenhausen, D. and Seitz, M.: *A Modular Approach for Solving the Peg-In-Hole Problem with a Multifingered Gripper*, Proc. Int. Conf. on Robotics and Automation, Nagoya, Japan, 1995.
28. Weigl, A., Schwartz, M. and Bettenhausen, K.: *A Flexible Motion Control Structure for Autonomous Robotic Disassembly*, Proc. Int. Conf. CAD/CAM Robotics and Factories of the Future (Gill, R. and Syan, C.S., Ed.), Middlesex University Press, London, UK, 1996.
29. Seitz, M.: *Untersuchungen zur Nutzung von Bildverarbeitung für Manipulationsaufgaben in der Robotik*, Phd Thesis, Darmstadt University of Technology, Shaker Verlag Aachen, 1996.
30. Weigl, A.: *Exemplarische Untersuchungen zur flexiblen automatisierten Demontage elektronischer Geräte mit Industrierobotern*, Phd Thesis, Darmstadt University of Technology, Shaker Verlag Aachen, 1997.
31. Kleinmann, K.: *Lernende Regelung eines Mehrfingergreifers*, Phd Thesis, Darmstadt University of Technology, 1996.

# 4

# THE ICa APPROACH TO INTELLIGENT AUTONOMOUS SYSTEMS

R. SANZ, F. MATÍA and E. A. PUENTE
DISAM-UPM
*Universidad Politécnica de Madrid*
*José Gutierrez Abascal 2*
*28006 Madrid, Spain*

## 1. Introduction

Our main interest is the identification of engineering methods for building intelligent machines for real -industrial type- applications. These machines are capable of operation in non-controlled conditions by means of advanced control capabilities, interacting with uncertain environments and performing complex work in collaboration with humans.

For us is of extreme importance achieving immediacy in the task of building an intelligent system for a special purpose. Mars *Sojourners* should be a piece of technology instantiation as immediate as a SCARA robot or a motorcycle. Immediacy means that these systems should be built with reduced effort and increased trustability. In this chapter we present an infrastructural software technology to build intelligent systems: the *Intelligent Control Architecture* (ICa). ICa forms part of a research project on constructability of autonomous systems (ASys). The scope of this chapter is restricted to the engineering approach promoted by ICa. Information on ICa implementation can be found in the ICa manuals.

The inherent complexity of the tasks assigned to such intelligent systems, lead to the need of using quite innovative technologies in their implementation. This motivates the traditional *prototype-like* approach to the construction of these systems.

However, the *one of a kind* issue is not exclusive of intelligent machines. In other areas of technology, the *one of a kind* system construction is the normal engineering way. Moreover, these systems are not prototypes, but well engineered, reliable pieces of technology. Look for example at bridges, or ferries or even skyscrapers.

We want to simplify and make more reliable the process of building complex behavior machines. Machines that can solve complex problems in co-operation with other machines and human beings, sensing their environments, focusing their attention, making decisions and building action plans. We want to build methods to construct artificial minds for them.

In this construction we need to solve several high level problems related with five main elements (with some overlapping):

- **Task**: The task the robot must perform.
- **Environment**: The environment of the robot and the task.
- **Beings**: The behaving entities in the environment[1].
- **Body**: The body of the robot.
- **Mind**: The mind of the robot.

What we want to build are artificial minds for artificial bodies. In dealing with this engineering purpose we envision four parts for our minds:

- The *task solving system*, in charge of generating control actions to achieve the objectives of the intelligent system.
- The *perception system*, in charge of data gathering from the environment and from the self.
- The *social system*, in charge of relations with other beings. Composed of three parts: interaction with humans, interaction with peers and interaction with aliens[2].
- The *self*, in charge of the robot body and mind management; it is a reflective [1] system.

## 2. The Concept of Intelligence and Autonomy

What is an *intelligent autonomous system*? What are its features and capabilities? Where is the frontier between autonomous and not autonomous or between intelligent and not intelligent?

These questions are challenging those big gurus of the artificial intelligence community. We, as simple engineers, have a clear idea of what we say when applying the adjectives *autonomous* or *intelligent* to some system. The main issue here is to be aware that these

---

[1] For these systems several terms have been employed, mainly *actors* and *agents*. For us *actors* are behaving entities that take a fake role and *agents* behave on behalf of others. In these category we include only those external active entities that can interact through a dialogue with our robot. Other entities, passive or not dialoging, are included in the environment.

[2] For the intelligent being, humans are also aliens, but for us, in our GOD role, humans are worth a special interaction mechanism; because machines are our slaves (at least in the present).

terms are not talking about classic -boolean valued- concepts. The concepts of *autonomy* and *intelligence* are pure fuzzy concepts. They refer to the degree of presence[3] of some property in a specific, behaving system. But these properties should be always analyzed from the point of view that they have only sense in relation with task accomplishment of the system.

This means that *autonomy* and *intelligence* are not properties but relations of systems that perform activities; and they are not absolutely measurable properties of an isolated system. What we can really say is that one system behaves more intelligently or more autonomously than other. Intelligence is related with the capability of solving complex tasks. This means that intelligence is considered greater of those systems than can solve more complex tasks. Autonomy is related with the capability of performing the tasks in an independent, helpless way.

When two systems A and B are requested to accomplish task T, the relative intelligence of the systems is the relation of the quality of the results of each one. When confronted with different tasks, task complexity should also be taken into account. We will say that system A is more intelligent than system B if its results are better or if the task solved is more complex. This can lead to inconsistencies in the more-intelligent-than relation, because A can get better quality than B in task T1 and worse in task T2. More, system A can solve a task T3 that is more complex than task T4 than can be accomplished by B; and the reverse can hold for two other tasks T5 and T6.

We conclude that intelligence relations should be established as a function of quality of results for specific tasks. So we should talk about intelligence in performing specific tasks (finding objects, controlling arms, ordering boxes, perceiving targets, etc.) and not as global properties of systems.

The same type of analysis can be done for autonomy. We must say that A is more autonomous than B performing task T and not that A is an autonomous system.

The use of intelligence and autonomy as properties of a unique system performing a specific task can only be done if there exist some metric of result quality. So we can equate intelligence level to quality given by the metric. This is what is done with IQ tests. If there is an upper bound of quality (i.e. there exist a maximum quality or task perfect accomplishment) we can say that the system obtaining that maximum quality is intelligent without referring to any degree of it. But always in relation with a task.

When we use the expression *intelligent autonomous systems* we mean systems that solve more complex tasks than others, that obtain better results and that do need less help from external entities to achieve their results.

From now on, we will use the acronym IAS, to refer to a system that should perform a complex task with limited interaction with external resources.

---

[3] And not *presence or absence*.

## 3. A domain for Intelligent Autonomous Systems

Intelligent autonomous systems (IAS) have two parts: a body and a mind. IAS' body designs are strongly related with the physical properties (shape, size, speed, etc.) of the tasks they are intended to perform. IAS' mind designs are related with the complexity of the task.

When we use the term intelligent autonomous system, we are mostly concerned with the IAS mind (mainly due to the task scale independence mentioned earlier). In fact, in our autonomous systems laboratory, we are addressing the application of these technologies in two quite different domains:

- ***Complex continuous processes***: These systems are characterized by three factors: scale, money and, sometimes, risk. They are big systems with lots of elements and variables an their control must achieve economically good results whithin safety levels; this means that no exploratory control can be done and that bad actions are absolutely forbidden.
- ***Autonomous vehicles***: In this case, complex perception is the main issue. The mobile vehicle must construct an internal view of the world he is moving within[4].

It can be thought that it is difficult to reach a common point between these two domains; that the fashionable *embodiment* concept cannot be easily accomplished with such different bodies. But our experience as control engineers tell us that control loop designs, independently of their complexity or technology, can be reused in such different domains. The same fuzzy controller can be in charge of controlling an oxygen flow in a kiln or a robot trajectory in a plant floor (obviously with modifications). These insights lead us to devising a modular approach to mind construction.

What is common to several domains is the abstract structure of the problems to solve; problems that require in many cases heterogeneous solutions to achieve specific performance levels.

## 4. A Constructive Approach to Artificial Minds

Our research is based in a simple hypothesis: that foundations of intelligent behavior can be independent of the task at hand. We *believe there are mechanisms for minds that can be reused in several tasks.* The level of task independence can vary from mechanism to mechanism, but we believe that they can be constructed with modularity and composablility properties.

Autonomous systems, or better, elementary mechanisms of mind classification, can be managed with a design space approach in which different orthogonal properties are used to classify mind designs. Some of the dimensions of this design space are:

---

[4] This does not mean that there exist an explicit representation of the world inside the robot controller [17], but that some form of coherent perception arises, perhaps directly in the form of robot behavior.

- **Analysis vs design**: Some artificial mind designs are based on models of biological systems; they are built using an analytical approach. From the other side, the type of minds we want to address are designed minds; models of minds that do not resemble known models of biological systems, but that have been designed for a specific purpose.
- **Top-down vs bottom-up**: Some minds are built using high-level cognitive approaches –for example expert systems- and others are built using subcognitive approaches –for example neural networks. In the last case, some form of learning or adaptation is necessary to achieve specific functionality.
- **Simplicity vs complexity**: It is not the same a mind for a welding robot that a mind for a submarine exploration robot. Complexity, as mentioned before, is the main issue of intelligence.
- **Constructive vs evolutive**: Minds can be built from their parts producing a somewhat static final structure or can be grown up from stupid stuff using some form of stuff evolution.

The following table shows a draft evaluation of placement of technologies and approaches into the design space.

| Technology | Complex | Constructive | Bottom-up | Designed |
|---|---|---|---|---|
| Computational Neuroscience | + | - | - | + |
| Artificial Life | - | + | + | + |
| Real Life | +++ | --- | +++ | --- |
| Neo Modularity | ++ | ++ | + | + |
| ICa | ++ | +++ | - | +++ |
| Neural Networks | - | -- | ++ | + |
| Genetic Algorithms | - | -- | ++ | - |

We want to address the issue of constructiveness of complex, design based minds. We want to use a globally top-down designed mind architecture, using a constructive approach to the mind component integration problem. We believe that the constructive way is the most effective way to achieve really complex artificial minds, in terms of money spent in their construction.

This does not mean that we neglect the use of evolutionary or bottom-up models. We think that this models are extremely useful –and even better in some cases- to build specific mind components, but their usefulness decrease when the global mind complexity increases; obviously due to task complexity.

Our proposed method to autonomous intelligent system construction is based on an almost classical sequence of steps:

- Specification of complex task
- Analysis and decomposition in canonical tasks
- Design of task solving components
- Design of integration components
- Construction of task components
- Construction of integration components
- Component integration

Our approach is near to what Stein calls modular functionalism, because we think that evolutive or not modular approaches does not scale well from the toy problems in which they are usually employed [2].

In this chapter, we address an almost pure software problem; the problem of providing adequate software infrastructure to build minds for autonomous systems. This infrastructure is addressed at the core modularization and integration issues.

## 5. A Reuse-Based Construction Model

The model we propose for artificial minds is based in the agency concept. We have selected this model not because it resembles the human mind [3] but because it offers a whole bunch of engineering alternatives, based on its intrinsic flexibility.

When searching for a technology to implement IAS, we must identify the requirements posed to that technology. IAS requirements are so varying that no simple technology can comply with this. Flexibility, extensibility and scalability are desirable properties for any type of software. In the case of background software for IAS, they are not desirable but indispensable properties.

In order to provide a coherent agent implementation framework we have developed ICa. It offers technology to adapt generic architectural design patterns to specific implementations, tailoring artificial intelligence based architectures to the concrete tasks posed to the IAS. The IAS can then be built by means of composition of functional modules atop the basic integration infrastructure. The modules are constructed by instantiation and adaptation of task oriented object frameworks to implement the collection of interacting agents that will constitute the IAS mind.

What is clear is that from system to system, a N-layered (or whatever design pattern you prefer) architecture needs to be refurbished to comply with the requirements of the new system. Direct, clean reuse is not possible if the problems are so varying. What we need is not reusability but *refurbishability*.

We have suffered this problem during the last years of work on intelligent control applications [4]. We have started from scratch many projects because specific requirements limited our ability to reuse pieces of software from previous projects.

Architectural reusability means effective use of previous work in software architecture. Under the ICa framework, the types of reusable items we perceive are:

- Analysis
- Design
- Objects
- Components[5]
- Agents

These different reusable items are commented below, analyzing how they are reused and the degrees of reuse that can be obtained.

## 5.1 ANALYSIS REUSABILITY

Analysis reusability may sound quite strange. Analysis is viewed as the process of studying a problem and identifying the ways to the solution. So, analysis should be centered in a specific problem, in a specific domain, in a specific context. Too much specific to be reusable.

Analysis reuse is impossible except if you have the same problem to solve again. This is true, but not *absolutely true*. The real thing is that problems are quite complex. Control problems for which intelligent control is viewed as a possible solution are really wicked control problems. Autonomous system controllers are among the most complex controllers (from the point of view of uncertainty in the environment). In most cases, these problems arise from secondary or tertiary complexities [5][6], due to the uncertain interaction of large quantities of not well known subsystems.

The natural approach to analysis reusability will be some form of *analysis encapsulation*. Encapsulated analysis will be reusable in the same sense as encapsulated code. The analysis work should be in many cases a pattern matching process: identifying the problem (matching against a problem collection) and adapting the coupled solution.

## 5.2 DESIGN REUSABILITY

The patterns movement [6] is one of the most serious efforts to bring design to the category of engineering from the category of art[7].

The term *design* is usually employed in two ways:

- As a verb, meaning devising something with some purpose in mind.
- As a noun, meaning the *how* of the designed thing.

---

[5] Including *architectural components*.

[6] Primary = size (number of things), secondary = interactions, tertiary = uncertainty.

[7] Well, in fact the terms should be *engineering art* and *fine art*, because design, as the term is usually understood, means creating something *new*, and this is always a form of art (See [Winograd 96] for more on the art view of software design).

Obviously in our case *design reusability* refers to the second meaning; because the first meaning means work (it's a verb) and work cannot be reused. Only the results of the work can be reused.

Design reuse means use for a new application a design that was used for a previous one. Design can be easily reused if the applications are equal at least in an encapsulated part (as is the case of analysis reuse). However, design can be reused from an application to a different but analogous one.

This design by encapsulated analogy is what the pattern movement promotes. Identify a section of the system to build that matches a system previously known (this matching is the though point of the process) and adapt the solution (the design) to the new system.

To work in this way you need a knowledge base of design cases and methods to instantiate new designs from old ones: adaptation, fusion, integration, etc. The term *pattern language* has been coined to identify these *bodies of knowledge*. Authors write books and papers proposing new pattern languages to bring design knowledge to specific domains. An example strongly related with ICa is the book of Mowbray and Malveau [7], proposing a pattern language for CORBA based systems.

The contents of a pattern in a pattern language can have two main forms quite related with classical artificial intelligence knowledge representation. It is possible to specify generics and instantiate them by aspect completion or to specify prototypes[8] and derive from them by aspect change. The use of one or another is a matter of taste. From our opinion some persons work better with prototypes and other work better with generics. It is even a matter of domain.

Notwithstanding, the expression of a body of knowledge as generics, cause some repulsion to many people, because it sounds too vague and abstract. Therefore, in most cases generic bodies of knowledge include examples of use to reduce vagueness. This is the usual form in which design patterns are presented: collections of generics with accompanying examples.

For software design knowledge expression, we think that this form is better than a collection of prototypes, because of the inherent flexibility of software systems. Prototypes cannot encompass all the possibilities of a particular design. Prototypes suffer from the classical problem of the *one-of-a-kind* systems that are usual in the IAS community.

In a former paragraph we said that to obtain a new design, all you need is to *instantiate new designs from old ones using adaptation, fusion, integration, etc.* This is easy to say but no so easy to do. What these pattern languages lack is a clear way to perform the active part of *design* (the verb meaning). Pattern languages simplify design by reduction of the learning process of the designer. Their main purpose is the transmission of *static design knowledge* from designer (expert) to designer (novice).

Active design knowledge is needed, but is difficult to express. We think that it is possible to identify two types of it:

---

[8] In the sense of *model*, not in the sense of *unfinished*.

- Horizontal design knowledge, applicable in all domains.
- Vertical design knowledge, domain focused.

Obviously these sets of knowledge are *fuzzy sets*; there exist design knowledge that is partially horizontal and partially vertical.

## 5.3 OBJECT REUSABILITY

Lots of literature exist related with the issue of object reusability. In fact this is one of the selling points of object oriented programming. This is the reason for selecting object oriented technology for ICa implementation. Almost all elements in ICa will be implemented as C++ classes.

## 5.4 COMPONENT REUSABILITY

In ICa we call components to code frameworks used to develop problem solving code. They are artificial mind building blocks. Therefore, components don't appear in an application. What appear in applications are customized or instantiated components[9]. So they are not directly reused, but reused through work, as is the case of design reuse. These customized components are the building blocks of ICa applications, they will be named "agents" if they comply with the ICa Generic Agent structure or with a derived agent specification for an application.

These means that not all mind components need to be agents[10] because they will not need to comply with the application basic agent behavior. This leads to lightweight run-time components that have reduced requirements from the computational system.

In ICa we talk about two principal types of components:

- Problem solving components
- Architectural components

What ICa tries to promote is a building block approach to application development. This approach will lead to a loosely coupled development process of components. With loose coupling we mean that the interaction between component customization processes is strictly limited to a specific set of interfaces for the customized components.

This almost means not taking into account possible uses of a component when defining and constructing it. The identification and implementation of mechanisms and not use policies will lead to great level of reuse.

The reusability will be obtained if these building blocks are *future proof* in the sense used in [8]. This means that they will be useful in the future whatever the future will be.

---

[9] The exact form of this customization can vary from component to component. ICa does not force specific framework structures for components. What ICa imposes are behavior restrictions and architectural styles.

[10] Even when there is no ultimate definition of what an ICa agent is.

A sample component we have just finished is a fuzzy system component. It permits the construction of fuzzy processors to implement agents with soft computing capabilities. We have built fuzzy controllers, filters, validators, constraints, etc. using this component for several applications.

## 5.5 AGENT REUSABILITY

Agents are autonomous processes that perform autonomous activities and offer services to other agents. As said before they comply with the ICa Generic Agent Specification[11].

ICa agent reusability can vary depending of the actual implementation of them. Component frameworks are normally flexible enough to let developers put whatever they want in their applications. This is inevitable and only quality and reusability assurance policies during development can limit reuse-limiting constructs.

The ICa methodology will provide mechanisms to assure reusability in customized components. One of the main mechanisms is the actual independence of code and domain knowledge. This is the central concept of knowledge based systems (the so called *knowledge hypothesis*), in which the domain knowledge used by the agent has the form of data and not of code. These agents are inherently reusable due to the proposed structure for building them. Non-representational systems, however, have a great interest because they resemble better the behavior of biological systems.

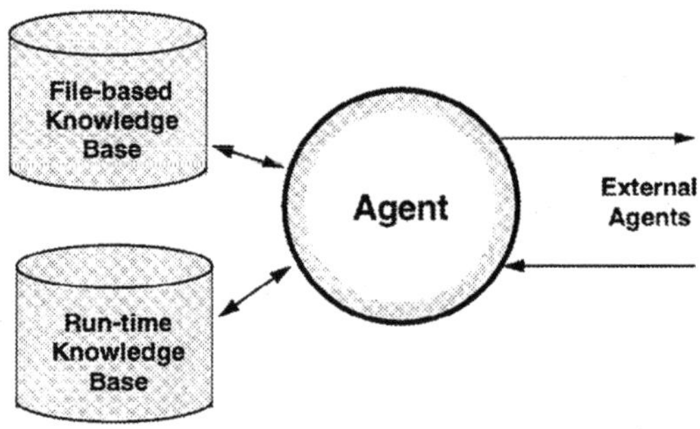


**Figure 1**: A standard structure that can be used for a broad collection of ICa agents (based on the KNOWLEDGEBASE pattern.

This form of application construction, this architectural design, is an example of a design pattern: KNOWLWEDGEBASE.

[11] Well, in this case they will be called *strictly compliant agents*. Other type of agents are possible, because there will be parts of ICa architectural components that will be available in non-strict compliant ICa platforms. By example, client-only agents don't require support of POSIX threads from the underlying operating system.

Most ICa Agents will follow the KNOWLDEGEBASE pattern, and due to this, the main code of them will be reusable from application to application; only new knowledge bases will need to be developed for new applications (See Figure 1).

The reusability of other agents will be also high if developed with enough flexibility to accommodate requirements of new applications as the KNOWLEDGEBASE pattern has.

## 6. The MetaArchitecture Concept

*Meta*Architecture means something that talks about architecture. Something that is used to specify a specific architecture. Something that can be used to store it in a transmissible form. The ICa MetaArchitecture is an abstraction of *possible* architectures in intelligent process control applications.

What we try to find is something that simplifies the process of architecture development for a particular application. Some form of an architectural handbook for intelligent process control designers. We saw three main stages:

- The final stage of an application *architecture implementation*, generated by a process of software engineering from an architecture specification.
- A stage of *application architecture* specification generated using the ICa architectural methodology using as inputs the application requirements, the available software technologies and a body of knowledge about architectures for intelligent process control applications.
- This body of knowledge is the third stage: ICa MetaArchitecture.

In some sense a specific application architecture is an instance of the ICa MetaArchitecture. What we want to obtain is a straightforward process, flexible enough to let final applications satisfy requirements with ample scope.

### 6.1 DESIGN PATTERNS

From our point of view one good way to capture the architectural expertise represented by the MetaArchitecture is the use of design patterns [9] [10]. There are lots of ongoing work and interest on design patterns, by example, the Winter volume of the IEEE Computer Society Software Engineering TC Newsletter offers a quite interesting list of Web sites on patterns.

Using design patterns the process of instantiation an application architecture from generics is simpler and repeatable (It will be more engineering and less art). We can even think in searching an advance in the CMM[12] level for our intelligent control process.

Design patterns offer solutions to architectural problems ranging from the basic class implementations, to planet-wide corporate infrastructure. There are different classification schemes for patterns proposed by different authors. Gamma et al. [9] distinguish three types: Creational, Structural and Behavioral. Mowray and Malveau [7]

---

[12] CMU SEI Capability Maturity Model.

talk about the importance of the scale: objects, microarchitecture, framework, application, system, enterprise and global.

The MetaArchitecture will be a collection of patterns suitable for intelligent process control applications. In particular they will be applicable to the construction of IAS.

## 6.2 PATTERN LANGUAGES

The use of such a collection of patterns can be quite difficult. As mentioned in the design reuse section static design knowledge needs to be worked out to be productive. The active design knowledge will form part of the ICa Methodology, that will include architectural methods based on the set of patterns.

To be really productive, the MetaArchitecture pattern set should have the form of a pattern language; a set of interrelated patterns that can be used in cooperative or integrative ways to specify complex architectural designs.

The ICa Pattern Language (ICA PL) will be a *pattern language for intelligent control architectures*. Pattern languages for other purposes have been described, focused on specific areas; there are even examples in the automatic control area. See by example the work on the G++ pattern language [11] for computer integrated manufacturing applications. The first pattern this language proposes is the LAYERS pattern, one of the main patterns that appear everywhere; some comments on it will be included later (See page 17).

The content and extent of the pattern language is dynamic, as are its main characteristics. The generativity of the pattern language is not an objective of the project, even when we intend to provide a broad coverage of the possible applications in the ICa domain [12].

## 6.3 THE DEVELOPMENT PROCESS REVISITED

Following these ideas, the specific stages in system development can be tagged with more specific information. These tags can be seen in Figure 2.

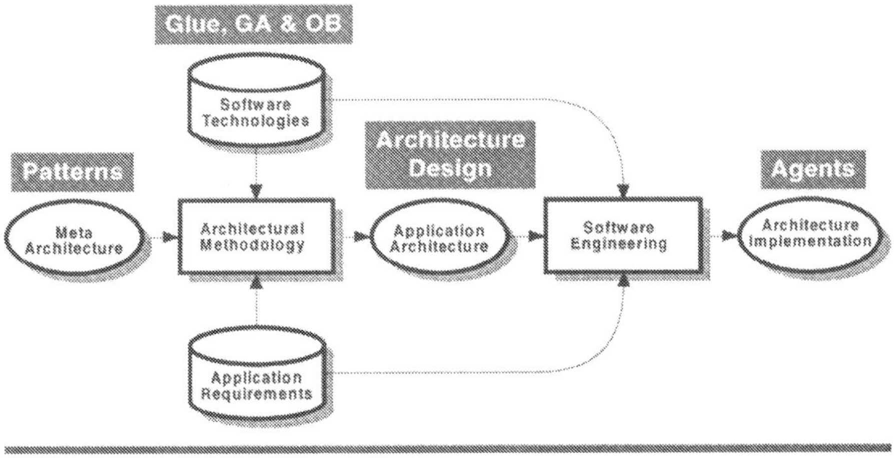


**Figure 2:** Overview of the architectural process with some contents identified. Glue, GA and OB are some basic components of ICa. They will be explained later.

The *metaarchitecture* is a pattern language; the *software technologies* are those provided by ICa architectural components (more on this later) and other problem solving components; the *application architecture* is an agent based design and the final *implementation* has the form of an agency (collection of interacting agents).

The ICa methodology will provide support for all these phases, even for the requirements specification or the selection of software technologies

## 6.4 PATTERN TYPES

From one point of view we can classify the patterns for control depending on the scale:

- Class level
- Framework level
- Application level
- System level
- Corporate level

Other classification we will use is to distinguish between generic patterns and control specific patterns. The classification of patterns in the ICa pattern language will be done with the two criteria mentioned before: level and focus. The table below shows some examples.

| Level/Focus | Generic | Control |
|---|---|---|
| Class | SINGLETON | VARIABLE |
| Framework | LOCK | FUZZYFICATOR |
| Application | LAYERS | FILTERPID |
| System | BROKER | RISKMAN[13] |
| Corporate | WEB | - |

In the level dimension we will differentiate between the mentioned levels. In the focus dimension we will only differentiate between generic and control patterns. Obviously some patterns will be usable in several niches; the division is only for organizing purposes, it does not try to establish a definitive taxonomy of patterns for process control systems.

Generic patterns are patterns useful for all type of computer applications, not only control and supervision. It is the same case for control patterns, even when the names indicate their intended use they can be used in other type of applications not related with process control.

As a result of the ICa project, the contents of the ICa pattern language will be enough to reissue applications like those demonstrated in ICa. But the pattern language will be extensible and, for sure, the proposed patterns will evolve in a natural way, following the *forces*[14] of the process control environment: process technology, safety regulations, efficiency requirements, pollution control, economic operation, etc.

A pattern language is a collection of interrelated patterns that can be used to describe systems architectures. A generative pattern language is a pattern language that can express all possible architecture. The objective of the ICa PL is to provide an extensible framework to express application architecture in the restricted domain of autonomous controllers for complex systems. It is not intended as a general pattern language for software architecture description even when most of the patterns are generic enough to be applied in other domains.

The ICa PL offers a collection of patterns to address common problems in the implementation of advanced controllers for autonomous systems. This means that the patterns will be generic to be reusable but will be specific to determinate problems or tasks that appear in this type of systems. So, in most cases, control specific terms or roles will be employed even when the patterns are not restricted to such domain. This pattern

---

[13] Real applications can be converted in patterns by means of a flushing process; classical expert system shells are an example of this. This is the main way to create application and system level patterns.

[14] In the patterns terminology *force* is a term used to represent things to have in mind when designing, i.e. some form of requirement or constraint to the design.

language is being constructed as an evolving architectural handbook, so it is planned to suffer continuous modifications based on the experience gained in the use of the patterns. The ICa PL however, has a dramatic difference with other pattern languages. It has a direct implementation in the form of a object-oriented framework suitable for developing applications based on the ICa Core software. This means that the distance between the pattern and the real implementation in smaller in the ICa PL than in other pattern languages.

Some of the patterns that are under development in the ICa Pattern Language are listed in the following sections.

### *6.4.1 Generic Patterns*

- LAYERS: Structure for maintainability.
- BROKER: Simplify access to a set of resources.
- SINGLETON: Coordinate work using a single window approach.
- WRAPPER: Make things appear as new.
- STUB: Providing interaction mechanism to separate entities.
- LOCK: Control access to a shared resource.
- AGENT: Get autonomous behavior.
- UPDATER: Gets data from some place and put it in other place.
- MONITOR: Looks at information to detect situations.
- COORDINATOR: Get a good global behavior from interacting components.
- SUPERVISOR: Makes comments about the state of things. Usually they are suggestions to humans.
- The LAYERS pattern: Basic structuring pattern for all types of systems. One of the most generic patterns for all software engineering tasks.
- The BROKER pattern: Mediating between service requesters and providers. A basis for transparent relocation, fault tolerance, maintenance, etc.
- The KNOWLEDGEBASE pattern: The separation of generics from specifics is one of the basis for reuse.

### *6.4.2 More Control Related Patterns*

- PID: Proportional-Integral-Derivative controller.
- FILTERPID: PID with input signal filtering.
- SENSORFUSION: Get a coherent perception coming from several sensors.
- STATEFEEDBACK: A classical type of controller.
- AUTOTUNNER: Evolve a controller with the plant.

- REACTIVEBEHAVIOR: Do work as immediate response to perception.
- BEHAVIORFUSION: Get complex behavior by fusing elementary ones.
- EXPERTCONTROL: Put operator knowledge into the computer.
- $H_\infty$: Robustness is a good feature of a controller.
- DIAGNOSER: Identify where is the problem.
- CONTROLLER: Basic one-loop controller pattern. A basis for other patterns.
- AGENCY: Artificial minds as collection of interacting agents.
- FILTER: Cleaning data coming from noisy sources is the best way to success.
- STATEFEEDBACK: A basic multivariable control pattern.
- SENSORFUSION: The basis of good perception is using several sensing systems.
- MODELBASEDREASONING: Providing behavior based on the type of reasoning about systems that engineers do.
- SUBSUMPTION: A behavior structuring pattern for simple minded systems.
- PERCEPTION: How to go from raw magnitudes to knowledge.
- CONTROLTUNER: Adjusting behavior of lower level controllers. A specific form of the layers pattern.
- FUZZYCONTROLLER: A basic pattern for knowledge based, multivariable, strongly non-linear controllers.
- MRAC: A classical pattern for adaptive control based on a reference model.
- NEUROMRAC: Using a neural network as model in an MRAC.
- REMOTE: Teleoperation is of extreme importance in uncertain environments.
- PLANTSTATE: Managing a digital *photograph* of the status of the plant under control.
- IMPLICITINVOCATION: How to inform to people you don't know about.
- HISTORY: Knowing the past is the basis for a better future.
- SAMPLER: Getting information from a changing world in an adequate way for the future use of the data.
- SUPERVISION: Hierarchical control with humans atop.
- CLIENTSERVER: A basic software pattern for multicomponent systems.
- DCS: A typical pattern of a distributed control system.
- PID: The most classical control pattern: proportional, integral and derivative control.

## 6.5 A SAMPLE GENERIC PATTERN

In the pattern languages literature, pattern schemas (or templates) are used to present the information in a homogeneous way. Our impression is that patterns tend to be quite heterogeneous and forcing all of them to a specific description schema makes no sense. Forcing a pattern to a template makes incomprehensible some sections of the template mainly because they have no application to that specific pattern.

On the other side, there are clear sections, like pattern name, examples, etc. that make always sense. What we will try to do is to provide a template focused in the process control domain; template that will be used freely to describe patterns. Some sections will be dropped when its contents are difficult to identify.

In the literature on system[15] architecture a recurring schema appears almost everywhere; it is the LAYERS pattern (See [10]). This pattern is so basic that, from our point of view, is used in all software with both complexity and future.

The LAYERS pattern pervades all areas of software. From operating systems, trough middleware to application software. It is even used beyond the scope of the pure code. This pattern can even be applied not to a specific piece of code but to a conceptual model of a broad class of software designs.

### *6.5.1 Name*

LAYERS

This pattern has been extensively described in the literature using the same name. But in each case the focus is slightly different due to the orientation of the intended use. The foundation, however, remains the same in all cases.

### *6.5.2 Aliases*

Layered control, hierarchical control, control loop nesting.

### *6.5.3 Example*

Cement kiln control is a quite complex problem because there are some ill conditions. The two principal are:

- The sensing system is not as good as it should be because there are unmeasurable magnitudes (for example the most critical one: *burning zone temperature*) or the sensing process suffers great delays[16].
- The controllability of the kiln is quite limited. For example, raw material input composition can only be approximately set. Responsiveness of the kiln to control actions is also limited.

---

[15] Intelligent or not.

[16] It is necessary to perform chemical analysis of the clinker -the kiln product- that can only be done after the material leaves the kiln.

Control of kilns is done by a mixture of simple controllers (usually PLCs) in charge of maintaining basic process magnitudes, and a human operator that takes all decisions concerning the setpoints of the basic controllers.

Operator experience has a critical influence on the state and performance of the kiln. Continuous attention to critical magnitudes (for example contents of $NO_x$ in exhaustor gases) is also of extreme importance.

### 6.5.4 *Task*

Provide increasingly complex -intelligent- behavior without sacrificing performance and/or reliability.

### 6.5.5 *Problem*

The problems that LAYERS tries to solve are the complex integration issues that can appear when control making components have different cognitive capabilities. Some authors used to talk about the *principle of increasing precision with decreasing intelligence.* In our case, we should talk about not only precision, but speed and trustability [5].

### 6.5.6 *Context*

Plants with control problems that are not addressed adequately using a single approach. Heterogeneous control systems are required with different levels of intelligence.

### 6.5.7 *Solution*

Each cognitive level interacts only with the adjacent levels. A complete gradation of authority is specified and controllers attain to it. Interactions with upper layers are *perception* interactions; interactions with lower layers are *action* interactions.

Additional advantages are obtained from the use of LAYERS. The clear decomposition simplifies development of the components (the layers) of the systems, because interaction and dependency are limited.

### 6.5.8 *Structure*

Interaction between layers takes the form of perception/action pairs. Perception goes upward toward increasing intelligence. Actions go downwards towards the real plant under control.

Layer $L_N$ get information about the state of the virtual plant it sees from the immediately lower layer. For layer $L_N$ the plant under control is the real plant plus all the layers up to it:

$$Plant_N = Plant + L_1 + L_2 + \ldots + L_{N-1}$$

The unique control actions available to $L_N$ are those control inputs admisible for $L_{N-1}$.

### 6.5.9 Dynamics

There are two typical layer behaviors in these type of systems:

- *Active layers* that actively get state information from the lower layer and perform actions based on this information.
- *Passive layers* that perform their activity only upon reception of an activity request.

The two typical activation policies are all-active layers and event driven activity triggered by state reception from lower layers.

### 6.5.10 Implementation

Layers are usually implemented as sets of processes. Couplings between layers are simple enough to be implemented using whatever type of integration mechanism is available. In the example TCP/IP message based interaction between processes was used.

### 6.5.11 Timing

It depends clearly on the application and the layers involved. Lower layers are usually faster than upper layers. There are some typical timings for continuous process plants ranging from milliseconds in sensing and basic control loops to hours in strategic decision making. In the mobile robotics area, times can be the same. In faster applications (avionics, machine control, etc) times can be one tenth or hundredth of these.

### 6.5.12 Example application

The example for this pattern is the CONEX [13] application. CONEX is an agent based, distributed intelligent control system developed for a cement kiln. This system uses the LAYERS pattern to implement a multilayered control system based on heterogeneous control technologies. The objective of the controller is the optimization of clinker burning conditions (temperature, time, oxygen, etc). The system was designed to maximize control capability to maintain kiln conditions under strict margins.

The control layers implemented in the system were conventional control, pattern based reactive control, expert control, model-based control and optimization and human supervision. The agents implemented in the kiln controller were:

- Process interface: In charge of interaction with the plant; data capture and action application. It included filtering of actions to prevent misapplication of control actions.
- Low level control: A quasi-conventional control layer based on classical PID controllers, state feedback controllers and fuzzy controllers.
- Reactive control: The reactive control layer detected patterns of behavior in the plant and reacted to them acting on the lower layer. The objective of the pattern based controller was the selection of low level controllers and the adjustment of them; mainly through the setting of values (parameters, setpoints, fuzzy membership functions, etc) .

- Expert control: The expert control was a classical expert system built atop G2 to perform decision taking regarding combustion state.
- Model based control: Based on a complex model, was devised to perform kiln optimization.
- Operator interface: A user interface for the whole application.

All these layers were implemented using independent active agents. Each agent was composed by two or more VMS processes using message-based interactions over TCP/IP.

This approach made possible the execution of trustable (in logical and temporal behavior) agents in one computer and non-trustable agents in other computer. This division was of extreme importance to provide 365 × 24 availability of the lower layers.

### 6.5.13 *Variants*

Minor deviations are usual in the application of this pattern. In fact it is usual to find superimposed patterns in the same system, because each design provides solutions to specific problems.

A variant that is included as another pattern is the CONTROL TUNER pattern[17].

In the CONEX example, an agent that somewhat violated the LAYERS guidelines was the *model and simulator* agent. It is a three tiered model (numerical, qualitative and knowledge based) that was developed (only partially) to support the whole range of activities of the rest of the system.

### 6.5.14 *Known uses*

LAYERS is used in all types of control system.

### 6.5.15 *Care with*

- Layer dependencies
- Modification of interfaces
- Excessive performance penalty

### 6.5.16 *The magic of 3*

The layering concept is quite natural from the engineering point of view. What is not so immediate is that in many cases the proposed number of -relevant- layers is *three*. 3 seems to be a magic number for layered intelligent control, why?

Some examples at hand of these 3 tier layered architectures are:

- HINT: Heterogeneous Integration Architecture [14]. The three intelligent levels specified for the HINT blackboard are:

---

[17] It can be considered not really a variant but a specific application of the Layers pattern.

    - Operational: In charge of continuous state maintenance.
    - Tactical: In charge of task level related activities (by example problem solving).
    - Strategic: In charge of attaining high level objectives: production optimization, quality, safety, etc.

- The reasoning and deciding component (RDC) of the COSY agent architecture is decomposed in three layers:
    - Strategic: What to do.
    - Tactical: How to do it.
    - Executional: When to do it.
- $_3T$ is a three tier (so the name) architecture for intelligent robots [15]. The layers are:
    - A dynamically reprogrammable set of reactive skills coordinated by a skill manager
    - A sequencing capability to activate or deactivate skills to accomplish specific tasks.
    - A deliberative planning capability with deep reasoning.
- INTERRAP proposes an agent architecture with three layers:
    - A behavior based layer.
    - A local planning layer.
    - A cooperative planning layer.
- The theory of Hierarchically Intelligent Control (HIC) systems of Saridis [16] proposes the following layers:
    - An execution level: hardware and software controllers.
    - A coordination level.
    - A knowledge based organization level.

The contents of the layers seem quite similar in most cases but without a clear relation between them. The triple layering seems as an externalization of our modes of behavior:

- *Reaction*: A reaction follows immediately a perception.
- *Task*: A sequence (possibly a tree) of terns *perception→action→follow-up.*
- *Plan*: Neither real perception nor real action. Making task trees for the future.

When going down to real applications in the industry, the layering concept is still maintained as a principle. But in the final implementation, the layering is not clear. It is not easy to say neither the number of layers nor their relations. In many real implementations, engineers solve real problems patching the clean layered design, implementing ad hoc solutions to specific sections of the system under development.

## 7. References


[1] Son, S.H. (Ed): *Advances in Real-Time Systems.* Prentice-Hall, 1995.

[2] Ferrel, C.B. and Kemp, C.C.: An Ontogenic Perspective to Scaling Sensorimotor Intelligence.

[3] Minsky, M.: *The Society of Mind..* Simon and Schuster, 1985.

[4] Sanz,R., Galán, R., Jiménez, A., Matía, F., Velasco, J.R. and Martínez, G.: Computational Intelligence in Process Control *ICNN'94, IEEE International Conference in Neural Networks.* Orlando, USA, 1994.

[5] Sanz, R.: *Arquitectura de Control Inteligente de Procesos.* PhD Thesis. Politechnical University of Madrid. 1990.

[6] Schmidt, D.C., Johnson, R.E. and Fayad, M.: Software Patterns. Communications of the ACM, Vol. 39, No. 10, October 1996.

[7] Mowbray, T.J. and Malveau, R.C.: *CORBA Design Patterns.* Wiley, 1997.

[8] van der Linden, F. and Müller, J.: Creating Architectures with Building Blocks. IEEE Software 1995.

[9] Gamma, E., Helm, R., Jhonson, R. and Vlissides, J.: *Design Patterns. Elements of Reusable Object Oriented Software.* Addison-Wesley, 1994.

[10] Buschmann, F., Meunier, R., Rohnert, H., Sommerlad, P. and Stal, M.: *Pattern Oriented Software Architecture. A System of Patterns.* Wiley, 1996.

[11] Aarsten, A., Brugali, D. and Menga, G.: Designing Concurrent and Distributed Control Systems. *Communications of the ACM*, October 1996.

[12] DIXIT Consortium *DIXIT Product Definition.*, 1997.

[13] Sanz, R., Jiménez, A. and Galán, R.: CONEX: A Distributed Architecture for Intelligent Process Control. *World Congress on Expert Systems*, Orlando, U.S.A. December 1991.

[14] HINT Consortium. *HINT Manual for System Developers.*, 1994.

[15] Bonasso, P., Kortenkamp, D., Miller, D.P. and Slak, M.: Experiences with an Architecture for Intelligent, Reactive Agents. In *Intelligent Agents II.* Michael Wooldridge, Jörg P. Müller and Milind Tambe. Springer Verlag 1996, pp. 187-202.

[16] Lima, P.U. and Saridis, G.N.: *Design of Intelligent Control Systems Based on Hierarchical Stochastic Automata.* World Scientific, 1996.

[17] Brooks, R.A.: *Intelligence without Reason.* MIT AI-TR 1293. 1991.

# 5

# INTELLIGENT ROBOTIC AGENT COMBINING REACTIVE AND COGNITIVE CAPABILITIES

W. JACAK, S. DREISEITL
*Johannes Kepler University of Linz*
*Institute of Systems Science and*
*RISC Research Institute for Symbolic Computation*
*A-4040 Linz, Austria*, jacak@sea.uni-linz.ac.at

## 1. Introduction

The design of intelligent and knowledge-based autonomous systems (agent type) that learn by themselves to perform complex real-world tasks is a still-open challenge for the fields of system and control theory, robotics and artificial intelligence.

In this chapter we present the concept of a framework for an *autonomous robotic agent* that is capable of showing both *machine learning-based* and *reactive behaviour*. Past experience has shown that neither purely reactive nor purely machine learning-based approaches suffice to meet the requirements imposed by real-world environments.

In multi-agent robotic systems, one is primarily interested in the behaviour and interactions of a group of agents, based on the models of the agents themselves. The idea of the reactive model is to specify an agent by its behaviour, i.e., the way the agent reacts to certain environment stimuli. With every perceptual input, one associates a certain action that is expressed in the form of rules or procedures that calculate the reaction of the agent. Reactive systems have no internal history or long term plans, but calculate or choose their next action based solely upon the current perceptual situation.

On the other hand, machine learning-based models are motivated by the representation of the system's knowledge. The adaptation of symbolic AI techniques has led to the introduction of believes and intentions into the reasoning processes of the system. Such cognitive models permit to use more powerful and more general methods than reactive models; this, however, makes them inadequate for many real-time applications.

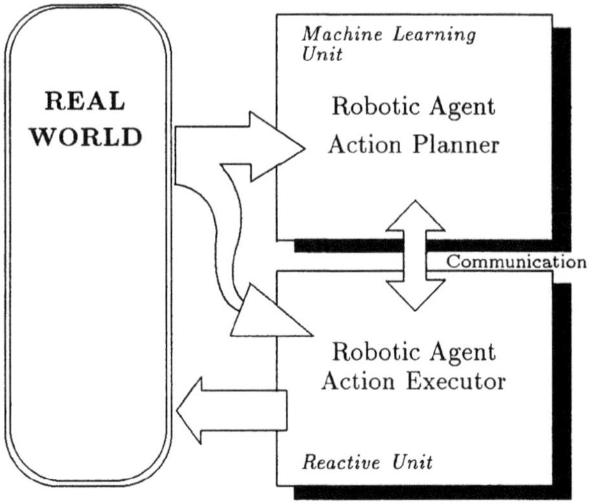


Figure 1: Intelligent robotic agent components.

Usually an agent has only partial information about the world state obtained by its perception system. Machine learning aims to overcome the limitations such as *knowledge bottleneck, engineering and tractability bottleneck* [25], by enabling an agent to collect its knowledge on-the-fly, through real-world experimentation. Learning from real-world experiments can be simplified by considering systems that face whole collections of control learning problems over their entire lifetime. Processing, storing and using information obtained during several task executions is called *lifelong learning.*

In our concept, the structure of a robotic agent consists of two independent subsystems: the *action planning system* and the *action execution system.* These components are shown in Figure 1.

## 1.1. LIFELONG LEARNING-BASED ACTION PLANNER OF INTELLIGENT AGENT

In classical approaches [3] we assume that the full model of the agent and its influence on the surrounding world is known. More exactly, the world state set $S_{World}$ is given and the agent's action effect model is specified by two functions: the action function **do**, and the observation function **see**:

$$\mathbf{do} : S_{World} \times A \to S_{World}$$

$$\mathbf{see} : S_{World} \to O.$$

Here, $A$ is the action (control) set and $O$ is the observation set obtained from the perception subsystem of the agent. For such a specification of the agent's model there exist many methods, based on graph searching and resolution procedures [3], for finding the sequence of actions $a^* = (a_1, a_2, ..., a_n)$ (called the *fundamental plan*), that transforms the agent from the current state into the given goal state. One such approach, based on a finite state machine formalism, is presented in [24].

In contrast to classical approaches [3, 22, 21, 23], we assume that an agent has no prior information about its environment state, so it cannot establish the world state observation function **see** : $S_{World} \to O$. Since the observation space of the agent's perception subsystem $O$ consists of the set of all possible vectors that are the collections of appropriate sensors signals, this space will be mapped to (interpreted as) a "conceptual" state space $S$ of the real world. To this aim, one has to construct the generalization function

$$\mathbf{gen} : O \to S, \tag{1}$$

which maps any observation vector $o \in O$ to a conceptual world state class $s \in S$ in such a way that the closer points are in $O$, the closer their corresponding classes will be in $S$, with neighboring points in $O$ being mapped to the same conceptual world state class. This property of the mapping is called *local generalization*, [4].

Having found a generalized state in the agent's conceptual space, one has to find an action sequence leading the agent toward its goal. This can be done during the action planning stage, when the next action $a_{i+1}$ can be determined from the previous action $a_i$ and the conceptual state $s_{i+1}$ with the function

$$\mathbf{plan} : \; S \times A \to A \tag{2}$$

as the action that maximizes the value function $\mathcal{Q}(s, a)$: $a_{i+1} = \mathbf{plan}(s_{i+1}, a_i) = \text{argmax}_{a' \in A} \; \mathcal{Q}(\mathbf{do}(s_{i+1}, a_i), a')$, where:

- The value function $\mathcal{Q}(s, a)$ is a function that returns scalar utility values. These values have to be learned over the whole lifetime of the agent while it is acting in the same environment. This learning can be accomplished with the Q-learning method [25] that is a popular method for learning to select actions from delayed or sparse reward.

- The function **do** models the effect of an agent's behaviour on the conceptual state space as
$$\mathbf{do} : \; S \times A \to S. \tag{3}$$

One can see that the machine-learning based behaviour of an autonomous agent can be completely characterized by the three functions **gen, do** and $\mathcal{Q}$. Generally, the agent consists of five main parts:

- the *conceptual world state generalization* module that generalizes the agent's sensor system observations,

- the *action planner* that finds sequences of agent actions realizing desired goals,

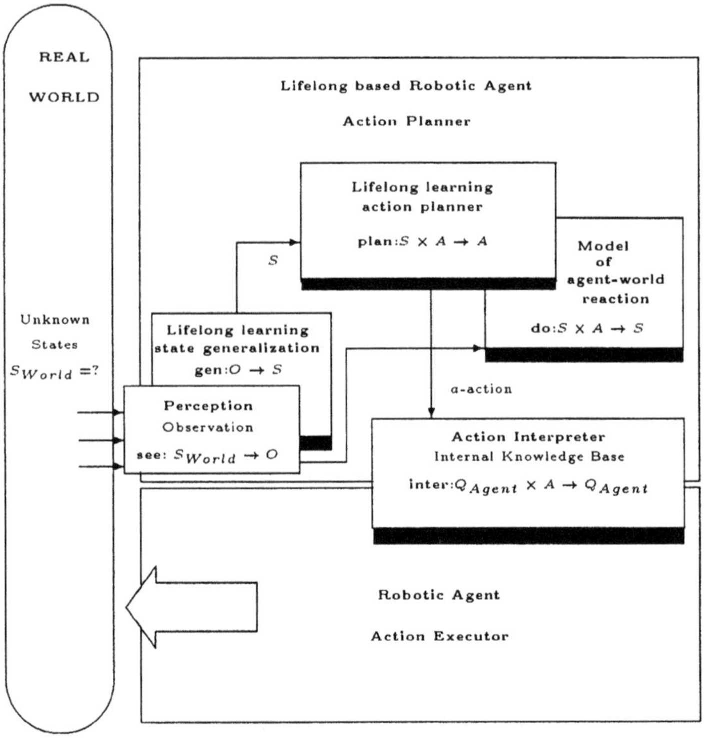


Figure 2: Structure of autonomous agent based on lifelong learning approach.

- the *modeller of the agent behaviour on the environment* that models the interaction between the agent and the conceptual world,
- the *action interpreter* that transforms elementary agent actions to goal states in its joint space, and
- the *action executor* that calculates controls that drive the agent toward its goal state.

The structure of such an intelligent agent is shown in Figure 2. The components of the autonomous agent that implement the reactive behaviour (the action interpreter **inter** and the action execution controller **contr**, with on-line motion and trajectory planning) are described in Section 4..

## 2. Neural Modelling of Robot Action Effects

One of the main components of the machine learning layer in an autonomous robotic agent is the module for building a function that gives conceptual states from sensor data. Our goal is to construct a world state generalization function **gen:** $O \to S$ that forms conceptual states as categories of observation vectors. This leads to a clustering of the observation vectors, with vectors in the same cluster being mapped to the same conceptual state by the function **gen**.

### 2.1. NEURAL IMPLEMENTATION OF GENERALIZATION ALGORITHM

The problem of obtaining conceptual world states from sensor input in such a way as to preserve proximity information among the sensor vectors can be solved by a neural network that forms clusters in its input space, and produces a "good" representative of this cluster as output. The network is trained in an unsupervised manner, so no other input but the sensor vectors are necessary. We implemented this generalization network incorporating ideas from Kohonen networks [18] as well as fuzzyART and fuzzyARTMAP algorithms [1].

For the training process, the robotic agent is first operated in an *active mode*, i.e., sensor readings are taken with the robot in positions that can be chosen by the user.The *passive mode* of learning takes place when the robot performs a task in its environment.

Our clustering algorithm is based on measuring the similarity of an input vector with the category information that is stored, for each category, as the weights of the corresponding output neuron. When an input vector is presented to the network, all the output neurons calculate their similarity (i.e., the distance of their weight vector to the input) in parallel. The neuron with the smallest distance is the winner neuron. If the observation vector is close enough to the winning neuron's weights (according to a given distance measure), then these weights are adapted to reflect the new entry in this category by moving into the direction of the observation vector. If the observation vector is too far away, it is not similar enough to any of the current categories to be included in one of them, and a new category (in the form of a new output neuron) has to be created. It is clear that the number of new categories created will be inversely proportional to the distance allowed for two observations to be included in the same category. If the measure of similarity is loose, only few category neurons are needed to cover the whole input space.

To prevent a proliferation of output neurons, we include a pruning step that cuts an output neuron when it is either not the winning neuron for some time, or when the number of observations that it encodes is small compared to the number of observations encoded by the other neurons.

The network presented above produces conceptual states from observation vectors. These conceptual states are then used as inputs for the **do** function that models the effects of the agent's behaviour on the conceptual states space. The network that implements this function is presented next.

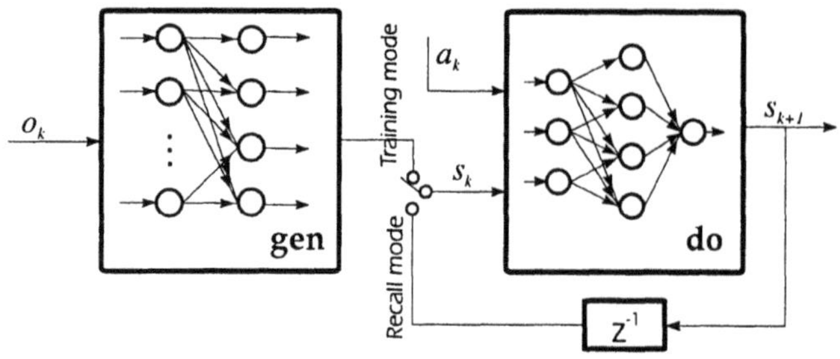


Figure 3: Structure of the network modeling the function **do.**

## 2.2. NEURAL IMPLEMENTATION OF ACTION EFFECTS MODELLER

The function **do** can be seen as the state transition function of a dynamical system with states in $S$ and inputs in $A$. As such, it can be implemented by a multilayer feedforward network that learns and refines the mapping as the robot operates. The neural network runs in two different modes:

- The *training mode*, when input patterns $(s_i, a_i)$ and desired outputs $s_{i+1}$ are presented to the network, and the network is trained on this static mapping; and

- the *recall mode*, when the conceptual state information is not obtained from the **gen** function, but from memory cells that store the previous state output of the **do** network and feed it back as input with the next action.

The structure of such a network, operating on state information from the **gen** function in the training mode and on state information from its own delayed output in the recall mode is shown in Figure 3. The multilayer feedforward network that implements the static function **do** can be trained with the backpropagation algorithm.

### 2.2.1. Extensions of the algorithm

The algorithm presented here can be extended in a variety of ways to increase the accuracy of the modeling. Two possible extensions are to *change the number of hidden neurons* in the action effect modeling network or to *extend the number of previous actions and conceptual states* used as inputs for the action effect modeling network. The second is more intricate, instead of using only the current conceptual state $s_i$ and the current action $a_i$ to predict the next conceptual state $s_{i+1}$, a "history" of states and actions before $s_i$ and $a_i$ could be taken into account as well. The number of previous states and actions that should be taken into consideration cannot be determined beforehand and depend on the complexity of the robot environment. We propose to use a genetic algorithm to find those past states

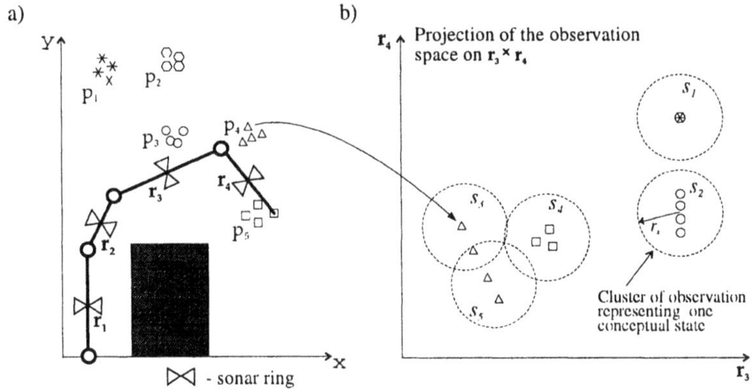


Figure 4: a) Sensor readings taken during active mode of learning. b) Conceptual states of sensor readings on last two links.

and actions that can contribute to determining the next state, while eliminating those that are not relevant. Such a genetic algorithm was developed in a different context [9, 10] and will be implemented in this work in the future.

Simulation results, presented in the next section, show that the methodology presented in this chapter can operate in a simple scene without the extensions mentioned above.

### 2.2.2. Simulation results

A two-dimensional 4-DOF manipulator was placed next to an obstacle in a simple scene. We assume that sensors were mounted on both sides in the middle of each link. Sensor readings are scaled from 0 (contact) to 1 (time out at infinity). Since there are no objects on the left of the manipulator, we disregarded the sensors on that side since they always showed a distance of infinity, reducing the observation space dimensionality to 4. The training of the two networks for the **gen** and **do** function was performed in two steps: first, the active mode of learning for the generalization network, and then a simulated normal operation of the robot when both the generalization and the state transition network were trained.

During the active learning of the generalization network, we placed the manipulator into five poses $p_1, \ldots, p_5$ that were far apart in the manipulator's configuration space. Each of these poses was changed slightly four times and sensor readings were taken (see Figure 4 a)). Using this training data of 20 observation vectors for the generalization network resulted in five conceptual state categories (see Figure 4 b)). However, these five categories were the result of one category being formed for two poses that were identical in sensor space (both now represented by conceptual state $s_1$), and the sensor reading of one pose being split into two distinct conceptual states ($s_3$ and $s_5$), because the small changes caused one

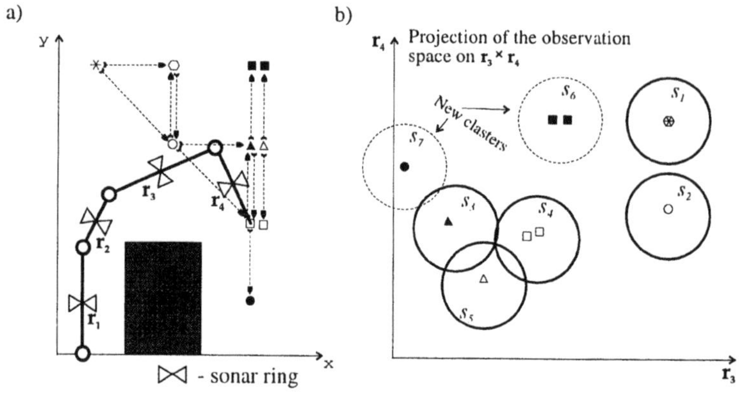


Figure 5: a) Sensor readings taken during normal operation of robot. b) Conceptual states of sensor readings on last two links.

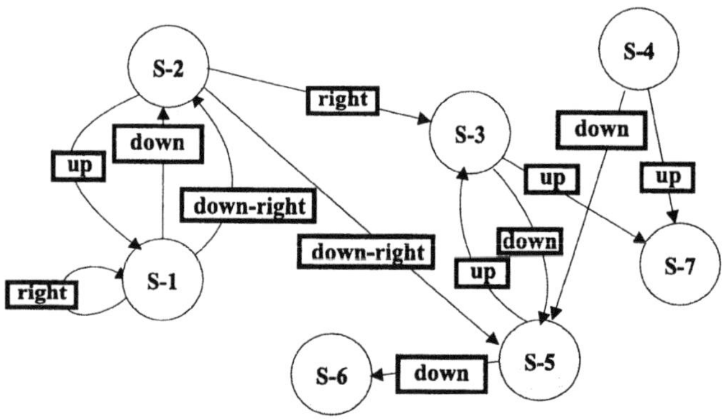


Figure 6: State transitions graph obtained during simulation.

sensor not to see the obstacle.

During the simulated normal operation of the robot, the actions (i.e. movements of the effector end) up, down, left, and right and sensible combinations thereof were allowed. Actions were performed with the manipulator being in one of the five original poses that were used for training the generalization network (see Figure 5 a)). The states reached during operation of the robot are shown in Figure 6.

## 3. Q-learning-based Intelligent Robot Action Planner

The inadequacy of classical planning algorithms for real-time applications and lack of full knowledge about of the surrounding world has motivated the use of learning approaches in order to quickly calculate the behaviour of the agent. A lifelong learning agent might acquire general knowledge about itself and its environment, or acquire generally useful skills that can be applied to speed up learning a new task. Moreover, lifelong learning promises to reduce the number of training examples required for successful learning.

We use the Q-learning method to synthesize the action planner [25]. Q-learning is a popular method for learning to select actions from delayed and sparse reward. The goal of Q-learning is to learn the strategy for generating whole action sequences that maximize an externally given *reward function.* The reward may be delayed and/or sparse, i.e., reward is only received upon reaching the goal of the task or upon total failure.

Based on the observation $o$ and the corresponding state $s = \mathbf{gen}(o)$, the agent picks an action $a \in A$. As a result of applying this action, the world state changes. The trainer receives a scalar *reward value*, denoted by $r(s, a)$, that measures the action performance. This reward can only be received upon reaching a designated goal or upon total failure, respectively.

The Q-learning method should find an action strategy mapping $\pi : S \to A$ from conceptual world states $S$ to actions $A$ that, when applied to action selection, maximizes the so-called *cumulative discounted future reward* $R = \sum_k \gamma^k r(s_k, a_k)$, where $\gamma \leq 1$ is a discount factor.

To quickly find the best action in the current state $s$, the key of Q-learning is to learn a *value function* for picking the actions. Such a value function

$$\mathcal{Q} : S \times A \to \mathcal{R}$$

maps conceptual states in $S$ and actions in $A$ to scalar utility values.

In the ideal case $\mathcal{Q}(s, a)$ is, after learning, the maximum cumulative reward one can expect upon executing action $a$ in state $s$. The function $\mathcal{Q}$ schedules actions according to their reward. The larger the expected cumulative reward for applying action $a$ in the current state $s$, the larger is its value $\mathcal{Q}(s, a)$.

The value function $\mathcal{Q}$ is used to select optimal actions for the current state $s$ as $\pi(s) = \mathrm{argmax}_{a' \in A} \mathcal{Q}(s, a')$. The values of $\mathcal{Q}(s, a)$ have to be learned over the whole lifetime of the agent in these same environment. The function $\mathcal{Q}$ will be realized as a neural network.

Initially, all values $\mathcal{Q}(s, a)$ are set to zero. During learning, values are incrementally updated, using the following standard recursive procedure [26]:

$$\mathcal{Q}^{new}(s_i, a_i) = \begin{cases} +N & \text{if } a_i \text{ final action, goal reached} \\ -N & \text{if } a_i \text{ final action, agent failed} \\ \gamma\ [(1-\lambda) \cdot \max_a \mathcal{Q}(s_{i+1}, a) & \text{else} \\ \quad +\lambda \cdot \mathcal{Q}^{new}(s_{i+1}, a_{i+1})] & \end{cases}$$

Such Q-learning learns individual strategies independently, ignoring the opportunity for the transfer of knowledge across different tasks [26]. In order to transfer knowledge, the planner should learn the predictive action model **plan** : $S \times A \rightarrow A$. Action models indirectly describe the effect of the agent actions. Such a model can be constructed by using the state transition function **do** and the value function $Q$.

The function **do** is apriori unknown and has to be learned based on observed state transitions over the whole lifetime of the agent. This dynamical model is implemented as an artificial neural network with memory cells [9, 10]. Once trained, the neural network model can be used to analyze observed episodes by providing explanation. The explanation allows to derive the slopes of the value function $Q$. The slopes are used for generalizing the episodes, and to generalize training instances in sensors space and conceptual state space, since they indicate how small changes will effect the target value function $Q$. They are extracted from the general action model, which is acquired and used over the entire lifetime of the agent.

## 4. Reactive Execution of the Action Plan - Reactive Motion Planner and Controller

While the robot is tracking a path in an unknown environment, some extraneous objects may enter the work space. These objects have to be detected with the help of sensors mounted on the links of the manipulator. For the perception system of a robotic agent, we propose the installation of ultrasonic sensors and touch-sensitive skin on each manipulator link. We can use neural networks to calculate the proximity of objects with each link. The vector of sensors signals (together with the distance to the goal position) establishes the point in sensor space. These points are used to construct the conceptual state space.

### 4.1. ACTIONS AND INTERPRETER

Let $\delta$ be a finite length of the displacement vector of the effector end in Cartesian space. Based on $\delta$, we can introduce the following set of standard displacement vectors of the effector end:

$$\delta p = \{(+\delta x, 0, 0), (-\delta x, 0, 0), (0, +\delta y, 0), (0, -\delta y, 0), (0, 0, +\delta z), (0, 0, -\delta z)\}$$

$$= \{\text{Forward, Backward, Left, Right, Up, Down}\}$$

Additionally, we should decide about the form of the geometric shape of the robot's manipulator: This point is most important if the number of degree of freedom is large. To describe the topological form of the manipulator shape, especially for the last arm links, we use the following set of terms:

$$\text{Wrist Shape} = \{\text{straighten out, arch up, arch down}\}$$

This set determines the shape of the wrist for a planar manipulator. For non-planar wrists, the Wrist Shape set has to be extended by the two terms "arch left" and "arch right". Based on these two sets, we can introduce the elementary action set $A$:

$$A = \delta p \times \text{Wrist Shape},$$

where the pair $(\delta, shape) \in A$ denotes the action of displacing the effector end while simultaneously changing the wrist shape. The action set is necessary to find the model of robot kinematics for calculating robot configurations. Let $Q_{Agent}$ denote the joint space of the robot. The kinematics model can be expressed as the discrete dynamical system in the form

$$\textbf{inter} : Q_{Agent} \times A \rightarrow Q_{Agent}$$

The main problem of action interpretation with this model of kinematics is to find the next state function **inter**. Recall that the function has to be calculated by solving under-determined kinematics equations of the manipulator. In order to choose one configuration from among many possibilities, we use the modified method proposed by Jacak in [6, 7].

We decompose the whole manipulator into two submanipulators and introduce two parameters calculated on the basis of two sets of joint numbers. Let $Q = \{1, \ldots, n-2\}$ denote the set of initial joints of the submanipulator, i.e., the sequence of the kinematic chain from joints 1 to $n-2$, and let $G = \{\nu, .., n\}$ be the set of terminal joints of the submanipulator. The parameter $\nu$ can be chosen arbitrarily. From these sets, we choose joints that decompose the manipulator into an *active* and a *passive* part.

The selection strategy of changing the configuration acts as follows:

$$(l, g) = \max_{i \in Q}(\min_{j \in G}\{C^i_{j,min} \leq c^i_j \leq C^i_{j,max} \ \& \ i < j-1\}),$$

where $C^i_{j,min}$ and $C^i_{j,max}$ denote the relative maximum (minimum) reach between joints $i$ and $j$, respectively, (i.e., the maximum respectively minimum size of the kinematic chain connecting the joints $i$ and $j$), and $c^i_j = ||t_i(q_k) - [t_j(q_k) + \delta]||$. The two parameters $l$ and $g$ thus determine the active and passive part of manipulator, with joints before $l$ being static and joints after $g$ moving.

The new position for each joint in Cartesian space is determined as

$$p_i^{new} = \begin{cases} t_i(q_k) + \delta & \text{for } i = g, \ldots, n \\ t_i(q_k) & \text{for } i = 1, \ldots, l \\ i\text{th joint position procedure} & \text{for } i = l+1, \ldots, g-1 \end{cases}$$

The joint position procedure determines the new position of a joint by transforming of inverse kinematic problem into an optimization problem [11, 12, 14]. For planar manipulators, this can be reduced to a simple algebraic algorithm [7]. The activation of the manipulator parts depends on the set $G$. We can introduce two

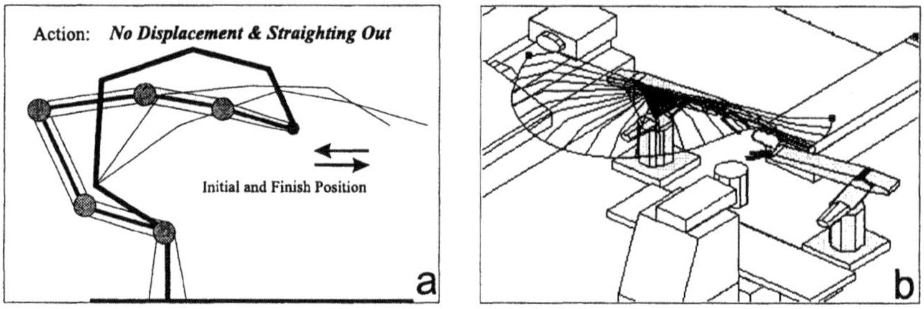


Figure 7: a) Action interpretation b) Collision free path

strategies: the *active terminal joints - reach motion strategy*, where $G = \{n\}$, and the *active initial joints - gross motion strategy*, where $G = \{3, \ldots, n\}$. In the first, a minimum number of links close to the effector end is moved, and in the second, a minimum number at the manipulator base.

As result of this interpreter function, we obtain the goal configuration of the motion step, realizing the given action. This configuration is the input for the reactive motion controller (executor). An example of interpreting a robot action is shown in Figure 7.

## 4.2. REACTIVE ACTION EXECUTOR

Dynamic obstacles entering the workspace of a robot are supposed to be detected with the help of sensors mounted on the links of the manipulator. The sensor readings are then analyzed to check for collision. The problem of finding the path of a robot equipped with sensors has been dealt with by Lumelsky and Sun in [21]. In their approach, the manipulator moves along a specified path and traces the outer boundary of the obstacle when one of the links touches an obstacle. In our case the objective of sensory manipulation is preventive collision avoidance.

We propose the installation of ultrasonic sensors and touch-sensitive skin on each manipulator link. The proximity of objects is calculated by a neural network. These distances are compared with the local virtual model of robot environment (created on the basis of the earlier experiments) to recognize a new dynamic obstacle. The structure of an intelligent reactive controller of the robot arm is presented in Figure 8. Based on the signals coming from the external sensors as well as its internal signals, the system produces the robot's input forces and torques which make the robot perform a prescribed task. The intelligent controller and planner consists of two main parts: the *on-line geometric motion planner* based on a multi-sensor local world-model builder and neural computed kinematics and the *reactive dynamic motion planner and controller* based on neurally calculated dynamics.

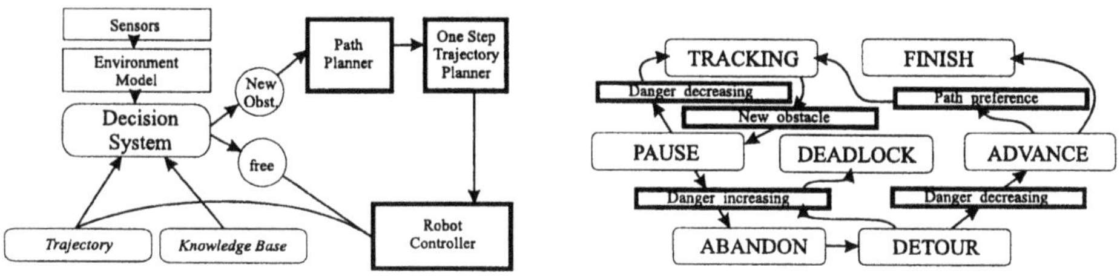


Figure 8: Structure of intelligent manipulator and decision system.

### 4.2.1. Sensing Process and Decision System

Instead of assuming only touch-sensitive skin on each manipulator link, we propose the additional installation of the rings of ultrasonic range finders on the links and use a neural network to calculate the proximity of the obstacles with each link.

The sensing process can be defined as the process by which measurements (sensor) data are obtained. The perception process, implemented in a neural network, can be defined as the process of deriving information about structures in the environment.

The inputs to the neural network for the perception process are signals from each ultrasonic sensor in the ring mounted on the $i$th manipulator link. The outputs of the neural network are vectors $d_i$, represented the minimal distances to objects in robot environment for each joint $i$ of the manipulator. The sensing process can be interpreted as a mapping of the state of the ultrasonic sensors and touch sensors into a set of multisensor images of much lower dimensionality.

In general, the preplanned geometrical and dynamical trajectory of motion together with the local model of the world can be expressed as

$$Traj(\text{move from A to B}) = (q(s), s(t), D(s))$$

where $q(s)$ is the parameterized track of motion in joint space, created by the path planner, $s(t)$ is the time realization of motion along the track, calculated by the off-line trajectory planner, and $D(s) = [D_1(s), D_2(s), \ldots, D_n(s)]$ is the sensor-image compatible local model of the path environment. Here, $D_i(s)$ represents the minimal distance vector from the $i$th joint to static obstacles. For unknown environments, we set $D_i(s)$ to the maximal range of the sensors.

During the motion, the local model is compared with the output of the neural network-based sensor-image processor, and the following external inputs for a discrete event decision system are created [8]: **Path_free, New_Obstacle, Danger_increasing, Danger_decreasing**.

The decision system is represented by an event-oriented system, the transition function of which is presented in Figure 8. The state **Tracking** denotes the state when the robot manipulator is tracking its currently assigned collision-free path. **Pause** is the state when the manipulator is stationary. **Detour** denotes the state when the manipulator is trying to circumvent an obstacle. **Advance** denotes the state when the obstacle has been circumvented and the manipulator approaches its immediate goal. **Abandon** denotes abandoning the assigned path and **Deadlock** is the state when the environment and the manipulator seem to be static.

If the dynamic obstacle comes dangerously close to a link, the manipulator is stopped or reconfigured to avoid touching the obstacle. In extreme cases, it might happen that the dynamic obstacle cannot be avoided if the effector end is to pursue its preplanned path. The manipulator must now be moved away to yield to the dynamic obstacle. Afterwards, the manipulator is made to go through a sequence of actions to bring the effector end back to its original path. To realize this task, the robot controller uses the event-based decisions as the strategy to change between manipulator motion along the preplanned trajectory and on-line planning of motion.

## 5. On-line Path Planning of Robot Movement

In the extreme case that the robot must abandon the preplanned trajectory, it has to find the next collision-free position. This position should be close to the preplanned path as long as possible.

On-line path planning and reactive control depends on the computation of robot kinematics, and is computationally expensive. Therefore it is attractive to develop a neural network that automatically generates robot kinematics. Neural networks can be used to reduce the computational complexity of the robot kinematics model. Additionally, neural networks can give robots the abilities to learn and self-calibrate.

Some neural network models used to attack this problem include the back propagation network [10, 20], the Hopfield network [19, 27], the context sensitive network [20, 17, 28], and the two layer counterpropagation network [5]. As the experiments show, the forward kinematics models based on these neural network types have many disadvantages. For all, the training time is too long for on-line relearning and these approaches can only provide approximations of the robot's positions in the base frame.

For this reason, we use a multilayer feedforward neural network with hidden units of sinusoidal activation functions [20]. To eliminate the training time and to increase accuracy, we apply hybrid techniques that automatically create the full network topology and weight values by symbolic computation of forward kinematics model components.

Each component $t_i(q)$ of direct kinematics can be represented in the form

$$t_i^m(q) = \sum_{j=1}^{k_i} l_j^m \sin(\mathbf{w_j^m}^T q) \tag{4}$$

for $i = 1, \ldots, n$ and $m = x, y, z$; here $t_i^m(q)$ defines the $(i, m)$ output representing the $m$th Cartesian variable of the $i$th joint position, $q$ is the joint state vector and $\mathbf{w_j^m} = (w_{j0}^m, w_{j1}^m, \ldots, w_{jn}^m)^T$ represents the weight vector of the $j$th sinusoidal function. This form, obtained from the symbolic level of network generation, represents the input-output function of a neural network with sinusoidal hidden units. The training of the feedforward network can be omitted, since the parameters $l_j^m$ and weights $\mathbf{w_j^m}$ can be calculated symbolically using technical data of the robot.

The model of manipulator kinematics must be able to generate a sequence of robot configurations that realizes the motion along the desired effector displacement and avoids the obstacles. To determine these configurations, the inverse kinematics of the manipulator have to be calculated.

The solution of the inverse kinematics problem can be obtained by combining a network of derivatives (see below) with the direct kinematics network to form a recurrent loop, such that, given a desired Cartesian pose, the feedback loop iteratively generates joint angle correction terms to move the output of the forward kinematics network toward the given pose. For obstacle avoidance, we can add additional contributions combining the information from the local model of the robot environment (the values of neural network outputs $d_i$) with the current position of each robot joint. A more detailed derivation is given next.

Let

$$e_n(q) = [p(s(k+1)) - t_n(q)]$$

denote the pose-error between the current position of effector end (in step k), calculated by forward kinematics and a desired next Cartesian pose $p(s(k+1))$ on the path. Obstacle avoidance can be achieved by introducing additional errors for each joint:

$$e_j = [p_j - t_j(q)] \quad \text{for} \quad j = 1, 2, \ldots, n-1,$$

where:

$$p_j = [t_j(q) - \eta \frac{d_j}{||d_j||}],$$

and $d_j$ is the minimal distance vector from the $j$th joint to obstacles given by the neural sensor-image processor, and $t_j(q)$ is the position of the $j$th joint in Cartesian space.

Then the inverse kinematic problem can be transformed to an optimization problem with constraints as follows: *For the given vector of Cartesian poses $P = (p_1, \ldots, p_n)$, find the joint configuration $q^*$ that minimizes the performance criterion*

$$v(q^*) = \frac{1}{2} \sum_1^n \lambda_i e_i(q)^T e_i(q) \rightarrow \min \tag{5}$$

*subject to $\sum_1^n \lambda_i = 1$ and $q^{min} \leq q^* \leq q^{max}$.*

The above problem can be transformed to a problem without constraints by introducing a penalty function $\omega$ to obtain the joint limits availability. We define additional errors for each degree of freedom: $\epsilon_i^m = q_i^{min} - q_i$ and $\epsilon_i^M = q_i - q_i^{max}$ for $i = 0, \ldots, n-1$. The penalty function is defined as

$$\omega(q) = \frac{1}{2}\epsilon_m^T \Phi_m \epsilon_m + \frac{1}{2}\epsilon_M^T \Phi_M \epsilon_M, \tag{6}$$

where $\epsilon_m = (\epsilon_1^m, \ldots, \epsilon_n^m)^T$, $\Phi_m$ is a diagonal matrix$\Phi_m = diag[\ \varphi_i^m]$, and $\varphi_i^m = 1/(1 + \exp(-\beta \cdot \epsilon_i^m))$ is the logistic function. The modified performance function is $\bar{v}(q) = v(q) + \omega(q)$. By using large values of $\beta$ in the sigmoid function $\varphi$, we activate the errors $\epsilon_i^m$ and $\epsilon_i^M$ only if $q_i < q_i^{min}$ and $q_i > q_i^{max}$, respectively.

The modified function $v$ can be interpreted as the Lyapunov function of a dynamical system. For convergence of $t_i(q)$ to $p_i$, error dynamics are involved. With the choice

$$\dot{q} = \eta \cdot \frac{\sum_1^n \lambda_i e_i(q)^T e_i(q) + \epsilon_m^T \Phi_m \epsilon_m + \epsilon_M^T \Phi_M \epsilon_M}{grad\ \bar{v}(q)^T \cdot grad\ \bar{v}(q)} grad\ \bar{v}(q) \tag{7}$$

the errors $e_i$ are minimized when the constraints are satisfied. The gradient of the function $\bar{v}$ has the form

$$grad\ \bar{v}(q) = -\sum_{i=1} 1^n \lambda_i J_i(q)^T e_i - \Phi_m \epsilon_m - D(\epsilon_m)\Phi'_m \epsilon_m$$
$$+\Phi_M \epsilon_M + D(\epsilon_M)\Phi'_M \epsilon_M$$

where $J_i(q)$ is the Jacobian matrix of $t_i(q)$, $D(\epsilon_m) = diag[D_i]$ is the $n \times n$ diagonal matrix with $D_i = \epsilon_i^m$ and $\Phi'_m = diag[\varphi_i'^m]$ is the diagonal matrix of first derivatives of $\varphi_i^m$ such that $\varphi_i'^m = \beta\varphi_i^m(1 - \varphi_i^m)$.

It can be seen that the time derivative of $\bar{v}$ is negative if $\eta > 0$, ensuring convergence of the Lyapunov function. The presented solution of the inverse kinematics problem uses only direct kinematics models, namely $t_i(q)$ and Jacobians $J_i(q)$. The neural implementation of the performance function and its gradient drastically reduces the computational burden. The Jacobians $J_i(q)$ can be implemented as networks of sinusoidal neurons in a similar way as the forward kinematics functions $t_i(q)$.

An example of on-line path planning is shown in Figure 7. When moving the manipulator from one point to another, the geometric motion planner [10] uses the local world model to check whether the point is reachable without collision and without violating joint limit constraints. If so, the point is passed to the reactive dynamic trajectory planner and controller. If not, a new collision-free configuration is calculated and the new point is the passed to the dynamic controller.

## 6. Reactive Dynamic Trajectory Planner

This controller part consists of the following components *neural network-based robot dynamics model, one-step trajectory planner* and *motion controller* [8]. The latter two units both use the neural model of robot dynamics.

## 6.1. HYBRID SYMBOLIC AND NEURAL NETWORK SYSTEM FOR DIRECT AND INVERSE ROBOT DYNAMICS

In order to fully utilize a robot's capability its dynamics should be taken into account. We follow the idea of exploiting maximum information of the problem domain (a structure of robot dynamics) and using neural network techniques to quickly compute any robotic expression.

A model of a nominal robot's dynamics has to be obtained based on technical data of a robot. It is assumed that the model of robot's dynamics takes the form

$$\sum_{j=1}^{n} D_{ij}(q)\ddot{q}_j + \sum_{j=1}^{n}\sum_{k=1}^{n} C^i_{jk}(q)\dot{q}_j\dot{q}_k + \sum_{j=1}^{n} R_{ij}\dot{q}_j + G_i(q) = u_i \qquad (8)$$

for $i = 1, \ldots, n$, where $D(q)$ is a $(n \times n)$ inertia matrix, $C(q)$ is a $(n \times n \times n)$ array of centrifugal and Coriolis terms, $R$ is a $(n \times n)$ diagonal matrix of viscous friction coefficients, $G$ is a $(n \times 1)$ vector of gravity terms, $q$ is a $(n \times 1)$ vector of generalized coordinates (configuration), and $n$ is a number of degrees of freedom.

We obtain the equations of a robot's dynamics by applying the Euler-Lagrange formalism, modelling the robot body as the set of elementary solids, and using the Steiner Theorem.

We can observe that having the fixed structure of robot's dynamics, the whole information about the dynamics is closed to its coefficients $G_i(q), D_{ij}(q), C^i_{jk}(q), R_i$ for $i, j, k = 1, \ldots, n$. So it is enough to transform the coefficients into sum-of-sines format [10] to get an NN representation of the robot's dynamics. All coefficients can be implemented as neural networks after a suitable normalization operation.

Active mode of learning The purpose of the active mode of learning is to obtain the real dynamics of a robot with zero payload (RD0). An active experiment relies on forcing the robot to follow prescribed trajectories and, during this process, measuring important robotic data. It is important to note that by choosing appropriate trajectories (points) in phase space, the problem of identifying the dynamic parameters of a robot can be decomposed into a set of simpler kinematic problems. This insight is of great importance when neural network techniques are used while deriving a robot's dynamics. By taking advantage of the possibility to freely choose a measuring (identification) point we can identify all the terms in (8).

Passive mode of learning The coefficients of robot dynamics depend on payload parameters and can vary when different payloads are carried. We tune the weights of the neural networks so that the robot can properly react to different payloads. When doing real work, we can only get information about coefficients by *passive experiments*, i.e. by following trajectories prescribed by a user.

## 6.2. NEURAL GAIN SCHEDULING CONTROL SYSTEM WITH ONE-STEP TRAJECTORY PLANNING

In our system, we use a one-step trajectory planner to control a robot arm along a given path. In an environment with uncertainties, only local information is available, so the whole time trajectory cannot be planned in advance.

It was assumed that the arm's motion is to be performed with piecewise constant acceleration in internal space. For modelling the robot dynamics, we use the neural network nominal dynamics model with zero payload. The task is to find functions parameterized by time that describe changes of joint accelerations, velocities and positions such that they start from the manipulator's current state and go through the desired position without violating the acceleration and velocity constraints.

The trajectory planner performs each of the following calculations in every iteration step.

Joints limits computation Since the task is only to *estimate* the limit values, and for typical situations consecutive points on the joint path lie close to one another, the limits obtained with this assumption are very similar to real ones. We use the NN model of robot dynamics in this step.

Computation of duration of movement Having computed limits on joint velocities and accelerations we are can find the time period within which the manipulator will be able to perform the desired movement. We do this by computing a lower bound of movement realization time $t_{d\,\min}$, checking whether each joint can be moved in this time, increasing $t_{d\,\min}$ if one joint cannot be moved in this time, and repeating this process until each joint can be moved in time $t_{d\,\min}$.

Trajectory segment computation Because for every manipulator joint the computation is performed in the same way, from now on we will continue the description for one joint only.

From the previous steps of calculation we obtained the final time $t_d$ within which the movement must be realized, and joint velocity and acceleration limits in the form

$$-\dot{q}_{\min} \leq \dot{q} \leq \dot{q}_{\max}, \quad \ddot{q}_{\min} \leq \ddot{q} \leq \ddot{q}_{\max}.$$

There are typically infinitely many possible trajectories satisfying the admissibility conditions, so an optimization criterion is needed to select only one trajectory. Two criteria are proposed: the *jerks minimization criterion* in the form

$$Q_j = \int_{t_0}^{t_d} |\dddot{q}(t)| = |\ddot{q}_1 - \ddot{q}_0| + |\ddot{q}_2 - \ddot{q}_1|, \tag{9}$$

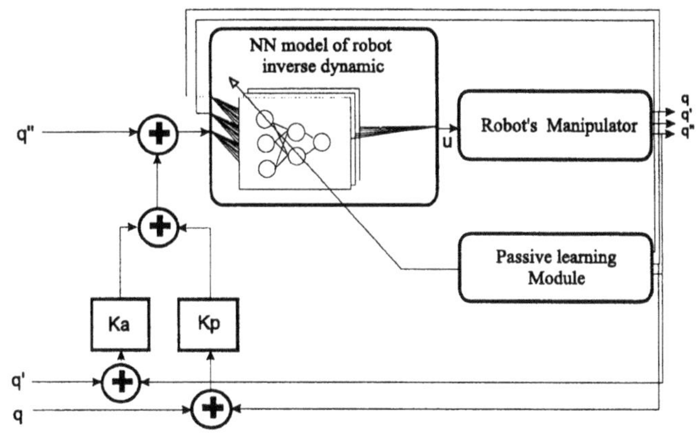


Figure 9: Computed torque controller with NN model of inverse dynamics.

and the *overshoots minimization criterion* that can be formulated in two different ways:

$$Q_o = \|q(t) - \tilde{q}(t)\| = \begin{cases} \max\limits_{t_0 \le t \le t_d} |q(t) - \tilde{q}(t)| \\ \\ \displaystyle\int_{t_0}^{t_d} (q(t) - \tilde{q}(t))^2 \mathrm{d}t. \end{cases} \tag{10}$$

We solve this polyoptimization problem by transforming it to the monooptimization problem

$$Q' = \lambda \tilde{Q}_j + (1 - \lambda)\tilde{Q}_o$$

where

$$\tilde{Q} = \frac{Q - Q_{\min}}{Q_{\max} - Q_{\min}}.$$

Because this method uses a parameter $\lambda$ that has to be set according to which criterion is more desirable, we use a fuzzy logic-based self-tuning method to establish this value [16].

## 6.3. MOTION CONTROLLER WITH NEURAL DYNAMICS MODEL

The trajectory planned by the one-step trajectory planner is executed by the controller. We apply the typical computed torque controller with PD loop. The CT controller is shown in Figure 9. The neural network dynamic model is used to compute the inverse dynamics. In order to improve the controller performance, the neural network model is trained to fit the actual dynamics during the control process. To speed up the learning, we use network updating based on either internal controller signals or external signals. The neural networks dynamic models

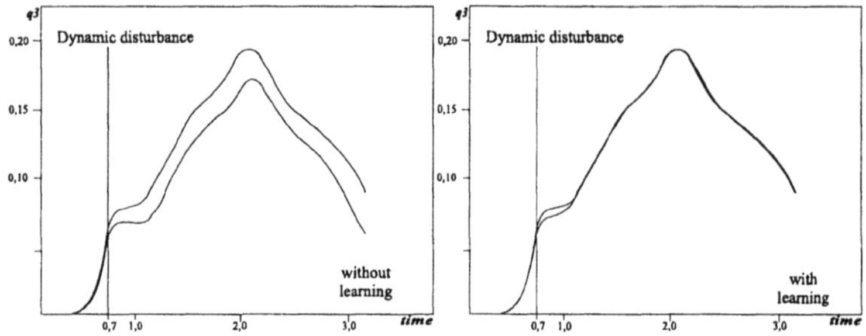


Figure 10: Joint position obtained by applying the CT controller with NN model of robot dynamics and real dynamics disturbance at $t = 0.7$. without (left) and with (right) passive learning.

weights are stored for several different manipulators payloads and in the case of obtaining an external signal the proper set of the weights is updated.

The performance of the system has been tested for a Puma-type robot manipulator. After deriving the robot dynamics equations in the closed form the calibration was performed.

The purpose of this test was to create a basis for comparing the performance of the typical computed torque controller (in which a nominal dynamics model without any calibration and adaptation is applied) with the performance of the controller proposed above. Figure 10 shows how the computed torque controller with nominal dynamics model works while driving the real manipulator without any payload, slightly different from nominal one. It is easy to see that such a controller is unable to control the robot properly.

To avoid such tracing errors, passive learning of the neural network can be applied. The robot gripped a payload at time 0.7. The performance of the system for disturbed robot dynamics is shown in Fig. 10. A large improvement in the system properties can be observed. There is a small error in the trajectory tracing just after the dynamics disturbance occurred, but because of the fast adaptation of the dynamics model to the new conditions it decreases rapidly.

## References

[1] G.A. Carpenter, S. Grossberg and D.B. Rosen. Fuzzy ART: Fast stable learning and categorization of analog patterns in an adaptive resonance system. *Neural Network*, **4**, pp. 759–771, 1991.

[2] I. Dulęba, W. Jacak and R. Muszyński. Hybrid computing approach to robot control in the presence of uncertainties. In *Proceedings of the IFAC workshop on Motion Control*, Munich, Germany, pp.664–671, 1995.

[3] M.R. Genesereth and N.J. Nilsson. *Logical Foundations of Artificial Intelligence*, Morgan Kaufmann Publishers, 1987.

[4] M. Han, B. Zhang. Control of robotics manipulators using a CMAC-based reinforcement learning system. In *Proc. of IEEE Conference on Intelligent Robots and Systems IROS'94*, Munich, Germany, 1994.

[5] R. Hecht-Nielsen. Counterpropagation networks. In *IEEE First Inter. Conf. on Neural Network*, Vol.2., 1987, pp. 19-32.

[6] W.Jacak. Discrete Kinematic Modeling Techniques in Cartesian Space for Robotic System. In *Control and Dynamic Systems: Advances in Robotic Systems*, ed. M.C. Leondes, Academic Press, New York, San Diego, London, Chapter V., 1991,

[7] W. Jacak. A Discrete Kinematic Model of Robots in the Cartesian Space. *IEEE Transactions on Robotics and Automation*, RA-5, No. 4, 1989, s. 435-444.

[8] W. Jacak and G. Jahn. Distributed Intelligent Control of an Autonomous Robotic Cell. In *Intelligent Autonomous Systems IAS'95*, ed. R. Dillman, IOS Press, Amsterdam, Oxford, Karlsruhe, 1995.

[9] W. Jacak and S. Dreiseitl. Hybrid Evolutionary Programming - the Tools for CAST. In *Proc. of Intern. Conf. EUROCAST'95*, Innsbruck, Austria, 1995.

[10] W. Jacak and S. Dreiseitl. Genetic Algorithm based Neural Networks for Dynamical Systems Modeling. In *Proc. of IEEE Intern. Conference on Evolutionary Computing ICEC'95*, Perth, Australia, 1995.

[11] W. Jacak. Hybrid Evolutionary approach to control of intelligent arm. In *Proc. of II Inter. Symposium on Methods and Models in Automation and Robotics MMAR'95*, Szczecin, Poland, 1995.

[12] W. Jacak, S. Dreiseitl, T. Kubik, and D. Schlosser. Distributed Planning and Control of Intelligent Robot's Arm Motion based on Symbolic and Neural Processing. In *IEEE Internat. Conference on Systems, Man and Cybernetics SMC'95*, Vancouver, Canada, 1995.

[13] W. Jacak, B. Buchberger, I. Duleba, and R. Muszynski. Symbolic Computation based Synthesis of Neural Network Dynamics and Controller for Robots. In *Proc. of IEEE Intern. Conference on Neural Networks ICNN'95*, Perth, Australia, 1995.

[14] W. Jacak, T. Kubik, and R. Muszynski. Neural Processing based Robot Kinematics Modelling and Calibration for Pose Control. In *Porc. of XII Conference on Systems Science*, Wroclaw, Poland, 1995.

[15] W. Jacak, B. Buchberger, and S. Stifter. Intelligent Systems Combining Reactive and Learning Capabilities. In *Proceedings of the European Meeting on Cybernetics and Systems Research*, Vienna, Austria, 1996.

[16] W. Jacak. Fuzzy Rules Based Control of Intelligent Robotic Workcell. In *Cybernetics and Systems'94* , ed. R. Trappl, World Scientific, Vienna, Singapore, 1994, pp. 1451-1459.

[17] M. Kawato, Y. Uno, M. Isobe, and R. Suzuki. A hierarchical model for voluntary movement and its application to robotics. In *IEEE First Inter. Conf. on Neural Network*, Vol.4., 1987, pp. 573-582.

[18] T. Kohonen. *Self-Organization and Associative Memory.* Springer Series in Information Sciences, Springer Verlag, third edition, 1989.

[19] S.Y. Kung and J.N. Hwang. Neural network architectures for robotics applications. *IEEE Trans. Robotics Automation*, RA-5(5), 1989, pp. 641-657.

[20] S. Lee and G.A. Bekey. Applications of neural network to robotics. In *Control and Dynamic System: Advances in Robotic Systems*, ed. C.T. Leondes, Academic Press Inc., San Diego, New York, 1991, pp.1-70.

[21] J.C. Latombe.: *Robot motion planning*, Kluwer Ac. Pub., Boston, Dordrecht, 1991 .

[22] D. McDermott. A temporal logic for reasoning about processes and plans. *Cognitive Science*, vol. 6, 1982.

[23] M. Spong and M. Vidyasagar.*Robot Dynamics and Control.* MIT Press, Cambridge, 1989.

[24] I. Sierocki. Solving the State Identification Problem by Experimentum Crucis. In *Cybernetics and Systems 94*, ed. R. Trappl, World Scientific, 1994.

[25] S. Thrun. A lifelong learning perspective for mobile robot control. In *Proc. of IEEE Conference on Intelligent Robots and Systems IROS'94*, Munich, Germany, 1994.

[26] S. Thrun and T. Mitchell. Integrating inductive neural network learning and explanation-based learning. In *Proc. of IJCAI'93*, Chamberry, France, 1993.

[27] C.M. Wu, B.C. Jiang, nad C.H. Wu. Using neural networks for robot positioning control. *Robotics and Computer Integrated Manufacturing*, Vol. 10(3), 1993, pp. 153-168.

[28] D.T. Yeung and G.A. Bekey. Using a contex-sensitive learning network for robot arm control. In *Proc. of IEEE Intern. Conf. on Robotics and Automation*, 1989, pp. 1441-1447.

# 6

# REACTIVE PLANNING WITH LEARNED OPERATIONAL CONCEPTS

VOLKER KLINGSPOR
*Universität Dortmund*
*Lehrstuhl Informatik VIII*
*44221 Dortmund, Germany*

## 1 Introduction

In the near future, services will be provided more and more by robots. The potential market for these service robots is expected to exceed that of current industrial robots until 2010 [1]. Currently, the first service robots are used in hospitals, factories, and to help the disabled [2]. Since users of service robots are usually no specialists in robots programming, new aspects for controlling robots should be topics of current research. The first topic we want to mention in this chapter is:

*How communicates the user with the robot?*

Communication comprises two direction, from the user to the robot to specify the task the robot should perform, and vice versa, from the robot to the user to report about what the robot has actually done in reality.

By *task specification,* we refer to the process of providing the robot with information about the task to be accomplished. An example of task specification via natural language is the system KANTRA [3]. The user can tell the robot which action it should perform with which object, for example "Grasp the red screw driver". In this approach, perception and action are separated. Planning is done in a classical way, creating a detailed sequence of actions, to be broken down to the lowest level of action and executed until some problem arises. Problems of this approach are: real situations are computational intractable, planning is inadequate for worlds with unpredictable events, plans have to be very detailed, and plans do not relate to concrete situations [4].

Agre and Chapman view plans as one among several sources of information for deciding what to do next [4]. Instead of defining a sequence of fixed and deterministic operators, plans just help to decide what's good to reach a given goal. Therefore, interpreting and executing a plan is more complex than in classical planning, because the interpreter itself must be able to solve the problem at least in a rudimentary way. To execute a complex sequence of actions, such a system must be accompanied by a user/instructor giving instructions like: "Head left now, ..., now rotate to the right, ..., slow down, ...". It cannot plan ahead to decide on what to do in the future in order to reach a goal.

The project "Situated artificial communicators", similarly to [4], aims to embed commands into sequences of actions that are jointly perceived by the user and the system [5]. The whole system integrates this planning and communication part with a natural language interface and an object identification system. Each command triggers a single behaviour like "grasping", and relates this behaviour to an object. Like above, the commands are not used to define complex goals.

All the previously described approaches do not really combine knowledge for planning with knowledge for control. Either the system is reactive and thus has no planning capabilities. Or the system can communicate with the user, but operation execution and planning are separate and hence control is not reactive. Very few approaches combine reactive control with planning, e.g., [6]. These approaches, unfortunately, do not use *learned* concepts for communication. Learning, however, as we argue later on in the chapter, is essential for building a common language for communication [7].

*Execution supervision* is the second topic of human robot communication. It is necessary to report about actions performed by the robot. That requires a classification of the perceived situation into categories that are useful and understandable for the user. Even for a human domain expert it is difficult to specify appropriate mappings from sensor patterns to monitoring conditions, failure classifications, failure explanations, and recovery strategies. Providing explicit examples of erroneous behaviour is even more difficult and is especially sensitive to errors in the user's model of the robot and the task [8].

Supervision as described in [8], tries to find faulty operations and reports about these errors to the user. Therefore, the system must be able to classify faulty situations into different classes of errors. Our approach differs from this. Instead of classifying error situations, we try to classify all situations. That means, during execution all perceptions and actions are analyzed and categorized. At each time the robot can tell the user in terms of perceived concepts what it is currently doing, i.e. it can communicate the result of applying an operator. In addition, if the perceptions differ from what the robot expects, explicit notifications can be send to the user.

Our goal for communication is close to human navigation in unknown environments. It is based on events which can be reached during the act. An example of a sequence of commands is: "move through the door, rotate beside the door, move along the wall, and stop at the cupboard." Because the goal is not to develop a sophisticated natural language interface, the communication entities are PROLOG predicates. The statement above can then be translated into a sequence of PROLOG facts linked by their arguments (denoting time instances):

```
move_through_door(T1, T2) &
rotate_in_front_of_door(T2, T3) &
move_along_wall(T3, T4) &
approach_cupboard(T4, T5).
```

The same concepts that are used by the user to define tasks can be used by the robot to report about its activities. Therefore, the robot can use different levels of abstraction of our representation hierarchy (described in [9]) to enable the user to find reasons for the failure of a mission. E.g.:

> "mission failed because
> move_through_door failed, because
> through_door failed.
> Instead, move_along_door was perceived."

where through_door is the perception of the robot when crossing a door, and move_along_door is the whole concept (including action and perception) of passing a door. In this way, the user has access to the state of the robot's mission at all times.

A further important research topic for service robots is

*What and how should service robots learn?*

Learning has different aspects, too. First, learning is essential to enhance the applicability of the robot. If service robots are to become a really large market, it is impossible that the salesperson configures each robot exactly to the demands of the customer. Instead, the customer must be able to make the robot understand how to detect relevant objects and which actions it must execute at or with these objects. Therefore, the robot must be able to learn from user given examples of actions. This approach is different from those of learning single behaviours [10, 11], where the robot learns to enhance perception-action-rules. It is also different from learning plan skeletons [12, 13] or heuristics for increasing planning efficiency [14, 15]. Closer to our learning approach is programming by demonstration [16, 17]. Here, a teacher demonstrates the system an action, and the system learns to reproduce this action in similar situations, i.e. it learns skills. The learning result of skill acquisition, indeed, is different from our learning result, because skills cannot immediately be used as operators for planning, and they cannot not be used to report about performed activities.

Learning is not only essential because of the flexibility of applications for service robots, but for better communication, too. Whereas the vocabulary used for communication must be the same for user and robot and the extension of the concepts described by the used words, the intensional description of the concepts might be very different. The sensors of a robot and the senses of a human differ very much, and hence, robot and human perceive the same situation very differently. Hence, the description of the same situation, and thus of the same concepts, must be very different [7]. Because of this difference, one needs very much knowledge about the sensors to generate concept descriptions. This task can be solved automatically if the robot learns concept definitions by its own, based on examples of the concept provided by the user [9].

In the remainder of this chapter, we first describe the representation of operational concepts. We will not describe once more in detail, why these concepts are suitable for communication, and how they can be learned – learnability together with executability requires a different representation than executability alone. The reader can find publications about these topics in [18, 9, 7]. The main topic of this chapter is, how learned concepts can be used for planning and reactive plan execution. This approach differs from other approaches in the way that it brings together three important aspects of service robots, namely:

- the capability to communicate,
- the capability to learn, and
- the reactive operation in the real world.

## 2 The Knowledge Used for Reactive Planning

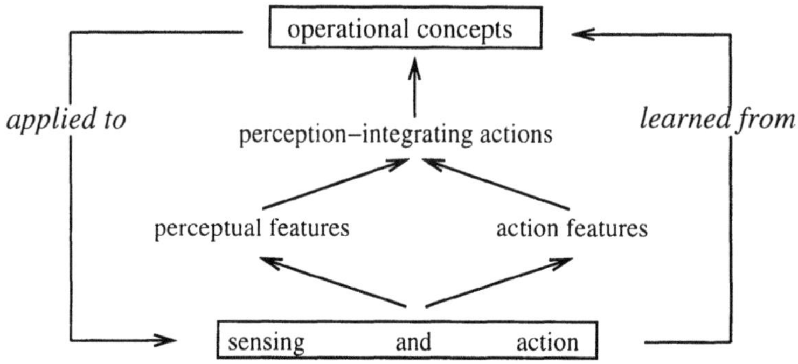


Figure 1: The Representation Hierarchy

In previous papers, we have argued intensivly, why a relational representation formalim like first order logic is well suited to represent the data necessary to control the robot, if communication and learning are further aspects in the whole system [9, 7]. In a nutshell, the reasons are the following:

- It allows an easy handling of time and time relations.
- It allows to combine different sensors easily.
- Arguments may remain variable until the actual situation determines their values.
- Background knowledge can be integrated.
- Perception and action can be combined.
- There exists a lot of learning algorithms within the framework of inductive logic programming [19] and a theory about learnability (e.g. [20]).

In our representation, several levels of abstraction link the level of raw sensor and motion data with the level of operational concepts. The hierarchy of more and more abstract representations comprises perceptual features at several levels which are all related with actions as well as action feature which become integrated with perceptual features at a higher level. Figure 1 shows the representation hierarchy in a simplified version.

*Perceptual features* are features describing time intervals or single time instances, where the robot perceived an object. This perception depends on the action the robot performed, and hence the name of the perception consists of action and perception like "along_door". Nevertheless, the action itself is not represented at this level. Perceptual features are described by two 6-ary PROLOG-predicates tip (*time interval perception*) and tpp (*time point perception*).

tip(Trace, $\text{Time}_1$, $\text{Time}_2$, Perception,
p_info(PerceptionDirection, RobotSide), MovementRelation).
tpp(Trace, $\text{Time}_1$, $\text{Time}_2$, Perception,
p_info(PerceptionDirection, RobotSide), MovementRelation).

Trace is an identifier for learn and test purposes, indicating a unique identifier for a data set. $\text{Time}_1$ and $\text{Time}_2$ define the time interval during which the predicate holds. For time point perceptions, usually these two are equal. Perception is the acronym for the situation the robot perceived, like "along_door" in the example above. p_info is a 2-ary functor describing necessary perceptual information. Currently, this information is the direction, in which the object was perceived (e.g. "right_side"), and which side, a small or a wide one, of the robot is oriented towards this direction (e.g. "small_side"). We used a functor for this information to make the predicates tid and tip independent of the needs of the current robot. E.g. for circular robot, the robot_side-information is senseless, but maybe, other data must be gathered to control the robot. In that case, only the functor changes, but not the arity or the structure of the perceptual features. MovementRelation defines in which way the robot moved relative to the object, e.g. "straight_towards" or "parallel" along the object.

Of course, perceptual features are not calculated immediately from the sensor data, but there are some intermediate levels, defining less abstract features. Since this chapter's topic is not object identification, we refer to [9, 7].

At the same level of perceptual features, one find *action features*. These denote the action, the robot performs, independent of what it perceives. An action feature is a 4-ary predicate

BA(Trace, $\text{Time}_1$, $\text{Time}_2$, a_info(Speed, MovementDirection)).

BA is a dummy name for the actual predicate name. Currently, predicate names for action features are stand, move, and rotate. The first three arguments are like above. a_info is again a functor, denoting additional action information, like the robot's speed (a discrete value) and the direction of the motion (e.g., to the right). Perceptual features and action features at these levels define the *lowest level of interest* (LLOI) for planning and plan execution.

At the next higher level, perception and action are combined by *perception integrating action features* (PIAF). PIAFs are 7-ary predicates with the following structure:

PIAF(Tr, $\text{Time}_1$, $\text{Time}_2$, Perception, p_info(...), a_info(...), add_info(...)).

Again, PIAF is a dummy name for the predicate. The real predicate name combines information of the movement relation and the action, e.g. parallel_moving. PIAFs receive their arguments from the defining perceptual features and action features. The last argument, add_info, is sometimes used for planning, it is not determined from the sensor information. An example for a clause defining the PIAF parallel_moving is:

tip(Tr, $\text{Time}_1$, $\text{Time}_2$, Perception, p_info(...), parallel) &

move(Tr, $\text{Time}_1$, $\text{Time}_2$, a_info(...))

→ parallel_moving(Tr, $\text{Time}_1$, $\text{Time}_2$, Perception, p_info(...), a_info(...), add_info(...)).

A fact with this predicate denotes, e.g., that the robots moves in Trace Trace during the time interval [$\text{Time}_1$, $\text{Time}_2$] in a parallel track.

The highest level data are *operational concepts*. They are 3-ary predicates with arguments trace identifier and time interval. Operational concepts are defined by three properties: precondition, action itself, and end-of-action condition:

standing(Tr, T1, T2, in_front_of_door, p_info(PDirect, small_side),
a_info(_, _), add_info(_)) &

parallel_moving(Tr, T2, T3, through_door, p_info(right_and_left, wide_side),
a_info(Speed, PDirect), add_info(_)) &

standing(Tr, T3, T4, in_front_of_door, p_info(rear, small_side),
a_info(_, _), add_info(_))

→ move_through_door(Tr, T1, T4).

The rule describes the operational concept of crossing a doorway. It combines the recognition of the doorway with the action of moving through it on a parallel track. Action and perception are linked by the perceptual features occurring as arguments in the action predicates.

The first premise of the rule states that the robot stands during a time interval [T1, T2] and senses the perceptual feature in_front_of_door from the sensors mounted at a small side. This premise denotes the precondition for crossing a doorway on a parallel track.

The second premise states that in a following time interval the action parallel_moving should be executed, measuring the perceptual feature through_door by the sensors at the right and the left side of the robot. Note, that the particular values for the time intervals do not matter for the operational concept but are instantiated by the specifics of a particular path. Only the time relation is important. This makes the operational concept independent of the particular depth of a doorway and the robot's particular rate of advance.

The third premise describes the end of the movement through the doorway: the robot stands in front of the door, now sensing with its rear sensors mounted at its small side the feature in_front_of_door after having perceived the feature through_door. PDirect, the direction in which this door is perceived at the beginning (e.g., left_side, right_side, front, rear) is not fixed by the rule. It is only relevant that the orientation should not change during the parallel move and that with respect to this orientation the doorway is sensed in the rear after the movement. Because we define the direction in which the robot moves as its front whatever the absolute orientation was at the beginning of the action, the direction in which the door is perceived at this time is with start of the movement defined as being front so that after completing the action the doorway is sensed by the opposite, the rear sensors.

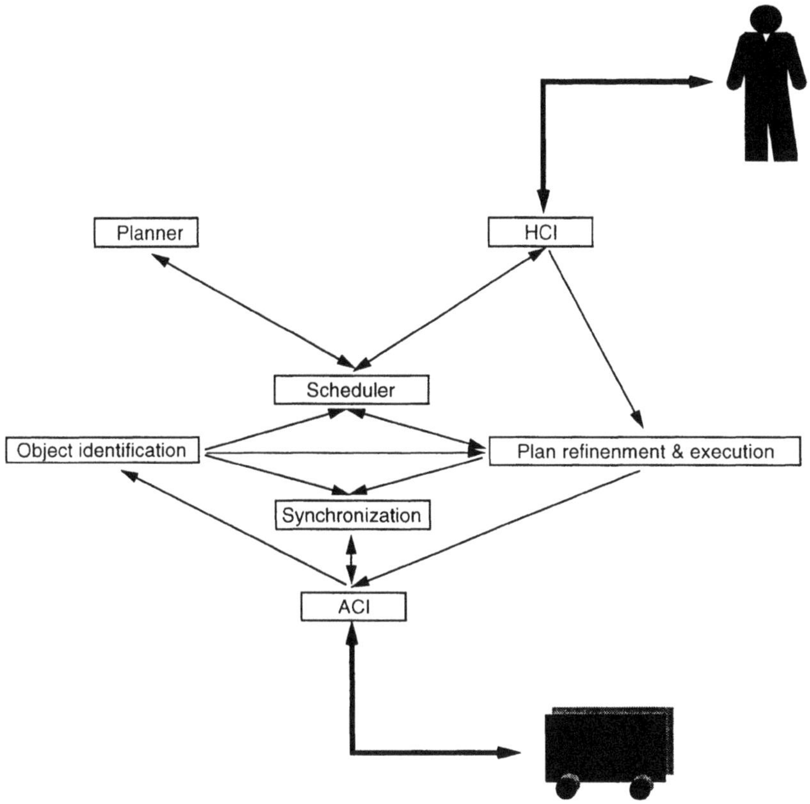


Figure 2: The SHARC performance system

## 3 The Architecture of the Performance System SHARC

After having introduced the knowledge used for reactive planning, we now present the system that uses this knowledge for controlling the robot. Remember, the goal of this system is a reactive but user understandable and user controllable system, that uses knowledge learned in a previous learning step.

The performance system SHARC[1] consists of different modules, communicating with each other using the inter process communication system PVM (Parallel Virtual Machine) [21] (Fig. 2). Each of these modules consists of at least a single process. Especially the *object identification* module, however, consists of many more small processes (currently, for the robot navigation with PRIAMOS[2], around 80 processes). The different processes

[1] SHARC can be seen as acronym for "System for Human Agent interactive Reactive Concept interpretation".

[2] The mobile robot PRIAMOS is a development of the Institute for Real-Time Computer Systems & Robotics, University Karlsruhe [22]. During the ESPRIT project BLearn-II and afterwards, all real world experiments are made with their platform.

can be distributed over an arbitrary network with different computer architectures and different operating systems, as long as the respective process runs on the target platform. We now sketch the different modules, before we come closer to the planning and plan execution process.

**Human Computer Interface (HCI):** The HCI is the interface between the user and the SHARC. It allows the user to define goals the robot has to solve and to track the course of the mission. We do communication without a natural language interface, because this is a research topic of its own. The used symbols, however, build a human understandable language. The HCI receives from Scheduler and PREM (see below) plans at different abstract levels for visualization, and it sends the user's commands to the Scheduler.

**Object Identification Module:** This module interprets the robot's sensor data[3]. Specialized inference engines generate perceptual features, action features, and the higher level features via forward inferences, using the learned rules at different levels of abstraction. Features higher than or equal to the LLOI are send to Scheduler and PREM. The object identification module and the different inference engines are described in detail in [23, 24].

**Planner:** The Planner generates an abstract plan based on operational concepts of the highest level. This plan cannot be executed immediately, but is has to be broken down to an operational level. The Planner receives the demand to plan from the Scheduler and sends the found plan or a message about failed planning back to the Scheduler.

**Scheduler:** The Scheduler distributes the different tasks. It communicates with the HCI, triggers the Planner and the PREM, and it receives real world data from the object identification module.

**Plan Refinement and Execution Module (PREM):** The PREM refines abstract plans generated by the Planner and reactively executes the refined plan. It receives the current plan via the Scheduler from the Planner. To visualize the process of plan execution, information about the current state is sent to the HCI.

**Agent Computer Interface (ACI):** The ACI is the interface between the agent (here the robot, but software agents like Internet agents are imaginable as well) and the SHARC. It receives low level sensor data from the agent and sends them to the object identification module, and it receives action features from the PREM and transforms them to the robot's elementary operations.

**Synchronization:** This module is used to start the whole system, and to synchronize agent, object identification, and plan execution.

---

[3]Data about the position of the robot are seen as sensory data as well (they can also be seen as perceptual data), because we only use relative coordinates which easily can be determined from the wheel encoders.

## 4 Reactive Planning

**Definition 1 (Reactive Planning)** Reactive Planning *is the process of guiding an agent from a current situation to a user-given target situation.*[4]

With this definition, planning is immediately embedded into plan execution, because the goal of reactive planning is not to find an operator sequence, that theoretically should transpose a start state into a target state and this plan is executed later on by an complicated control system knowing very much about how to execute the operators even in cases of unpredictable events. Instead, the goal is to bring the real robot itself into the target situation.

In a nutshell, the idea of our approach is to search for a set of operator sequences possibly reaching the goal from different start situations. At least one of these plans has the current situation as precondition of the first operator. The shortest plan that is executable in the current situation, is carried out as long as the actual perceptions match the expected perceptions according to the plan. In cases, where the current plan fails, another plan out of the pre-calculated set of plans is chosen and executed in the same way. If no other plan is applicable, the set will be extended.

In the following subsections, we describe how we generate sets of plans, managing efficient jumping between different plans, and how the robot actions are triggered. Planning appears at different abstract levels. The more abstract the level, the less the plan changes during execution. We now explain the different levels of plans we generate and we describe in detail, how plans are generated and executed.

### 4.1 ABSTRACT PLANNING

The most abstract level of planning is the level of operational concepts. Abstract plans are sequences of concepts, where the postcondition of a step in this sequence is unifyable with the precondition of its successor. The start situation, i.e. the robot's current situation, is defined by the current perceptions. The perceptions are not necessarily deterministic, because the different learned rules may infer different perceptual features simultaneously. The start situation is therefore a *set* of PIAFs. The target situation is defined by the user. This situation is deterministic. The user cannot give different goals at the same time.

To find a sequence of operational concepts, guiding the robot from start to target situation, breadth-first search backwards from the target to the start situation is applied. In this way, the shortest operator sequence can be found, transferring *one* of the current perceived situations into the goal situation.

This approach to take the shortest possible path corresponds to our aim of optimistic planning. Instead of calculating all possible sequences a priori, we hope that one sequence guides the robot to the target. In cases of differences between current perceptions and planned sequence, the current plan is aborted and the systems tries to find another, again shortest path from the new current situation to the goal.

[4]We know that the term "reactive planning" is used by [25], too. They extend operators by explicit test for error states to be able to react immediately to these states.

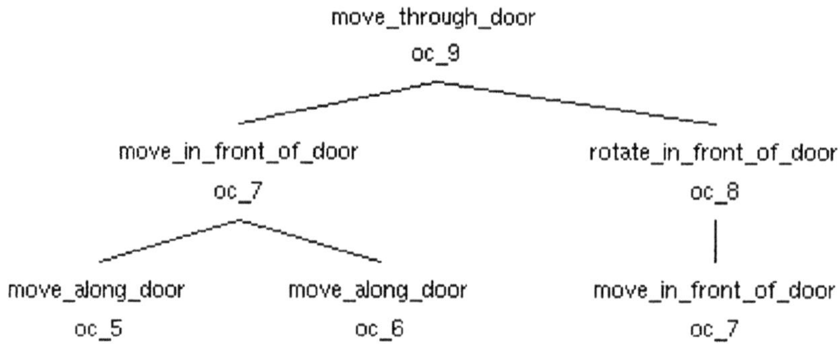


Figure 3: An example of an abstract plan tree

To make this plan change as fast as possible and to make plan execution as flexible as possible, we do not only store the sequence found, but the whole tree spanned by the breadth-first search as well. Each path between a node of this tree and the root is a valid operator sequence guiding the robot from the precondition of that node to the target node. In cases where the suggested plan fails, a lot of different plans are provided by this plan tree without any effort to re-plan. Only the correct node must be found, i.e., the current perception must match the precondition of that node. Now we give the formal definition of an *abstract plan tree*.

**Definition 2 (Abstract Plan Tree)** *Let $G$ be a planning goal, a PIAF. $W$ is a set of currently inferred perceptions, which are PIAFs, too. An* Abstract Plan Tree *for $G$ and $W$ is a tree with the following properties:*

- *Each node is assigned an operational concept.*
- *The root of the node is a assigned a concept whose postcondition is unifyable with $G$.*
- *For each edge $(V_1, V_2)$, the postcondition of $V_1$ is unifyable with the precondition of $V_2$. The variable of both clauses assigned to $V_1$ and $V_2$ are substituted by the unificator.*
- *At least one leaf of the tree is assigned a concept whose precondition is unifyable with an element of $W$.*

Figure 3 shows an example for an abstract plan tree[5]. The goal concept of this tree is move_through_door. The current perception is standing(..., parallel_in_corner, p_info(_, large_side), ...). This perception is precondition of the concept move_along_door. In this

[5]This figure, as well as Figure 5, is a screen-shot of parts of the HCI.

figure, each node is not only attached a concept name, but a unique identifier of the rule defining the concept.

The definition of the abstract plan tree shows, that unification of pre- and postcondition defines the search space. In Figure 4 the planning algorithm is shown. The algorithm performs standard breadth-first search. However, for each node it is checked, whether this node occured anywhere else in the tree, because the search may then be pruned at these nodes. Since the arguments of the preconditions may not be full instantiated, different cases of occurency must be distinguished:

- The precondition of the new operator $O_n$ is more special than another precondition $O_a$ in the tree or $O_n$ and $O_a$ are equivalent. In the earlier case, the subtree below $O_n$ is more special than the subtree below $O_a$, in the latter case the two subtrees are equal. In both cases, $O_a$ appears earlier in the search than $O_n$, so that the start node will be found earlier in the subtree of $O_a$. Search below $O_n$ can therefore be pruned without loss of completeness.

- If the precondition of $O_n$ is more general than the one of $O_a$, a solution that is in both of the subtrees below $O_a$ and $O_n$, is found earlier in the subtree of $O_a$. However, since $O_n$ is more general, there may exists more solutions for the planning problem in the subtree of $O_n$. Therefore, none of the two subtrees is pruned.

Furthermore, the call of background knowledge is restricted. A substitution is only valid, if it is deterministic. Otherwise, the algorithm would explode at this step, because all instantiations generate own subtrees. To handle this restriction efficiently, we require that for all background predicates of arity $n$, each $n-1$ arguments determine the remaining argument. Background predicates can thus be called, if at least $n-1$ of their arguments are instantiated.

**Complexity** The complexity of our planning algorithm is determined by two factors, the maximum size of the plan tree and the time needed to expand a node. Given a fixed signature, $Arg_i$, $i = 1, \ldots, t$ are the number of terms that can be instantiated at argument position $i$ of a PIAF. $t$ is the arity of all PIAFs, currently 7. If the argument $i$ can be substituted by a term *or* by a variable, $Arg_i$ is incremented by 1. Furthermore, $Arg_0$ is the number of different predicate names for PIAFs. $MaxArg := \max_{i=0,\ldots,t}(Arg_i)$. Then, the upper bound of the size of the plan tree is the number of different instantiations of PIAFs:

$$\prod_{i=0}^{t} Arg_i = \mathrm{O}(MaxArg^t) \quad (1)$$

Since trace is constant in any case, and $\mathsf{Time}_i$ is variable in any case, $Arg_1 = Arg_2 = Arg_3 = 1$. With a fixed number of arguments, in our case 7, the maximum size of the tree is $\mathrm{O}(MaxArg^5)$.

To estimate the effort for planning, we define the following values. $K$ is the maximal number of nodes in the tree (the upper bound is calculated in Equation 1). $Op$ is the number of all rules defining operational concepts. $W$ is the number of current perceptions. As

$Plan := \{\}$
$Open := [Goal]$
while $(Open \neq empty)$
  $Subgoal := head(Open)$
  $\forall Op \in operationalconcepts$
    $Op = (PreCond, \ldots, BKCall, PostCond)$
    if $\exists$ m.g.u. $\sigma$ of $Subgoal$ and $PostCond$
      $\sigma'$ := valid substitution, satisfying $BKCall$
      append$(Plan, Op\sigma\sigma')$
      if $\exists W \in CurrentState$ : $\exists$ m.g.u. $\sigma''$ of $PreCond\sigma\sigma'$ and $W$
        **leave** Planner with $Plan$
      if $\nexists PreCond' \in Plan : PreCond'$ subsumes $PreCond\sigma\sigma'$
        append$(Open, PreCond\sigma\sigma')$
**leave** Planner with $planning\_failed$

Figure 4: The planning algorithm[6]

before, $t$ is the arity of PIAFs. The time for unification and the subsumption test is linear in the length of the terms to be tested, $O(t)$ [26, page G–25]. For each node in the plan tree, the following operations are performed:

- All predecessor must be determined. Therefore, for each operator a unification is necessary: $O(Op \cdot t)$. Furthermore, for each valid operator, it has to be tested, whether its precondition can be unified with a current perception: $O(W \cdot t)$.
- For each node, before adding it to the tree, it has to be tested, whether it is subsumed by another node. This requires $O(K)$ subsumption tests, and hence $O(K \cdot t)$ time.
- Adding a node to the tree, i.e. appending it to the list of tree nodes, takes $O(K)$ time.
- Adding a node to the open list again takes $O(K)$.

All together, we get a planning complexity of

$$O(K \cdot (Op \cdot t \cdot W \cdot t + K \cdot t + K + K)) = O(K \cdot Op \cdot W \cdot t^2 + K^2 \cdot t)$$

Note, that the number of concept defining rules only linearly influences the complexity. This behaviour is important, because the influence of the learning result is rather small. The main complexity can be reduced by choosing a simpler and more problem tailored signature yielding in lower values for $K$ and $t$.

[6]The brackets in the second line denotes a list, "m.g.u." is an appreviation for "most general unifier".

## 4.2 PLAN REFINEMENT

The plan generated at the abstract level is not immediately executable. Even for reporting about the execution, it is too abstract, because the user is not able to find the real reasons causing a failure of the plan execution. Two less abstract levels of planning are used to execute the abstract plan:

- operational plan trees and
- execution lists.

**Operational Plan Trees** *Operational plan trees* are plans at the level of PIAFs. They immediately correspond to the abstract plan trees described before, they are generated by backward chaining from operational concepts to their definitions by PIAFs. Thus, an operational plan tree defines again a set of possible plans to reach the given goal, where at least one plan starts with a currently given perception. A representation at this level combines compactness and concreteness. It is the lowest level at which action and perception are represented together.

Operational plan trees are calculated from abstract plan trees by deductive inference from operational concepts to PIAFs. Each node of an abstract plan tree corresponds to three nodes in an operational plan tree, namely precondition, action to be performed, and postcondition. Because postcondition and precondition of each two successing operational concepts meet each other, they are represented in an operational plan tree by a single node. An abstract plan tree with $K$ nodes therefore corresponds to an operational plan tree with $2K + 1$ nodes. The "+1" is due to the fact that the postcondition of the root node meets no other precondition. Figure 5 shows the operational plan tree corresponding to the abstract plan tree shown in Figure 3. The PIAFs, attached to the nodes are not shown in all detail, only instantiated arguments are printed to reduce the space needed to display the graph.

To generate the operational plan tree, for each node in the abstract plan tree, the clause defining the operational concept is searched for. This consists of three PIAFs and some calls to background knowledge. Each clause is scanned from PIAF to PIAF to gather the background knowledge calls. This way, for each abstract plan tree node, a linear branch of three nodes is generated. The root of each of these branches is marked by the keyword postcond. This keyword is used to put together the different branches of the operational plan tree in a post-processing step. This step is simple and only of technical relevance (and hence not shown in the algorithm). It avoids the need for following a special order in processing the abstract plan nodes. The algorithm is shown in Figure 6.

**Executable Lists** The lowest level to which operational plans are broken down to be executed are *executable lists*. In contrast to abstract and operational plan trees, executable lists are not computed a priori, but they are generated on demand. At each moment of plan execution, exactly one executable list exists. An executable list is the definition of a PIAF broken down to the predicates of the *lowest level of interest* (LLOI). In our case, these are perceptual features, action features, and background knowledge predicates. Executable lists are calculated by deductive inference from the PIAF backward to the predicates of

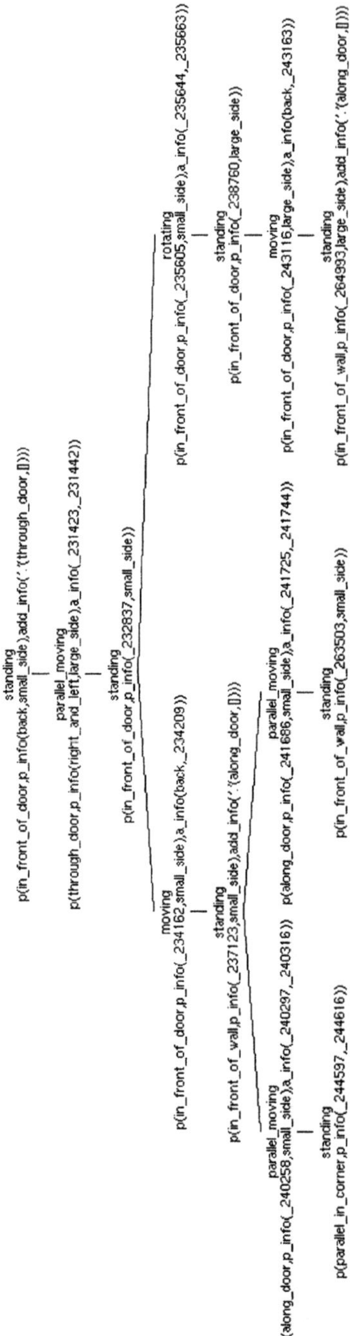


Figure 5: An operational plan tree

```
assert_edges (+ListOfPrems, +OC_Node, –PremNodeID)
  if (ListOfPrems is empty)
    PremNodeID := postcond
  else
    FirstPIAF := head (ListOfPrems)
    FurtherPrems := tail (ListOfPrems)
    get_bk_call (FurtherPrems, NextPIAF, RestPrems, BK_Call)
    PremNodeID := generate_new_node_id ()
    NextNodeID := assert_edges (RestPrems, OC_Node)
    assert_prem_node (FirstPIAF, NextPIAF, BK_Call, PremNodeID, NextNodeID, OC_Node)

get_bk_call (+ListOfPrems, –NextPIAF, –RestPrems, –BK_Call)
  BK_Call := []
  RestPrems := []
  while (head(ListOfPrems) is not a PIAF)
    append(BK_Call, head(ListOfPrems))
    ListOfPrems := tail(ListOfPrems)
  NextPIAF := head(ListOfPrems)
  RestPrems := tail(ListOfPrems)
```

Figure 6: The plan refinement algorithm[7]

the LLOI. This step is a standard technique, so we abstain of presenting the algorithm in more detail.

Suppose, the current perception of the robot is standing(..., parallel_in_corner, ...) (lower left corner in Figure 5). The next PIAF to be executed, the father node in the tree, is the PIAF parallel_moving(..., along_door, p_info(..., small_side), ...). This PIAF is broken down to the following executable list:

[move(..., back), tip(along_door, p_info(right, small_side), ...)].

That means, to execute the next PIAF in the tree, parallel_moving(..., along_door, p_info(..., small_side), ...), first a move with the specified arguments must be executed, and along_door must be perceived. How the action is triggered and the perception is monitored is explained in the next section.

## 4.3 REACTIVE PLAN EXECUTION

Plan execution is a perception driven process, receiving data from the object identification module, and sending data to different other processes. Corresponding to the different types of data that might come in —PIAFs, perceptual features, or actions features— different actions are performed:

[7] In this figure, + denotes input parameters, – output parameters. "Operational concept" is appreviated by "OC", "background knowledge" by "bk".

**PIAF:** Is the received PIAF unifyable with the current node of the operational plan tree?

- **yes:** The current operator of the plan is correctly executed, and so the next operator can be performed.
  - Call the background knowledge attached to the current node, to instantiate variables that are currently not instantiated. If there exist different possible instantiations, only use the first one. Mobile robots can not really be set back to a backtracking point, hence we have to be optimistic that the current choice is the right one.
  - Set the next node, the father node in the operational plan tree, as current node. This node is the one to be executed now.
  - Throw away the current executable list. Since PIAF and executable list correspond, it is just a timing reason why the executable list has not been executed completely — the PIAF has been received earlier than the LLOIs. Generate the executable list for the new current node. Is the first item of the list is an action feature, execute this action (see below).
- **no:** The received PIAF is not the expected one. However, the expected one might be received someone later in the near future. Therefore, the PIAF is stored in the list of currently perceived PIAFs. Nothing other happens.

**perceptual feature** or **action feature:** Is the perceptual feature or action feature unifyable with the first element of the current executable list?

- **yes:** The current perception matches the expected perception, all works as it should.
  - Delete the first element of the executable list. This element has been verified.
  - As long as the first element of the list is a background knowledge predicate, call this predicate and delete it from the list.
  - If now an action features is the first element of the list, this action is the next one to be executed. The action feature is send to the agent computer interface, which transforms it to an elementary operator of the robot. The action feature is not deleted from the executable list, the system waits for the action feature to be perceived by the object identification unit. That ensures that the suspected actions are really performed, and eases the handling of time delays.
  - If the executable list becomes empty, the corresponding PIAF is completely executed. The same steps as if an expected PIAF had been received are executed.
- **no:** Is the received data a perceptual feature?
  - **yes:** It is thrown away. It is not necessary to verify the executable list. Jumps in the plan are only performed at the level of PIAFs.
  - **no:** The received item is an action feature and this action does not correspond to the expected one. That means, the current executable list and the corresponding PIAF failed. In contrast to perceptions, which are

indeterministic and several different perceptions can be delivered at the same time, actions are deterministic and if the action is the wrong one, the right one can never appear. If the PIAF failed, this path in the plan tree cannot be executed further. We have to find another plan that might be better. Because we have generated not only a single plan, we can now try to find an alternative plan starting with the current perceptions, (stored in the list of currently perceived PIAFs, see above). Therefore, we first test all nodes in the abstract planning tree (starting at the root), whether they are unifyable with one of the current PIAFs. Again, we are optimistic, the first matching PIAF is the winner, it becomes the new current node and plan execution starts at this point. If no node matches a current perception, a new abstract plan has to be generated.

We call the plan execution strategy opportunistic and optimistic. Opportunistic, because as soon as the current plan fails, and another plan might lead the robot to the goal, the system changes the current plan. Optimistic, because every time the system takes the shortest plan to the goal, and it hopes, it is the right one. This strategy allows an easy handling of action cycles, where the number of cycles is not known a priori. If the robot should leave a room through the door, it first must find the door (this corresponds to the given example). To find a door, it first moves to the next wall, moves along this wall into the next corner, rotates in this corner, moves along the new wall into the next corner, ..., until it found a door opening while moving along a wall. Whenever the robot reaches a corner, moving along a door opening failed, and the plan starts again with the PIAF describing standing in a corner. No re-planning is needed for this case, only a jump in the plan tree.

## 5 Conclusion

In this chapter, we presented the reactive planning and plan execution system SHARC. This system brings together three different aspects. The first aspect is to use learned operators for planning, where not only the action part of the operators are learned, but also the perceptual parts. The second aspect is to have a system reactive to its perceptions, which is opportunistic by swapping plans, whenever other plans seems to be better, and which is optimistic, by searching for the shortest valid plan instead of searching for all alternatives. And third, the system is able to communicate with its user. Currently, this communication is based on PROLOG predicates. The plan execution process is visualized by marking the nodes of the plan trees with different colours, depending on whether a node has been executed with success, or whether it is the node that is currently in execution.

### 5.1 EXPERIMENTS

We have performed different experiments with parts of the system. The object identification unit has been tested in a real environment with the robot PRIAMOS. It has been shown, that the learned rules are able to classify real world concepts, but that the classification process is still not perfect. In addition, the experiments have shown, that the parallel inference system used to infer operational concepts from sensor data is fast enough to work in real time [23].

Experiments with the SHARC have been performed only in simulation. The perceptions are not produced by a real robot, but by hand. Therefore, we only tested how planning, plan refinement, plan execution, jumping between different plans, and, also, visualization of the plan execution process succeed, but this is clearly no real world proof for the system.

## 5.2 FURTHER WORK

**Emergency And Default Plans** In the previous sections, we described how to handle plans that guide the robot from a currently perceived situation to a goal situation. While executing plans, two exceptional situations may arise:

1. The robot moves into a critical situation, i.e. critical for itself (e.g. moving towards downwards going stairs) or critical for something/somebody else (e.g. moving towards a human).

2. The robot perceives nothing relevant, that can be used for planning.

In the first case, the robot must immediately perform an action bringing it out of the critical situation. This action must override all planned actions. In the second case, the robot must perform some action, that is likely to bring it into a situation that can be classified and used for planning. These operations are only applied, if no other action is sensible. Therefore we have in mind to enhance our system of operational plans in the following way in the future.

Instead of having a single operational plan, the PREM manages multiple operational plans at three different priority levels. The highest level is the one of emergency plans. Emergency plans consists of critical preconditions and solutions, i.e. action to be performed in these situations. The medium level is the level of plans described in Section 4. This plan guides the robot to the goal situation. Plans at the lowest priority level are plans for unclassifiable situations. They consists of, more or less, empty preconditions and default actions to be performed in this situation. Clearly, emergency plans and default plans are only small plan fragments, their only task is to bring the robot into save or classifiable situation, respectively.

Having these different levels of priority, the control structure of the PREM changes a bit. For each perceived PIAF, the PREM tests successively from the highest priority plans to the lowest priority plans whether it matches the precondition of a plan. In this case, this plan is executed.

In principle, proceeding in this way can be extended to arbitrary many different levels of priority, to distinguish plans with different prospects to proceed. However, our representation of operational concepts and our planning algorithm do not immediately suggest how to implement goal oriented plans with different weights. So, this last point is more a vision than a short-term goal.

**ACKNOWLEDGEMENTS** We would like to thank Katharina Morik for her discussions, ideas, remarks, and all the time she spent for supporting this work. Furthermore, thanks to Stefan Sklorz, who developed previous versions of the representation of action parts operational concepts and a first planner. We thank all members of our research group and the group at the University of Karlsruhe for the fruitful discussions about preliminary versions of the system, especially Peter Brockhausen and Michael Kaiser. Many thanks to Ralf Klinkenberg for proof reading this chapter.

## References

[1] M. Hägele. Service Roboter – ein Beitrag zur Innovation im Dienstleistungswesen. IPA – Frauenhofer Institut für Produktionstechnik und Automatisierung, 1994.

[2] J. Evans. HelpMate: An autonomous mobile robot courier for hospitals. In *IEEE/RSJ Intern. Conf. on Intelligent Robots and Systems (IROS '94)*, Germany, 1994.

[3] T. Längle, T. Lueth, G. Herzog, E. Stopp, and G. Kamstrup. KANTRA – a natural language interface for intelligent robots. In *Intelligent Autonomous Systems IAS-4*, pages 365–372. IOS Press, 1995.

[4] P. E. Agre and D. Chapman. What are plans for? In Pattie Maes, editor, *Designing Autonomous Agents*, pages 17–34. MIT Press, 1990.

[5] S. Förster, H. Lobing, and K. Peters. Architekturkonzeption eines teilautonomen Montageroboters. Report – Situierte Künstliche Kommunikatoren 95/10, Universität Bielefeld, 1995.

[6] A. Saffiotti, K. Konolige, and E. H. Ruspini. A multivalued logic approach to integrating planning and control. *Artificial Intelligence* **76** (1995), 481–526.

[7] V. Klingspor, J. Demiris, and M. Kaiser. Human-robot-communication and machine learning. *Applied Artificial Intelligence* **11** (1997), 719–746.

[8] L. Seabra Lopes and L.M. Camarinha-Matos. Planning, training and learning in supervision of flexible assembly systems. In *Procs. of Conf. on Balanced Automation Systems*, 1995.

[9] V. Klingspor, K. Morik, and A. Rieger. Learning concepts from sensor data of a mobile robot. *Machine Learning* **23** (1996), 305–332.

[10] L. P. Kaelbling. Foundations of learning in autonomous agents. *Robotics and Autonomous Systems* **8** (1991), 131–144.

[11] J. Millán and C. Torras. Learning to avoid obstacles through reinforcement. In *Procs. of the 8th Intern. Workshop on Machine Learning*, pages 298–302. Morgan Kaufmann, 1991.

[12] A. Segre. *Machine Learning of Robot Assembly Plans*. Kluwer, Boston, 1988.

[13] J.G. Carbonell, C. Knoblock, and S. Minton. PRODIGY: An integrated architecture for planning and learning. In K. VanLehn, editor, *Architectures for intelligence*. Lawrence Erlbaum, 1991.

[14] S. W. Bennett. Learning uncertainty tolerant plans through approximation in complex domains. Technical Report UILU-ENG-89-2204, University of Illinois at Urbana-Champaign, January 1989.

[15] K. Zercher. *Wissensintensives Lernen für zeitkritische technische Diagnoseaufgaben*. infix, Sankt Augustin, 1992.

[16] A. I. Cypher. *Watch what I do – Programming by Demonstration*. MIT Press, Cambridge, Massachusetts, 1993.

[17] M. Kaiser and R. Dillmann. Building elementary robot skills from human demonstration. In *IEEE International Conf. on Robotics and Automation*, Minneapolis, 1996.

[18] V. Klingspor and K. Morik. Towards concept formation grounded on perception and action of a mobile robot. In *IAS–4, Proc. of the 4th Intern. Conf. on Intelligent Autonomous Systems*, pages 271–278, 1995. IOS Press.

[19] S. Muggleton, editor. *Inductive Logic Programming*. Academic Press, 1992.

[20] J. U. Kietz. *Induktive Analyse relationaler Daten*. PhD thesis, Technische Universität Berlin, 1996.

[21] A. Geist, A. Beguelin, J. Dongarra, W. Jiang, R. Manchek, and V. Sunderam. *PVM: Parallel Virtual Machine – A Users' Guide and Tutorial*. MIT Press, 1994.

[22] R. Dillmann, J. Kreuziger, and F. Wallner. PRIAMOS – an experimental platform for reflexive navigation. In *IAS-3: Intelligent Autonomous Systems*, pages 174–183. IOS Press, 1993.

[23] V. Klingspor. On-line inference of off-line learned operational concepts. In Henry Hexmoor and Lisa Meeden, editors, *ROBOLEARN '96: An Intern. Workshop on Learning for Autonomous Robots*, pages 54–63, 1996.

[24] A. Rieger. MP: An efficent method for calculating the minimum Herbrand model of chain Datalog programs. In W. Wahlster, editor, *Procs. of the 12th European Conf. on Artificial Intelligence*, 1996.

[25] M. Beetz and D. McDermott. Declarative goals in reactive plans. In *Procs. of the First International Conference on AI Planning Systems*. Morgan Kaufmann, 1992.

[26] Quintus Corporation, Palo Alto, Califonia. *Quintus Prolog, Release 3.1*, 1991.

# 7
# SENSORS USED FOR AUTONOMOUS NAVIGATION

E. M. NEBOT
*Australian Centre for Field Robotics*
*The University of Sydney*
*NSW 2006 Australia*

## 1. Introduction

Navigation and guidance are defined as the processes of determining and controlling the position of a vehicle. An autonomous system will have to include these two basic procedures to perform any useful task. The information generally required for navigation is the direction, speed, and position of the vehicle. The basic navigation techniques were designed several years ago to navigate vessels across the ocean. In the past the position of ships was estimated with dead reckoning techniques using speed and heading observations. With this information the trajectory of the ship was predicted with errors that accumulated over time. The heading information was obtained with the position of the sun or other stars and the velocity was evaluated with a log compensating for the effect of local currents. Absolute information was also used to correct the position estimation. These fixes were done when well-known natural landmarks were recognised. The Mediterranean was the testing place of all these techniques. It was appropriate since the water currents are light in most places and many natural landmarks can be seen along most possible journeys.

A completely different problem arises when the navigators of the fifteen-century needed to cross the ocean. Even with the invention of the compass dead reckoning became useful for short periods of time due to difficulty in obtaining accurate measurements of the velocity. The difficulties in compensating for local currents made dead reckoning useless after a period of a few hours. The navigators used absolute information to correct their dead reckoning estimation. In the open sea natural landmarks are scarcely available and hence new methods or new sensors were needed to obtain absolute position information. Methods to determine Latitude were developed in the early 1550s by the Portuguese. Latitude information could be found with relative accuracy (30 miles) by the observation of the meridian altitude of the sun and certain stars, such as the Pole star. The determination of Longitude was not an easy problem, and in fact it took another 300 years to be solved. In 1714 England's parliament, offered a significant prize to anyone who could formulate a reliable method to obtain absolute information in the open sea. Two approaches were mainly followed at this time. One method was based on accurate prediction and observation of the moon position. Difficulties in obtaining the lunar orbit prediction with enough precision and the accuracy required of the instrumentation made this approach almost impossible to implement at that time. The other important approach was based on knowing the time with enough accuracy to estimate the Longitude of a particular place. John Harrison developed a chronometer to keep time at sea with accuracy not available on any land clock of this time. By 1770 both techniques were mature enough and were used in many journeys across the ocean. By the end of the century, navigation in the open sea became predictable and safe, making possible the growing of business involving commercial transactions from all over the world.

In today's application the approach to navigation is very similar. The navigation techniques are still designed based on dead reckoning and absolute information. There is considerable research into dead reckoning and absolute sensors that can work reliably in different environments and weather conditions. Different sensors suites are required for each application, such as navigation in underground and open cut mining, underwater vehicles, ships and space vehicles. Each system will require a particular combination of dead reckoning and absolute sensors.

In many of these applications the reliability is a very important issue. Such is the case in autonomous operation of large machines. In these applications the navigation system has to be composed of more than one navigation loops, each based on independent failure mode sensors. For system integrity each loop needs to be designed with dead reckoning and absolute sensors based on different physical principles.

This chapter will present a description of a variety of dead reckoning and absolute sensors that are applicable for autonomous operation. Inertial sensors are described in detail since they have recently become available with the required specification and at affordable price for commercial applications. A variety of external or absolute sensors are also described.

## 2. Dead Reckoning Sensors

Dead reckoning sensors measure quantities such as position, velocity, acceleration and orientation. Most autonomous machines use wheels to move around in the environment of operation. The most direct approach to track their position is to measure the angular velocity of the wheels and the orientation of the steering mechanism. This information is then used with a kinematic model of the vehicle in order to predict position.

The velocity of the wheel is usually measured by attaching a rotation sensor to the shaft of the motor or the axle of the wheel. The final sensor selection will depend on the resolution required, the maximum speed expected and economic considerations.

Some of the most common dead reckoning sensors are potentiometers, incremental and absolute encoders (optical, inductive, capacitive, magnetic), tachometer, synchros, resolvers, Linear Variable Differential Transformers (LVDT), compass and inertial sensors such as gyroscopes and accelerometers. The following section presents a description of the most common dead reckoning sensors with emphasis on inertial sensors.

### 2.1. Potentiometers

This type of sensor is based on a variable resistor whose value changes according to the position of its shaft. There are many types of potentiometers available including single and multiple turns. The position is sensed measuring the analog voltage between the moving contact and one of the extremes of the potentiometer. The resolution of these sensors depends on the quality of the device, the electronic noise level present and the number of digit used by the analog to digital converter. These devices are in general categorised as low precision sensors and are used in low cost applications or to preposition incremental high-resolution sensors.

## 2.2. Encoders

These are one of the most common sensors used to measure the velocity or position of a rotating device. Encoders are typically attached to a shaft to measure their rotational velocity. The frequency characteristics of these sensors are band-pass. To perform dead reckoning, wheel rate encoders and steer angle encoders are passed through the vehicle's forward kinematic equations in order to obtain a prediction of the position and orientation. One problem with encoder based dead reckoning is that the vehicle model is usually derived in two dimensions, so the further the terrain is from a plane, the less accurate the solution. Another important problem is that encoder based dead reckoning cannot detect wheel slip, [1].

### 2.2.1. Incremental Encoders

Sensors of this type are based on a variety of different principles. They usually have one or two channels measuring or counting the number of revolutions of a shaft. The higher the resolution of the device the more sensitive it to very small rotations of the shaft. They usually produce square wave signals that are proportional to the number of marks detected on the shaft.

One channel encoders give information on velocity but it can not detect direction. This problem is solved by adding another channel that generates an additional square wave with appropriate phase shift.

These sensors have very good medium and high frequency responses but can produce large errors at very low frequency. The system is usually based on the sensor itself and the decoder that transforms the square wave output into a digital count.

The most common available encoders are based on optical principles. These sensors have a transparent disk with a number of lines painted to obstruct the light of two optical devices. These devices are mounted such that the square wave output obtained has a phase shift of 90 degrees and is usually referred to as a quadrature output. An index pulse is also generated to mark the beginning of a revolution.

These sensors are capable of measuring the relative position of a shaft with respect to an initial position. The resolution can be increased by adding more lines to the transparent disk. The maximum number of counts per revolution is a function of the diameter of the encoder's disk and the quality of the light transmission. It is possible to have up to 10000 counts per revolution with an encoder of less than 3 inches in diameter. BEI and MVC are a couple of companies that supply a large variety of these sensors.

There are other encoders that are based on magnetic, inductive and capacitive principles. These encoders are available for outdoor operation since they are constructed with a heavy-duty cases and are environmental sealed to comply with IP66 norms.

The decoding system is usually built around a standard HCTL20XX integrated circuit. This chip provides the necessary logic to process the quadrature outputs and to keep track of the position with a 16 to 20 bit counter.

There are also some more sophisticated encoders control integrated circuits. These include the HCTL1100 [Hewlett Packard] and the LM628 [National]. The HCTL1100 has a 24-bit encoder and provides a controller based on a lead-lag transfer function. The pole, zero and gain are programmable as well as other parameters based on the desired trajectory profile. The LM628 provides a 32-bit encoder counter and a controller based on a standard PID algorithm. The proportional, integral and derivative gains are programmable as well as many other control and trajectory parameters. These ICs are a very cost effective solution for decentralized control systems since the central computer only needs to send the desired set points, leaving the computational intensive control task to the dedicated processors.

### 2.2.2. Absolute Encoders

These devices return absolute information on the position of the shaft. They can be constructed with resolutions in the order of 8 to 10 bits per revolution. They are used in applications where high accuracy is not necessary and where loss of reference from power outage is not acceptable. They can be manufactured with the same environmental specification as incremental encoders although high-resolution devices are much more expensive than their counterparts. They provide digital outputs in serial or parallel form following Gray, natural binary or BCD codes. These sensors are best suited to measure the orientation of low frequency devices, such as a steering wheel and antenna orientation. Unlike incremental encoders, upon power up they do not require any reference for calibration purposes.

## 2.3. Linear Variable Differential Transformers (LVDT)

The Linear Variable Differential Transformer or LVDT, consists of a series of inductors in a hollow cylindrical shaft and a solid cylindrical core. The LVDT produces an electrical output proportional to the position of the core. The LVDT is constructed with two secondary coils placed symmetrically on either side of a primary coil contained within the hollow cylindrical shaft. Movement of the magnetic core causes the mutual inductance of each secondary coil to vary relative to the primary, and thus the relative voltage induced from the primary coil to the secondary coil will vary as well. The lack of friction between the hollow shaft and the core prolongs the life of the LVDT and enable very good resolutions. The normal operating range of these sensors is in the order of 6 to 300 mm. These devices can be used to observe the orientation of the wheel by measuring the displacement of a particular joint of the steering mechanism. The output is usually an analog signal proportional to the displacement. Although they provide absolute information, they may need initial calibration and sometimes on line re-calibration due to instabilities with temperature variations.

## 2.4. Synchros and Resolvers

These devices are used to transmit angular data electrically from one location to another, where a high degree of accuracy is required. They are essentially variable transformers in which the coupling between the winding varies with the rotor position. Resolvers are primarily used for the solution of trigonometrical problems since they provide two phase outputs that are related by sine and cosine functions to the angular position of the rotor. The synchros are similar devices but they provide three sine wave outputs with a phase lag of 120 degrees. Typical applications of these devices include:

- Range and elevation calculation for radar equipment and gunnery

- Remote digital measurement of angle via suitable signal converters
- Conversion of geometrical coordinates
- Remote positioning of low torque mechanisms
- Remote control of servo mechanisms

These sensors require an appropriate analog to digital signal converter to transform the voltage output of the different phases to a digital count. Integrated circuits such as the HSR1056 from Natel that convert Synchro/Resolver analog outputs to a 14/16 digital number. These devices are usually much more expensive than the single analog to digital converters of the same precision.

## 2.5. Compass

These are devices that measure the direction of the earth's magnetic field. This physical principle has been used since the 1200s to navigate ships across the ocean. A simple device consisted of an iron needle floating in water. The compass information needs to be corrected because of the discrepancy between the true north and the direction of the earth's magnetic field. Another correction is also required due to vertical magnetic forces. This correction depends on the position of the vehicle, being negligible at the equator and maximum at the poles.

The main drawback of these devices is that the quantity measured, the intensity and direction of the magnetic field, can be deformed in the presence of metals and other electrical or magnetic fields, such as power lines and transformers. Very often, this problem makes these devices very unreliable and therefore not desirable for autonomous land vehicle applications.

Flux gate compasses with good resolution can be obtained from several companies such as KVH and Watson. Some sensor packages are available which overcome the limitation of the standard Flux gate compass by adding gyros to the flux gate sensors. One example is the Watson unit FGM-G100HS that combines a gyroscope sensor with a flux-gate magnetometer. The integrated sensor includes its own processor which generates an output that uses the high frequency component of the gyro and low frequency information given by the compass. With appropriate frequency weighting schemes fast changes in compass information due to unexpected magnetic field modification can be rejected. At the same time, the low frequency drift of the gyro is also rejected. The system is still not absolutely reliable since there are many situations where the compass can be constantly perturbed for long periods of time translating into a biased heading output.

In general these sensors are recommended for autonomous applications only in very few applications where it can be guaranteed that the environment will be free of all these disturbances.

## 2.6. Inertial Sensors

Inertial sensors make measurements of the internal state of the vehicle. A major advantage of inertial sensors is that they are non-radiating and non-jammable and may be packaged and sealed from the environment. This makes them potentially robust in harsh environmental conditions.

Historically, Inertial Navigation Systems (INS) have been used in aerospace vehicles, military applications such as ships, submarines, missiles, and to a much lesser extent, in land vehicle applications. Only a few years ago the application of inertial sensing was limited to high performance high cost aerospace and military applications. However, motivated by requirements for the automotive industry, a whole variety of low cost inertial systems have now become available in diverse applications involving heading and attitude determination.

The most common type of inertial sensors are accelerometers and gyroscopes. Accelerometers measure acceleration with respect to an inertial reference frame. This includes gravitational and rotational acceleration as well as linear acceleration. Gyroscopes measure the rate of rotation independent of the coordinate frame. They can also provide 3-D position information and, unlike encoders, have the potential of observing wheel slip.

The most common application of inertial sensors is in the use of a heading gyro. Integration of the gyro rate information provides the orientation of the vehicle. A good quality gyro will have zero or constant bias, and small noise variance. There are many types of gyroscopes but few in the price range that can be afforded in commercial mobile robot applications. Fiber optic and vibratory gyroscopes have been recently released with satisfactory specifications and reasonable price ranges for the applications under considerations, [1].

Another application of inertial sensors is the use of accelerometers to measure the attitude of the vehicle. The inclination of a platform can be evaluated with two orthogonal accelerometers. This requires the knowledge of the gravity value in the region of operation. There are tilt sensors available for this purpose that can provide accurate information while the vehicle is stationary, such as the CXTILT02 manufactured by Crossbow. This device can provide bank and elevation information with an accuracy of up to 0.1 degrees. These sensors can not be used when the platform is not stationary since they are sensitive to translational and rotational accelerations.

Pendulum accelerometers are also used to determine the bank and elevation angles. These devices use the principle that a pendulum with an equivalent radius of the earth will always point to the vertical irrespective of the acceleration. This effect can be approximated with appropriate feedback loop compensators. Although they are able to obtain some rejection of the undesired acceleration due to the movement of the platform they can only be used under very low vehicle accelerations.

The following sections present some of the most common type of gyroscope and accelerometers used in land vehicle applications. An analysis of the error generated by these devices is also presented to allow the user to specify the quality of the inertial sensor required for a particular application. Finally the problem of working with a full six degree of freedom INS is addressed presenting the necessary algorithms to perform alignment and calibration and to predict attitude, velocities and positions in 3-D.

### 2.6.1. Accelerometers

The accelerometers measure the inertia force generated when a mass is affected by change in velocity. This force may change the tension of a string or cause a deflection of beam or may even change the vibrating frequency of a mass. The accelerometers are composed of three main elements: a mass, a suspension mechanism that positions the mass and a sensing element that returns an observation proportional to the acceleration of the mass. Some devices include an additional servo loop that generates an apposite force to improve the linearity of the sensor. A basic one-degree of freedom accelerometer is shown in Figure 1 . This accelerometer is usually referred to as an open loop since the acceleration is indicated by the displacement of the mass.

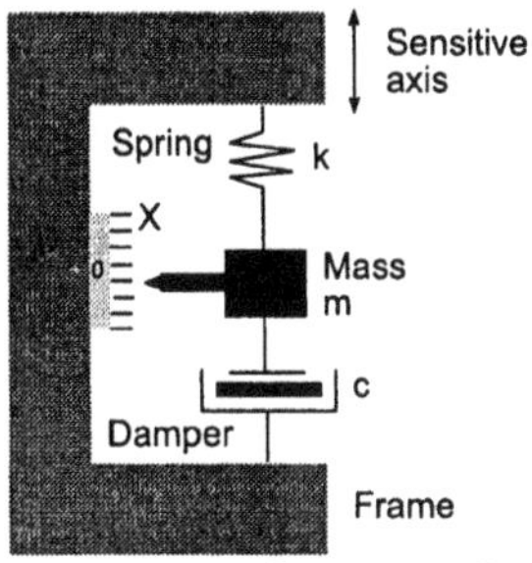


**Figure 1 Basic components of an open loop accelerometer.**

The accelerometer described in this example can be modelled with a second order equation:

$$F = m\frac{d^2x}{d^2t} + c\frac{dx}{dt} + k\,x \qquad (1)$$

where F is the applied force to be measured. The damping can be adjusted to obtain fast responses without oscillatory behaviour. Many of the actual commercial accelerometers are based on the pendulum principle. They are built with a proof mass, a spring hinge and a sensing device. These accelerometers are usually constructed with a feedback loop to constrain the movement of the mass and avoid cross coupling accelerations. There are accelerometers that implement variation of this principle such as the Sundstrand "Q-Flex" design and the Analog Device silicon micro-machined accelerometer.

One important specification of the accelerometers is the minimum acceleration that can be measured. This is of fundamental importance when working with large machines, where the acceleration expected is usually in the range of 0.1-0.3 g. Figure 2 shows the acceleration measured when travelling with a car at low speed. A standard accelerometer for such applications is usually capable of measuring acceleration of less than 500 μg. The dependence of the bias with temperature and the linearity of the device are also important specifications.

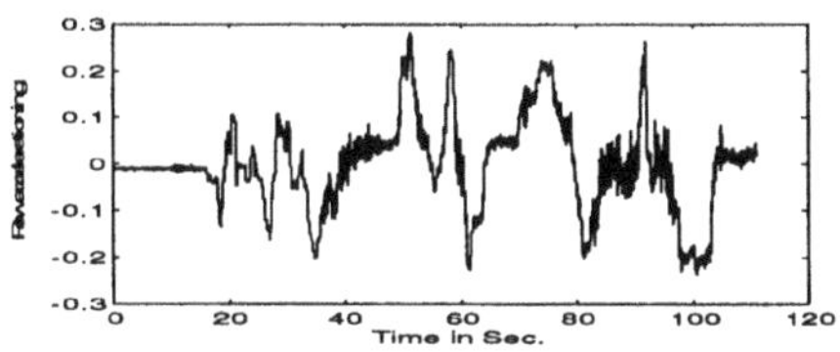


**Figure 2. Raw Acceleration**

### 2.6.2. Gyroscopes

These devices return an output proportional to the rotational velocity. There is a large variety of gyroscopes that are based on different principles. The price and quality of these sensors varies significantly. The following sections present some of the most common types of gyroscopes available for industrial applications.

### 2.6.2.1. Vibratory gyroscopes

These types of gyroscopes can be manufactured in very different forms but they are all based on a basic principle. The device can be modelled by a simple mass-spring system as shown in the Figure 3.

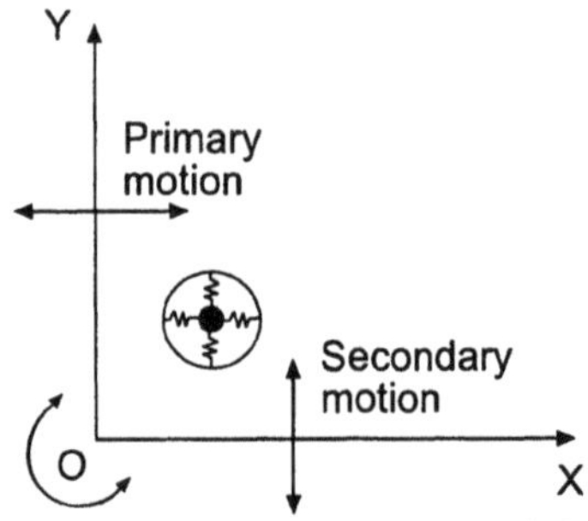


**Figure 3. Mass spring model**

The purpose of the gyroscope is to measure the angular velocity of the particle that is supposed to be rotating about the axis OZ. With this design it is necessary to make the particle vibrate with constant amplitude along the axis OX. This motion is usually referred to as primary motion and is controlled by an embedded circuit that maintains the oscillation at constant amplitude. Under rotation the mass will experience a coriolis inertia force that will be proportional to the applied rate of turn and will have a direction parallel to the OY axis. This motion is referred to as secondary motion and its amplitude is measured to provide information proportional to the angular rotation. Some devices provide an additional feedback mechanism for this secondary motion to increase the linearity of the sensor. Murata manufactures a gyroscope model Gyrostar ENV05 that is based on this principle. This device presents very good high frequency characteristics but poor low frequency performance. The drift expected is in the order of 1 degree / minute. Noise and significant bias temperature dependence also affect the performance of this particular sensor. Better performance can be expected from the Vibrating Structure Gyroscopes from BAE and the quartz rate sensors build by Systron Donner. These devices have a drift of less that 0.15 degree/minute and the linearity is better than 0.05 % of the full range.

### 2.6.2.2. Fiber Optic Gyros

The fiber optic gyros are based on the Sagnac effect discovered by Georges Sagnac in 1913. This effect can be easily explained assuming two waves of light circulating in apposite direction around a path of radius R. If the source is rotating at speed $\omega$, the light travelling in the opposite direction will reach the source sooner than the wave travelling in the same direction, as shown in Figure 4. The wave travelling with the rotation will covers a distance $D_1$ in a transit time $t_1$, while the other signal covers a distance $D_2$ in a time $t_2$.

$$D_1 = 2\pi R - R\omega\, t_1$$
$$D_2 = 2\pi R + R\omega t_2 \tag{2}$$

Making the waves travel the path N times, the difference in transit time becomes:

$$\Delta t = N(t_2 - t_1) = \frac{4\pi N R^2 \omega}{c^2} \tag{3}$$

It is important to relate this time difference with a phase shift at a particular frequency.

$$\phi = 2\pi\Delta t f \tag{4}$$

For a given rotation ω the phase shift will be

$$\Delta\phi = \frac{8\pi^2 RN}{c}\omega \tag{5}$$

Most low cost implementations of these devices work in an open loop manner. The maximum phase shift that can be evaluated without ambiguities is 90 degrees.

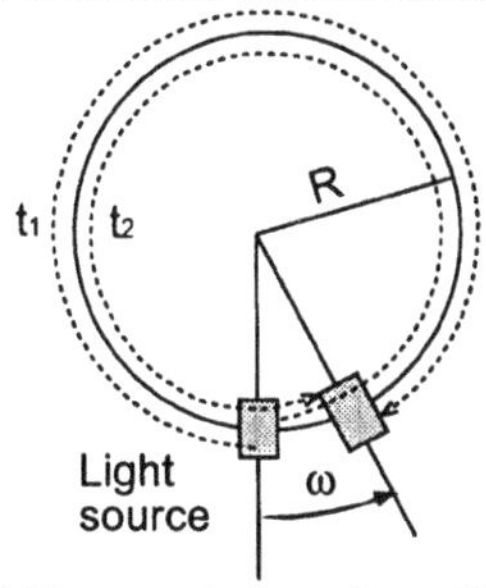


**Figure 4 Transmission time difference**

There are commercial laser gyros such as the Andrew model 2030 and Autogyro, which are capable of drifts as low as 0.036 and degree 0.12 per minute respectively. Closed loop optical gyros are also available but they are more expensive

### 2.6.3. Accelerometer and Gyro errors

The accelerometers measure the absolute acceleration with respect to an inertial frame. We are interested in the translational acceleration hence the algorithms used should compensate for all other accelerations. For practical purposes it can be considered that gravity is the only other acceleration present. In Figure 2 it can be seen that the average acceleration obtained from a land vehicle is smaller that 0.1 g. This implies that the orientation of the accelerometer has to be known with very good accuracy in order to compensate for gravity without introducing errors comparable to the actual translation acceleration. This has been the main limitation that has prevented the widespread use of inertial sensors being used as position sensors in low cost implementations.

The orientation of the accelerometer can be tracked using a gyroscope. These sensors provide an output proportional to the rotation speed but are always contaminated by noise and drift. For short periods of time the drift can be approximated by a constant bias. The orientation is evaluated with a single integration of the gyroscope output:

$$\theta_m = \int \dot{\theta} + b + v \, dt \quad = \quad \theta + bt + \int v \, dt \tag{6}$$

The integration given in equation 6 returns the rotation angle with two additional undesired terms. A random walk due to the noise and another term that grows with time proportional to the gyro bias. The maximum random walk error is proportional to the noise standard deviation and the square root of time. This error has to be considered in cases where this value is comparable with the expected drift of the gyro.

As shown in figure 5, the gyro drift will introduce an error in acceleration, velocity and position given by:

$$ax = abt \quad vx = \frac{1}{2}abt^2 \quad posx = \frac{1}{6}abt^3 \tag{7}$$

It is important to remark that a constant gyro bias introduces an error in position determination that grows proportional to $t^3$.

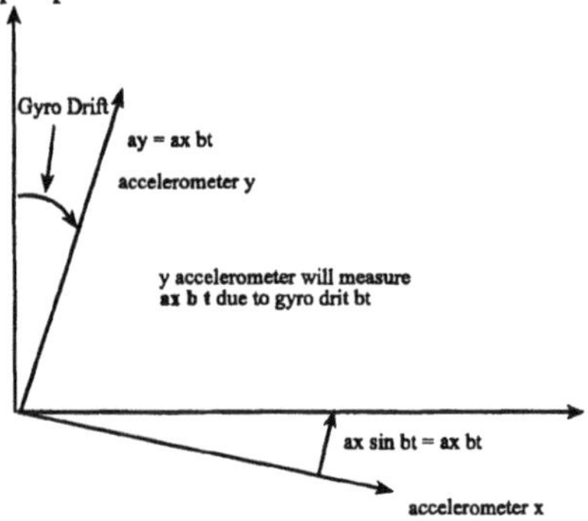


**Figure 5. Gyro drift errors**

For standard low cost, good quality gyros the bias expected is in the order of 1-10 degrees / hour. Without calibration, the bias could introduce an error of approximately 140 meters due to the incorrect compensation of gravity after only 2 minutes of operation

Another aspect that needs to be considered is the resolution of the analog to digital converter used, especially when working with low drift gyroscopes. The original noise standard deviation of the gyro noise could be incremented selecting an inappropriate digital to analog converter. A gyro that can work with in a range of 100 deg/sec and with stability of the order of 1 degree per hour will require a 16-bit converter to operate under its specifications.

The bias in the accelerometer will also increase the error in position and is proportional to the square of time:

$$vx = b_a t \qquad posx = \frac{1}{2}b_a t^2 \tag{8}$$

The error introduced by the accelerometer is important enough not to be neglected. The calibration procedure must incorporate the identification of gyro and accelerometer biases.

### 2.6.4. Full six degree of freedom Inertial Navigation systems

A full inertial navigation system (INS) consists of at least three (triaxial) accelerometers and three orthogonal gyroscopes that provide measurements of acceleration in three dimensions and rotation rates about three axes. An INS system assembled from low cost solid-state components is almost always constructed in a "strap-down" configuration. This term means that the gyros and accelerometers are fixed to a common chassis and are not actively controlled on gimbals to align themselves in a pre-specified direction. This design has the advantage of eliminating all moving parts. The strap-down construction, however, needs substantially more complex algorithms in order to compute and distinguish true linear acceleration from angular acceleration and body roll or pitch with respect to gravity, [3]. Once true linear acceleration has been determined, vehicle position may be obtained, in principle, by double integration of the acceleration. Vehicle orientation and attitude may also, in principle, be determined by integration of the rotation rates of the gyros. In practice this integration leads to unbounded growth in position errors with time due to the noise associated with the measurement and the non-linearity of the sensors. Consequently, INS information is useful for determining position and orientation only over short periods of time. How long one can rely on INS information to provide a measure of position and orientation depends ultimately on the magnitude of the measurement error or "drift rate". The predicted trajectory will mainly be a function of the initial calibration and alignment of the platform. By calibration is meant the determination of the biases on the accelerometers and gyros. The initial calibration aspect and on line calibration has been addressed in a number of works, [4][5]. Some of these works consider very high quality inertial units. The cost of this hardware makes them prohibitive for standard commercial systems. The alignment process consists of determining the initial orientation of the platform. It is also of fundamental importance since the dead-reckoning algorithm will use this initial orientation to update the attitude information and then evaluate the accelerations in the navigation frame. The following section will present the algorithms required to perform initial calibration and alignment and to obtain the inertial attitude, velocity and position indications with a standard Inertial Measurement unit (IMU).

### 2.6.4.1. Coordinate Systems and Transformations.

When designing a multi-sensor navigation system it is necessary to relate the information from the different sensors to a single navigation coordinate. The obvious coordinate choice for land vehicle applications is a local geographic frame n, with axes $^n\{N, E, D\}$, (North, East and Down). The inertial measurement unit is mounted on the vehicle constituting a new frame. This is called the body frame "b'', and has axes $^b\{R, P, Y\}$, (Roll , Pitch and Yaw). This frame will be in constant rotation with respect to the "n" (navigation) frame. The velocity of this rotation is measured by three near-orthogonal gyros. The transformation matrix that relates "b" and "n" coordinate frames can be obtained with the following integration:

$$\dot{C}_b^n = \Omega_{bn} C_b^n \qquad \Omega_{bn} = \begin{bmatrix} 0 & -\omega_Y & \omega_P \\ \omega_Y & 0 & -\omega_R \\ -\omega_P & \omega_R & 0 \end{bmatrix} \tag{9}$$

where $\Omega_{bn}$ is the anti-symmetric velocity matrix and $\omega_{i's}$ are the rotation velocities measured by the gyros in the body frame. In real time applications the integration is implemented with the following approximation

$$C_b^n(k+1) = C_b^n(I + \Delta\theta)\ , with\ \Delta\theta = \begin{bmatrix} 0 & -\Delta\theta_Y & \Delta\theta_P \\ \Delta\theta_Y & 0 & -\Delta\theta_R \\ -\Delta\theta_P & \Delta\theta_R & 0 \end{bmatrix} \tag{10}$$

where $(I+\Delta\theta)$ is the small angle direction cosine matrix relating the frame at time k and the rotating frame at time k+1. This approximation is valid for small angles, which constrains the minimum sampling time of the gyros such that the transformation matrix may be obtained with reasonable accuracy. This sampling time will be a function of the severity of manoeuvres expected from the vehicle. In many land vehicle applications the rotation velocity expected is usually less than 25 degrees/sec. With a sampling time of 100 Hz the maximum angle variation will be less than 0.25 degree satisfying the small angle approximation. Vehicles with faster dynamics will require a faster sampling time to be able to track the transformation matrix appropriately.

### 2.6.4.2. Calibration and Alignment algorithm

To start the inertial navigation loop it is necessary to calibrate and align the Inertial Navigation Unit. The unknowns are the gyro and accelerometer biases and the actual orientation of the platform, that is the bank, elevation and heading angles. The inclination (tilt) of the platform could be obtained, in principle, from the accelerometers x and y, but the unknown biases prevent its accurate determination. At steady state, additional sensors, such as inclinometers usually provide more stable information, especially when working with small tilt angles. This information is used to obtain two of the three Euler angles (Roll & Pitch) necessary to evaluate the initial Cosine Matrix. The other angle (Yaw) is usually obtained from the compass information. The Direction Cosine matrix can be defined considering three successive rotation of angles $\theta$, $\beta$,$\gamma$ around the x, y and z axis respectively and in that order. The order of the rotation is of fundamental importance since these operations are not commutative. The initial cosine matrix is:

$$\begin{bmatrix} a^{nx} \\ a^{ny} \\ a^{nz} \end{bmatrix} = \begin{bmatrix} \cos\beta\cos\gamma & -\cos\theta\sin\gamma+\sin\theta\sin\beta\cos\gamma & \sin\gamma\sin\theta+\cos\theta\sin\beta\cos\gamma \\ +\cos\beta\sin\gamma & \cos\theta\cos\gamma+\sin\theta\sin\beta\sin\gamma & -\sin\theta\cos\gamma+\cos\theta\sin\beta\sin\gamma \\ -\sin\beta & +\sin\theta\cos\beta & \cos\theta\cos\beta \end{bmatrix} \begin{bmatrix} a^{bx} \\ a^{by} \\ a^{bz} \end{bmatrix} \tag{11}$$

This matrix relates the acceleration in the body and local navigation frame, that is

$$a^{ni} = C_b^n a^{bi} \tag{12}$$

The bias present in the measured acceleration $a^{bi}$ can be estimated with the inclinometer information. These devices can provide bank and elevation angular information of high quality at very low frequencies. The acceleration in x, y and z are corrected according to the following equations:

$$\begin{aligned} a_{bias}^{bx} &= a_{meas}^{bx} - \sin(elevation) \\ a_{bias}^{by} &= a_{meas}^{by} + \sin(bank) \\ a_{bias}^{bz} &= a_{meas}^{bz} - (1-\cos(elevation) * \cos(bank)) \end{aligned} \tag{13}$$

During initial alignment, the velocities and accelerations of the platform can be considered zero except for gravity. With this assumption, the acceleration measured in the navigation frame is:

$$a^{ni} = C_b^n a^{bi} \;\; with \;\; a^{ni} = \begin{bmatrix} 0 & 0 & g \end{bmatrix}^T \tag{14}$$

The measured acceleration in the body frame can be evaluated with:

$$a^{bi} = C_n^b a^{ni} \;\; with \;\; C_n^b = \left[C_b^n\right]^{-1} = \left[C_b^n\right]^T \tag{15}$$

Since the transformation matrix C is orthogonal, the inverse is just its transpose:

$$\begin{bmatrix} a^{bx} \\ a^{by} \\ a^{bz} \end{bmatrix} = \begin{bmatrix} \cos\beta\cos\gamma & \cos\beta\sin\gamma & -\sin\beta \\ -\cos\theta\sin\gamma+\sin\theta\sin\beta\cos\gamma & \cos\theta\cos\gamma+\sin\theta\sin\beta\sin\gamma & \sin\theta\cos\beta \\ \sin\gamma\sin\theta+\cos\theta\sin\beta\cos\gamma & -\sin\theta\cos\gamma+\cos\theta\sin\beta\sin\gamma & \cos\theta\cos\beta \end{bmatrix} \begin{bmatrix} 0 \\ 0 \\ g \end{bmatrix} \tag{16}$$

With the tilt information (bank and elevation angles), the value of the measured body acceleration can be predicted. These are the values of the element of the third column of the transformation matrix. Then the Euler angles θ and β can be obtained solving:

$$\sin\beta = \sin(elevation)$$
$$\sin\theta\cos\beta = \sin(bank) \tag{17}$$
$$\cos\theta\cos\beta = \cos(elevation)\cos(bank)$$

From equation 17 the Euler angles θ and β can be evaluated using

$$\beta = elevation$$
$$\theta = \sin^{-1}\left(\frac{\sin(bank)}{\cos(-elevation)}\right) \tag{18}$$

The remaining Euler angle γ can be evaluated with the provided heading data.

$$\gamma = heading\ \ angle \tag{19}$$

With the Euler angles θ, β and γ the elements of the initial transformation matrix $C_b^n$ can be evaluated using equation 11.

#### 2.6.4.3. Implementation issues and Experimental results.

Figure 6 presents the tasks required to work with a full inertial measuring unit.

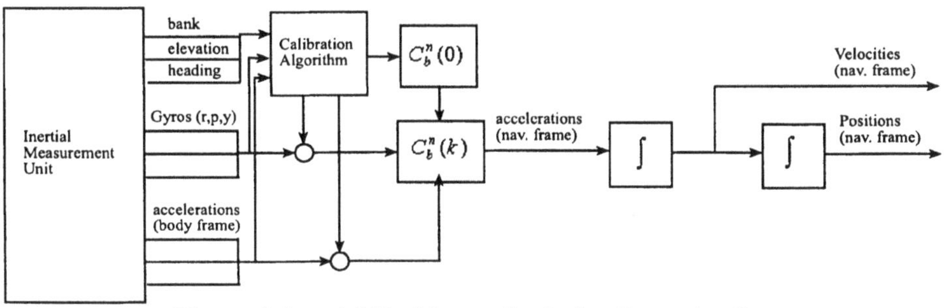


**Figure 6. Inertial Position and velocity determination**

The bank, elevation and heading information is used to calibrate and align the unit while the vehicle is stationary. Once the initial transformation matrix $C_b^n(0)$ is obtained, it is updated at fast rates with the gyroscope information to obtain $C_b^n(k)$. This matrix is used to transform acceleration in the body coordinates to the navigation coordinates. Then the single and double integration is performed to obtain inertial velocity and position.

An example of this application was implemented using a Watson IMU retro-fitted in a land vehicle. The vehicle was run for 3 minutes in a trajectory of approximately 1 km long. Initially the vehicle was at rest, with the engine on, for a period of approximately 15 seconds. This stationary data was used for calibration and alignment purposes. Figure 7 shows the acceleration and pendulum gyro data used for calibration. A 15-second average was used in order to filter the engine vibration information. After the estimation of the biases and attitude information, the raw accelerations were compensated and the initial 40 seconds are shown in Figure 8. It can be clearly seen that the compensation is almost perfect since the acceleration in the x and y directions are zero while the vehicle was stationary.

Figures 9 presents the velocity and position predictions compared with the true values obtained with a DGPS unit. It can be seen the INS prediction, black, is following the true trajectory with small error for approximately 100 seconds.

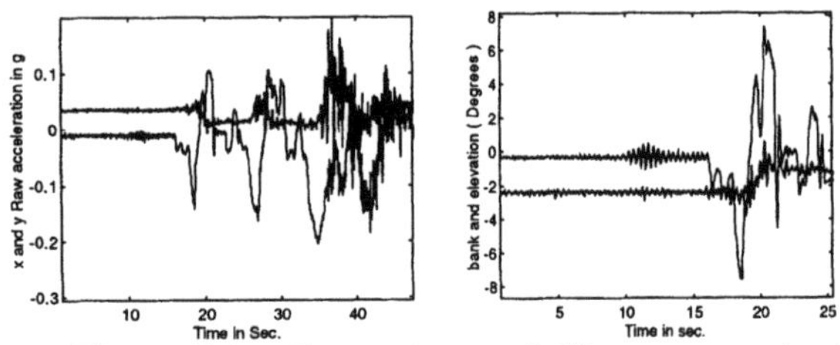


**Figure 7 Accelerometers and tilt sensors output**

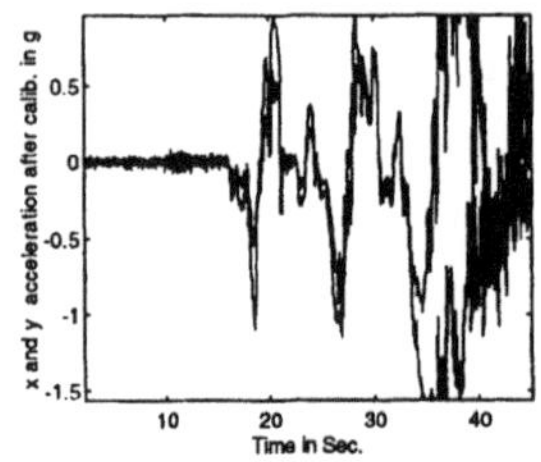


**Figure 8 Acceleration**

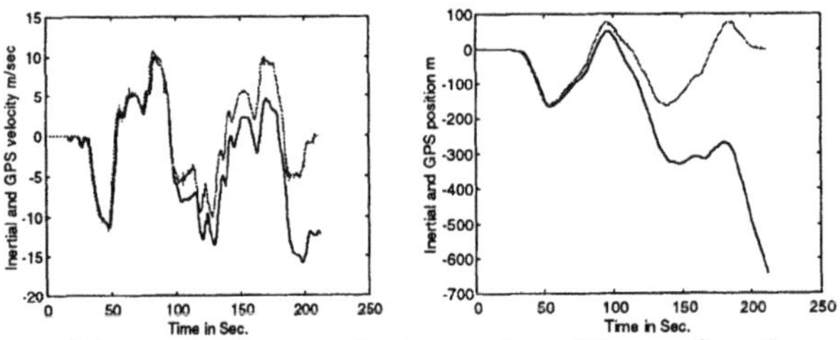


**Figure 9 North velocity and position estimation**

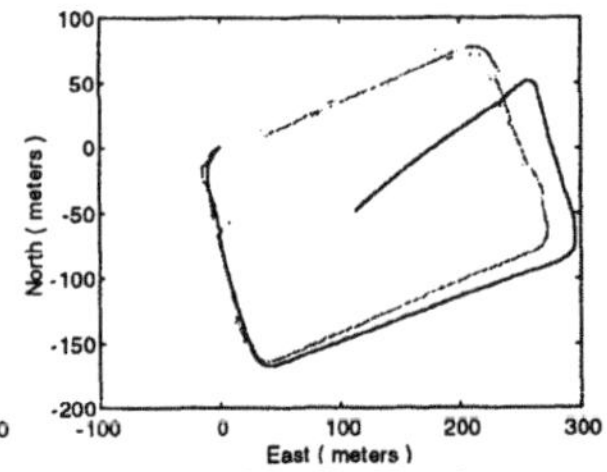


**Figure 10 2-D Trajectory**

The 2-D trajectory is presented in Figure 10 . An enhanced view of the first 160 meters is also shown in Figure 11. During this period the DGPS system experienced some multipath errors that can be rejected when fused with the INS information [6]. Finally Figure 12 represents the history of the Euler angles information during the run. The yaw angle is changing 360 degrees per turn but roll and pitch have very small variations. Errors in roll and pitch are very important since they are responsible for the gravity compensation. From these results it can be seen that the INS position and velocity indication can be efficiently used in a complementary manner with another external low frequency sensor such as GPS.

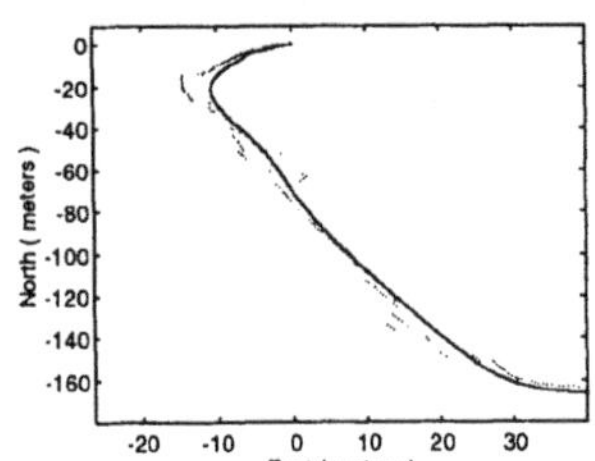


**Figure 11 2-D trajectory**

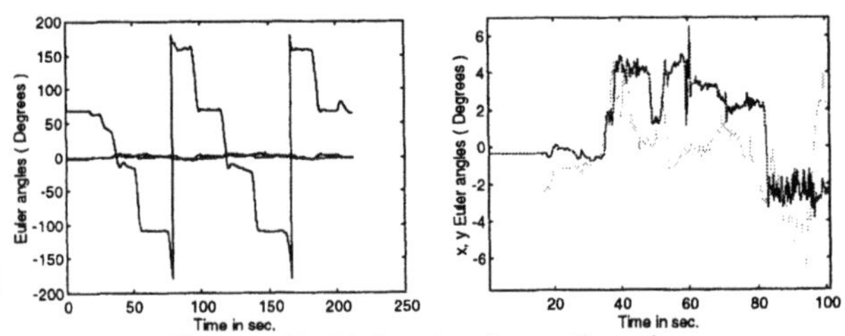


**Figure 12 Euler Angles estimation.**

# 3. External Sensors

External sensors provide positional information through the direct observation of landmarks, either natural or artificial. These sensors are essential in providing absolute position information. They measure range or bearing to landmarks located at known positions. These sensors are typically less robust and more expensive than dead reckoning devices. They can be categorised as having low pass frequency characteristics. Their bandwidth is limited by the computing time required to detect and process all the landmarks. Sensors such as **M**illi-**M**eter-**W**ave-**R**adars, laser-beacons and GPS are examples of such devices. In the following sections the most common external sensors that are applied to vehicle navigation are presented.

## 3.1. Bearing-Only Laser Scanner

This type of sensor is commonly used for navigation of industrial autonomous guided vehicle (AGV) systems. The basic principle of operation is very simple. A pencil laser beam is mechanically scanned in azimuth around the environment. Passive reflectors, occasionally bar codes, are placed around the environment at known surveyed locations. When a scanner mounted on a vehicle moves through the environment it detects the presence of these reflectors or beacons. An encoder mounted on the scanner is used to record the azimuth angle or bearing at which a reflector is detected. In some cases it can also return the width of the target making range measurements also possible. Thus as the vehicle moves through the environment, a sequence of bearing measurements, to a number of fixed and known landmarks are made. If the locations of these reflectors are known to the vehicle navigation system, then the absolute location of the vehicle may in principle be computed from the sequence of bearing measurements made.

Figure 13 shows a bearing only laser. These scanners are manufactured by GEC Electrical Projects, now GCS Ltd, and by NDC Netzler & Dahlgren Co AB, SWEDEN. The GEC scanner uses an interlocked 30mW laser diode modulated at 5MHz to eliminate effects of ambient lighting. The scanner rotates at approximately 1Hz. The bearing resolution is in the order of one milli-radian. The maximum range is approximately 20m (theoretically 50m is possible). The reflectors used for indoor applications are made with standard reflective tape.

These sensors are widely used for indoor autonomous guidance vehicles. Some applications use standard triangulation techniques to estimate the vehicle position. When the environment becomes complicated more beacons are needed and a full filter implementation fusing dead reckoning and bearing information is usually required. Even in the case where few beacons are present, a filter is also necessary to identify spurious reflections. In some cases the reflection may not be due to a beacon but to other infrastructure with similar reflective indexes. The navigation algorithm must track the position of the vehicle, validate the observation and determine the beacon identity. If no ambiguities are present then the known position of the beacon is used to update the vehicle state. The next chapter will present a navigation loop design based on this type of sensor.

## 3.2. Laser Range Sensors

These devices return range information measuring the time of flight or phase differences. There are 3-D Laser range sensor devices but they are still very expensive. Over the past few years 2-D version of laser range sensors have become available with

very competitive prizes. The PLS is a laser ranger manufactured by Sick opto-electronics of Germany. It does not require reflectors and works on the time-of-flight principle. The PLS was designed primarily as a safety device for the protection of danger areas and as an electronic, non-contact 'bumper bar' for AGVs. The PLS has a single moving part, the rotating mirror. There are also outdoor versions that are capable of attenuating the effects of snow and rain with a proprietary software algorithm.
Typical specification of this device are presented in the following table:

| | |
|---|---|
| Range: | 50m |
| Range resolution: | Approximately 70 mm |
| Field-of-view: | 100-180 degrees |
| Maximum angular resolution: | 0.25-1 degree |
| Scan time: | 40ms |
| Computer interface: | RS232 or RS422 (9600- 500000 baud) |

These sensors are mainly used to provide warning signals when an object is detected inside the programmable security region. These devices can also be used for many other applications since they are provided with a communication protocol based on telegrams to read the laser sensed data. Figure 14 shows the indoor version of the Sick PLS laser. In relation to navigation, the sensors become very useful in tasks such as map building and obstacle avoidance. These sensors are also very robust and can work in harsh environments, such as underground mining [1], where very good quality real time maps of the tunnels were obtained while mounted in a Load, Haul and Dump machine.

**Figure 13 Bearing only laser.**

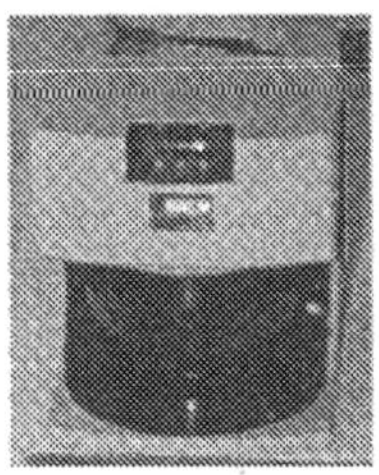

**Figure 14. Range and bearing laser**

## 3.3. Ultrasonic Sensors

Ultrasonic transducers are used as range sensors for map building and collision detection in AGV applications. They typically use a time-of-flight measurement to determine the distance to acoustically reflective objects. When the speed of sound is accurately known, they provide very good range data, however due to their wide beam, the azimuth information is poor. Interpretation of the return echo can be difficult, with a number of problems arising, which are related to the physics of sound propagation. The sound waves suffer specular reflection, diffraction, and multi-path reflection from oblique specular surfaces, and absorption by porous surfaces. Some researchers have attempted to model and simulate the physical properties of the ultrasonic beam and its return echo, [7]. Significant problems arise from multiple reflections in the beam path, firing the sensor at oblique angles at plane specular surfaces and side-lobe interference.

The most common ultrasonic sensor device is the one manufactured by Polaroid. This ultrasonic ranging unit comprises of an electrostatic transducer and an ultrasonic circuit board. Together these components are capable of detecting the distance to objects within a range of approximately 30 cm to 10 meters. When activated, the transducer emits a pulse which is a high frequency, inaudible "chirp", lasting for 1 ms and consisting of fifty-six pulses at four carefully chosen ultrasonic frequencies; 60 kHz, 57 kHz, 53 kHz and 50 kHz. By using four different frequencies, it overcomes the possibility of having an echo cancelled because of certain topographical characteristics. Figures 15 and 16 show the actual response of the sensor with different angles and the angle model uncertainty commonly assumed for the sensor, [8]. It can be appreciated that the model considers the return echo coming from the first lobe only. With some surfaces this may also present a problem giving return echoes from objects outside the cone.

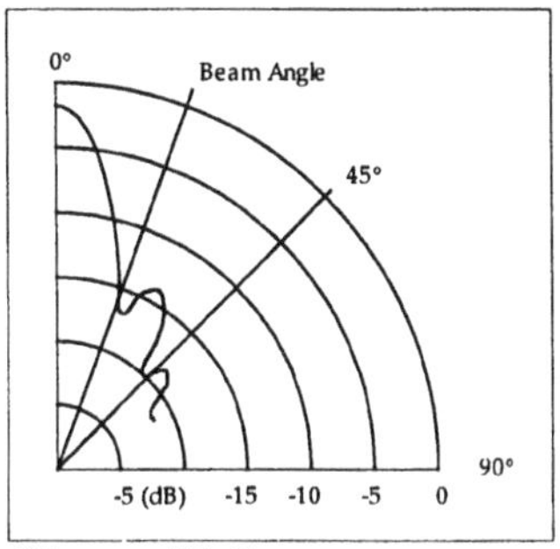


**Figure 15. Sensor response**

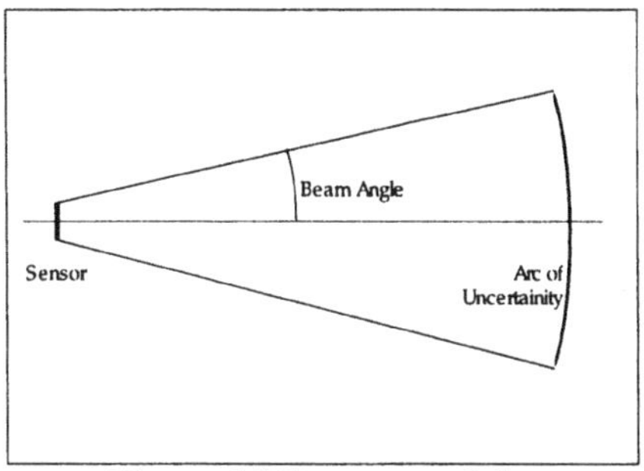


**Figure 16 Sensor model**

There are some more robust implementations of ultrasonic sensors that permit its use to rugged environments such as underground mining. These are usually ceramic devices operating at a frequency of 140kHz. High frequency sonar has the advantage of being more accurate and less susceptible to noise than low frequency devices. However, higher frequency devices have much lower range.

### 3.4. Global Positioning Systems

The Global Positioning System (GPS) is a world wide positioning system based on 24 satellites circulating in 6 different orbits around the Earth. It has been developed to solve the military needs of determining position of land, air and sea vehicles anywhere on earth and under all possible weather conditions. The satellites transmit information at frequencies L1=1575.42 Mhz and L2=1227.6 Mhz. They are modulated with a pseudo random C/A and P codes and with a navigation message. The Precision Positioning System (PPS) is provided by the P code included in the L2 signal. This code is only available to authorized users. The Standard Positioning System (SPS) is based on the C/A code that is included in both frequencies. The accuracy of each system is presented in Table 1.

The C/A code is a repeating 1 Mhz pseudo random binary sequence. There is a separate C/A code for each satellite in operation. The GPS receiver has an internal copy of all the possible 32 codes. The code is used by the receiver to identify the satellite and to obtain range information. During the initial satellite search a shifted version of each code is correlated with the incoming code. A maximum correlation is only obtained with the correct satellite code. The time shift that achieves the maximum correlation is proportional to the distance to the satellite. The P code is a very long (1 week) 10 MHz pseudo random sequence. This code has been encrypted since January 1994 and is called Y code.

**Table 1. Global Position system precision**

| | Horizontal Accuracy | Vertical accuracy | Time accuracy |
|---|---|---|---|
| **SPS** | 100 meters | 156 meters | 167 nano-sec. |
| **PPS** | 17.8 meters | 27.7 meters | 100 nano-sec. |

The navigation message is present in both the L1 and L2 frequencies. It is a 50 bit per second code sent in a thirty-second data frame consisting of five 300-bit sub-frames. Each sub-frame contains information about the precise satellite ephemeris, clock correction parameters, and low precision ephemeris data for the other satellites. They also include parity bits and allow for some error detection and corrections. The 1500 bit data frame contains the complete set of parameters that precisely describe the trajectory of the satellite. After 30 seconds of continuous tracking the GPS receiver is able to use the tracked satellite for position determination. After 12.5 minutes of continuous satellite tracking the navigation message is completed. This message includes many parameters of interest such as low precision orbit parameters for the rest of the satellites and parameters for ionospheric correction.

### 3.4.1. GPS Observables

Two types of observations are of main interest. One is the pseudo-range which is the distance from the satellite to the vehicle plus additional errors due to clock drifts, the ionosphere, the troposphere, multi-path and Selective Availability, (SA). SA is a deliberate error introduced by the US Department of Defence and consists of additional noise included in the transmitted satellite clock and satellite ephemeris [10]. Another observation used by the GPS receiver is Doppler frequency information. As the vehicle and the satellites are in constant motion with respect to each other the receiver signal will experience a change in frequency proportional to their relative velocities. This information, usually referred to as delta range, can be used to generate very accurate velocity estimation. Although most GPS receivers generate position and velocity information, in only a few of them this information is not correlated. Only the more sophisticated systems generate velocity from Doppler observation making it independent of position information. This is an important consideration for the data fusion algorithm since position and velocity will not be correlated.

The precision of the solution is affected by two main factors: the geometry of the actual relative positions of the satellites with respect to the receiver, usually referred as PDOP (position dilution of precision), and secondly, the precision in range determination. The 3-D position of the receiver can be evaluated using standard triangulation techniques with range observations from at least four satellites. The position of the satellites is predicted with the ephemeris information. With time stamp range and satellite position information the following set of non-linear equations can be stated to obtain the 3-D position of the rover x, y and z, and the GPS receiver clock drift $\Delta t$.

$$
\begin{aligned}
r_1 &= \sqrt{(x-x_1)^2+(y-y_1)^2+(z-z_1)^2}+c\Delta t \\
r_2 &= \sqrt{(x-x_2)^2+(y-y_2)^2+(z-z_2)^2}+c\Delta t \\
r_3 &= \sqrt{(x-x_3)^2+(y-y_3)^2+(z-z_3)^2}+c\Delta t \\
r_4 &= \sqrt{(x-x_4)^2+(y-y_4)^2+(z-z_4)^2}+c\Delta t
\end{aligned}
\qquad (20)
$$

More complex GPS receivers include all the satellites in view to obtain a position fix. They usually implement a least square or Kalman filter algorithm to estimate position and velocity of the receiver. The vehicle location is obtained in the coordinate frame of reference used by the satellites. For the GPS, this reference is the Earth centred coordinate system using the WGS-84 datum (World Geodetic System 1984). The datum defines the ellipsoid that approximates the curvature of the earth.

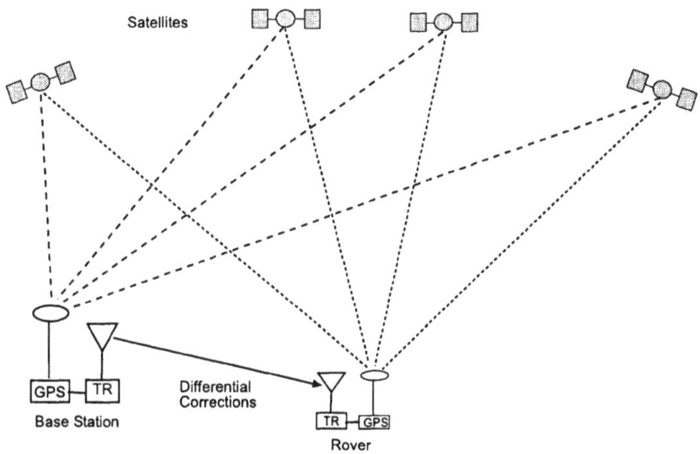


**Figure 17. Differential implementation of GPS**

The errors in range determination can be significantly reduced with the use of another station placed at a known fixed location. This station processes the information from all satellites available in the region and determines the errors in range information. These errors are then broadcast with a radio modem and used by the mobile receivers (rovers). This is called differential GPS (DGPS) and is shown in Figure 17. It eliminates almost all the errors due to atmospheric and ionospheric delays and to a great extent the ones due to SA. Due to the latency involved in the correction the effect of SA can not be completely eliminated in real time implementations. These systems are capable of generating position with an error of the order of 0.5-2m Circular Error Probable (CEP).
Modern DGPS units also use phase carrier information to improve the position determination. Differential carrier phase tracking consists of measuring the phase shift between the same signal received at the base and the rover stations. This phase shift is proportional to the distance between these two stations. Much more complicated hardware and software algorithms are needed for this implementation since these measurements are subject to phase ambiguities. Sophisticated software techniques are used to solve for the ambiguities. Using phase carrier information on a single frequency,

real time errors of the order of 20 cm CEP can be achieved using a minimum of 5 satellites. Using carrier phase information from both frequencies the errors can be reduced up to 2 cm requiring at least 6 satellites in view. A complete review of these techniques is presented in [9].
Two main problems affect the accuracy of DGPS position determination. The most notorious one is satellite availability and shadowing. The rover needs to see at least 4 satellites to determine standard 3-D DGPS position with range corrections and 5 or 6 in order to get the solution generated with phase carrier implementations using single and double differences respectively. In many cases buildings and large equipment may obstruct the signals from one or more of the satellites and a position fix will not be generated. Another important problem is multi-path reflection. The range information is evaluated measuring the travel time of the signal from the satellite to the rover. Sometimes this path may be obstructed but the signal will still reach the rover antenna through an indirect path by multiple reflections. The receiver will obtain erroneous range and phase carrier difference information and an incorrect position fix will be generated. This problem is also present in non-differential implementation of GPS but it is overshadow by SA that is responsible for errors two orders of magnitude larger than multi-path. The multi-path problem can be reduced with an appropriate selection of the antenna, GPS receiver, and accepting observations only from satellites with a minimum elevation angle. Since multi-path is due to multiple reflection, satellites with lower elevation angles are more prone to generate these undesired problems. The antenna can also be provided with a choke-ring ground plate to eliminate low angle reflection. Finally some GPS receivers include algorithms and extra hardware to detect and eliminate some multi-path problems. Although this technology has been making enormous progress in the past few years it will be almost impossible to guarantee a given GPS accuracy 100% of the time when working in application such as mining or stevedoring. Large pieces of equipment, such as cranes and trucks will almost certainly cause important perturbation to the system for a limited period of time.
While it makes sense to use GPS, due to the absolute precision obtained, it needs to be combined with other information, both to improve position precision and to provide a degree of resilience to outages and performance degradation.

### 3.5. Milli-Meter Wave Radar (MMWR)

These devices can be used to determine range and bearing to a set of beacons placed at known locations. These targets return the incident radiation beam directly back along the incident path over the capture angle of the beacon. Such a sensor has been used in [11], to autonomously navigate a vehicle of gross weight 80 tones to centimeter levels of precision at speeds of up to 6 $ms^{-1}$. This sensor, although more expensive than a comparable GPS system, is more reliable since one has total control of the number and the location of the beacons in the area of operation. Unlike a laser units, which typically return range and bearing to the closest reflector the radar returns the power spectra along each azimuth. These features make this sensor one of the most attractive for reliable outdoor navigation.

#### 3.5.1. Principle of operation

A typical radar as shown in Figure 18 operates in the 77 GHZ band frequency. The device has a rotor which rotates a deflector to obtain 360 degree-azimuth range information. A control unit maintains a constant rotational velocity of the deflector plate positioned above the aperture and reads azimuth of each radar sweep. With this configuration the beam angle in the horizontal direction is 1 degree and the vertical angle is approximately 6 degrees to make the radar insensitive with respect to roll and pitch of the vehicle. The vertical angle is made wider incorporating a slight curvature to the deflector plate. The bearing information is obtained with an encoder mounted on the shaft of the rotating plate. This radar is designed for beacon detection and for obstacle avoidance tasks. To satisfy this dual purpose the radar is designed to enable the beacons to be identified from all other object reflections. This is achieved though appropriate design of the beacons and polarization of the radar transmitted signal. The radar generates a vertical polarized signal that is transformed into a circular polarized signal by a $\lambda/4$ plate. This plate is designed to produce a 90-degree phase shift in one of the orthogonal axis. The beacon's design incorporates a vertical-polarising grid that will reduce the effective reflected area of the vertical signal. This signal will be significantly attenuated with respect to the horizontal component that will enter the corner shape beacon. The returning signal will have horizontal polarization producing a circular polarization back in the receiver after the $\lambda/4$ plate. Any other polarisation will make the output somehow elliptical. The only polarisation that will split into equal power on the two orthogonal channels will be the one generated from the horizontal polarization signal returned by the beacons. This is the feature that is used to detect beacons from all other targets in the environment.

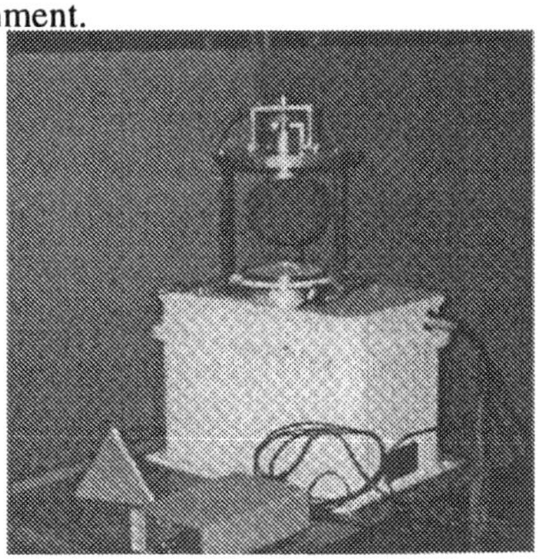

**Figure 18. Radar system**

The time between transmission and reception gives a measure of the distance to the reflecting surface. Due to the limited maximum range it is not practical to measure such small time intervals. The carrier wave is frequency modulated over 600 MHz in a 1 msec interval. The transmitted and received signals are then mixed and the frequency difference extracted and processed to determine the position of the reflectors. A frequency spectrum of the frequency difference is obtained and then the distance to the reflector is determined. The system can obtain range information up to 250 meters with a resolution of 0.25 meters.

The plate rotates at 1 Hz and 4096 samples are taken from both channels 250 times per second. The Discrete Fourier Transform of these sequences needs to be evaluated in less than 4 msec. This requires substantial processing hardware and necessitates the use of state of the art DSP units. Although relatively expensive this sensor becomes very attractive for outdoor localisation purposes. There are other units under investigation in the application of highway vehicles that promise to bring down the price of this technology in the near future.

## 4. Conclusion

This chapter presents a description of the most common sensors used for navigation with emphasis to land vehicle applications. Dead reckoning and absolute sensors are described enhancing their main capabilities and drawbacks. This is of fundamental importance for high integrity navigation systems. Such systems require a combination of sensors that must not fail under any possible situation. A sensor can be operating correctly but it may return erroneous information. Such is the case of the encoder's information obtained from a wheel that is slipping. In this case an inertial based sensor can be the ideal complement in order to detect the fault. The same things applies to external sensors such as GPS, radar and lasers where simultaneous failure can be of very low probability with appropriate allocation of beacons and reflectors.
There are some other types of internal and external sensors that provide dead reckoning and absolute information in different environments and applications. Depending on the required reliability of the application the designer will need to explore the right combination of sensors based on their failure modes.

## 5. References


1. Scheding, E. M. Nebot , M. Stevens and H. Durrant-Whyte, J. Roberts and P. Corke, . Cunningham and B. Cook, "Experiments in Autonomous Underground Guidance", IEEE Conference on Robotics and Automation, 1997, pp. 1898-1903.
2. Lawrence A., "Modern Inertial Technology", 1993, Springer Verlag.
3. Britting K., "Inertial navigation System Analysis", Wiley Interscience, NY, 1971.
4. Itzhack Y., "Minimal Order time Sharing filters for INS in flight alignment", Journal of Navigation, Vol 5, #4, July 1982, pp.396-402.
5. Saab S., Gunnarssin T., "Automatic Alignment and Calibration of an Inertial Navigation System", Proc. of Pos. Location and Navigation , 1994, pp. 845-852.
6. Nebot E. , Sukkarieh S., Durrant-Whyte H., "Inertial Navigation aided with GPS information", pp 169-174 Proc. of M2VIP-97, Australia, September 1997.
7. Barshan B., Kuc R., "Differentiating sonar reflections form corners and planes by employing an intelligent sensor", IEEE Trans. Pattern Analysis and machine Intelligence, 12(6):560-569, 1990.
8. Pagac D., Nebot E., Durrant-Whyte H., "An Evidential approach to probabilistic map-building", pp. 745-750Proc. of the IEEE ICRA, 1996, Minnesota.
9. Leick A., Global Positioning System, Satellite Surveying", John Wiley & Sons, Inc. 1995.
10. Nebot E., Scheding S., Durrant-Whyte H., "Frequency domain Modelling of Aided GPS with Applications to High-Speed Vehicle Navigation Systems". Proc. IEEE ICRA, New Mexico, April 1997, pp. 1892-1897.
11. Durrant Whyte H., "An autonomous guided vehicle for cargo handling applications", Int. Journal of Robotics Research, 15(5), pp. 407-441, 1996.

# PART 2

# MOBILE, WALKING AND SNAKE-LIKE ROBOTS

# 8

# DESIGN, CONTROL, AND APPLICATIONS OF AUTONOMOUS MOBILE ROBOTS

D. FLOREANO, J. GODJEVAC, A. MARTINOLI,
F. MONDADA and J-D. NICOUD
*Micro-computing Laboratory*
*Swiss Federal Institute of Technology in Lausanne*
*LAMI-INF-EPFL, CH-1015 Ecublens*
*Switzerland*

## 1. Introduction

An autonomous robot is a machine that operates in a partially unknown and unpredictable environment. In contrast to robots used in manufacturing plants, where the environment is highly controlled, autonomous robots cannot always be programmed to execute predefined actions because one does not know in advance what will be the universe of required sensorimotor transformations required by the various situations that the robot might encounter. Furthermore, the environment might have dynamic characteristics that require rapid online modifications in the robot behaviour. For these reasons, in the last ten years several researchers have looked at novel methods for setting up autonomous mobile robots.

The basic idea behind most approaches is to break down sequential top-down programs into a set of simple, distributed, and decentralised processes that have direct access to sensors and motors of the robot. The first formalisation of this approach is the subsumption architecture proposed by Rodney Brooks at MIT [2, 3] where several local behaviours (sensorimotor modules) continuously operate in parallel using only local available information. The emergent overall behaviour is flexible, robust against environmental noise and mechanical failure, and based on compact modular codes. Another key feature of these novel approaches is bio-inspiration, that is the attempt to implement mechanisms of biological adaptive behaviour [27, 26, 4, 21]. Adaptation, combined with a decentralised bottom-up approach, is often seen as a solution to the problem of generating and maintaining stable behaviours in partially unknown and dynamic environments.

In this chapter we give an overview of recent work in autonomous mobile robotics done at our laboratory, covering hardware methods and design, adaptive control, collective autonomous robotics, and conclude with considerations on industrial applications. Other related and complementary approaches can be found in two special issues of the journals *Robotics and Autonomous System* [14] and *IEEE Transactions on Systems, Man, and Cybernetics - B* [5]. In the next section we shall introduce some important methodological issues in setting up a research tool for autonomous robotics, such as miniaturisation, modularity, and flexibility which are behind a mobile robot developed at our laboratory and currently used by several hundreds research centres. We shall then describe two experiments using neuro-fuzzy adaptive control, a methodology which provides an easy interface between user knowledge and robot autonomous operation. Some of these experiments will be later used as a starting point for introducing an evolutionary approach where genetic algorithms are used to automatically develop neurocontrollers without human intervention in a range of different environments. In a later section, we shall address hardware and control issues in autonomous collective robotics. Here, global behaviour of a team of robots emerges from dynamical interactions among several robots programmed to operate using only local information. Finally, we provide some practical considerations on industrial applications of autonomous mobile robots and conclude with a comparative summary and outlook for future developments.

## 2. Issues in hardware design

Simulation studies have for long time been considered a valid investigation methodology to develop autonomous mobile robots, both for research purposes and industrial applications. Recently, a large number of researchers [2, 3, 6, 12, 36, 38], including the authors of this chapter, have stressed the importance of using real robots for the development and validation of novel solutions to fully capture important interactions between the robot and the environment. Despite a widespread agreement, real mobile robots are not yet always employed during the initial research stage. We believe that there are some practical reasons behind this fact. Most of the researchers working in the field of artificial intelligence are software engineers or academics coming from various disciplines like biology, psychology, or anatomy. These people are not very keen to deal with mechanical and electronical problems. Mobile robots were often very unstable devices, built by hobbyists, and needing specific know-how and special care to operate them. Only few laboratories in the world have the possibility to build a robot and program it to achieve interesting research results. Furthermore, most of the commercial products available were made by roboticians for roboticians, resorted to very specific programming languages and thus were not accessible to outsiders. This situation is improving, but a majority of robots still needs specific know-how. Finally, in universities, where software writing is considered "without costs", a real robot seems often too expensive in comparison to simulations.

Lack of simple, efficient, robust and low cost robotic tools prompted us to develop the Khepera miniature mobile robot, initially designed in 1991 by E. Franzi, A. Guignard, and F. Mondada [30], based on ideas of J-D. Nicoud. Given the interest of our laboratory in innovative and multidisciplinary approaches (some of which will be outlined in the following sections), the Khepera robot has been intended right at the beginning as a tool that could be easily used by a variety of people with different backgrounds and needs, such as biologists willing to test theories of adaptation, neurophysiologists developing models of neural networks, psychologists implementing models of learning and cognition, and engineers evaluating different control strategies for specific applications. In the following subsections, we describe the essential design elements of this robotic tool.

### 2.1. MINIATURISATION

Miniaturisation of a mobile robot brings some advantages to the researcher who works with it, but it is a presents several challenges for the engineer who has to develop it. Khepera represents a viable compromise, featuring 55 mm in diameter and a variable height; it is sufficiently small to easily move on a desk and sufficiently big to be built with standard electronic and mechanical components. This size brings several important advantages. **I.** The experimenter can build complex environments on a limited surface. For a miniature robot like Khepera, indeed, a normal office desk of 0.9m x 1.8 m represents a working surface equivalent to that of a tennis court for a standard-size robot having a diameter of 55 cm. Additionally, one can use a thin suspended cable for power supply without disturbing the robot movements and also place a camera with normal lenses on the ceiling for monitoring the whole environment. **II.** Fundamental laws of physics give higher mechanical robustness to a robot of this size. In order to intuitively understand this physical phenomenon, compare a robot of 50 mm in diameter crashing against a wall at 50 mm/s with a robot of 1 m in diameter crashing against the same wall at 1 m/s. The miniature robot will resist the collision, the other robot will probably report serious damages. **III.** The price of a miniature robot like Khepera is lower than that of a robot with comparable performance, but larger size. This is due to smaller parts, smaller circuits, smaller infrastructure to build and test the robots, and simpler mechanics which allow us to use electrical part (connectors, for instance) for both electronic and mechanical functionalities (body structure, for instance). Miniaturisation brings also some drawbacks, like difficulty of mounting large devices on the robot (ultrasonic sensors, laser range finders, etc.).

## 2.2. HARDWARE AND SOFTWARE MODULARITY

Hardware modularity enables different possible configurations and experiments using the same basic components. It means also possible extensions and, globally, cheaper equipment. Software modularity means flexibility and possibilities for extensions, which enables the software developer to write only parts of the program required for the specific application. Khepera is based on this concept of modularity, both in hardware and software. At the hardware level, Khepera has an extension bus that makes it possible to add turrets on the top of the basic configuration, depending on the needs of the experiments to be done. This modularity is based on a parallel and a serial bus. The parallel bus can be used for simple extensions directly under control of the main Khepera processor. The serial bus implements a local network for inter-processor communication. Using this second bus, other processors can be connected to the main one in order to build a multi-processor structure centred on the Khepera main processor. This kind of structure has the advantage that one can employ additional computational devices on extension modules, thus keeping the main processor free for global management of the robot behaviour.

At the software level, modularity is needed to support the multi-processor structure of the robot. It consists of a flexible protocol that recognises all added extension modules when the robot is powered, informing the main processor about all functionalities available in each extension as well as the procedures for activating these functionalities. The BIOS of the Khepera, which includes all basic procedures for robot management, is also based on a modular structure. Motors, sensors, and timing functions are grouped into distinct modules to simplify management of the robot and improve software robustness. The main software also supports remote control and down-loading of specific applications through a serial cable. Such software structure simplifies the task of the user, who can easily add her own software to the management modules already implemented on Khepera.

## 2.3. FROM SIMULATIONS TO APPLICATIONS

Several researchers in the field of autonomous mobile robots belong to computer science, artificial intelligence, neuroanatomy, biology, psychology, and many other related domains. The tool needed by this community to move from simulation models to the real world is a simple robot that can be easily

*Figure 1.* The miniature mobile robot Khepera in its basic configuration.

controlled from the desk computer in the same way in which a simulated robot is controlled.

Khepera is situated between simulations and real-world applications. If on the one hand it keeps a level of simplicity and operation modality similar to simulation tools, on the other hand it cannot reach the complexity of real-world applications. However, being a physical robot, it introduces most of the characteristics of robots used for real-world applications. This position between simulators and applications is confirmed by several Khepera users who have moved from simulations to the miniature robot and have highlighted the advantages of playing at these two levels [33, 19, 28]. Recent construction of the new larger Koala robot, which is software compatible with Khepera, enables the transfer of developments made on the Khepera to a more complex platform which can be used for real-world applications.

## 2.4. KHEPERA CHARACTERISTICS AND EXTENSIONS

The Khepera robot consists of a basic platform and a number of extension turrets. Some of them are described below and are used in the experiments described in the following sections, while other modules are under development.

In its basic configuration (figs. 1 and 2), Khepera consists of two layers corresponding to two main boards: the sensory-motor board and the CPU board. The motor system consists of two lateral wheels and two pivots on the front and back. This configuration is very good for facing complex geometric obstacles because the robot can turn in place without lateral displacement. The sensory system available in the basic configuration is placed on the lower board, consisting of 8 infrared-light proximity sensors distributed around the body, 6 on one side and two on the other (this asymmetry can be used to establish the front and back of the robot). These sensors can detect the presence of objects by emitting and measuring re-

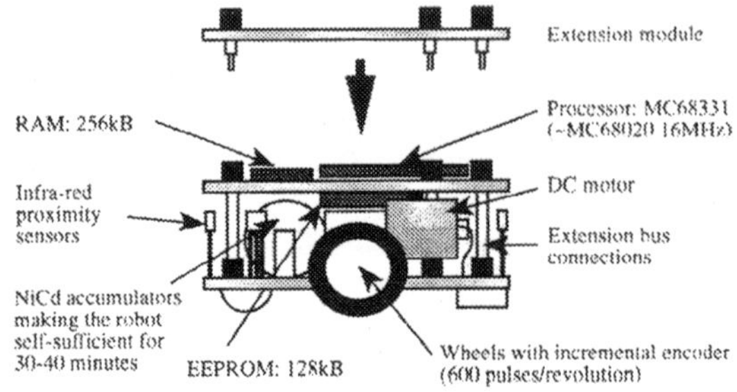


*Figure 2.* Khepera robot structure and extension possibilities.

flected light and can also be used as simple passive infrared light sensors. On the sensorimotor board are also placed NiCd batteries with a capacity of 110 mAh which allow the robot to be self-sufficient for approximately 30-40 minutes. The CPU board encloses the robot's main processor (a Motorola MC68331 with 128 K-bytes of EEPROM and 256 K-bytes of static RAM). An A/D converter allows the acquisition of analog signals coming from the sensory-motor board. An RS232 serial line is also available on the board via a miniature connector. On this same connection, a wire can also provide continuous power supply from an external source.

The electrical link between sensory-motor board and CPU board is done by connectors that also provide mechanical support. These connectors pass through the boards and form an extension bus for addition of further modules to the robot. Possible extensions are, for example, the gripper module and the linear vision module illustrated in figure3. The gripper module (figure 3a) is connected on the extension bus and has its own processor which communicates with the main Khepera processor using the local network. The gripper has two degrees of freedom: elevation and grasping. Sensors inform about elevation of the arm, position of the gripper fingers, presence of an object inside the gripper (optical barrier), and electrical resistivity of the grasped object. The K213 linear vision, which can be placed on the basic platform (figure 3, b)) or on the top of the gripper (figure 3, c)), has its own local processor which uses the local network to share information with the main processor. This module has a horizontal linear camera composed of 64 photo-receptors giving a gray-level image spanning a 36-degree visual field, and an auto-iris function for automatic adaptation of image sensitivity to changing light conditions which also provides information on global light intensity.

a) b) c)

*Figure 3.* From left to right: Khepera with a gripper extension, with a "linear vision" extension, and with both modules simultaneously.

## 2.5. IMPROVEMENTS OF THE ACTUAL DESIGN METHODOLOGIES

Widespread use of the Khepera robot (more than 500 users spread in more that 200 research centres and schools all over the world) and publication of the scientific results have clearly improved replicability and comparison of research results. However, despite this progress, it is still necessary to move in the direction of real applications, which means larger robots and more complex environments. This kind of experimentation needs considerable investments of time and money, and it cannot maintain the standardisation achieved with Khepera. If a simple Khepera environment can be easily reproduced, a larger and much more complex office environment will never be reproduced by other research groups to compare the efficiency of the control algorithms. Therefore, new tools are required.

A project currently in progress at LAMI[1] aims at testing a possible solution, based on the growing communication networks, to this methodological problem. The basic idea is to make available to the scientific community one or more complex mobile robotic set-ups through computer networks like Internet or ATM. With this equipment, every university or research group in the world could access in real time, through a network, the same robotic set-up, download algorithms into the robot, monitor the robot behaviour, or modify the environment, just as if the robot was in the next room. Comparison between different approaches in the field of autonomous mobile robotics can be made only when all conditions are perfectly identical. This means that not only the robot has to be exactly the same, but also

[1]This project, called "Sharing of Unique or Expensive Equipment for Research and Education: The Remote Manipulation Paradigm" is part of the Swiss Priority Programme SPP ICS of the Swiss National Foundation.

that the environment has to be identical, in all details from light conditions to colour and type of floor. A unique and shared robotic set-up meets these requirements and could bring a new dimension to scientific research.

In the following two sections, we will shift our attention to different types of adaptive control applied to the Khepera robot.

## 3. Fuzzy control

Fuzzy logic offers the possibility to express and implement human know-how in the form of linguistic if-then rules which can be applied for the control of nonlinear systems, such as mobile robots [39]. Every rule has two parts: the antecedent part (premise), expressed by **If...** , and the consequent part, expressed by: **then...** . The general form of a linguistic if-then rule is:

> **If** a set of conditions is satisfied **then** a set of consequences can be inferred.

The antecedent part is the description of the state of the system which is used to activate one rule, while the consequent part is the action that the operator who controls the system must take. The process states and control variables are called linguistic variables. A linguistic variable can take several linguistic values. For example, figure 4a shows a set of seven values for the linguistic variable *speed*, labelled as: *NEGATIVE BIG*, *NEGATIVE MEDIUM*, etc. These linguistic values are expressed by continuous functions, called membership functions which represent fuzzy sets. Each membership functions is defined by two parameters: its centre and width.

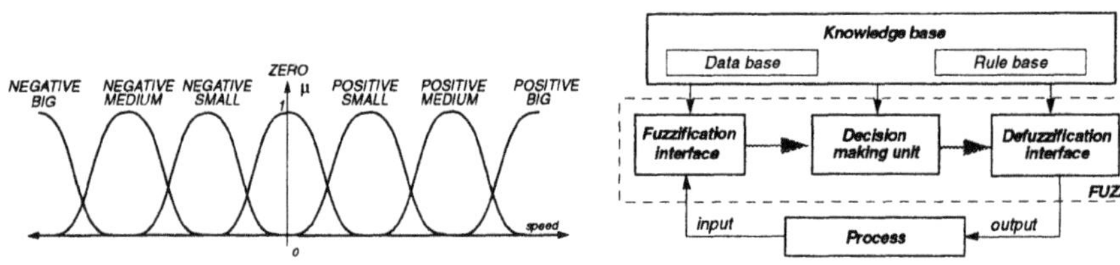


a) b)

*Figure 4.* a) Universe of discourse for linguistic variable *speed.* b) Block diagram of a fuzzy controller.

A fuzzy controller is composed of four principal modules, as shown in figure 4b. The *fuzzification interface* performs the transformation of crisp values into fuzzy sets. The *knowledge base* supplies the fuzzification module, the inference engine, and the defuzzification interface with necessary information (parameters of membership functions and rules) for their proper functioning. The *decision making unit*, or inference engine, computes the

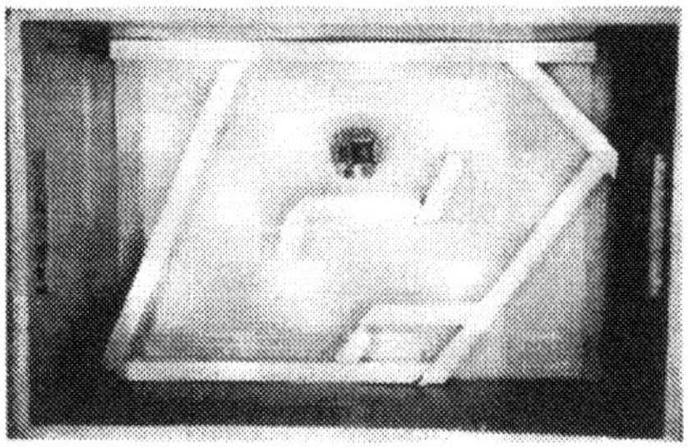

*Figure 5.* Circular maze employed for navigation experiments described in sections3.1 and 4.1.

meaning of the set of linguistic rules. The *defuzzification interface* transforms the union of fuzzy sets (individual contributions of each rule in the rule base) into a crisp output.

Although one can implement a simple controller for obstacle avoidance on the Khepera with few rules, the main effort is that of designing the appropriate membership functions and choosing the rules.

### 3.1. NEURO-FUZZY CONTROL FOR OBSTACLE AVOIDANCE

We have applied an automatic method for the design and the analysis of a fuzzy controller for two different behaviours: obstacle avoidance and wall-following. Since this method is based on a supervised learning procedure developed in the frame of neural networks, we will call it a *neuro-fuzzy* controller.

The first step in the design of a neuro-fuzzy controller is to establish a set of initial rules and parameters of the membership functions. The second step is the implementation of the controller. During the system operation, a supervisor guides the robot through the environment while the parameters of the membership functions and of the rules are adjusted by a learning algorithm based on stochastic approximation method [15]. After learning, the parameters define a new set of rules which can be used for autonomous operation of the robot. Additionally, the learned rules can be extracted in a form intelligible for a human being.

The training environment used for the two experiments consists of a circular maze with several sharp corners (figure 5). The results of the experiments are shown in figure 6. The chosen neuro-fuzzy controller has 4 linguistic variables as inputs: *distance from the left* $D_l$, *from the front* $D_f$, *from the right* $D_r$, and *from the back* $D_b$ of the robot. These are calculated

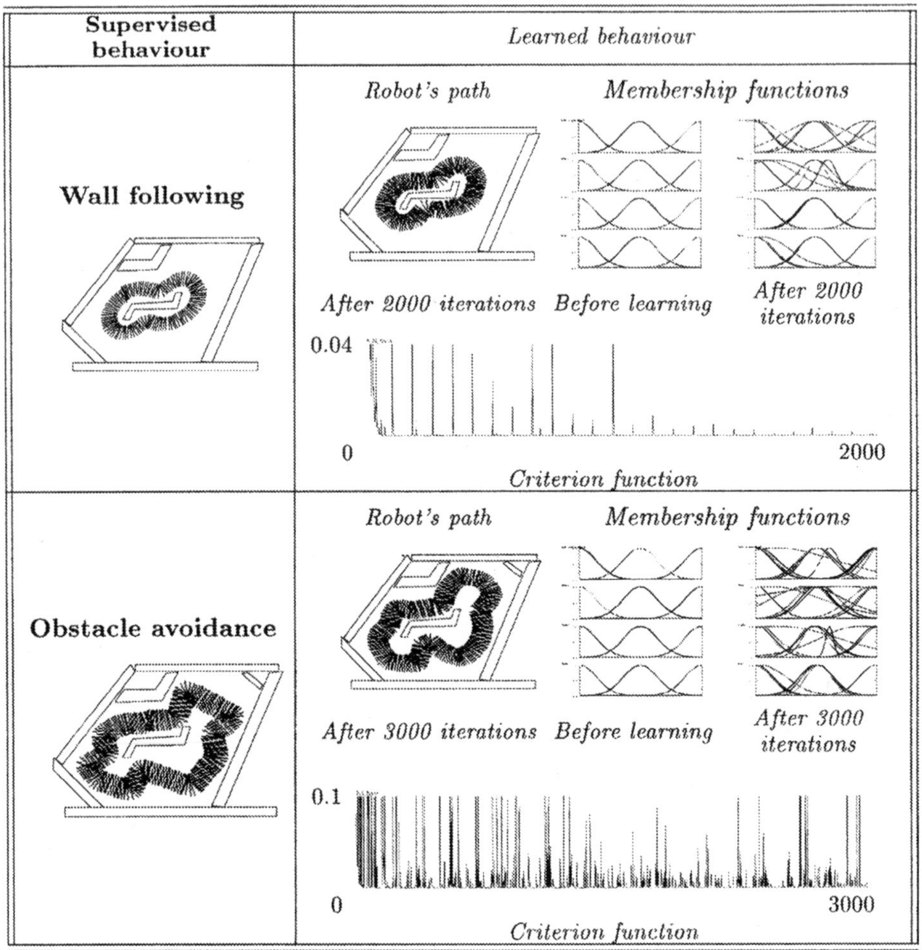


*Figure 6.* Results of adaptive neuro-fuzzy experiments.

in the following way:

$$D_l = S_0,\ D_f = \frac{S_2 + S_3}{2},\ D_r = S_5,\ D_b = \frac{S_6 + S_7}{2}, \tag{1}$$

where $S_i$ are the sensor values normalised within the interval [0,1]. For each input variable, we defined three linguistic values in order to classify the distances as: *Big*, *Average*, and *Small*. These values are sufficient to capture the various states of the sensors for this environment. The parameters of the membership functions are initialised so that they uniformly cover the input space. The controller also has two output linguistic variables: *left motor speed* and *right motor speed.* They can take seven linguistic values.

The first column of figure 6 plots a trace of the robot's path during the learning phase while the robot is guided to avoid obstacles or follow the central wall of the environment shown in Figure 5. The Khepera's path is displayed as a series of segments that link the contact points of the wheels on the floor. The motion is computed by odometry using the internal data from the incremental sensors of the wheels. Since the sampling period is constant, closer lines mean slower robot motion. The second column displays the robot paths after learning without human control, the membership functions before and after the learning phase, and error values during learning.

In the wall-following experiment depicted in the first row, learning is successfully accomplished after 1000 iterations (sensorimotor loops), whereas in the obstacle avoidance experiment learning can be considered satisfactory after 1500 iterations. These experiments show that the learning method can be successfully applied to a real robot with noisy and imprecise sensors. The robot learns only the real state vectors corresponding to those situations which are encountered in its real environment. An interesting feature of adaptive neuro-fuzzy control is that learning can start from a set of rules that the engineer might think are important (instead of starting from random parameters) and modify them as required by the environment. Seldom encountered situations need long time to be learned, as shown by the error peaks for the obstacle avoidance experiment after 1500 iterations. This drawback is the price that we pay for using on-line learning instead of memorising learning vectors which would take large memory and computational resources.

The modified membership functions can be used for off-line extraction of linguistic rules for analysis purposes. The possibility to obtain a linguistic representation of the transfer characteristics is an interesting properties that several other learning methods -such as artificial neural networks- do not offer. When extracting the rules associated with the learned parameters, one not only learns about the fuzzy controller, but also about the sensorimotor transformations required by the robot for a specific task.

## 4. Evolutionary Robotics

Evolutionary Robotics is a technique for automatic creation of control systems for autonomous robots that is inspired upon the Darwinian principle of selective reproduction of the fittest individuals. Compared to the neuro-fuzzy system described above, an evolutionary approach does not requires less human knowledge.

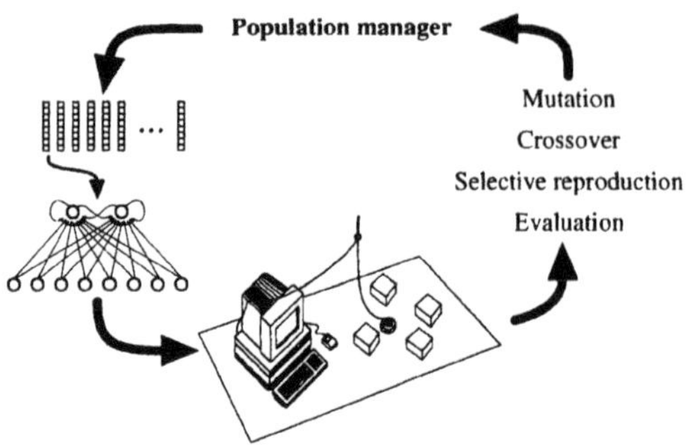


*Figure 7.* Evolutionary robotics on a single robot. Each individual is in turn decoded into a corresponding neurocontroller and the robot is let free to move while its fitness is automatically evaluated.

A population of different artificial chromosomes, each encoding the instructions to build a different neurocontroller, is decoded and tested on the robots. While each robot freely interacts with the environment according to the decoded control system, a "fitness function" automatically assesses its performance. New populations of increasingly better individuals are created by repetitively applying selective reproduction, crossover, and mutation to the evaluated chromosomes for several generations [18] (figure 7).

The role of the engineer in evolutionary experiments is to design a fitness function that measures the performance of the robots with respect to the desired behaviour or task. Artificial evolution automatically develops suitable control systems while the robot autonomously interacts with the specific environment where it is situated incorporating the physics and dynamics of the environment which often are not available to the engineer. In the next subsections we shall describe some results in evolutionary robotics that show how different environmental constraints automatically shape significantly different behaviours without requiring major changes in the fitness function of an autonomous robot trying to keep itself in a viable state.

### 4.1. A SIMPLE EXPERIMENT

As a simple comparative example, let us consider evolution of straight navigation and obstacle avoidance for the Khepera robot in the same environment (figure 5) already used for the experiments with fuzzy logic described in section 3.1. The fitness function $\Phi$ to be maximised is based on three variables which are measured at each time step using sensor and motor activations, as follows,

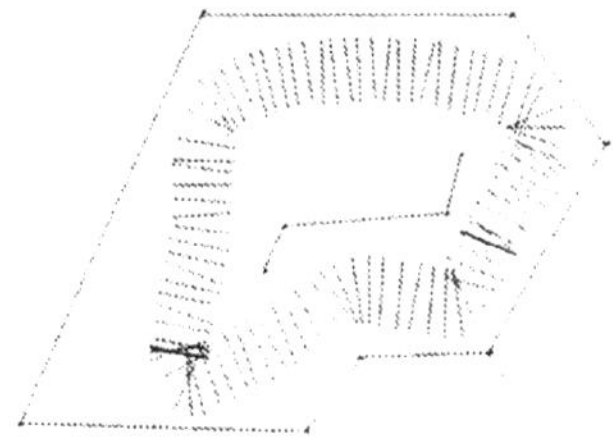

*Figure 8.* The trajectory performed by one of the evolved robots measured using a laser positioning device. Segments represent successive displacements of the axis connecting the two wheels. The direction of motion is anti-clockwise.

$$\Phi = V\left(1 - \sqrt{\Delta v}\right)(1 - i) \tag{2}$$

where $V$ is a measure of the average rotation speed of the two wheels, $\Delta v$ is the absolute value of the algebraic difference between the signed speed values of the wheels (positive is one direction, negative the other) and $i$ is the normalised activation value of the proximity sensor with the highest activity. The fitness values are accumulated during the "life" of each individual and then divided by the total number of actions performed. The function $\Phi$ has three components: the first one is maximised by speed, the second by straight direction, and the third by obstacle avoidance. Each artificial chromosome encodes the synaptic weights and unit thresholds of a neurocontroller with eight input units clamped to the Khepera's sensors and two motor units with recurrent connections, each controlling one motor of the robot. All the 80 chromosomes of the initial population are randomly generated and each individual is separately tested in the environment. The whole procedure is entirely automated and we can observe the population statistics on our workstation a few offices away.

After approximately 45 generations, during which the average population fitness steadily increases, an individual is born which exhibits smooth navigation around the maze (figure 8) [7]. Despite the simplicity of this experiment, which is merely intended to explore the feasibility of the evolutionary approach, there are at least three interesting emergent properties of the evolved system which were not predefined by the engineer. The first concerns the development of appropriate recurrent connection strengths on the motor neurons which prevents the robot from getting stuck in situations when contralateral sensors are equally activated, therefore generating behaviours which are more efficient than a feed-forward controller [29]. The second properties is the direction of motion. Although the robot is perfectly circular and the fitness function does not specify the direction of motion, the evolved controller always moves in the direction with higher sensor density which gives a better resolution of encountered obstacles. Finally, the evolved controller displays an optimal cruising speed that, although it is not the maximum available speed, it well matches the geometry of the environment, the sensor characteristics, and the sensor update rate. If the robot moved faster, it would crash into a wall before having the possibility of detecting it.

## 4.2. COMPLEX ENVIRONMENTS GENERATE COMPLEX CONTROLLERS

The experiment described above indicates that evolution can automatically develop a set of smart and efficient solutions tailored for the environment where the robot operates. Here we describe another experiment where some changes in the environment and in the robot characteristics lead to different behavioral strategies without making the fitness function more complex and/or changing the evolutionary algorithm. The new environment consists

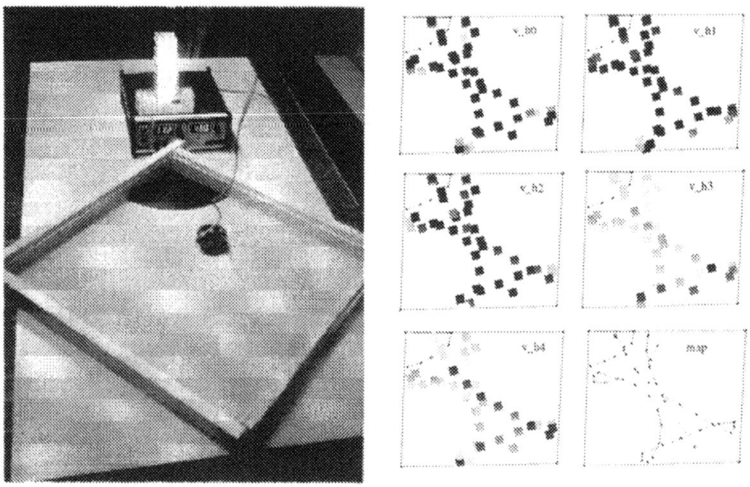

a) b)

*Figure 9.* a) The environment of the experiment on battery recharge. The light tower is positioned in the far corner over the recharging area which is painted black. There are no other light sources in the room. b) Activation of five internal neurons plotted every 100 ms while the robot navigates in the environment with the robot starting on the lower right corner; darker squares indicate higher activation values. The last frame plots only the trajectory of the robot.

of a 40x45 cm arena delimited by walls of light-blue polystyrene (figure 9a), as in the previous experiment. A 25 cm high tower equipped with 15 small DC lamps oriented toward the arena is placed in one corner and the room does not have other light sources. Under the light tower, a black-painted sector is intended to simulate the platform of a battery charger under construction. When the robot happens to be over the black area, its simulated battery is instantaneously recharged. Beside the eight infrared sensors, the neurocontroller receives input also from two ambient light sensors, each on one side of the body. Additionally, another ambient light sensor is placed under the robot platform, pointing downward, to detect when the robot arrives on the black-painted sector. The robot is provided with a simulated battery characterised by a fast linear discharge rate (max duration: approx. 20 seconds), and with a simulated sensor giving information about the battery status.

The fitness function is a simplified version of that used in the previous experiment, without the second component of equation 4.1 which accounted for straight navigation. The fitness function is accumulated at every time steps. Therefore, robots that could learn to discover the position of the battery charger and periodically return to it would survive longer. It should be noticed that when the robot is on the battery charger, the fitness value is zero because the robot is very close to the walls ($i = 1$).

After 240 generations of continuous fitness progress without human intervention, a robot is born which could navigate in the environment and periodically return to the charging station. In order to maximise the fitness function, the robot returns to the charging station just 1 second before battery failure. By analysing the evolved neurocontroller dynamics and correlating the neuron activation with the robot position while it is freely moving in the environment (thanks to a laser positioning device), it was found that some neurons are responsible for obstacle avoidance and others for homing to the recharger. One of the latter units displays a coding of the environment structure and charger position which is used by the robot to locate its own position, correlate it with remaining energy, and decide whether it is necessary to return to the station [8]. Localisation of the charging station, straight trajectories, self-localisation, and exact timing of the recharge phase are all emergent properties which evolved out of a very simple fitness function in order to satisfy the constraints posed by the environment in which the robot operated.

## 4.3. CO-EVOLUTIONARY ROBOTICS

In the effort to make the environment more complex and dynamic, and at the same time attempt to reduce human design even further (which up to now had been the formulation of simple fitness functions), we turned our attention to co-evolution of competitive robots. In the simplest scenario of two co-evolving and competing populations (for example, predator and prey), fitness progress of one species is achieved at disadvantage of the other population's fitness.

*Figure 10.* Right: The Predator is equipped with the vision module described in section 2.4. Left: The Prey has a black protuberance which can be detected by the predator everywhere in the environment, but its maximum speed is twice that of the predator. Both Predator and Prey are equipped with 8 infrared proximity sensors (max detection range was 3 cm in our environment).

For this experiment we employ two Khepera robots, a predator and prey, as depicted in figure 10 which evolve in parallel within a square arena of 47 x 47 cm. Each individual of one species is tested against the best individuals of the competitor species. A competition ends either when the predator touches the prey or after 500 sensorimotor updates. The fitness function $\Phi_c$ for each competition $c$ does not require any sensor or motor measurement, nor any global position measure; it is simply *TimetoContact* normalised by the maximum number of sensorimotor updates $TtC$ for the predator $pr$, and $1 - TtC$ for the prey $py$.

After evolution, each best individual is tested against all the best individuals of the other species; the average fitness of this "Master Tournament" for each best individual across generations is plotted on top of figure 11). Such a Master Tournament tells us two things: At which generation we can find the best prey and the best predator, and at which generation we are guaranteed to observe the most interesting tournaments. The first aspect is important for optimisation purposes and applications, the latter for pure entertainment. The best individuals are those reporting the highest fitness when also the competitor reports the highest fitness (marked by letters A and B in the graph). Instead, the most entertaining tournaments are those between individuals that report the same fitness level, because these are situations where both species have the same level of ability to overcome the competitor.

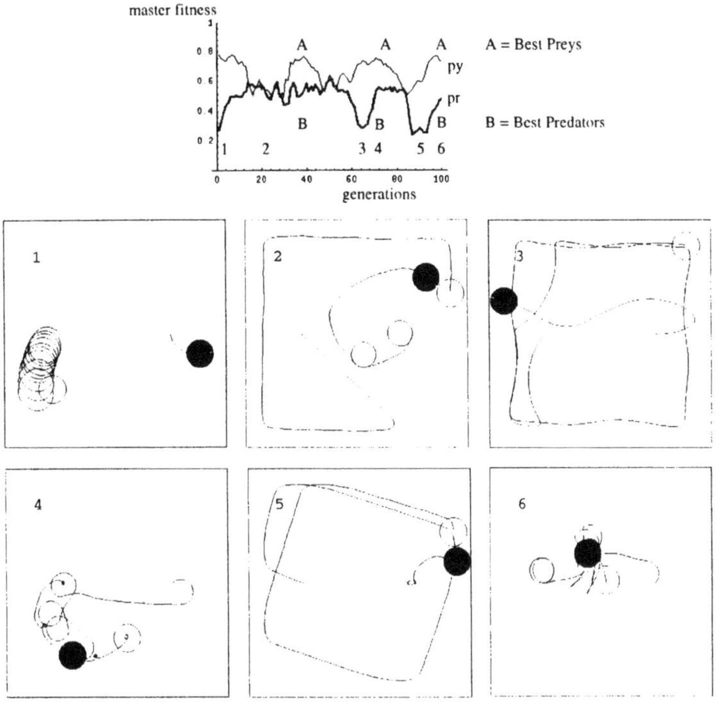


*Figure 11.* Top: Fitness of best individuals in Master Tournament for the two species. Letters indicate position of best prey and best predators. Numbers indicate position of individuals whose tournaments are displayed below. Bottom: Behaviours recorded at interesting points of co-evolution, representing typical strategies. Black disk is predator, white is the prey. See text for a description.

In the lower part of figure 11, behaviours of best competitors at critical stages of co-evolution, as indicated by Master Tournament data, give a more intuitive idea of how pursuit-evasion strategies are co-evolved.

Initially, the predator tends to stop in front of walls while the prey moves in circles (box 1). Later, the prey moves fast at straight trajectories avoiding walls while the predator tracks it from the centre and quickly attacks when the prey is closer (box 2). In box 3, the predator intercepts the prey which accelerates causing the predator to crash against the wall. Around generation 75, we have a typical example of the best prey (box 4); it moves in circles and, when the predator gets closer, it rapidly avoids it. Prey that move too fast around the environment sometimes cannot avoid an approaching predator because they detect it too late (IR sensors have lower sensitivity for a small cylindrical object, like another robot, than for a white flat wall). Therefore, it pays off to wait for the slower predator and accurately avoid it. However, some predators become smart enough to perform a small circle once they have missed the target, and re-attack until, by chance, the prey displays a side without IR sensors. As soon as the prey begin again moving around the environment, the predator develops a "spider strategy" (box 5): it slowly backs until it finds a wall where it waits for the fast-approaching prey. However, this strategy does not pay off when the prey stay in the same place. Finally, at generation 99 we have a new interesting strategy (box 6): the predator quickly tracks and reaches the prey which quietly rotates in small circles. As soon as the prey senses the predator, it backs and then approaches the predator (without touching it) on the side where it cannot be seen; consequently, the predator quickly turns in the attempt to visualise the prey which rotates around it, producing an entertaining dance.

These experiments indicate that competitive co-evolution is a promising technique for automatic gradual evolution of complex behaviours without effort in fitness design [11, 10]. In the following section we shall turn our attention to the study of systems that include more than two autonomous robots.

## 5. Issues in Collective Autonomous Robots

*Collective autonomous robotics* deals with teams of several autonomous robots which are involved in a shared mission. Design and control of the robot group requires the analysis of several collective mechanisms such as communication, interference, and cooperation. One way to tackle the problem is to take inspiration from collective intelligence displayed by social insects [1]. *Bio-inspired collective robotics* favours decentralised solutions and focuses on robot-robot and robot-environment interactions which can potentially lead to robust, goal-oriented, and emergent group behaviours.

Suitable team-behaviours can be achieved both by explicit programming or by adaptation. If the control solution is decentralised, explicitly programming each single robot is easy, but can become prohibitively difficult as the complexity of the desired team behaviour increases. Evolutionary or learning techniques, such as genetic algorithms or reinforcement learning methods, can help the engineer to select adequate behaviours for individual robots in the team. We believe that the integration of adaptive methods with explicit programming can strongly contribute to design a team of self-programming robots for a predefined task.

If only team performance is measured, robots are faced with a *credit assignment problem*, that is the problem of deciding to what extent their own behaviour has contributed to the team overall score [40]. Two ways for bypassing this problem have been proposed. One consists of exploiting

a) b)

*Figure 12.* a) Three Kheperas equipped with different combinations of modules on the energy supply board (from left to right): gripper and infrared modules, infrared and radio modules, and gripper and radio modules. The active seeds complete the set-up picture. b) A closer look at the Khepera with infrared and radio modules.

*global communication* among teammates [35] so that each robot is aware of the action effects of its fellows on the environment. However, this is not a completely decentralised solution and global communication among the robots is not always possible or does not scale well to large team sizes. Another solution consists in measuring the performance of each individual robot instead of team performance [25]. A potential drawback of this approach is that it might force collective behaviour to be the sum of identical individual behaviours, which is not necessarily the optimal strategy for every task constraints. We can achieve *real team solutions* only at the price of dealing with the credit assignment problem. Attempts in this direction in simulated environments have been recently published [40, 31].

At our laboratory we have conducted experiments with real robots who behave according to local programs [23, 22, 24] and are currently investigating the efficiency of genetic algorithms to adapt team behaviour for certain task boundary conditions.

### 5.1. SPECIFIC HARDWARE TOOLS FOR AUTONOMOUS COLLECTIVE ROBOTICS

In evolutionary or learning single-robot experiments, the robot is connected to a workstation through a cable which supplies required energy and supports intensive computing (see sections 2 and 4). However, it is impossible to use such connections for groups of more than two robots because the cables would become entangled.

Figure 12 shows the set-up developed at our laboratory for experiments in collective robots. To achieve extended autonomy we have developed a

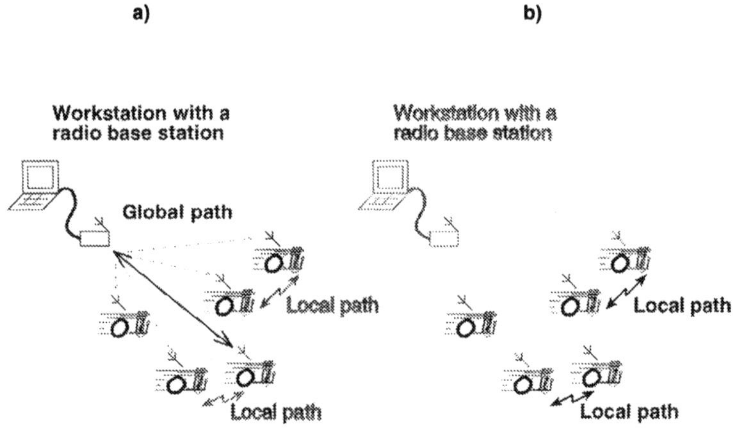


*Figure 13.* a) Global communication path. b) Local communication path.

special floor board as an interface between an external energy supply source and the robots. The required energy is acquired through electrical contacts placed underneath the robot platform, regardless of its position. Communication among robots is limited to neighbouring teammates. Furthermore, a communication link between workstation and robots enables supervision of the adaptive process. Therefore, we have a hierarchical communication strategy which optimises robot-to-robot (local path, via infrared link) and workstation-to-robot (global path, via radio link) communication (see figure 13). Finally, the infrared turret allows the Khepera to distinguish teammates from other objects in the environment.

The special pucks shown in figure 12, which we call "active seeds", introduce the possibility of increasing the environment complexity without increasing the complexity of the robot hardware in object-gathering experiments. The active seeds are capable of synchronously responding to the IR pulses of Khepera's proximity sensors. For instance, active seeds which respond with one pulse every two received pulses are seen by the robot as "blinking" objects. The blinking rate can be changed at pleasure, allowing the presence of several different detectable objects in the environment.

## 5.2. A SIMPLE BIO-INSPIRED COLLECTIVE EXPERIMENT

We present here a biologically inspired experiment concerned with clustering and gathering of scattered passive seeds (small wood cylinders). It is worth emphasising that in both experiments the robots operate completely autonomously using a local subsumption architecture [2] for each robot; there is no explicit communication (IR or radio link) with other robots or

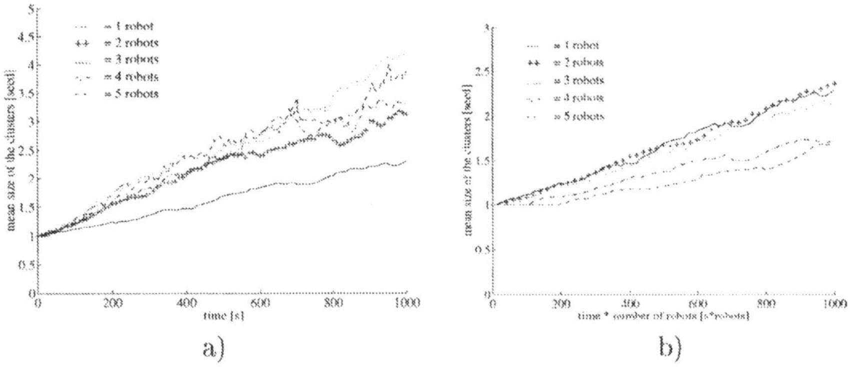


*Figure 14.* (a)Absolute performance of the group with increasing number of robots (1 to 5) on an arena of 80 × 80 cm and 20 seeds to be gathered. (b) Relative performance of a single robot within the group (1000 s of work using a single robot have to be compared with 500 s of work using 2 robots, with 333 s using 3 robots, and so on).

with the workstation. The only possible interactions among robots are reciprocal avoidance of collisions and modifications of the environment due to the displacement of the objects by other robots.

The control program that each robot executes can be described as follows. The robot moves on the arena looking for seeds. When its sensors are activated by an object, the robot begins the discriminating procedure. Two cases can occur: if the robot is in front of a large surface (a wall, another robot, or an array of seeds), the object is considered as an obstacle and the robot avoids it. In the second case, the object is identified as a seed. If the robot is not already carrying a seed, it grasps the seed with its gripper; if the robot is carrying a seed, it drops the seed close to the one it has found; then, in both cases, it performs a 180-degree rotation and begins resumes the search.

The experiments are conducted with a group of 1 to 5 Kheperas equipped with the gripper module and 20 scattered seeds in an arena of 80 × 80 cm. The measured team performance is the average size of the clusters created in about 17 minutes. Each experiment is repeated 5 times. The average performance of the 5 runs is plotted in figure 14. Typical patterns of seed scatter observed at the beginning and at the end of a longer experiment (120 minutes, repeated 3 times) are shown in figure 15.

Despite the simplicity of these experiments we can derive some interesting conclusions. The probability of incrementing the mean size of the clusters is always bigger than that of decrementing it. On the sole basis of simple local interaction rules, it is therefore possible to create clusters starting from a random placement of the seeds on the arena (see figure 14a).

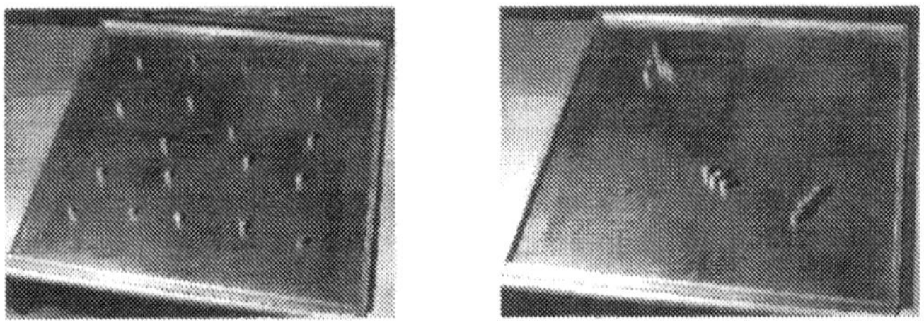

a) b)

*Figure 15.* Seed scattering a) at beginning of the experiment and b) after 2 hours, at the end of the longest experiment. The experiment has been conducted with the help of the extended autonomy tool.

Figure 14b clearly shows that there is no superlinearity in the team performance. On the contrary, groups with 4 and 5 robots show sublinear team performances due to interferences among teammates. After about one hour, the average cluster size reaches a saturation zone where interference and building gradient contributions are in equilibrium. We never observed all the seeds gathered in a single cluster.

The results indicate that in this kind of experiments with no explicit communication among robots and with no adaptation in robot control, a larger number of robots does not necessarily help to increase the team fitness. The introduction of adaptation could, for instance, allow individual robots to switch from an active phase to an inactive one when the ratio between the amount of work (in our case the seed finding rate) and the interference (in our case the encounter rate with other teammates) decreases under a given threshold. Similar mechanisms are supposed to play a crucial role in ant colonies [34].

## 6. Applications of Autonomous Robots

Despite the potentials of the approaches and algorithms described in the previous sections, industrial applications of autonomous mobile robots are not yet widespread. However, the conditions might be ready for an explosion of applications in the next few years.

Given the low complexity of current applications, a traditional engineering approach is typically employed to program basic behaviours (usually in C language) for the various situations that the robot might face. The control program is extensively tested on the robot in order to adjust poor reactions to unplanned situations. Most prototypes developed so far have been based on the layered subsumption architecture proposed by Brooks [2, 3]. These robots display behaviours similar to those described in the previous sections, such as obstacle avoidance, object search, but the environment around the robot is clearly defined and the behaviour is continuously verified.

At the moment of writing this chapter, we are not aware of any application where an autonomous robot automatically improves its own behaviour by modifying internal parameters according to experienced situations. Neural networks, genetic algorithms, and neuro-fuzzy approaches do not offer yet enough guarantee of stability in unsupervised situations for being accepted as a viable method in critical applications where risks and costs rank high. For real-world applications the robot must be sold in numerous copies to customers who will read a short set of recommendations, power the robot, and check from time to time that the work is being properly done. Customers are not willing to spend time in instructing their robot, letting it carefully explore the environment, and buy the risk of sub-optimal performance. Processors, interfaces, and sensors are sufficiently reliable nowadays, but not so is complex software. One possibility is that of combining deterministic and adaptive software, but much work in this direction remains to be done.

Another challenge for applications is energy autonomy which is currently supported by solar cells or on-board accumulators, but could be extended by including a behavioural module that guides the robot to a recharging station when necessary. Power requirements are negligible for sensor reading and processing, but become considerable when it comes to move actuators, such as wheels and grippers. Current limitations in energy autonomy naturally favour "white-collar" applications of autonomous robots, such as surveillance.

## 6.1. FULLY AUTONOMOUS APPLICATIONS

A fully autonomous, widely marketed robot is the Husqvarna lawn-mower for flat and prepared terrains. The robot is 15 cm high and has a surface of 80 cm by 40 cm covered by solar power cells which let it work for several months when the sun is high over the sky (a small battery back-up is used by the processor when sun light is not strong enough to power the robot). The robot moves randomly, exploiting small irregularities of the terrain, while checking for an electric wire (solar powered too) positioned on the perimeter by the owner. In case the robot gets stuck in unexpected situations, it starts beeping and waiting for human help. Lawn-mowing is a simple navigation task where random walk seems acceptable; furthermore, since the wheels move faster when the lawn is cut, the robot tends to spend more time on areas not yet cleared. Pool cleaning robots share some characteristics with the autonomous lawn-mower. Although several types are available on the market (e.g., see http://h2o-marketing.com/aquabot/aqua.html), they generally perform a random walk on the bottom and on the sides while scrubbing, vacuum cleaning, and filtering the pool. They are generally powered via a cable hanging from the centre of the pool and can also be remotely controlled, if necessary. A more systematic cleaning of the pool inner surface can be achieved by pressure sensors which exploit the regularities of tiles.

In both the applications described above, the environment is simple and stable enough to expect reliable operation. Unfortunately, not the same applies for home vacuum cleaners. Here the robot cannot rely on a regular environment, on wires delimiting perimeter and obstacles, on uniform surfaces, and on solar cells or a cable. Energy requirements are very stringent and batteries with a reasonable size might last only 5 minutes before needing a lengthy recharge. Currently, autonomous vacuum cleaning robots are restricted to specific large environments, such as airport lounges (Narita airport in Japan, e.g.). A prototype robot in the Paris metro was designed to follow a line buried in the ground, whereas the recently completed CLEAN Eureka project (nr. EU-1094 on the Eureka database: http://www.eureka.be) has attempted to develop a robot for cleaning hyper-market surfaces by exploiting pre-positioned active landmarks (see the concept of active seeds for extended sensors described in section 5.1).

Autonomous mobile robots seem a very suitable application for searching anti-personnel landmines [32]. However, several factors make this application very difficult: the terrain is often impracticable for wheeled robots (such as tall vegetation on abandoned fields), there are not yet powerful mine sensors available, and an almost systematic search is required to guarantee that no landmines are left behind (for military applications, such as breaching, not all mines must be neutralised: human and material losses are weighted against the strategic value of crossing the field). Recently, a legged water-proof robot has been developed for landmines positioned on the surf zone [17], which is a rather regular and defined terrain. All these robots are supposed to blow up mines by hitting them (and being destroyed in the meanwhile).

## 6.2. PARTIALLY AUTONOMOUS APPLICATIONS

The Mars Sojourner (http://mpfwww.jpl.nasa.gov/default.html) is a popular example of a robot with full energy autonomy, but limited behavioural autonomy. Since it moves very slowly, a rather small area of solar cells is sufficient to power the robot. It receives instructions from Earth on its destination, but it has to get there autonomously.

Another application with similar behavioral requirements is the autonomous wheelchair [16]. Several handicapped persons find it difficult to steer precisely their own wheelchair to get around corners or passing through doorways. By supplying these chairs with additional sensors and appropriate control systems that support semi-autonomous navigation, the owner can instruct the chair on the desired destination and let it get there autonomously.

Semi-autonomous mobile robots have a large potential market, from rescue robots to robots for maintenance of nuclear plants, and have several military applications, such as reconnaissance flying drones. An application that has attracted the interest of several industries, research institutes, and funding agencies is a semi-autonomous vehicle capable of navigating in daily traffic as well as on rough terrains. A well-known example is NavLab [20], developed by Carnegie Mellon University. The Swiss Serpentine [37] is a urban semi-autonomous vehicle designed for accommodating several standing persons which follows an inductive track providing power, self-localisation, and general directives on the task to be accomplished.

As it can be seen from the examples reported here, safety compliances, costs, stability, and risks are important issues in industrial applications where often a traditional engineering approach is still preferred over advanced learning abilities. However, certain niches of the market seem ripe for new forms of autonomous robot featuring some form of self-organisation. Entertainment and companion robots have recently attracted the interest of several small and large companies [13]. Here, the learning aspect is important both for adapting the robot to the requirements of the owner and for providing it with novel behaviours added to the old ones, which would otherwise become soon boring. Similarly, space exploration and asteroid mining will profit from robots equipped with higher autonomy and decision abilities.

## 7. Conclusion

In this chapter we have provided an overview of some important issues in mobile autonomous robotics, emphasising the directions explored at our laboratory from hardware to software development. At the hardware level, we have insisted on the development of appropriate research tools to investigate approaches and applications on an incremental and comparative fashion. The success of the concept behind the miniature mobile robot Khepera is indeed its strategic position between simulations and applications, between research development and industrial application, and between roboticians and people from other disciplines.

At the software level, three different approaches have been described here in further detail: neuro-fuzzy, evolutionary, and collective systems. Each of these fields of investigations has its own interest, advantages, and drawbacks, which become evident when all approaches are tested and compared on the same robot platform. Neuro-fuzzy systems offer the unparallelled feature of combining a low-level learning algorithm which operates on a set of knowledge-based rules. The result of the learning phase can then be transformed back into easily interpretable rules, as opposed to the often unintelligible pattern of synaptic connections in neural networks. However, the presence of a human supervisor during the learning phase limits somewhat the autonomy and applicability of the method. Evolutionary systems are powerful algorithms to develop interesting behaviours without much human intervention and knowledge. A clear drawback, however, is that they require considerable training time during which power must be somehow externally supplied. Collective systems, a relatively unexplored terrain, introduce new complexities which require a reconsideration of several issues at hardware and software level. Study of decentralised control in colonies of autonomous robots can provide interesting surprises, such as those outlined in section 5.2, which can be appropriately used as a starting point when developing applications requiring several interacting robots. Despite the obvious research interest and biological relevance, it is not yet clear whether application potentials can match hardware costs, both at the development phase and during final operation.

As outlined in the previous section on applications, it is clear that research on autonomous robots capable of self-organisation is still far ahead of industrial applications. We are still at a crucial investigation phase where it would be deleterious and short-minded to declare what are the most promising directions and what are those that should be abandoned. Probably, some combinations of these and other techniques will reciprocally overcome their own respective drawbacks. Some investigations are already underway on the combination of evolution and fast learning algorithms [9], on fuzzy logic to analyse and control the result of autonomous learning, and on adaptive local methods embedded in collective systems.

## References

1. E. G. Bonabeau and Theraulaz G. *Intelligence Collective.* Hermès, Paris, France, 1994.
2. R. A. Brooks. A robust layered control system for a mobile robot. *IEEE Robotics and Automation*, RA-2:14–23, March 1986.
3. R. A. Brooks. Intelligence without representation. *Artificial Intelligence*, 47:139–59, 1991.
4. D. Cliff, P. Husbands, J. A. Meyer, and S. W. Wilson, editors. *From Animals to Animats: Proceedings of the Third International Conference on Simulation of Adaptive Behavior*, Cambridge, MA, 1994. MIT Press/Bradford Books.
5. M. Dorigo. Special issue on learning autonomous robots. *IEEE Transactions on*

*Systems, Man and Cybernetics-Part B*, 26:361–364, 1993.
6. M. Dorigo and U. Schnepf. Genetic-based machine learning and behavior based robotics: a new synthesis. *IEEE Transactions on Systems, Man and Cybernetics*, 23:141–154, 1993.
7. D. Floreano and F. Mondada. Automatic Creation of an Autonomous Agent: Genetic Evolution of a Neural-Network Driven Robot. In D. Cliff, P. Husbands, J. Meyer, and S. W. Wilson, editors, *From Animals to Animats III: Proceedings of the Third International Conference on Simulation of Adaptive Behavior*, pages 402–410. MIT Press-Bradford Books, Cambridge, MA, 1994.
8. D. Floreano and F. Mondada. Evolution of homing navigation in a real mobile robot. *IEEE Transactions on Systems, Man, and Cybernetics-Part B*, 26:396–407, 1996.
9. D. Floreano and F. Mondada. Evolution of plastic neurocontrollers for situated agents. In P. Maes, M. Mataric, J-A. Meyer, J. Pollack, H. Roitblat, and S. Wilson, editors, *From Animals to Animats IV: Proceedings of the Fourth International Conference on Simulation of Adaptive Behavior*, pages 402–410. MIT Press-Bradford Books, Cambridge, MA, 1996.
10. D. Floreano and S. Nolfi. Adaptive behavior in competing co-evolving species. In P. Husbands and I. Harvey, editors, *Proceedings of the 4th European Conference on Artificial Life*, Cambridge, MA, 1997. MIT Press.
11. D. Floreano and S. Nolfi. God save the red queen! competition in co-evolutionary robotics. In J. Koza, K. Deb, M. Dorigo, D. Fogel, M. Garzon, H. Iba, and R. L. Riolo, editors, *Proceedings of the 2nd International Conference on Genetic Programming*, San Mateo, CA, 1997. Morgan Kaufmann.
12. N. Franceschini, J.-M. Pichon, and C. Blanes. Real time visuomotor control: From flies to robots. In *Proceedings of the Fifth International Conference on Advanced Robotics*, pages 91–95, Pisa, June 1991.
13. M. Fujita and K. Kageyama. An open architecture for robot entertainment. In *ACM Conference on Autonomous Agents*, pages 435–442, Marina Del Rey, February 1997.
14. P. Gaussier. Special Issue on Animat Approach to Control Autonomous Robots interacting with an unknown world. *Robotics and Autonomous Systems*, 16, 1995.
15. J. Godjevac. *A Method for the Design of Neuro-Fuzzy Controllers; an Application in Robot Learning*. PhD Thesis N° 1602, École Polytechnique Fédérale de Lausanne, 1997.
16. T. Gomi and K. Ide. The development of an intelligent wheelchair. In *Intelligent Vehicles Symposium*, Tokyo, 1996.
17. H. Greiner, C. Angle, T. Freed, J. Jones, P. Ning, R. Elsley, and G. Bane. Autonomous legged underwater vehicles for near land warfare. In *Autonomous Vehicles in Mine Countermeasures Symposium*, pages 14–22, Monterey, CA, 1995.
18. J. H. Holland. *Adaptation in natural and artificial systems*. The University of Michigan Press, Ann Arbor, 1975.
19. N. Jakobi. Half-baked, ad-hoc and noisy: Minimal simulations for evolutionary robotics. In P. Husbands and I. Harvey, editors, *Proceedings of the 4th European Conference on Artificial Life*, Cambridge, MA, 1997. MIT Press.
20. A. Kelly and A. Stenz. Analysis of requirements for high speed rough terrain autonomous mobile robots. In *IEEE Conference on Robotics and Automation*, pages 3318–3333, Albuquerque, 1997.
21. P. Maes, M. Mataric, J-A. Meyer, J. Pollack, H. Roitblat, and S. Wilson, editors. *From Animals to Animats: Proceedings of the Fourth International Conference on Simulation of Adaptive Behavior*, Cambridge, MA, 1996. MIT Press/Bradford Books.
22. A. Martinoli, E. Franzi, and O. Matthey. Towards a reliable set-up for bio-inspired collective experiments with real robots. In *Proc. of the Fifth International Symposium on Experimental Robotics ISER-97*, Barcelona, Spain, June 1997. In Press.

23. A. Martinoli and F. Mondada. Collective and cooperative group behaviours: Biologically inspired experiments in robotics. In *Proceedings of the Fourth International Symposium on Experimental Robotics*, Stanford, U.S.A., 1995. Springer Verlag.
24. A. Martinoli, M. Yamamoto, and F. Mondada. On the modelling of bioinspired collective experiments with real robots. In *Proceedings of the Fourth European Conference on Artificial Life ECAL-97*, Brighton, UK, July 1997. `http://www.cogs.susx.ac.uk/ecal97/present.html`.
25. M. J. Matarić. Learning in multi-robot systems. In G. Weiss and S. Sen, editors, *Adaptation and Learning in Multi-Agent Systems*, volume 1042, pages 152–163. Springer Verlag, Lecture Notes in Artificial Intelligence, 1996.
26. J. A. Meyer, H. L. Roitblat, and S. W. Wilson, editors. *From Animals to Animats: Proceedings of the Second International Conference on Simulation of Adaptive Behavior*, Cambridge, MA, 1993. MIT Press/Bradford Books.
27. J. A. Meyer and S. W. Wilson, editors. *From Animals to Animats: Proceedings of the First International Conference on Simulation of Adaptive Behavior*, Cambridge, MA, 1991. MIT Press/Bradford Books.
28. O. Michel. Evolutionary neurogenesis applied to mobile robotics. In M. Patel and V. Honovar, editors, *Advances in Evolutionary Synthesis of Neural Systems*. MIT Press, Cambridge, MA, 1997.
29. F. Mondada and D. Floreano. Evolution of neural control structures: some experiments on mobile robots. *Robotics and Autonomous Systems*, 16:183–195, 1995.
30. F. Mondada, E. Franzi, and P. Ienne. Mobile robot miniaturization: A tool for investigation in control algorithms. In T. Yoshikawa and F. Miyazaki, editors, *Proceedings of the Third International Symposium on Experimental Robotics*, pages 501–513, Tokyo, 1993. Springer Verlag.
31. A. Murciano and J. del R. Millán. Learning signaling behaviors and specialization in cooperative agents. *Adaptive Behavior*, 5(1):5–28, 1997.
32. J. D. Nicoud. Vehicles and robots for humanitarian demining. *Industrial Robot Journal*, 24(2):168, 1997.
33. S. Nolfi, D. Floreano, O. Miglino, and F. Mondada. How to evolve autonomous robots: Different approaches in evolutionary robotics. In R. Brooks and P. Maes, editors, *Proceedings of the Fourth Workshop on Artificial Life*, pages 190–197, Boston, MA, 1994. MIT Press.
34. S. W. Pacala, D. M. Gordon, and H. C. J. Godfray. Effects of social group size on information transfer and task allocation. *Evolutionary Ecology*, 10:127–165, 1996.
35. L.E. Parker. The effect of action recognition and robot awareness in cooperative robotic teams. In *Proceedings of IEEE International Conference on Intelligent Robots and Systems IROS-95*, volume 1, pages 212–219, Pittsburgh, PA, August 1995. Springer Verlag.
36. R. Pfeifer. Cognition – perspectives from autonomous agents. *Robotics and Autonomous Agents*, 15:47–70, 1995.
37. B. Saugy and R. Rovira. The Serpentine, a step forward in intelligent transport system. In *Proceedings of the Conference on Robotics and Intelligent Systems (IROS97)*, Grenoble, France, September 1997.
38. L. Steels. The Artificial Life Roots of Artificial Intelligence. *Artificial Life*, 1:75–110, 1994.
39. M. Sugeno and K. Murakami. An Experimental Study on Fuzzy Parking Control Using a Model Car. In Michio Sugeno, editor, *Industrial Applications of Fuzzy Control*, chapter 8, pages 125–138. North-Holland, 1985.
40. C. Versino and L. M. Gambardella. Ibots learn genuine team solutions. In M. Van Someren and G. Widmer, editors, *Proceedings European Machine Learning ECML-97*, pages 298–311, Kyoto, Japan, 1997. Springer Verlag. Lecture Notes in Artificial Intelligence.

# 9

# INTELLIGENT AUTONOMOUS SYSTEMS: VISUAL NAVIGATION FUNCTIONALITIES FOR AUTONOMOUS MOBILE VEHICLES

E. STELLA and A. DISTANTE
*Consiglio Nazionalle delle Ricerche*
*Istituto Elaborazione Segnali ed Immagini*
*Via Amendola, 166/5 - 70126 Bari, Italy*

## 1. Introduction

The main reason behind the lack, of a real commercial development of Autonomous mobile vehicles (AMV) is due to the difficulty for these kind of systems to satisfy three important constraints: cheapness, speed and uncertainty.
Cheapness means that intelligent systems should became comparable to a commercial washing machine in order to increase their diffusion among families and small commercial activities. The costs depend on both technological constraints and the level of "intelligence" that systems should have.
Moreover , regarding the second constraint, systems should be fast, compared to the environment where the AMV must operate. Like cheapness, speed should be a compromise between technological advances and "intelligence" complexity.
The last constraint, uncertainty, is probably the most important one. The intelligent systems must operate safetly. A tollerance of 10%, often, means that AMV can cause damages. Furthermore. uncertainty depends on technological aspects but, mainly, on "intelligence".
While, it is intuitive to understand what technological constraints are, what does "intelligence" mean? When a system can be considered an intelligent one?
A good definition of "intelligence" could be the capability of the system to learn autonomously as to associate behaviours to sensorial inputs.When more behaviours are possible, the systems should be able to choose among the most appropriate actions to perform. Probably, this kind of definition could be unrealistic compared to real needs.
A more pragmatic meaning of "intelligence" could be to restrict the previous definition considering the constraints of the environment where the systems should operate and the aim where the systems were designed for. So a system should be intelligent if it operates according to the desire of its designer. This assumption could be restrictive if we identify "the desire of the designer" as a predefined specialized task (such as to grasp an object from a bin, weld two parts together), but a full significance of the definition is obtained when the aim of the system becames more general (such

as to move safe in an unknown environment). In the last scheme many examples of "intelligent systems" can be found in literature. Usually, they represent prototypes of mobile vehicles resulting from scientific laboratory studies and not commercial solutions.
A classification of those mobile vehicles is not absolute but depends on different criteria:

- Generality: how much flexibility the system has to work on different applications.
- Architecture: functional (horizontal) or behavioural (vertical) decompositions are used [1].
- Context: indoor or outdoor environment.

However it is possible to identify some tasks enough to permit an evaluation of the performances of AMV. Considering an indoor environment where most of the vehicles are designed for, AMVs should have many functionalities. The most important can be considered the **navigational** one.
Navigation is the capability of an autonomous mobile vehicle to estimate its position in the environment.
In [2] a review of several prototypes of AMV and most of the sensors used, are described. Dead-reckoning sensors (odometers and gyro) have the main advantage to require no modification of the environment but they are affected by an incremental unbounded error.
Active sensors such as ultrasonics, infrareds and lasers do not suffer error growth problems but they need to modify the environment. In fact active beacons must be placed in the environment and the AMV should be able to sense them in order to estimate their position by triangulation. Moreover ultrasonic and infrared sensors do not have large ranges so beacons must be located not very far from the AMV. Furthermore, lasers cannot freely used in environments populated by human beings.
Valid alternative are **vision based** sensors. There are many reasons. The main advantage is that vision techniques are passive so no emission is necessary. On the other hand vision methods, are slow on conventional hardware, so a high performance computer (specialized processor or parallel architecture) should be used in order to increase performances. Though, vision based techniques are less precise than active or dead-reckoning sensors, they are characterized to have bounded uncertainty that can be a-priori modelled. Both, bounded uncertainty and passivity make vision based sensors more flexible and worthy to be considered. In the context of mobile vehicle navigation, visual methods can be roughly divided into two classes : **absolute** and **relative** techniques. Absolute navigation, also referred as self-location, consists of those techniques that estimate the absolute position and orientation of the vehicle determining the location of the Center-of-Projection (CP) and the orientation of the TV camera placed on the vehicle, referring to a world reference system. Generally, it is performed referring to absolute landmarks. The use of artificial beacons as landmarks

is more suitable but the possibility to refer to natural landmarks is preferred since no modification of the environment is necessary. In par. 2, self-location techniques are analyzed. In particular a technique to determine the location and orientation using three cheap visual beacons is described [3]. Moreover, the possibility to substitute artificial beacons with natural landmarks is described and related problems are analyzed.
The second class of visual navigation methods consists of relative ones. Relative navigation consists of determining the motion of the vehicle estimating the motion of the TV camera set on. The motion is estimated respect to a starting known position assumed as origin of a reference system. In par. 3 these kind of techniques are analyzed and a novel method is described.

## 2. Absolute visual navigation

In [4] a good review of the visual self-location methods is done. Most of the techniques use artificial beacons instead of recognized natural features. The reason is that artificial beacons are easy to detect without a burdensome image processing technique. On the other hand, the use of artificial landmarks means to modify the environment that in some applications could be limitative. Alternatively, the flexibility introduced by the use of natural landmarks is limitated by the necessity to have absolute 3D information about landmarks. This can be performed or by using a ranging device or building a 3D world model (such as a CAD model). Obviously, it is possible to recover information in the world model by visual sensors (TV cameras) placed on the AMV. In the sense that AMV must be able to match information in the world model with image data acquired during its route. This is not a simple task, considering that visual data are affected by both perspective and scale transformation. Furthermore, the management of image data can be performed in realtime only using specialized expensive hardware. The expensiveness is not only in the cost of the hardware but also in the complexity in its programmability.
The following paragraph describes a visual self-location method useful for AMV navigation. The method is tested on artificial beacons in order to verify performances and effectiveness. Successively, a technique, to select and recognize under scale and perspective transformation natural landmarks, is described with the aim to substitute the artificial beacons in robot navigation.

### 2.1 SELF-LOCATION OF TV CAMERA BY ARTIFICIAL BEACONS

Let $P_1(X_1,Y_1,Z_1)$, $P_2(X_2,Y_2,Z_2)$ and $P_3(X_3,Y_3,Z_3)$ be three landmarks in a world reference system $O_w$ (fig. 1a). Let $V(X_v,Y_v,Z_v)$ be the unknown coordinates of the center-of-projection (CP) of the camera. In the context of autonomous robot navigation the determination of the location of the CP in the $X_w$-$Z_w$ plane is all important, so the location problem is reduced to a two-dimensional one. Let $L_1(X_1,Z_1)$, $L_2(X_2,Z_2)$, $L_3(X_3,Z_3)$ and $CP(X_{cp},Z_{cp})$ be the projections onto the plane $X_w$-$Z_w$ of $P_1,P_2,P_3$ and $V$ (fig. 1b). If the **visual angle** $\alpha$ is known, then the triangle with vertices $L_1,L_2$ and CP can be circumscribed by a circle whose equation can be determined assuming that:

1. the circle passes through landmarks *L1* and *L2*
2. its radius is given (**generic triangle theorem**):

$$d/\sin\alpha = 2R$$

where d is the distance between landmarks $L_1$ and $L_2$.

So the three conditions to estimate the circle equation are completely defined. Summarizing, the equation of the circle $C_1$ (fig. 2) circumscribed to the triangle whose vertices are $L_1$,$L_2$ and CP is determined by solving the following system of equations:

$$\begin{cases} (X_1 - X_C)^2 + (Z_1 - Z_C)^2 = R^2 \\ (X_2 - X_C)^2 + (Z_2 - Z_C)^2 = R^2 \\ \dfrac{\sqrt{(X_1 - X_2)^2 + (Z_1 - Z_2)^2}}{2\sin\alpha} = R \end{cases} \qquad \text{(EQ1)}$$

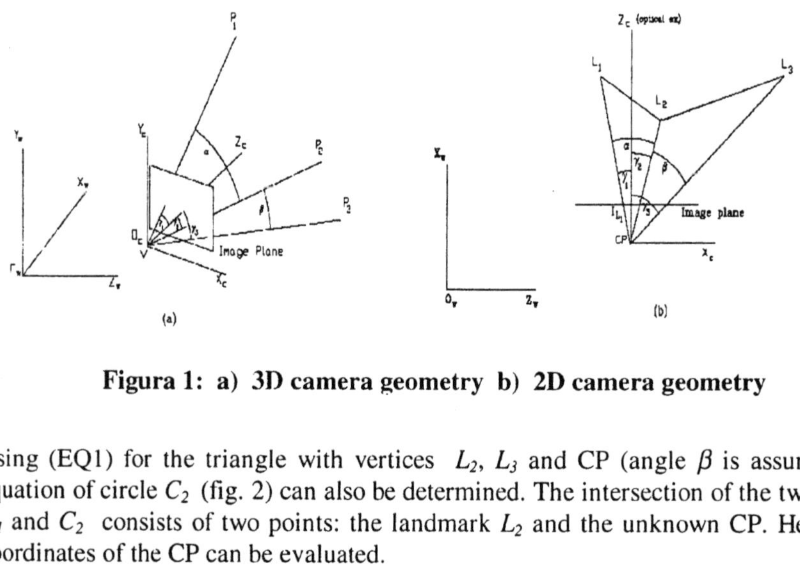


**Figura 1: a) 3D camera geometry b) 2D camera geometry**

Using (EQ1) for the triangle with vertices $L_2$, $L_3$ and CP (angle $\beta$ is assumed) the equation of circle $C_2$ (fig. 2) can also be determined. The intersection of the two circles $C_1$ and $C_2$ consists of two points: the landmark $L_2$ and the unknown CP. Hence, the coordinates of the CP can be evaluated.

The visual angles $\alpha$ and $\beta$ can be estimated using the excellent angular resolution of the CCD cameras. In fig. 2, the lines obtained from the projection of the landmarks to the CP form the angles $\gamma_1$, $\gamma_2$ and $\gamma_3$, so that:

$$\gamma_1 = a\tan\left(\frac{u_1}{f}\right) \qquad \gamma_2 = a\tan\left(\frac{u_2}{f}\right) \qquad \gamma_3 = a\tan\left(\frac{u_3}{f}\right) \qquad \text{(EQ2)}$$

then

$$\alpha = \gamma_2 - \gamma_1 \qquad \beta = \gamma_3 - \gamma_2 \qquad \text{(EQ3)}$$

where $f$ is the focal length of the camera and $u_1$, $u_2$, $u_3$ are the distances on the $x$-axis (on the image plane) of the projections of the landmarks from the **piercing point** (intersection of optical axis and image plane). Considering that, $u_1$, $u_2$, $u_3$ in (EQ2), are pixel coordinates referred to the CCD chip (not to the frame grabber), their estimation needs a **scale factor**. The piercing point, focal length, scale factor and lens distortion are camera intrisinc parameters that can be recovered by a camera calibration technique [5,6,7,8]. Our experiments have been performed, however, without the use of a camera calibration and we have adopted: the focal length suggested by the lens manufacturer, the piercing point at the center of the image plane (as suggests in [5]), a scale factor equal to 1 because we have
used a TV camera and a frame grabber with the same sampling frequency and no lens distorsion.

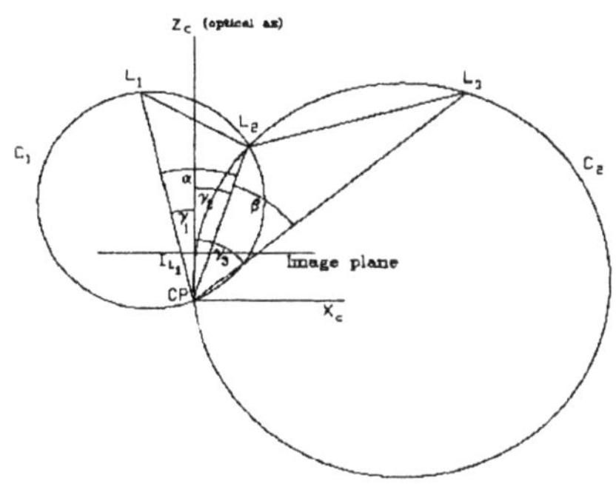


**Figura 2: CP location as intersection of two circles**

Considering (EQ1) again, it is possible to estimate the circle equation between two fixed points having a fixed radius. Normally, solving (EQ1) for $L_1$, $L_2$ and visual angle $\alpha$, it gives two possible solutions shown in fig. 3a. Two solutions exist for landmarks $L_2$, $L_3$ and visual angle $\beta$, hence, as shown in fig. 3b, four circles must be considered: $C_1$, $C_1'$, $C_2$, $C_2'$. $C_1$ intersects circles $C_2$ and $C_2'$ at 4 points: two points coincide with landmark $L_2$ and the others are two possible candidates to be the CP. If circle $C_1'$ is also considered two others candidates for the CP are obtained. We have defined two rules in order to select the correct CP from the four candidates.

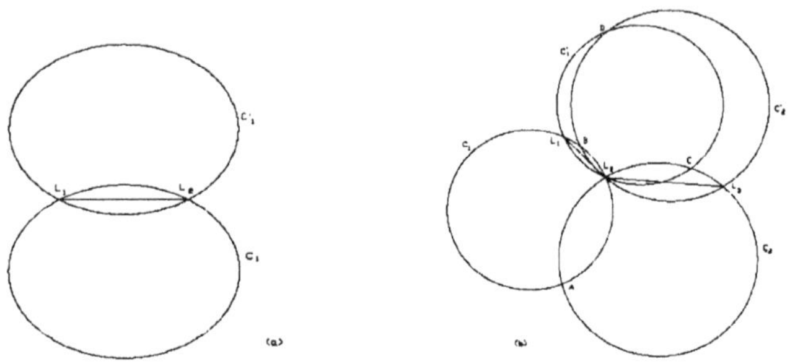

**Figura 3: a) Circles passing between two points. b) All circles to consider**

**Rule 1**

Let $C_1$ be one of the two circles defined for landmarks $L_1$, $L_2$ and visual angle $\alpha$. Let $C_2$ and $C_2'$ be the two circles defined for landmarks $L_2$, $L_3$ and visual angle $\beta$ (fig. 3b). $C_1$ intersects $C_2$ and $C_2'$ at 3 points: $L_2$, $A$ and $B$. Because, $L_2$ is the known landmark $A$ and $B$ are the two possible candidates for the CP. Considering the angle $L_1AL_3$ to be the visual angle when $A$ is chosen as CP and, alternatively, $L_1BL_3$ when $B$ is considered the CP, the correct CP between A and B is that one whose visual angle is equal to that one estimated on the image plane between the landmarks $L_1$ and $L_3$. This angle, referring to (EQ2) and (EQ3), is equal to ($\alpha + \beta$). In [9] the proof that only one of the angles $L_1AL_3$ and $L_1BL_3$ is equal to the visual angle estimated by the camera is given. Rule 1 applied to both circles $C_1$ and $C_1'$ reduces to two the possible candidates for the CP.

**Rule 2**

Let A and D be the two candidates for the CP after applying rule 1. The criterion to select between the two candidates consists of the following two steps:

1. Evaluating the equation of the optical axis of the camera in the world reference system. $P(X_p,Z_p)$ is taken as the generic candidate for the CP in the world reference system. The optical axis is the line which passes through $P$ and at an angle of $\gamma 1$ (see (EQ2)) to the line $L_1P$. So, using these two conditions:

$$\begin{cases} Z_P = mX_P + s \\ \cos(\gamma_1) = \dfrac{m'm}{\sqrt{1 + (m'm)^2}} \end{cases} \qquad \text{(EQ4)}$$

where $m'$ is the slope of the straight line between $L_1$ and $P$, $m$ and $s$ are, respectively, the unknown slope and the intercept of the optical axis. From (EQ4) two solutions for m are evaluated. The correct solution is selected by using the condition that the angle between the optical axis and the straight line between landmark $L_2$ and $P$ is $\gamma 2$ (refers to (EQ2) and fig.2).

2. Selecting between two optical axes given by step 1. Consider a local reference system $O_C$ centered on $P$ and a $z$-axis coinciding with the estimated optical axis (fig. 2). The equation of the image plane in $O_C$ is the straight line $i$: $z = f$, where $f$ is the focal length. The intersection $I_{Ll}$ between $i$ and the straight line between $L_l$ and $P$ represents the projection of the landmark $L_l$ on the image plane. So, considering $A$ and $D$, the point is selected as CP whose $I_{Ll}$ is equal to the $x$-coordinate on the image plane for landmark $L_l$.

Hence, the position of the CP and the orientation of the optical axis is univocally determined.
In order to verify the effectiveness of the method under real conditions, three artificial beacons have been used as landmarks. In par 2.2 a technique to select and recognize natural landmarks will be described.
Our landmarks are three semispherical infrared (most CCD cameras can see near infrared light) LEDs (Light Emission Diode). Fig. 4a shows how the landmarks are placed in the environment, while fig. 4b is an acquired image of landmarks.

**Figura 4: a) View of environment, vehicle and landmarks. b) Landmarks as seen from the robot**

The method for self-location has been tested on our mobile robot SAURO, shown in fig. 6b. SAURO is based on a TRC LABMATE mobile base controlled by an on-board VME-bus based computer. The imaging system is the VME board IPP ELTEC frame grabber. The processing is performed on a Motorola 68040 board. The TV camera used is a COHU 6710 mounted on a RS232 controlled pan-tilt motorized head. The acquired images are 576 lines by 768 columns.

In order to verify the reliability of self-location in navigation context, SAURO was asked to perform a path to a goal position. The path consisted of the segments shown in fig. 6a (solid line). The navigation strategy is the following:

1. The vehicle starts from an unknown position with an unknown orientation.

2. The landmarks are searched by using the pan-tilt head. Then, by self-location the mobile vehicle estimates its position and orientation in the world reference system.

3. In order to go to the next stop position, the rotation and translation are evaluated and then the vehicle moves on.

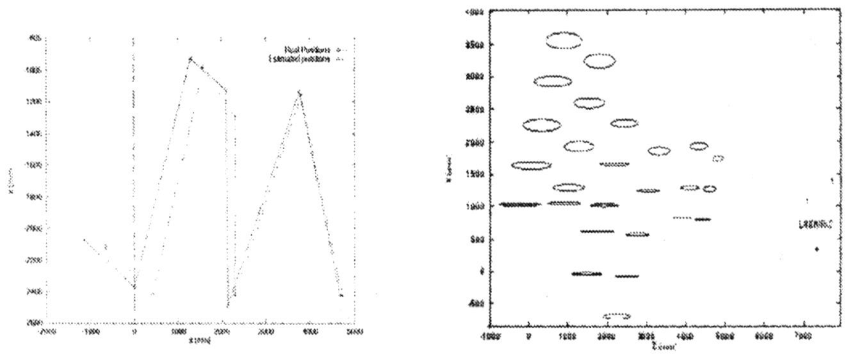

**Figura 5: a) Experimental results of self-location on a path. b) Self-location error respect to landmark position**

The described procedure is repeated until the vehicle reaches the goal. The positions estimated by self-location (dotted line) compared with the real positions (solid line) are shown in fig. 5a . The robot location was verified using a D-50 NIKON TOTAL STATION (Theodolite) that can estimate the coordinates of a point in space with an error of 5 mm. The D-50 was also used to determine the 3D locations of the landmarks in the world reference system. The results plotted in fig. 5a are tabulated in tab. 1 which shows that the error in self-location depends on the distance from the landmarks: nearer the landmarks are, less is the error in robot location. This result is evident in fig. 5b, where error is graphically represented by ellipses. The center of the ellipse is the real position of the TV camera, estimated by the D-50 station. The axis of the ellipse represent the standard deviation of the error by determining the camera location through the visual self-location method. Fig. 5b shows that the error grows with the distance from the landmarks. In fact, the distance modifies the appearence of the landmarks on the image plane of the camera, determining a growth of the unreliability in the visual angle estimation. In fact, using a zoom lens, hence modifying the appearance of the landmarks on the image, the error can be reduced.

Improvements in location estimation can be obtained using a motorized zoom lens in order to choose the best focal length depending on the distance from the landmarks. The self-location estimation time is about 50 ms, but the global time depends on the time that the pan-tilt head takes to move in order to permit the camera to see all landmarks. The speed of the head is 300 deg/sec. In order to reduce the global time, the head starts to move during the motion of the vehicle from a path position to another. This is possible because the path positions are a-priori known.
Furthermore, no a specific configuration of the landmarks is requested for self-location, so the LEDs are placed in the environment in order to have their maximum visibility.

**Tabella 1: Path performed by SAURO. (f-camera = 16mm).**

| Distance from landmarks (mm) | Real robot position (mm) | Estimated Robot position (mm) |
|---|---|---|
| 8009 | x = -1146.25 | x = -658.53 |
| | z = 2069.79 | z = 2118.55 |
| 7173 | x = 0.92 | x = 420.91 |
| | z = -2372.99 | z = -2413.45 |
| 6126 | x = 1286.88 | x = 1582.78 |
| | z = -926.99 | z = -982.14 |
| 5290 | x = 2099.26 | x = 2310.07 |
| | z = -1125.61 | z = -1295.10 |
| 5034 | x = 2137.45 | x = 2292.57 |
| | z = -2491.85 | z = -2421.27 |
| 3730 | x = 3754.05 | x = 3792.12 |
| | z = -1135.4 | z = -1160.02 |
| 2504 | x = 4673.89 | x = 4716.37 |
| | z = -2432.5 | z = -2426.00 |

## 2.2 NATURAL LANDMARKS: SELECTION AND RECOGNITION.

The use of natural landmarks instead of artificial beacons has the advantage to manage unforseen situations such as landmark occlusion or conditions where the distortion due to perspective introduces large error in location estimation. Those situations can be managed considering more landmark setups placed at different locations in the environment. So, the vehicle can decide which setup, at each time, is more suitable to be referred to vehicle position estimation. On the other hand, several setups determine many modifications of the environment. The straightforward solution is to use natural landmarks. The use of natural landmarks presents problems related both to the world model construction and to the recognition of landmarks.

The main problem in world model construction is that information must be represented in a way compatible with the sensory data. When visual data are used the world model construction is a very hard task because visual information depends on the viewpoint.
In [10] problems on the world model representation and construction are analyzed. For our aim, in order to simplify the world model construction step,we consider a set of real junctions (a-priori chosen) as a world model. Some, among the junctions in the world model, are used as landmarks for robot navigation. The world model is represented as a graph where junctions are the nodes, and relations among the junctions are the edges. That kind of representation permits to formulate the landmark recognition as a graph matching problem. Since, we manage visual data, the graph representation must be invariant to perspective and to scale changes. Graph invariance depends on the invariance of properties of nodes and on the relations chosen among the nodes. The proposed approach consists of two steps: an off-line and an online step.
During the off-line step, the world model graph representation is performed.
In fact a set of junctions in the environment is chosen as the world model and for each one of them, the 3D coordinates (considering an appropriate world reference system) are determinated. The 3D coordinates could be estimated by several methods: through sophisticated CAD representations or by the use of ranging devices. We have used a theodolite-like device (NIKON D-50 total station), that determines the 3D coordinates, simply, pointing each landmark by a telescope. Then, an image of the whole landmark context is acquired and it is processed (as described in section 2) in order to identify the junctions, label them and determine their relations.
The output of the off-line step is the world model graph. The online step takes place during the route of the vehicle in order to recognize the landmarks, to recover their image coordinates and to determine the location of the vehicle. In fact, a similar graph representation of the world model, is built from an image acquired by the camera on the vehicle during the route. The landmark recognition is performed as a graph matching problem and it is approached using the standard framework of finding correspondences between graphs [11].

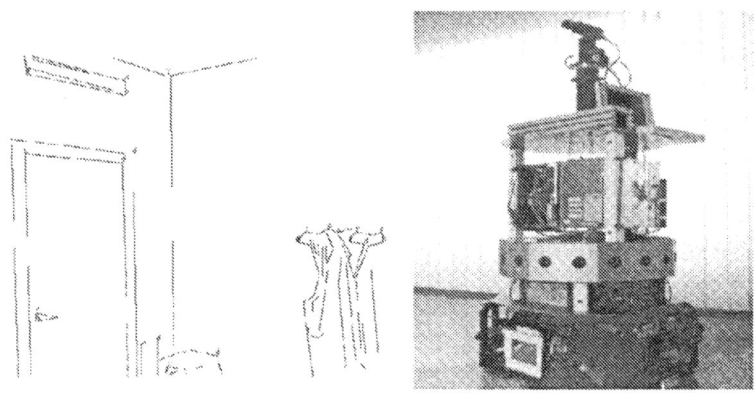

**Figura 6: a) Extraction and labelling of junctions. b) out vehicle SAURO**

The process to extract and to label junctions in an acquired image is similar to the one described in [12]. It consists of the following steps:

1. **Segment detection:** Canny's edge detection operator [13] is applied to the image in order to detect edges. Then, extracted edge are track-followed in order to convert raster edge into vectorial ones. A polygonal approximation [14] is applied to edge (vectorial form) to detect segments.

2. **Vanishing point detection:** the image coordinates of left,right and central vanishing points are determinated using the method described in [12].

3. **Junction labelling:** intersection points among segments converging to different vanishing points are considered junctions and are labelled into: "L","Y","T","X","V". For example, three segments converging each one to a different vanishing point identify a "Y" junction; two segments converging to two different vanishing points identify a "L" junction and so on. To simplify their management, the "L" junctions are divided into 4 categories ("a","b","c" and "d") depending on their orientation.

The output of junction extraction and labelling phase is an image in which each junction is properly labelled as shown in fig. 6a.
In order to construct a graph, it is important to define what is the nodes meaning and what are the relations among them. Junctions in our graph representation are the nodes. We have defined eight different types of nodes according to junction labels: "X","Y","T","V","a","b","c","d". The last four types identify the "L" junctions according to their orientation. The relations among nodes strongly depend on the use of the graph. In the proposed method, the graph must be used to recognize landmarks. Since,graphs are defined using information extracted from the images of the scene and visual information can change in terms of perspective and scale (depending on the viewpoint), the relations should be invariant to those changes.
The definition of invariant relations, like the definition of invariant measurements in terms cited in [15], is a difficult task. In order to overcome the problem, we have considered the so-called "partially" invariant relations. Namely, relations that can be considered invariant under *appropriate* changes of perspective and scale of the images.
In our approach we have considered three different kinds of relations:
UP-DOWN relation, RIGHT-LEFT relation and ADJACENCY relation. Considering two nodes $N_1$ and $N_2$ in the graph, whose related junctions have image coordinates, respectively, $(x_1,y_1)$ and $(x_2,y_2)$, the relations are so defined:

**UP-DOWN**

$N_1$ *UP-DOWN* $N_2$ *if* $y_1 >= y_2$

**RIGHT-LEFT**

$N_1$ *RIGHT-LEFT* $N_2$ *if* $x_1 >= x_2$

**ADJACENCY**

$N_1$ *ADJ* $N_2$ if the related junctions have a common segment.

Considering that our navigation context is *planar* in the sense that the vehicle moves on the ground and the TV camera is constrained to rotate around the $x$-axis (axis of the reference coordinate system parallel to the ground) and the $y$-axis (axis normal to the ground), while rotations around the $z$-axis (axis lined up to the optical axis of the TV camera) are not permitted. So, it is easy to demonstrate that the UP-DOWN relation is invariant to perspective and scale. While, it is not true for RIGHT-DOWN and ADJACENCY relations. In particular, ADJACENCY is strongly influenced by occlusion problems and sampling errors. In fact, common segments among junctions could disappear or could be divided into more parts.

The landmark recognition task has been formulated as a graph matching problem. In fact, similarly to the model, a graph can be constructed for the scene currently acquired by the robot TV camera. So, the recognition means to find the best subgraph to subgraph match between model and current graphs. A wide bibliography can be found on that problem [16]. An approach is to formulate the problem as a **maximum clique problem.** In fact, starting from two graphs a common graph, called the association graph (*AG*) can be built. A node, in the *AG*, is defined, if two nodes $N_1 \subset G_M$ and $N_2 \subset G_C$ have the same label (it means that corresponding junctions are of the same type). $G_M$ is the graph of the model and $G_C$ is the graph corresponding to the current image. Two nodes $A_1$ and $A_2$ in the *AG* are connected by an edge if the following condition is true:

let $N_1^1$ and $N_2^1$ be the nodes relative to $A_1$ and let $N_1^2$ and $N_2^2$ be the nodes relative to $A_2$. $A_1$ is connected to $A_2$

$$\textbf{if } \forall \textbf{ relation } R\textbf{:}\ N_1^1\ R\ N_1^2 \textbf{ and } N_2^1\ R\ N_2^2$$

The problem to find a subgraph-subgraph correspondence between two graphs can be formulated in terms of finding a **maximum clique** in the association graph. A **clique** is a subgraph full-connected in a graph. The cardinality of the clique is the number of its nodes. The **maximum clique** is the clique having maximum cardinality. The

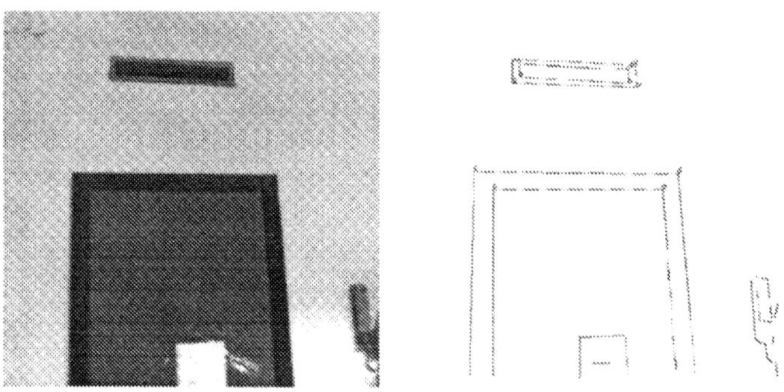

**Figura 7: a) Image of model. b) junctions identified on model.**

maximum clique problem has been extensively analyzed in literature from the theoretical point of view. A detailed review on the matter can be found in [11]. However, few references can be found in more specific computer vision literature on the formulation of feature matching problems in terms of graph matching. The first reference is the work described in [17] where authors formulated the problem to find correspondences between stereo images in terms of graph matching and resolved as a maximum clique problem. In [18], authors starting from [17] propose a method to speed up the original approach, however the proposed approach diverges from the graph matching framework towards a more classic hierarchical technique. The motivation for this lack of references is because the maximum clique finding problem is NP-hard: the techniques for finding the maximum clique are **enumerative** ones. They analyze exhaustively all cliques in a graph to determine the maximum one. So, the impracticability of computing time made the methods unable to manage applications where real-time is an important constraint. Recently, with the growth of computing power together with the introduction of optimization based methods for maximum clique problems, a new interest for graph matching formulation of real-time recognition problem is increasing. The optimization based methods [19,20] provide an approximated solution for maximum clique with a response time quasi-polinomyal and for many applications the approximate solution could be considered acceptable.

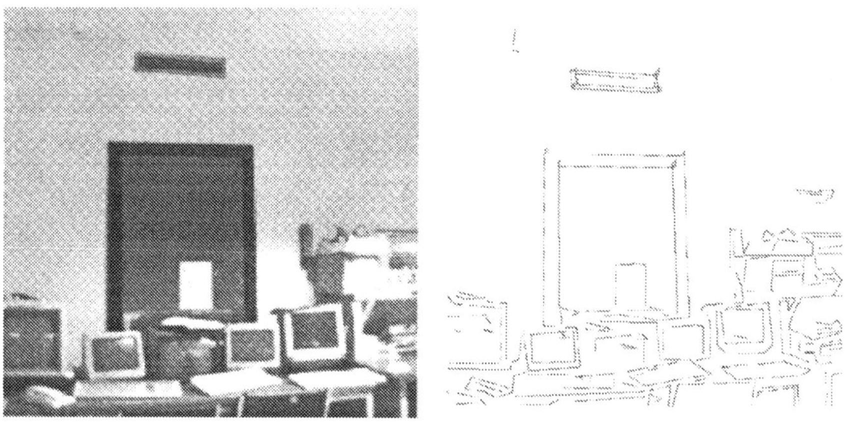

**Figura 8: a) Image acquired during the route of vehicle. b) matched junctions.**

During the world model construction, natural landmarks are chosen in the environment. For our experiments we have chosen to recognize as landmarks junctions relative to a door (fig. 7a). The 3D coordinates of landmarks are manually determined using a NIKON D-50 TOTAL STATION (theodolite-like device), while the graph model is built acquiring an image of the landmark context (fig. 7a). In fig. 7b junctions identified on image model are properly labelled. In fig. 8a, 9a, 10a three different views of the landmark context are shown. These views are acquired by the vehicle during its route. Fig. 8b, 9b, 10b show the results of the matching process. The recognized junctions are labelled on each image. Since, the used model (door) contains few junctions (about 20), both enumerative and optimization based techniques produce the same results in the same response time (<< 1 sec).

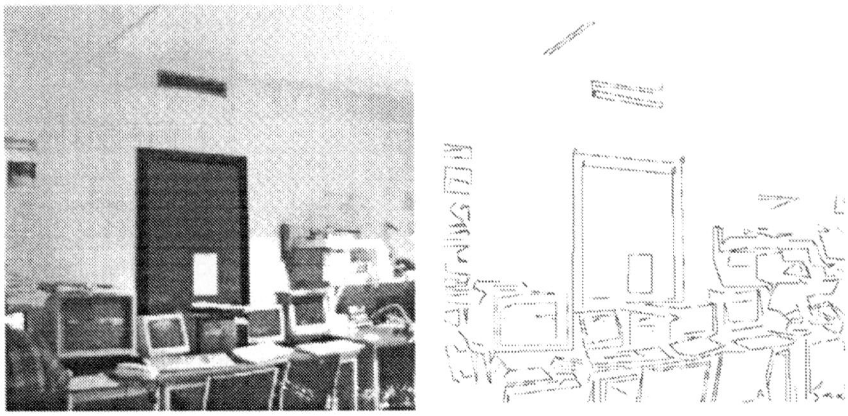

**Figura 9: a) Image acquired during the route of vehicle. b) matched junctions.**

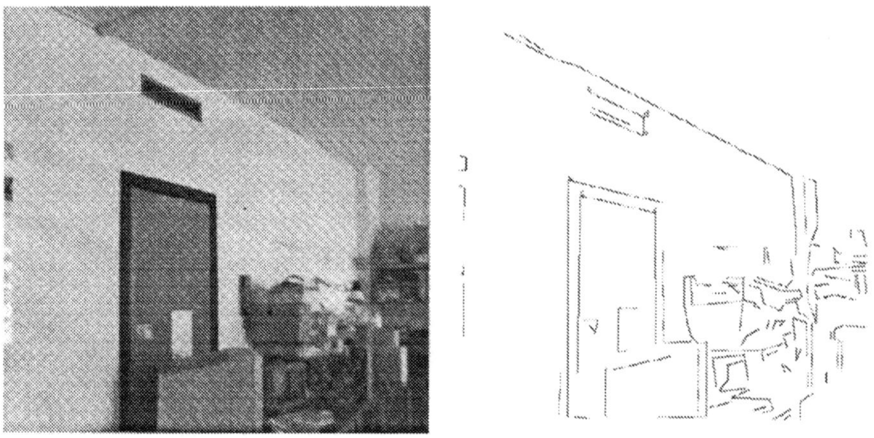

**Figura 10: a) Another image. b) Matched junctions.**

## 3. Relative visual navigation

Visual relative navigation is the capability to estimate the motion of an AMV respect to a known start position using image data. Approaches to relative navigation can be classified in two classes. **Photogrammetric approaches** are all techniques requiring a **camera calibration** task. Camera calibration is the process to estimate the intrinsic (focal length, scale factor, piercing point) and extrinsic (rotation and traslation)

parameters of one or more TV cameras matching 3D world points with their projection on image planes. In this class the **stereo navigation** methods are include [21,22,23]. These methods try to recover the translation and rotation of the AMV determining the rototranslation parameters of a set of 3D features estimated by a stereo pair of cameras placed on the vehicle. The difficulty of stereo navigation methods is twofold. Firstly the need to calibrate tv cameras. Secondly, the problem to estimate the motion parameters (rotation and translation) under stereo uncertainty [23]. Generally, the problem is formulated as a weighted least square problem whose solution can be recovered in close form only under specific constraints (such as gaussian modelling of stereo uncertainty) otherwise, it must be estimated by iterative standard technique with the known related problems (ill conditioning).
Another class of methods are the **optic flow** (OF) based ones. The estimation and interpretation of optic flow is a wide discussed topic of computer vision community. A good background about optic flow can be found in [24]. OF is a 2D vectorial field where each vector represents the apparent velocity of image pixels, originated by the motion of the TV camera. The basic problem in the intepretation of OF is to estimate the 3D camera motion parameters by 2D OF maps. The problem is ill-posed and can be resolved only imposing additional constraints.
Most of the applications related to AMV are interested to estimate more qualitative motion parameters such as **Heading** and **Time-To-Collision (TTC)**. Heading is the forward motion direction of the camera while TTC is the time to contact with the objects in the scene. The scene is considered static and the only moving thing is the AMV. Under additional but not limitative constraints (passive navigation and predominat forward translation motion) the estimation of those parameters is straightforward as described in the next sections.

### 3.1 ESTIMATION OF MOTION PARAMETERS: HEADING AND TIME-TO- CRASH

The idea regarding the estimation and interpretation of optic flow is the attempt to compute 3D translational motion parameters from the 2D optical flow obtained by projecting the 3D velocities on the image plane. Perspective projection models assume that a world point *P(X,Y,Z)* projects on the image point $(x,y)=f\left(\frac{X}{Z},\frac{Y}{Z}\right)$, where $f$ is the focal length. Longuet-Higgins and Prazdny [25] derived the following equations to describe the general rigid motion of an observer moving in a stationary world:

$$u = \frac{T_x - xT_z}{Z(x,y)} - xyR_x + (1+x^2)R_y - yR_z$$

$$v = \frac{T_y - yT_z}{Z(x,y)} - (1+y^2)R_x + xyR_y + xR_z \qquad \text{(EQ5)}$$

with $(T_x,T_y,T_z)$ the 3D translational velocity components, $(R_x,R_y,R_z)$ the 3D rotational velocity components, $(u,v)$ the projected velocity of a point $(X,Y,Z)$ on the image plane, $Z(x,y)$ the depth function.

Though various algebraic approaches have been proposed to resolve non-linear systems resulting from writing equations (EQ5).

in a suitable number of image points, the results are numerically instable due to the large number of equations to be solved and the noise in the $(u,v)$ velocity vector estimates.

We consider the application context in which the viewer translates on a flat ground and can rotate only around an axis orthogonal to the ground (**passive navigation**).

The resulting 2D motion field has a radial topology: on the image plane all 2D velocity vectors radiate from a singular point (that is the point where the flow vanishes) named focus of expansion (FOE).

The FOE is the projection on the image plane of the direction along which the observer moves. A rotation, occurring while the observer translates, will cause a FOE location shifting by preserving the radial shape.

The FOE location can be correctly estimated as the point where the translational motion $V_t$ vanishes:

$$V_t = \left( \frac{T_x - xT_z}{Z(x,y)}, \frac{T_y - yT_z}{Z(x,y)} \right) \qquad \text{(EQ6)}$$

hence,

$$(FOE_x, FOE_y) = \left( \frac{T_x}{T_z}, \frac{T_y}{T_z} \right) \qquad \text{(EQ7)}$$

If $T_z=0$ the FOE can be defined as a direction towards infinity defined by $(T_x,T_y)$ and all the flow vectors point in that direction.

Moreover, it can be shown that the time to contact defined as $TTC=\frac{Z}{T_z}$ can be easily computed from the OF in a small neighborhood of the FOE. From equations (EQ6) and (EQ7) we obtain the following equation for TTC estimation .

$$TTC = \frac{Z}{T_z} = \frac{FOE_x - x}{u} = \frac{FOE_y - y}{v} \qquad \text{(EQ8)}$$

The main problem in motion interpretation is due to the unknown depth $Z(x,y)$ depending on the different depth present in the scene. Though the FOE position is independent from the distances of world surfaces, the induced vector flow field depends on the unknown depths $Z(x,y)$ of the observed surfaces.

Generally this problem is overcome by making hypothesis of approximation with a planar surface. It is trivial to show that only the modulus of 2D velocity vectors depends on *Z(x,y)*, while the directions are independent.
3D motion interpretation (estimation of Heading and TTC) could be performed into two different ways. The first is to estimate FOE and TTC directly using a least-square or non linear optimization tecnique on optic flow vectors. In this case, a reliable optic flow map must be available. Alternatively, methods could try to estimate one between motion parameters (such as FOE) and use (EQ8) to determine the unknown one. The latter class can be associated the technique described in the next paragraph.

### 3.2 ESTIMATION OF FOE AND OPTIC FLOW

In our application context the resulting 2D motion field has a radial topology so the image plane all 2D velocity vectors radiate from FOE which is geometrically defined as the intersection of the heading direction of the TV camera with the image plane. A rotation, occurring while the observer translates, will cause a FOE location shifting on the image plane. FOE represents the main parameter to be estimated in planar passive navigation. Most of the techniques proposed in literature [26] for FOE estimation can be considered as a two level framework involving *2D motion field computation* [21], [22,28-34] through direct methods using local constraints or optimization methods using global constraints on such features and *FOE estimation*.
The FOE is the intersection point of all 2D displacement vectors computed for 2D sparse points in the image plane. Due to errors performed in estimating such 2D vectors, it is quite impossible for all intersection points to coincide with the FOE. In literature several methods have been proposed for FOE estimation [35-39], most of which use calibrated systems or require a preliminary decomposition of the optical flow field into its translational and rotational components, or propose to compute only a fuzzy region containing the FOE. Actually an accurate computation of FOE location from the 2D motion field seems to be a hard problem, mainly due to digitization errors, unreliable flow vectors and image noise. There exists a strict relation between the radial shape of optical flow generated by planar navigation and the FOE location: A correct 2D displacement vector estimates permit a correct FOE estimate, and information on FOE location can help in the 2D motion field computation.
Our idea is to consider 2D motion field computation and FOE estimation as a whole optimization process solved by minimizing an appropriate cost function containing constraints about radial shape and radiometric similarity. The input data are correspondences estimated by correlation for sparse features [21]. Optimal correspondences are estimated by updating iteratively the initial corresponding second image features until the energy function is minimized.
Once the energy function is minimized, all vectors will intersect in a single point which represents the FOE.
The estimation of FOE does not need a dense optic flow map, so displacement vectors are estimated only for features of "high" interest which are salient points that can be easily matched than others. These points generally occur at the corners and edges. We propose to use the interest operator introduced by Moravec to isolate points with

minimal autocorrelation values [21] in the first image of the sequence. The variance between neighboring pixels in *four* directions (vertical, horizontal and two diagonal) is computed over a window (the smallest value is called the interest operator value). Features are chosen where the interest measure has local maxima in windows of variable size. Moreover, constraints on relative distances between extracted high interest features are considered in order to obtain a spatial distribution on the whole image.

After the selection, raw matches, for each feature, are searched in the second image of the sequence using correlation similarity. A correlation similarity is maximized between windows centered on first image high variance features and second image candidate features. Actually, since we compute correspondences between regions, false matches are unavoidable: the correct match point can mismatch the center of the highest correlated window or several windows can have the same correlation.

Moreover, with a small amount of motion image features will scale up and correlation will fail because the patterns will enormously change. Matches computed through correlation can represent only an initial guess to improve through an optimization approach. In fact, raw matches are given as input to an optimization process to identify and correct bad measurements introduced by the correlation. The theoretical OF obtained by an image sequence, acquired by a TVcamera moving forward along its optical axis, has the typical radial shape with the FOE located at the same image coordinates of the principal point of the TV camera. Rotations occurring while the vehicle is translating around the axis normal to the ground plane do not affect the radial shape but the FOE location is not centered in TV camera principal point.

Our aim is to find the correct correspondence for each selected feature $p_i$ of the first image, starting from the initial guess $q_i$ estimated by the correlation in the second image, by solving an optimization process based on constraints related to **radiometric similarity** (each candidate match must meet a correlation similarity criteria) and **OF radial shape** (the correct matches must generate an OF map whose shape must be radial). The optimization problem is solved by minimizing an appropriate cost function through an iterative approach correcting the match positions in order to satisfy the imposed constraints. A radial OF is a map in which the direction of each flow vector (which represents a 2D velocity vector) seems to radiate from a common center (FOE). Let **i** and **j** be two flow vectors of the map and $\mathbf{P}_{ij}$ the image coordinates of the intersection point between the two straight lines representing the directions of each flow vector.

If two displacement vectors are correctly estimated their directions will intersect in the FOE location. Actually bad measurements generated from errors in match estimation and high uncertainty of the line interpolation give rise to a region of variable size where all intersection points are located. Feature locations in the second image must be updated iteratively in order to minimize the distance between all intersection points by always preserving the radiometric similarity between matched features.

Supposing that all intersection points $\mathbf{P}_{ij}$ have uniform probability distribution with mean

$$\mu = \sum_{i=1}^{N} \sum_{j=i+1}^{N} \frac{P_{ij}}{N_P} \qquad \text{(EQ9)}$$

where $N_P = \binom{N}{2}$ is the number of intersections generated by the directions of the $N$ available displacement vectors, and variance

$$\sigma^2 = \sum_{i=1}^{N} \sum_{j=i+1}^{N} \left(P_{ij} - \mu\right)^2 \qquad \text{(EQ10)}$$

To obtain a radial map means to vary the displacement vectors in order to minimize the variance $\sigma^2$ of their generated intersections. So, the matching process can be formulated as minimization of the following cost function:

$$E = \sum_{i=1}^{N} \sum_{j=i+1}^{N} \left(1 - S_i\right)\left(P_{ij} - \mu\right)^2 \qquad \text{(EQ11)}$$

where $S_i$ represents the similarity (correlation) between the selected feature in the first image $\mathbf{p_i}$ and the candidate match in the second image $\mathbf{q_i}$.

When $E$ is minimized the mean $\mu$ will represent the FOE. The norm $E$ will be minimized only when its partial derivatives with respect to all the points in the second image equal zero:

$$\frac{\partial E}{\partial q_i} = \left(1 - S_i\right) \sum_{j=1}^{N} \frac{\left(P_{ij} - \mu\right)}{\partial P_{ij} \partial q_i} = 0 \qquad \text{(EQ12)}$$

Satisfying the condition in (EQ12) for each of the $\mathbf{q_i}$ then a system of $N$ simultaneous equations in $N$ unknowns is generated. Thus, the solution which minimizes the squared norm of the difference vector $E$ amounts to find the set of points $\mathbf{q_i}$ so that the equation (EQ12) is solved.

Since the system (EQ12) is nonlinear we must use an iterative approach to compute the optimal solution. The approach we propose converges through iteration upon the desired correspondence points $\{\mathbf{q_i}\}$ by implementing gradient descent along the $E(\mathbf{q_i})$ surface, which expresses the quadratic cost function's dependency on all of the $\{\mathbf{q_i}\}$ points.

$$q_i = q_i + \frac{\partial E}{\partial q_i} \qquad \text{(EQ13)}$$

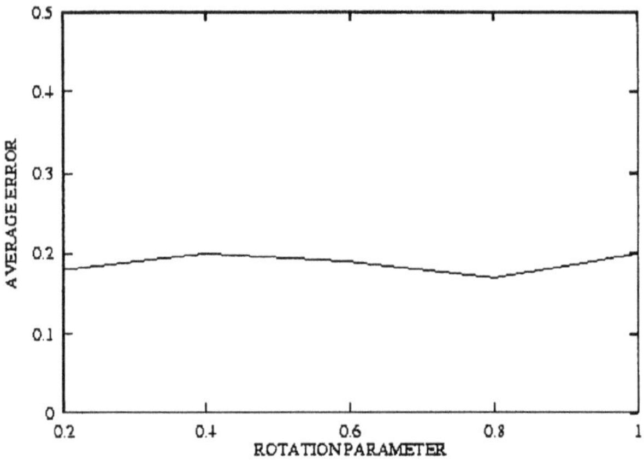


**Figura 11: Average error of the Heading estimated in the case of passive navigation varying the rotational velocity around Y-axis, computed on the theoretical flow fields with additional uniform noise with mean zero and variance 2.**

Starting from some approximate matches, the algorithm improves the solution until a predetermined convergence criterion is satisfied. The goal is to reject the noise introduced by the correlation measurements. Due to the nonlinearity of the system more than one solution can exist. The success to reach a global minimum, and not to be trapped in local minimum, depends on having a good first-guess for the solution. The algorithm will converge to an optimal solution if a number of input matches are correct. Correct measurements are not influenced from erroneous estimates, because the computed adapting signal control is zero due to the satisfaction of the imposed constraints. On the other hand, mismatches are influenced by correct matches determining the noise rejection.
Several experiments were performed on flow fields generated by setting randomly the HD and adding uniform noise with mean zero and variance 2, and varying the rotational parameter. All the results obtained are plotted in graphics in Fig. 11.
In order to verify the reliability and the performance of the proposed method, standard test image sequences [40] have been considered. In particular, we have used:
*div-tree sequence* (a synthetical image sequence where a TV camera is supposed to move forward along its optical axis against a tilted planar surface, TV camera focal length is 16 mm, the FOE is at the center of the image); *NASA sequence:* a real image sequence acquired in controlled indoor environment, motion is a primarily dilation, the forward speed is less than one pixel/frame; *rocket sequence:* a real outdoor image sequence, the sequence is acquired under a perturbed forward translational motion of the TV camera.
For the first sequence (div-tree) a reference true OF is available. In order to verify the performance of our technique an angular error between the true 2D velocity vector $\mathbf{v}_c = (x_c, y_c)$ and the estimated vector $\mathbf{v}_e = (x_e, y_e)$ has been computed as in [40]:

$$\phi_E = \arccos(v_c \bullet v_e) \qquad \text{(EQ14)}$$

So considering a pair of consecutive frames of the sequences, it is possible for each pair to estimate average angular error and its standard deviation. In figure (12e) a graphical representation of the average angular error and the standard deviation for the whole sequence is presented. In fig.(12a-b) two frames (the 1th and the 30th) of div-tree sequence are shown, while in fig. (12c) a typical OF map obtained by the correlation (used as a starting guess for optimization technique) between two frame of sequence is shown, while in fig. (12d) the corresponding final OF is drawn. In the matching process 30 features were selected.

For the other image sequences no reference flow is available, so, our resulting OF can be only compared with the OF obtained through correlation.

In figs.13a-b, 14a-b the first and the last frames of the NASA and rocket sequence are shown. In figs. 13c-d and 14c-d the OF obtained by correlation is compared with the OF estimated by our technique.

Finally, a set of experiments have been performed on images acquired in our laboratory with a TV camera mounted on a pan-tilt head installed on our vehicle SAURO.

The focal length of the TV camera is 6mm. In figs. 15a-b, 16a-b 17a-b images of the sequences acquired during a forward translational motion of SAURO are shown, and in figs. 15c-d, 16c , 17c the corresponding OF are shown.

All results seem to perform a good correction of the OF map obtained by correlation. The optimization process requires no more than 200 iterations to minimize the energy function. Though the proposed approach should recover the direction of motion in the presence of small amounts of rotations, in our experiments it is shown that even in the case of larger rotations shifting the FOE location far away from the image plane, correct estimates can be recovered anywhere.

## 4. Conclusions

In the chapter, two different methods to localize a mobile robot navigating in an indoor environment, are presented. The first method can be included among the so called absolute approaches that permit to estimate vehicle location into a world reference system. The method present a very low uncertainty level and it is suitable to perfom quantitative localization. An extension to select and to recognize natural landmarks is described making the self-location method more flexible to use in the context in which a real time performance is not required but a high precision is necessary. An alternative is of using a qualitative technique such as the one described, based on optic flow analysis. These kind of relative navigation methods provide a location information with an high level of uncertainty but this level can be reduced when the position estimation is quickly performed

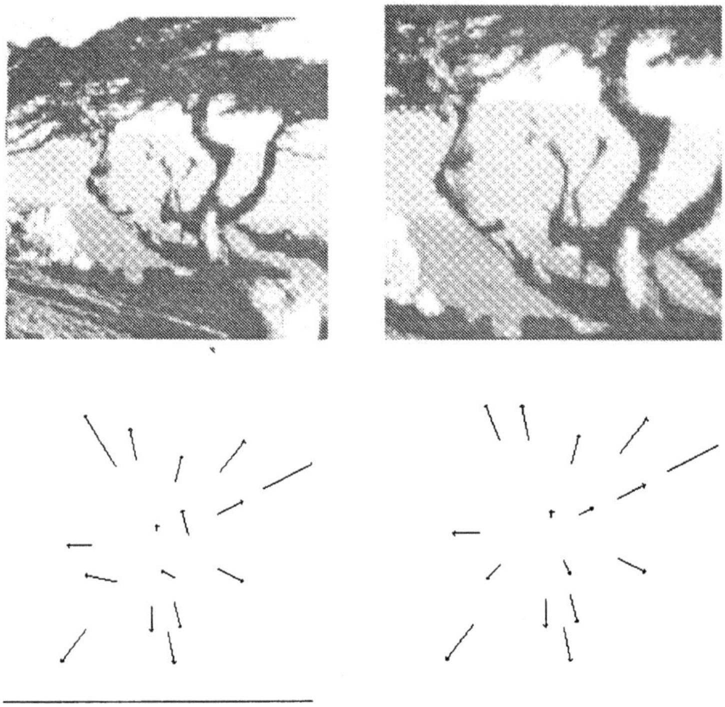

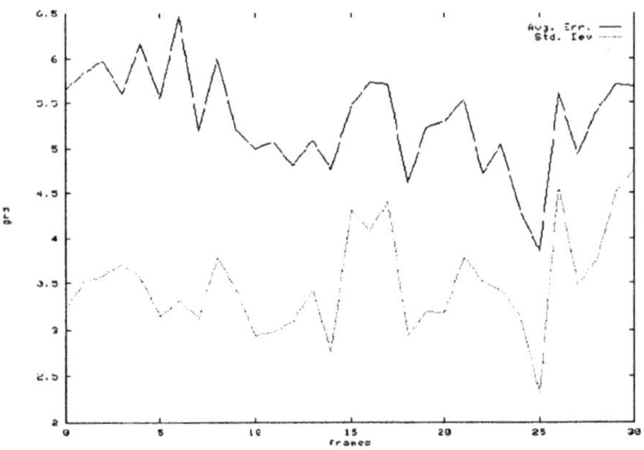


**Figura 12: Div-tree sequence: a) First frame. b) Last frame (30th) . c) OF by correlation. d) OF refined by optimization.**

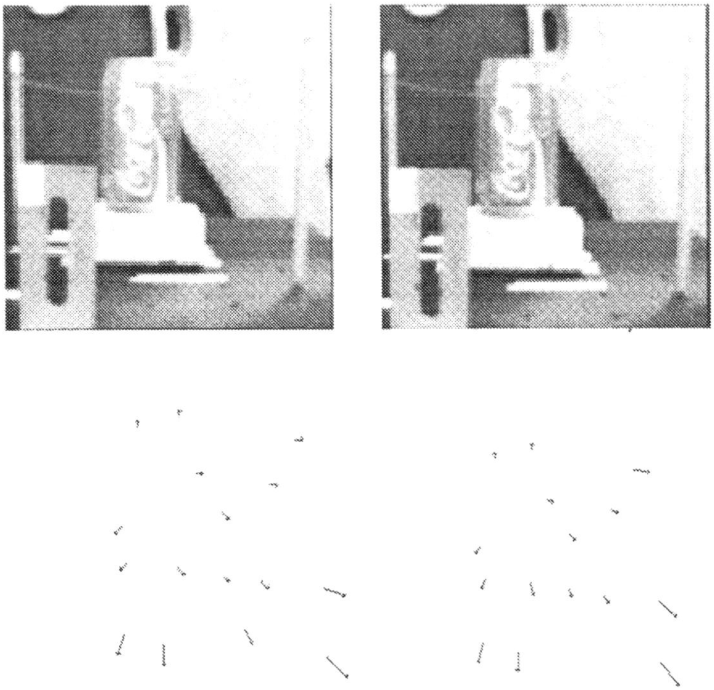

**Figura 13: NASA sequence: a) First frame. b) Last frame (20th). c) OF by correlation. d) OF refined by optimization.**

**Figura 14: Rocket sequence: a) First frame. b) last frame (20th). c) OF by correlation. d) OF refined by optimization.**

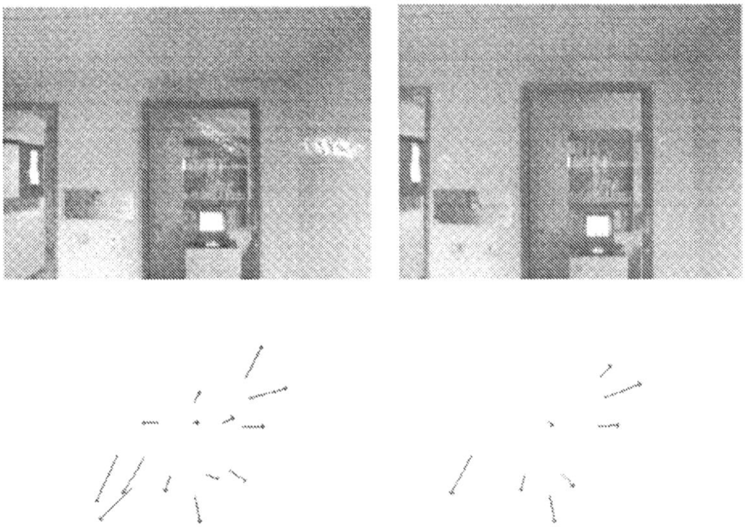

**Figura 15: SAURO sequence: a) First frame. b) last frame (20th). c) OF by correlation. d) OF refined by optimization.**

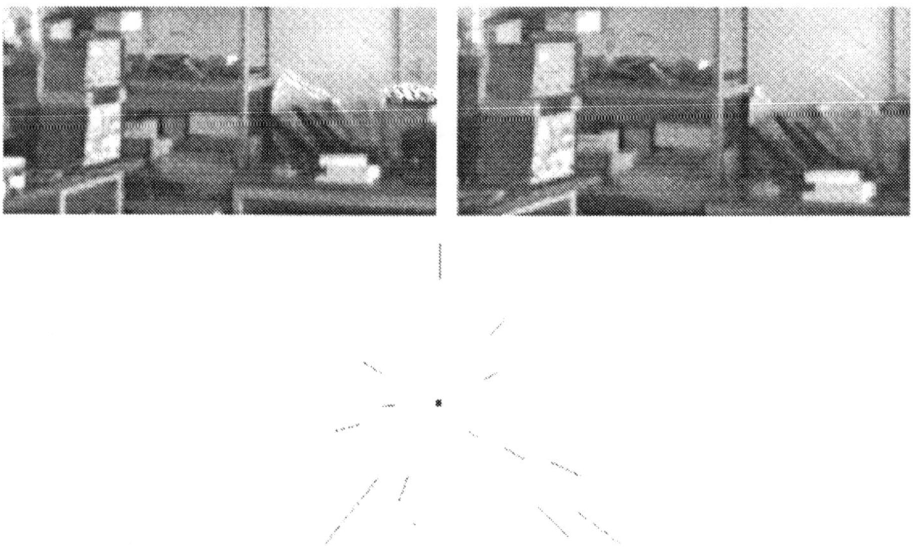

**Figura 16: Images have been acquired while the vehicle is traslating on rectilinear path; the distnce between frames is 500mm. Estimated FOE: x=10,y=15.**

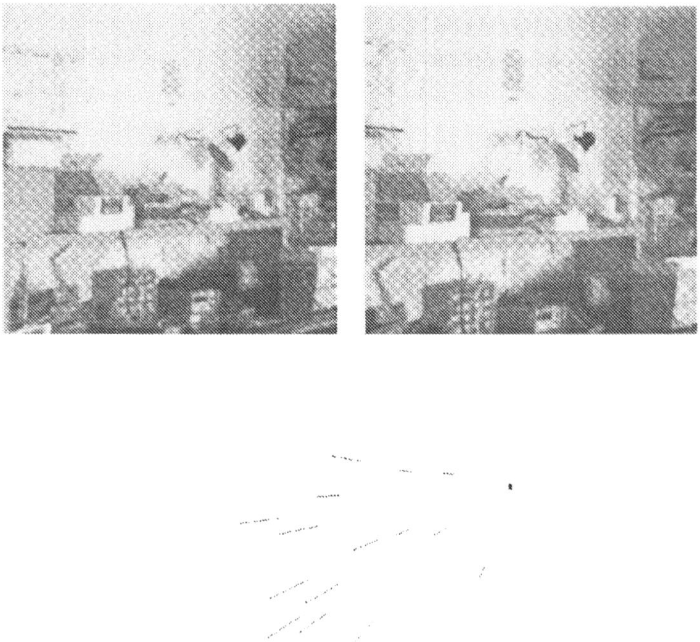

**Figura 17: Images have been acquired while the vehicle is moving on a curvilinear path; the distance between two frames is 500mm, the rotation angle is 1 deg. Estimated FOE: x=126, y=-1.**

and as much time as possible. The processing of visual data at an high rate is possible only using high performance hardware or specialized processors that today are not considered cheap. However the trend of development of technology is to make available a very powerful hardware with a low cost, so the use of visual techniques in IAS will become possible.

## References


1. Cameron, S., Probert,P. : *Advanced Guided Vehicles*, Wolrd Scientific, 1994.
2. Borestein, J., Everett, H.R., Feng, L. : *Navigating Mobile Robots : systems and techniques,* A.K. Peters LTD, 1996.
3. Stella, E., Distante, A. : Self-location of a mobile robot by estimation of camera parameters, *Journal of Robotics and Autonomous Systems - Elsevier,* **15** (1995)
4. Talluri, R, Aggarwal, J.K. : Position estimation techniques for autonomous mobile robots - a review, *Handbook of pattern Recognition and Computer Vision,* World Scientific, 1993.

5. Tsai, R.Y. : A versatile camera calibration technique for high-accuracy 3D machine vision metrology using off-the-shelf TV cameras and lenses', *IEEE J. Robotics Automat.,* vol. 3,No. **4**, (1987).
6. Weng, J., Cohen, P.,Herniou, M. : Camera Calibration with Distorsion Models and Accuracy Evaluation, *IEEE PAMI*, Vol.14, No.**10** (1992).
7. Lenz, R.K, Tsai, R.Y : Techniques for calibration of scale factor and image center for high accuracy 3D vision metrology, *Proc. IEEE Int. Conf. Robotics Automat.* (Raleigh, NC.), (1987).
8. Beardsley P.A., et al. : The correction of radial distorsion in images, *Technical report 1896/91*,Department of Engineering Science, Univ. of Oxford (UK) (1991).
9. Stella, E. : Visual Self location of mobile robot, *IESI-CNR Tech. Report RO20/94*, (1994).
10. Kosaka,A., Kak, A. : Fast Vision-guided mobile robot navigation using model-based reasoning and prediction of uncertainties, *Journal of Computer Vision, Graphics and Image Processing* : Image Understanding,Vol.56,no.**3**,(1992).
11. Paradalos, P.M., Xue, J. : The maximum clique problem, *Journal of Global Optimization*, **4** (1994).
12. Straforini, M., Coelho, C., Campani M.,Torre, V. : The Recovery and Understanding of a Line Drawing from Indoor Scenes, *IEEE PAMI*, Vol.14, no.**2**, (1992).
13. Canny, J. : A Computational Approach to Edge Detection, *IEEE PAMI*, vol. 8, no. **6**, (1989).
14. Ayache, N., Faverjon, B. : Efficient registration of stereo images by matching graph descriptions of edge segments, *Intern. Journ. Of Computer Vision*, **1** (1987).
15. Mundy, J.J., Zisserman, A. : *Geometric Invariance in computer vision*, MIT Press, 1992.
16. Ballard, B.D, Brown, C.M. : *Computer Vision*, Prentice Hall, Englewood Ciffs, NJ (1982)
17. Horaud, R., Skordas, T. : Stereo Correspondence through feature grouping and maximal clique, *IEEE PAMI*, **11** (1989).
18. Venkateswar, V., Chellappa, R. : Hierarchical Stereo and Motion correspondence using feature groupings, *Intern. Journ. of Computer Vision*, **15** (1995).
19. Pelillo, M., : Relaxation Labelling Networks for the maximum clique problem, *Journal of Artificial Neural Networks*, **2(4)** (1995).
20. Battiti, R., Protasi, M. : 'Reactive Local Search for the maximum clique problem, *ICSI- Berkeley, CA, TR-95-052,* (1995).
21. Moravec, H.P : The Stanford Cart and the CMU Rover, *Proc. IEEE*, **7(71),** (1983).
22. Ayache, N. : *Artificial Vision for Mobile Robots*, MIT Press , 1991.
23. Matthies, L. Shafer, S. : Error Modelling in stereo navigation, *IEEE Journal of Robotics and Automation,* **3(3)** (1987).
24. Sing, A. : *Optic Flow Computation,* IEEE Computer Society Press, 1991.
25. Longuet-Higgins, H.C., Prazdny, K. :The Interpretation of a Moving Retinal Image, *Proc. Roy. Soc. Lond. Ser. B 208* (1980).

26. Aloimonos, Y., Duric, Z. : Estimating the heading direction using normal flow, *Int. Journal of Computer Vision*, **13(1)** (1994).
27. Lee, J.J., Chang, J., Ho Ha, Y. : Stereo Correspondence using the Hopfield Neural Network of a new energy function, *Pattern Recognition*, **11(27)** (1994).
28. Clark J., Yuille, A. : *Data Fusion for sensory information processing systems*, Kluwer Academic Publishers, 1990
29. Pascual J.B., Starink, E. : Finding point correspondences using simulated annealing, *Pattern Recognition*, **2(28)** (1995).
30. Nasrabadi, N.M., Choo, C.Y. : Hopfield Network for Stereo Vision Correspondence, *IEEE Trans. Neural Networks*, **1(3)** (1992).
31. Zhou, Y.T., Chellappa, R. : A Neural Network for Motion Processing, *Neural Network Perception - Academic Press*, **1** (1992).
32. Ullman, S. : *The Interpretation of Visual Motion*, MIT Press,1979.
33. Convertino, G., Stella E., Branca, A., Distante A. : Optic flow estimation by an Hopfield neural network using geometrical constraint, Journal of *Machine Vision and Applications,* **10** (1997).
34. Marr, D., Poggio, T. : A computational theory of human stereo vision, *Proc. of Royal Society of London B .204.* (1979).
35. Burger, W., Bhanu, B. : Estimating 3D Egomotion from perspective image sequences, *IEEE PAMI,* **18(12)** (1990).
36. De Micheli, E., Torre, V., Uras, S. : The accuracy of the computation of optical flow and of the recovery of motion parameters, *IEEE PAMI,* 5(15) (1993).
37. Hummel, R., Sundareswaran, V. : Motion Parameter Estimation from Global Flow Field Data, *IEEE PAMI*, **5(15)** (1993).
38. Meyer, F.G. : Time to collision from first-order models of the motion field, *IEEE Trans. on Robotics and Autom.*, 6(10) (1994).
39. Prazdny, K. : Determining the Instantaneous Direction of Motion from Optical Flow Generated by a Curvilinearly Moving Observer, *Journal of Computer Vision, Graphics and Image processing*, **17** (1981).
40. Barron, J.L, Fleet, D.J., Beauchemin, S.S. : Performance of Optical Flow techniques, *Intern. Journal of Computer Vision*, **1(12)** (1994).

# 10

# MAPPING AND NAVIGATING IN TIME-VARYING OBSTACLE FIELDS

RAY JARVIS
*Intelligent Robotics Research Centre*
*Monash University, Wellington Road, Clayton*
*Victoria, Australia*

## 1. Introduction

Navigation schemes for autonomous mobile robots are critically dependent, in terms of functionality and robustness, on appropriately engineered and managed tight couplings between localisation, environmental mapping and path planning/execution sub-system components [ See Figure 1]. Various modalities of this scenario require more complex instrumentation and algorithmic considerations according to the degree of uncertainty which must be accommodated.

In situations where a detailed map of an unchanging environment is available initially, both localisation and path planning/execution are relatively simple since prior site preparation is usually acceptable in such cases. For example, under floor signal wires or on floor markings can be used to guide a robotic vehicle along near optimal pathways which have been determined in advance, the only decision points being at junctions of the path structure network. Simple odometry may suffice as the means of determining robot position since the intrinsic accumulative errors of such a localisation mode can be corrected at the junctions.

At the most complex end of the navigation mode spectrum, initial ignorance of the environmental map together with uncertainty arising from unknown time-varying aspects of the obstacle field require both sophisticated sensory instrumentation and algorithmic structures to cope.

In what follows various navigation modes from the simplest to the most complex will be introduced together with the instrumentation and algorithmic strategies needed to deal with them. A number of experimental case studies will be used to illustrate the methodology described. Before launching into the navigation problem itself various types of instrumentation and algorithmic tools currently available to support such tasks will be introduced under the classifications, localisation, environmental mapping and path planning.

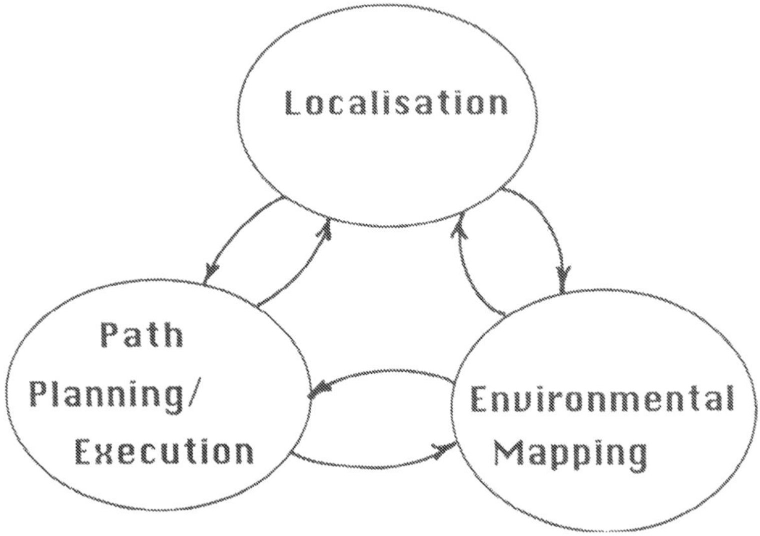


**Figure 1. Autonomous Navigation Sub-System Components**

**Figure 2. Odometry Wheel System**

## 2. Localisation

This term applies to the determination of the position and orientation of the robot relative to its working environment. In the simplest case this could be an (X, Y, θ) triplet in the context of a cartesian co-ordinate system encompassing the working environment.

The most direct way of knowing where a robot is and which way it is pointing is to restrict its movements along predetermined path segments specified by under-floor signal wiring (signals being detected electromagnetically) or on-floor markings (detected visually or magnetically, for example). Simple and reliable though such schemes are, little or no variation is possible without considerable inconvenience and/or expense. However, in some factory settings, such arrangements may be perfectly acceptable, given that factory layout changes are expensive in themselves and are usually avoided if possible.

Some degree of 'free-range' localisation can be accomplished using odometry wheels. Shaft encoders on a pair of wheels, one on each side of the robot, can supply distance travelled by averaging the rotation angle counts and orientation by accumulating rotation angle differences, but both measures are subject to accumulative errors due to surface undulation, wheel slippage, non-circularity of the wheels, changes in effective radius due to load shifts and inappropriate placement on the robot. All of these sources of error can be partially obviated. Ideally, the wheels should be large to minimise ground friction and improve smooth traversal of rough terrain, knife-edged to avoid effective radius change on angle shift (relative to ground), perfectly circular, unloaded to avoid distortion, and not be drive wheels (to avoid slippage on acceleration). In fact, if all these matters are carefully attended to, considerable accuracy can be achieved; ultimately, however, accumulative errors will become intolerable and some resetting/calibration will be needed (periodically). A reasonably accurate example of an odometry system is shown in Figure 2.

Absolute localisation can be achieved using beacons. These can be on-board, but are more usually at known positions in the environment. In indoor situations both ultrasonics [1] and optical [2] beacon systems can be used. Outdoor beacon systems include optical, radar [3] and GPS (Global Positioning Systems - based on satellites) types. From a geometric point of view, the fixing of a point relative to known beacons requires one distance and one angle, two angles, or two distances [ See Figure 3]. If more than a minimal set of such measurements are known, this redundancy can be exploited to improve the accuracy/reliability of the system by a combination of selecting the most reliable subset of measurements and averaging over competing estimations.

An optical (infra-red) rotating scanning bar code reader localiser made by Denning is shown in Figure 4. It is capable of identifying and measuring the bearing angles to up for ten bar code cards (approx. 20cm × 20cm) ten times per second. According to various strategies that can be applied to exploit the redundant information available, quite accurate localisation can be provided by such an instrument at the rate of 10Hz.

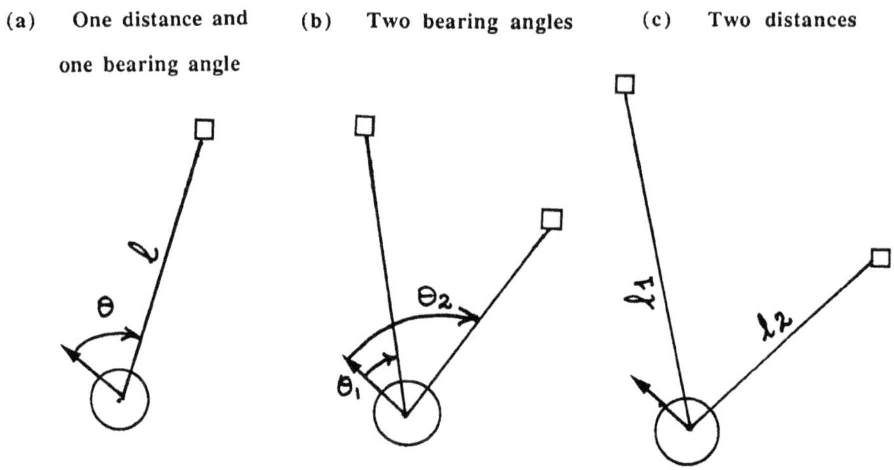


Figure 3. Beacon Localisation Geometries

Figure 4. Denning Bar Code Scanning Localiser

A more novel localisation scheme is shown in Figure 5. It consists of a differential ultrasonic panning head system designed to carry a video camera to follow a target identified by an ultrasound source. The panning head is fitted with a shaft encoder to indicate its position when the signal strength at the two receiver is approximately equalised. However, due to stiction, insensitivity and multipath possibilities this angle does not provide and accurate bearing to the target. If the ultrasound target is accompanied by an unambiguous, easily detected target like an illuminated globe, the analysis of the image provided by the camera on the panning head can provide the lower order bits on bearing angle as well as a good indication of range, which is inversely proportional to the square root of the area of the globe target. Thus only one localisation stations is needed (either on the robot or at a known fixed position) provided that the line of sight contact can be maintained. Should the visual target be monetarily lost due to it moving behind an obstacle, it can be recovered once it reappears and the failure will at least be known, since the lack of the visual target in the camera image will be easily detectable.

For outdoor localisation GPS (Global Positioning Systems) technology is now readily available at various levels of accuracy. Due to a deliberately introduced error (for security reasons) a single inexpensive (< $400) hand held unit, suitable for casual recreational use, can give a fix on the ground accurate to 50 metres r.m.s. (root mean squared) with at least three satellites in line of sight and can provide elevation data with at least four satellites in view. The effect of the deliberately introduced error can be minimised by using what is called 'differential mode'. A receiving station at a known, fixed location calculates an error vector of differences between the known distances to visible satellites (given in almanac data) and measured distances using the error prone signals themselves. This error vector is broadcast to receivers attached (or built-in) to mobile GPS units which use these signals to partially correct their estimated positions. Accuracies of ~ 10 metres r.m.s. are provided in this way. For many fairly open terrain operations (and open waterways for boats) this is sufficiently accurate for autonomous navigation purposes. However, care must be taken to provide back-up when an insufficient number of satellites are visible or other outages occur. The GPS system is very convenient when it is functioning correctly but can not be relied upon as the only source of positioning data. An even more accurate GPS mode is available, but is currently quite expensive (> $20,000). This is the carrier phase mode which uses the differentially corrected data as a starting point to refine position information by measuring phase shifts in the carrier waves of satellite signals using a complex iterative algorithm to resolve the 'phase window' ambiguity. Such systems can return a 1cm r.m.s. result after a number of minutes of operation. Often such systems are teamed up with INS [4] (Inertial Navigation Systems) to cover for 'outages' in GPS fixes.

For both indoor and outdoor localisation one of the most attractive approaches is to use natural landmarks as beacons, thus obviating the need for prior site preparation or the need of external systems such as G.P.S. (which of course would not be available on Mars). Whilst there are many ways of using natural landmarks, one can list a number of properties which can render them more useful for localisation purposes. In general they should be easily detectable, geometrically well distributed in terms of triangulation principles, unique, temporally stable, vertically sharp and viewpoint invariable in appearance. A good counter example is a colony of penguins. As visual and range sensing equipment becomes more sophisticated, accurate, reliable and inexpensive, the natural landmark mode of localisation will become more and more attractive, perhaps eventually eclipsing other methodologies in time. That less accuracy can be tolerated in sparce obstacle strewn environments where there are fewer natural landmarks and more accuracy required in cluttered environments where more landmarks are available to provided it is a very useful property of this modality of localisation.

## 3. Environmental Mapping

Knowing where everything is in a working environment not only provides a good basis for natural landmark based navigation but also provides the essential information for path planning and obstacle avoidance. This ideal can be met in full only when complete and unchangeable mapping data is provided beforehand or where the first task of the robot is to do a complete, once only, mapping for use in subsequent navigation tasks. What is more likely is that an initial map may not be available and some aspects can change (perhaps slowly). The hardest situation to manage is where the environment is initially unknown and subject to unpredictable and perhaps high speed variations (e.g. moving uncontrolled obstacles present). In all cases other than when complete mapping data is provided prior to navigation runs, vision and/or range sensors are needed to cope with the situation.

It is of interest to point out that the 'reaction' style of robot navigation championed by Rod Brooks of M.I.T. [5], in which a stimulus/response system constructed out of finite state machine elements is used for navigation without formal 'planning', does not use environmental maps in the normal meaning of that phrase. The 'planning' approach, on the other hand [6], does aim to have a complete map to reason about. As both approaches have matured conceptually and in experimental practise, it would seem that a 'hybrid' approach with both high level 'planning' and low level 'reaction' would be better than either, alone, in general. Sensors play a vital role for each, in any case.

Sensors used for 3D measurement and interpretation can be divided into three pairs of dichotomies [7] depending upon whether they are image based or not, whether they are active or not and whether they are single or multiple viewpoint systems. Figure 6 illustrates this classification scheme. For the purpose of this chapter, passive streopsis and active time-or-flight ranging methodologies will be chosen as examples of the types of sensors often used for environmental mapping purposes.

There are two forms of passive stereopsis [8], lateral and temporal. The lateral stereopsis [9] form exploits the inverse range relationship with image disparity between two or more images (acquired simultaneously) from cameras on base lines perpendicular to the ranging dimension. Temporal stereopsis [10] (often associated with optical flow analysis) exploits the image disparities evident in the looming out from the centre in images taken as a time sequence as a single camera moves along the ranging dimension into the scene. It is also feasible to combine both methods [11]. Optical flow studies [12] can extract this type of information, but usually at a very local intensity gradient related level. Larger scale processes of selecting features and then matching them across

Figure 5. Ultrasound/Vision Based Localiser

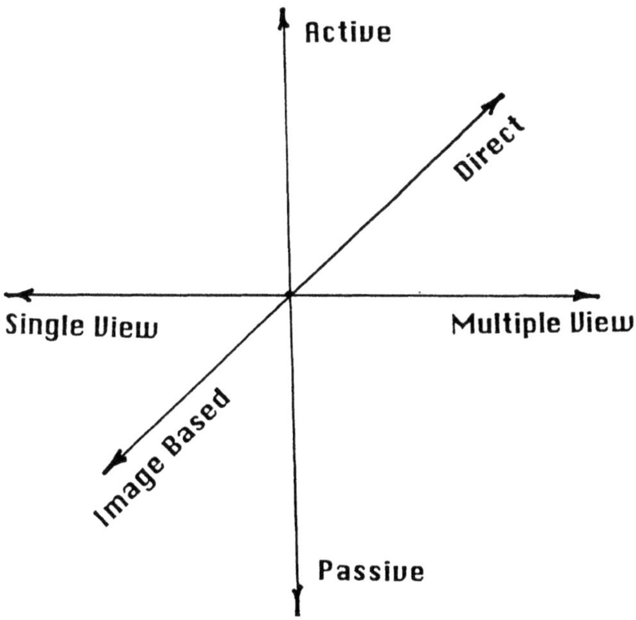


Figure 6. Range Sensor Classification

images are common for both modes of stereopsis and tend to be both computationally intense operations as well as suffer from mismatches and occlusion effects. Despite the intrinsic ambiguity of passive stereopsis ranging, this approach enjoys the advantages of low cost, low energy and eye-safety for a very wide range of applications.

Special systems [13] have been built to realised stereopsis based ranging at sufficiently high speed to render the method applicable to many robotic guidance tasks and particularly mobile robot navigation.

Active ranging methods in the time-of-flight class include ultrasonic, laser and radar systems, each with their own advantages and weaknesses. Ultrasonic ranging is relatively inexpensive but can suffer from specularity and wide beam width problems and is also limited to ranges up to approximately 15 metres due to severe signal attenuation in air. Sophisticated systems [14] which incorporate device physics into the analysis process are able to achieve high distance and bearing accuracy and can also categorise man made surfaces as being convex, concave or planar. A large number of mobile robot researchers use ultrasonics as their dominant ranging mode despite the short comings of most systems currently in use. In the laser time-of-flight range finder category a number of sophisticated but expensive systems are now available.For example,a system made in Germany (Dornier Ladar System) and costing approximately DM 80000, is capable of returning a 63x64 range image at 2 to 4 Hz rate for ranges between 8 and 100 metres.Radar ranging systems capable of ranging from 1/4 to 48 nautical miles are readily available for boating applications and can be purchased at a cost of between approximately 5 to 20 thousand dollars.

Less expensive and now quite widespread in use is the Erwin Sick scanning range finder(~ $7,000) which is capable of ranging up to 50 metres with a ± 3cm accuracy at 1/2 degree intervals over a 180 degree scan in the plane (usually horizontal) in two seconds. A unit itself and a single horizontal plane range profile are shown in Figure 7. An outdoor version capable of working in the rain but covering only 100° in sweep angle is available at a higher price. The Erwin Sick scanning range finders have iproven themselves very reliable and robust for autonomous mobile robot navigation support in recent years and have allowed quite reliable and simple systems to be demonstrated [15]; some of these are now commercially viable for general purpose application in cleaning, surveillance and healthcare.

## 4. Path Planning

Dealing first with a simple, unchanging 2D floor plan map based situation, clearly there are walls and obstacles to be avoided and clear areas which are navigable. In the 'planning' approach, optimal paths from a starting to a goal position are usually sought. Optimality is usually measured in terms of least path length, least time trajectory or least energy consumed units, sometimes all three being met at once. Whilst feasible rather than optimal paths are usually acceptable, particularly when safety and low computational complexity are factored in, optimality is a good starting focus when deriving robust algorithms to do planning. For a holonomic mobile robot configuration, a valid path consists of a set of (x, y, $\theta$) triplets through unoccupied (free) space. The working environment can be treated as continuous or tessellated.

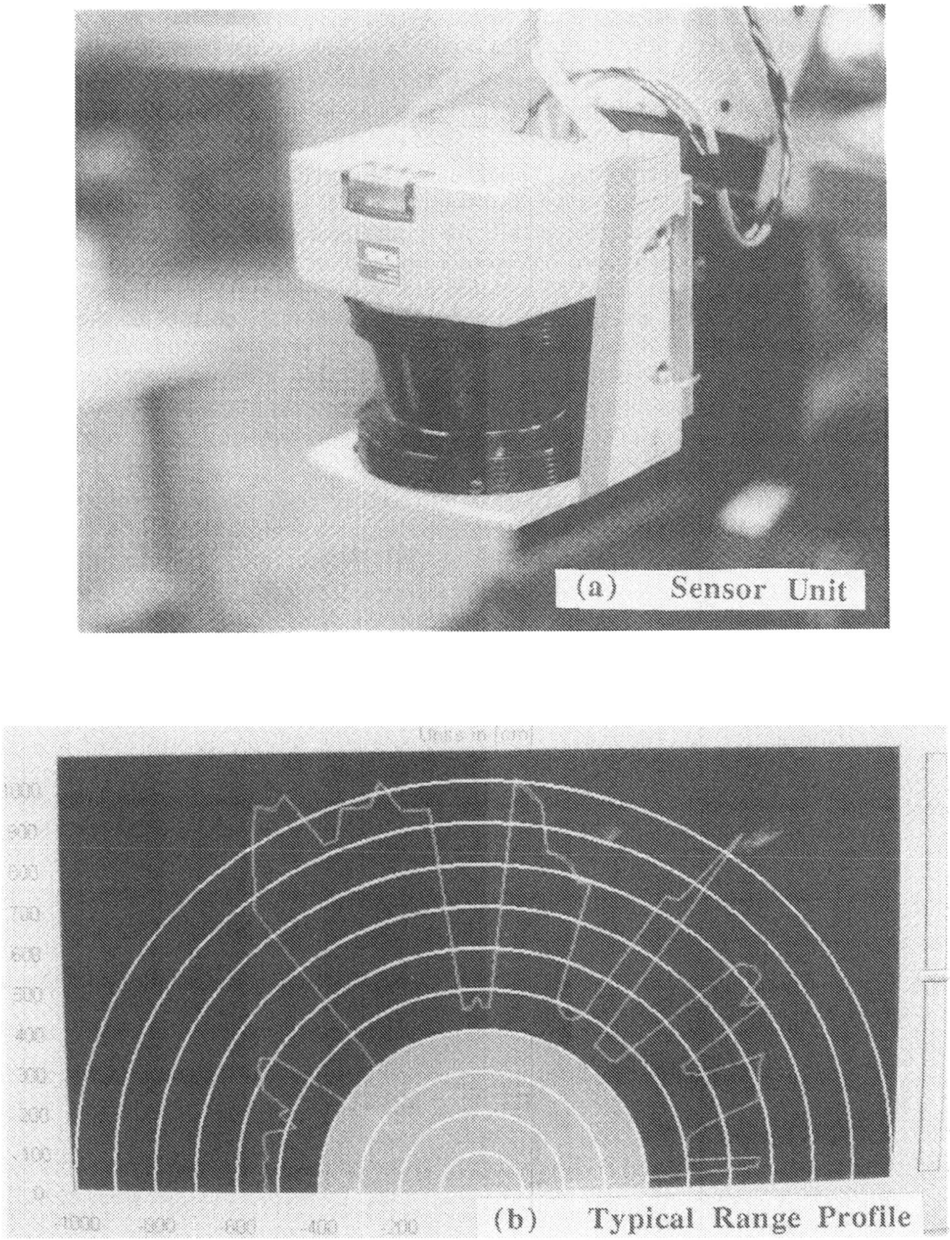


Figure 7. Erwin Sick Active Laser Time-of-Flight Range Scanner

A now classical approach to shortest path optimal path planning to apply the A * [16] search algorithm to the task, formulated in its simplest form as a point path through an obstacle field made up of convex polygonal objects in continuous space [17]. Figure 8 illustrates such a path. If the robot vehicle is itself represented as a convex polygon with a nominated reference point, the problem for a translation only (no rotation) solution, can be transformed to the point path solution by 'growing' the obstacles to compensate for the physical extent of the vehicle [18] . If both translation and rotation are considered, a set of grown obstacle fields, each for one orientation of the robot can be constructed typically for a quantised set of orientations from 0° to 360°. Moving through from one grown obstacle map to another represents pure rotation, moving in one map pure translation. Although continuity is compromised with this arrangement and transitions between adjacent maps must be allowed for (by defining a vehicle which includes the swept area over the rotation interval), acceptable paths can be readily computed. Elegant though this 'classical' approach is, its complexity growth over the number of vertices amongst the obstacles is very rapid [18] and this renders the approach computationally intractable in complex environments.

A tessellated space version of the optimal path planning problem allows complexity control at the cost of resolution. Using a regular rectangular tessellated space, various 'distance transform' methods derived from image analysis formulations [19] can be easily implemented. Details are provided in [20]. Whereas the original image processing formulations for the DT (distance transform) used the propagation of distance in integer steps inwards from the border of an image object to capture aspects of its shape, the path planning version grows distances outward from the goal (zero distance from itself) out through all reachable free space, flowing around obstacles. Multiple passes are necessary to infiltrate complex non-convex obstacle regions. Details are given in [20]. Region growing to allow for the physical extent of the robot can be carried out as for the A* approach, except in the quantised form imposed by the tessellation structure. The DT approach extends naturally to higher dimensional spaces to handle as many degrees of freedom as one wishes and thus can be applied to robot manipulator path planning [21] as well. Examples of the application of the DT algorithm are shown in Figure 9 for completely known and unchanging environments.

A variant of the DT method can be demonstrated in Delaunney tessellated space [22]. A quadtree structure [23] can also be adopted.

In both the rectangularity tessellated and Delaunney tessellated versions the DT first generates a distance function thoughtout free space. Then any steepest descent path from any point in free space to a goal represents a quantized minimal distance trajectory. The method is free of local entrappment unlike many potential field methods and can easily accommodate multiple start points and multiple goals as well as time/space paths, exploration tours and 'visit all' free cell modes with ease [20].

Not all free space need be equally accessible. In fact, entry into a cell from any of its eight neighbours could have its own cost associated with it. For example, a marshy or rough tract could be regarded as more heavy going than a dry, smooth one, the former areas being associated with higher costs. One way doors and traffic direction can be handled this way. By associating each cell with the eight costs of entry from its

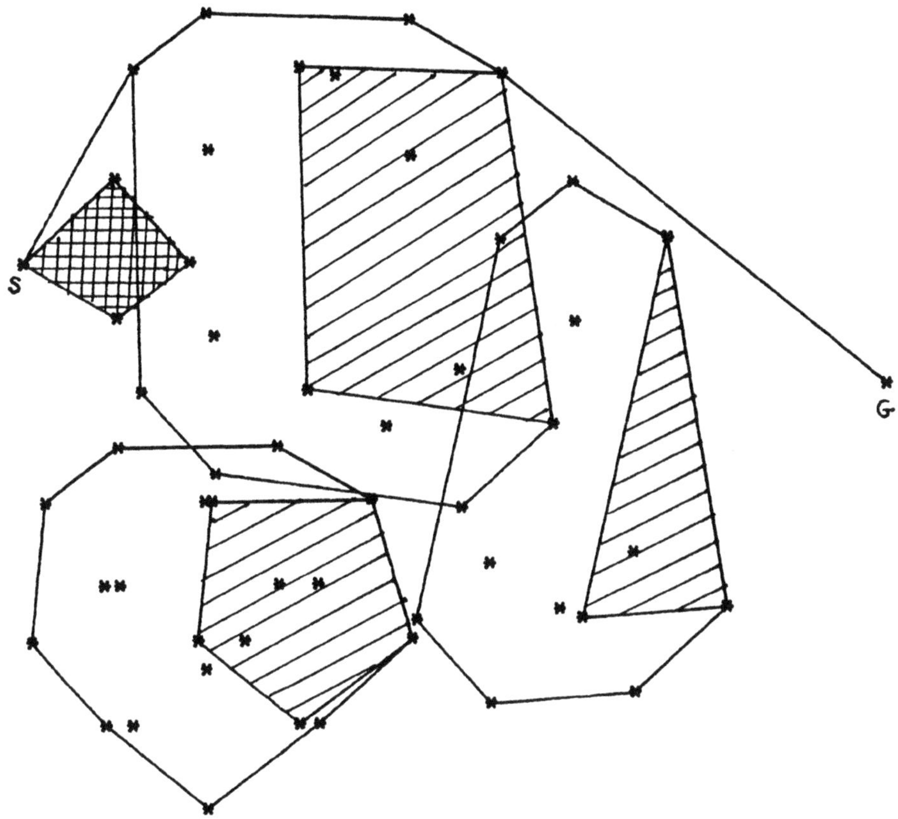


Figure 8. A* Based Optimal Path Planning
(showing grown obstacle field)

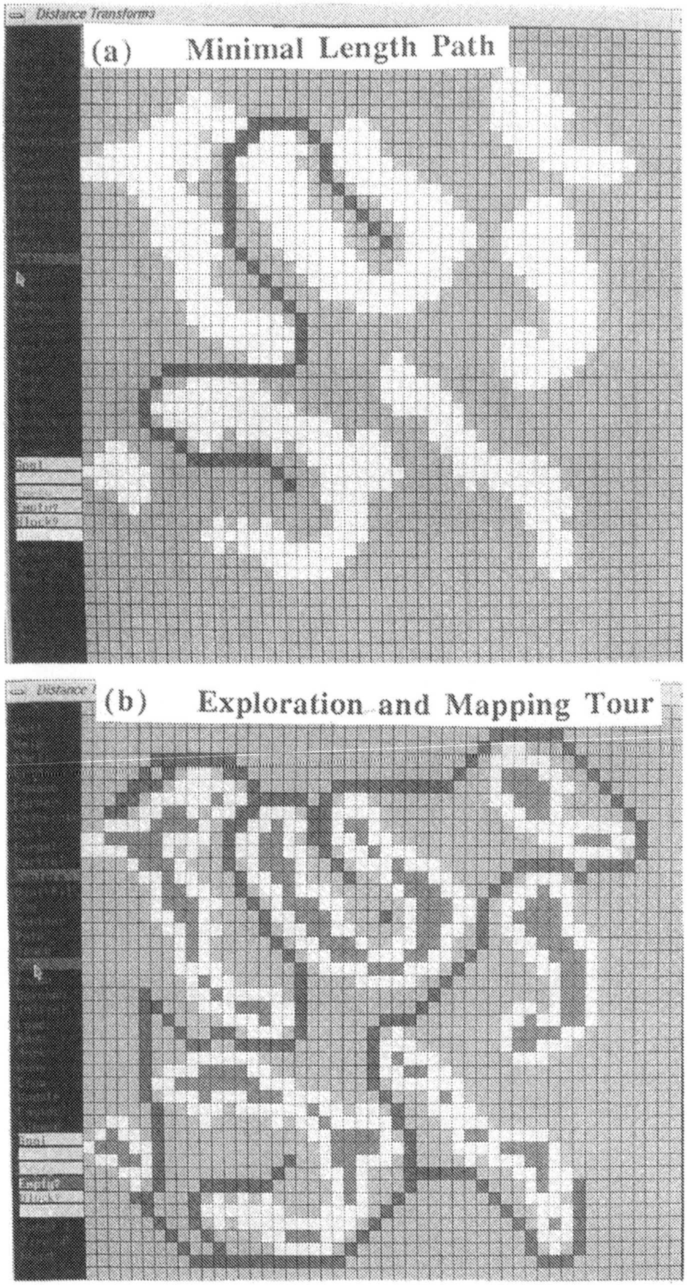


**Figure 9. Distance Transform Based Path Planning in Known and Unchanging Environments**

neighbouring cells the DT method can accommodate costs other than distance in a very simple way.

There are many other path planning methodologies [24,25,26] but what has been presented is sufficient to provide the background for the remainder of this chapter which deals with the whole navigation problem in ever increasing difficulty modes of uncertainty.

## 5. Autonomous Mobile Robot Navigation

A complexity hierarchy of autonomous navigation problems can be constructed where the complexity corresponds to degrees of uncertainty concerning the environment. Various nodes in the hierarchy will be dealt with, one at a time, in what follows. The appropriate DT variant will be used as a representative path planning method for illustrative purposes.

*Known Stationary Environments*

The simplest case is where a map explicitly indicating, in a regularly rectangular grid of appropriate resolution (after a cell the size of the robot is convenient to use), which cells are filled (forbidden or occupied) and which are empty (free) is provided. The obstacles, including the surrounding walls, are grown to account for the robot size (one cell more or less in the case where the cell size is that of the robot, or two be absolutely sure at the corners). A goal is specified, a DT algorithm propagates distance through connected free space and the steepest descent path from the starting position of the robot specifies the optimal (quantised shortest distance) path. The path is followed using some localisation scheme to check the trajectory. No environmental mapping sensors are needed unless natural landmark based localisation is being used. A smoothing of the path may be used if appropriate. If a safer, if less optimal (in shortest distance terms), path is required, access to free cells may be specified as more expensive in proportion to closeness to obstacle borders. In fact, the appropriate cost values can be generated by subtracting the distance from obstacle edges from a specified maximum value and resetting negative values to 1 (simple distance cost). The distances from obstacle edges may themselves be calculated by applying the DT outwards from all obstacle cells, each set as pseudo-goals ($\phi$), using the standard algorithm.

*Initially Unknown but Stationary Environments*

If it is known that an environment will not change but a map is not available, a mobile robot provided with appropriate localisation and environmental mapping sensors can complete a full mapping (within reactability or viewability constraints) before undertaking particular point to point tasks. The length of time taken for the mapping itself may not be significant, since it is a once off procedure which can produce results which are used for some considerable forward period of time during which no changes occur in the environment. If beacon localisation and range sensing (e.g. Erwin Sick scanning laser time-of-flight sensor) is used, the knowledge of the position and orientation of the robot allows observable obstacle edges to be marked in as forbidden cells in an initially blank map. An exploratory tour needs be taken which incrementally allows all free and occupied edge cells to be observed; the insides of obstacles need not be mapped since they are made up of inaccessible cells. Planning such a tour is easy using the standard DT algorithms with all unknown cells set initially as goal cells which become replaced by filled or free cells as they are observed by the range-finder.

All parts reachable from the start point of the tour will be mapped in this way. A thorough but not necessarily optimal mapping tour will result automatically. Once this is complete, continuing navigation tasks can take place in the now known environment.

*Initially Unknown but Possibly Slow Time-Varying Environments*

As an alternative to the strategy outline in the previous case, since the possibility of slow changes may render futile the notion of doing a complete exploratory tour (since the resulting map may be of only temporary validity), one may proceed under the assumption that all unknown space if free until proven otherwise through observation. This is known as the 'optimistic' assumption The standard DT algorithm is applied with forward planning, incremental movement and piecewise environmental mapping interleaved. Mapped obstacles cells may be retained until observed to be free again or be allowed to fade to the unknown category after a specified period of time, i.e. by the introduction of a 'forgetting factor' Point to point tasks can take place under such conditions with just a small amount of complexity added, as outlined above. The fade out to unknown category can be done gradually by incrementally lowering the cost of entering such a cell (during periods when it cannot be observed). Any ambiguity which arises from such procedures will be resolved when the robot revisits positions from which can observe the current state of affairs.

Again, strict path optimality cannot be guaranteed in the face of uncertainty of cell occupancy, but in most part acceptable and reliable behaviour results.

*Initially Known but Perfectly PredictableTime-Varying Environment* [27]

When all environmental changes are known exactly into the future, the DT algorithm can be used for optimal time/space trajectory planning, treating time as a third dimension of the search space. The time dimension differs from the other physical space dimensions in that time is irreversible. Oddly enough, this fact simplifies the propagation process which allocates cost to the cheapest goal through all time/space free cells. Goals are called 'rendezvous' since a cell in time space specifies both the place and exactly when a goal must be reached. If quantised time is considered to be specified by a third dimension , Z, say downwards into the future with each horizontal X, Y plane representing a time slice of physical environmental map showing free and filled cells, a vertical string of rendezvous cells represents a position which should be reached over a specified range of time intervals. Blobs of rendezvous cells indicate tolerances in space and time for goal attainment and so on. Both movement and time can be costed. The cost propagation algorithm works backwards from the represented future back to the present time, one time slice at a time. All rendezvous cells are marked as zero entry post and all obstacle cells infinite entry cost (i.e. impossible to enter).

All other cells are marked with a high but not 'infinite' value. A number of cost propagation strategies can be devised. A simple example is where a time wait of one interval without moving costs one unit, a unit x or y translation during that interval a cost of two units and translation of both a unit x and a unit y during a single time interval a cost of three. These costs are accumulated into the next level up in time (towards the present). The DT map at time t-1 can be derived from the map at time t and the accumulation of appropriate costs as described above. All cells entries at time t-1 can be calculated in parallel once the map at time t is known and so on. The propagation process is single pass backwards in time. When the propagation backwards in time reaches the present time slice (t = 0) all free cells are now loaded with accumulative costs to the least costly reachable rendezvous. Cells still containing their initial high values (or more) represent cells from which no goal can be reached within the remaining time . Steepest decent from any point in free space at the time $t = \phi$ level will follow an optimal time/space trajectory to the least costly rendezvous. This scheme works very well and, despite the seeming complication of the irreversibility of time, is actually simpler to implement than the regular (multi pass) DT algorithm and has considerable parallel processing potential.

*Initially Known but Unpredictable Time-Varying Environments-Fully Observable Obstacle Fields* [28]

Not surprisingly, unpredictability of the time-varying aspects of future obstacle fields adds considerable complexity to the path planning problem. Some general observations help to define this problem.

If the environment is changing unpredictably but slowly compared with the normal speed of the robotic vehicle, readjusting path planning and environmental modelling as an ongoing requirement during goal pursuance is the most direct approach to accommodating change. Parts of the map are altered when new information becomes available, forgetting factors fade out old information and global path planning is carried out more-or-less on a continuous basis, taking the current position as the new starting point at each update cycle. This type of environmental change situation does not essentially change the standard approach but merely requires a vigilant renewal of mapping/planning effort at short intervals.

At the other extreme, if the environment can change unpredictably and rapidly in relation to the normal speed of the robot, perhaps if obstacles are themselves moving rapidly, there is little sense in planning ahead in time and detailed global path planning may have to be abandoned. Aiming roughly at the goal and taking local obstacle avoidance reactive responses may well be the more appropriate strategy. However, no matter what approach is taken, collision may still occur since there is no guarantee that the robot can move out of the way of a rapidly moving obstacle quickly enough.

Thus, from a planning viewpoint, the intermediate situation of medium rate of change is the most interesting since the changes can not easily be accommodated simply by perpetual replanning nor need a global detailed plan be more-or-less abandoned as in the fast changing environment case. There is still a strong case for exercising control based on sensory data and path planning.

Observability becomes a critical issue, as does some assumptions about continuity of motion. If a medium rate of change obstacle field is fully observable, as might be the case with a high vantage point camera overlooking the working environment, and some assumptions on change continuity might reasonably apply, 'snap shots' of the obstacle field at a sequence of time instants can be used to estimate the future state of the obstacle field to aid forward global path planning. Tolerance can be introduced by generously expanding the moving obstacle cell into the future as its predicted short term path is plotted, thus allowing for errors in prediction and/or weaknesses of the continuity assumption. If no continuity assumptions can be reasonably adopted, each moving obstacle cell can be expanded at its maximum rate of movement in all directions; such an approach would soon render the whole map as a potential obstacle. Thus, depending on the speed of change, the time into the future that may be sensibly planned becomes limited.

Starting with the perfectly predictable case discussed earlier, the same framework of path planning in time/space can be adopted with some specific modifications when some degree of unpredictability needs to be accommodated. In the simplest version two 'snap shots' of the environmental map are taking with one time interval between them. If one is taken at the current time is $t = \phi$, then the other 'snap shot' most recently available is at $t = -1$. The differences between these time shots gives an indication of leading extensions of where space, previously free is becoming filled and where space, previously filled is becoming free. Both types of differences can be expanded in to provide an estimate of what the obstacle field might look like into the future some specified number of time intervals. The DT is used to propagate costs back from that future to the present time and a preliminary path plan is made. The robot moves along that path for one time interval and then the whole procedure is repeated. Obstacles can be expanded generously in the direction of predicted movement and new free spaces conservatively in their movement directions to allow for errors in prediction. This scheme provides a consistent way of accommodating uncertainty in forward path planning. Of course, longer sequences of contiguous 'snap shorts' can be used to refine predictions of future maps but two 'snap shots' are the minimum requirement..In the future beyond which predictions are quite unrelable, the 'optimistic' assumption is made (unknown space is considered free until shown otherwise).

*Initially Unknown and Unpredictable Time-Varying Environments - Partially Observable Obstacle Fields*

If an environment map can only be constructed using on-board sensors, then it can only be constructed incrementally because of the restricted view of the environment such sensors provide. When the environment is changing (at a medium rate) but unpredictably, the navigation problem is even more complex than the case discussed immediately above, since limited observability plus unpredictability makes for very hazardous operations. In human operations such as road traffic control, external regulation in terms of one way streets give way protocols, traffic lights and speed restrictions help reduce the hazards of moving in such environments. The cautious driver is encouraged to keep all the rules but not assume any one else will always keep them This is the so called 'defensive' approach to safe driving. In the autonomous mobile robot navigation situation, if all moving obstacles are other robots under the same control regime or humans who can be expected to react sensibly in the presence of moving robots, some degree of safety can be imposed by introducing a set of precedence rules which minimise the risk of collision. However, if obstacles are unpredictable in their movements and are themselves unable to react sensibly, all caution must be adopted by the robot itself. If the robot is slow the situation is even more difficult, since being collided into whilst trying to get out of the way may not be much of a consolation.

Research in this area of autonomous navigation is not yet mature but some researchers have adopted the stance that swarms of robots [29] working in co-operative yet autonomous modes are a good basis of studies of planning, obstacle avoidance and co-operation which might mimic certain classes of insect society and perhaps, ultimately, human society as well.

## 6. Experimental Case Studies

Three case studies of robot navigation systems capable of operating with slow unpredictable time-varying environments will be presented. Two are indoor systems and the third outdoor. Both the instrumentation and the algorithmic aspects of these systems will be briefly described. Details can be found in cited references.

*Natural Landmark Based Robot Navigation in Distance Transform Space [30]*

A three wheeled Denning robot [See Figure 10] capable of moving without body rotation (wheels are steered and driven in unison), fitted with an ESP Technologies scanning time-of-flight range finder [see Figure 11] and a Denning remote bar-code reading localiser [Figure 4], was used for this project. The localiser was used only to establish the initial position and orientation of the robot relative to an initially empty grid map and to gauge the effectiveness of the natural landmark based localisation strategy for experiment studies (but not for actual navigation). The ESP scanner proved to be very temperature and target surface albedo sensitive and could only reliably measure ranges up to 4 or 5 metres for light coloured obstacles. This meant that, within the 10 metre by 10 metre environment in which the system was operated, only part of the environment visible from the robot could actually be mapped by the range scanner.for any single robot location.

A $128 \times 128$ grid map was used for mapping the obstacle field but a reduced ($64 \times 64$) grid used for path planning calculations using DTs. Cells were loaded with numbers indicating their certainty of occupancy, multiple observations building up the numerical structure and failures to observe obstacles in cells (within the range scope of the robot sensor) which were previously thought to contain obstacles reduces these scores (eventually to zero). Any cell having a score exceeding $\phi$ was regarded as containing an obstacle for planning purposes. The obstacles were grown out an amount approximating the robot radius to convert the navigation problem to a point path task. A layer of two cells further out from the grown obstacles were given higher access cost weightings in the path planning map to discourage, but not prohibit, access close to obstacles.

Starting from a known position and orientation, the robot completes 360° range scans, marks observed obstacles into its environment map with some credibility rating increment, plans a global path to be goal using the 'optimistic' strategy in which

Figure 10. Denning Mobile Robot

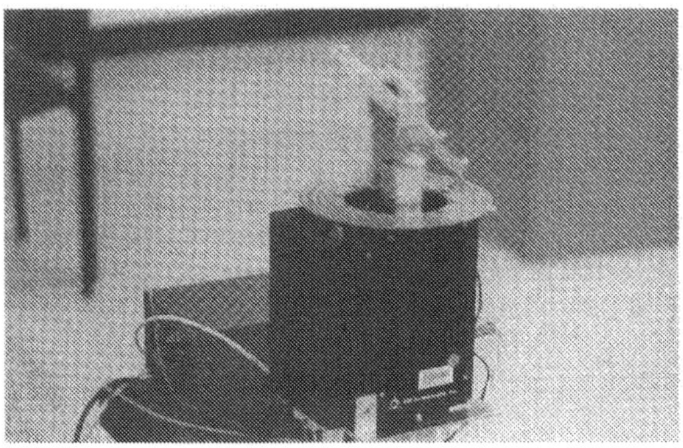

Figure 11. E.S.P. Technologies Scanning Time-of-Flight Range Finder

all unknown cells are regarded as free until shown otherwise, and takes the first straight line component of this path before repeating the range scan, environmental mapping, path planning and movement cycle, but with a natural landmark based localisation phase added for all cycles except the first.

Odometry on the robot gives a dead reckoning position and orientation fix; this is mixed with natural landmark based corrections which are applied in proportion to the reliability of the matching of current range data with the pre-existing environment map data within the range scope of the robot. Lateral and rotation shift correlation match searches are used in the proximity of the current position/orientation estimate to guage the x, y and $\phi$ localisation errors and these are used to adjust the previous localisation estimate according to the quality of the correlation match. The process is slightly more complicated than outlined above but these details can be found in [30].

Despite the simplicity of the above approach to natural landmark based navigation, the system operated fairly reliably, the main weakness being the relative poor quality of the ESP range data. Note that specific identifiable landmarks were not used but all range data within the range scope of the robot used collectively as landmark data. If, alternatively, a specific landmark based method was to be applied, the delectability and the identity of individual landmarks would have to be dealt with and this would have increased the complexity of the system considerably. Had there been very few obstacles which could have been used (within the range scan of the robot sensor) in the landmark matching process adopted, the odometry based incremental shifts from the previous adjusted estimates would have dominated the new localisation fix. Whilst the localisation error may be high, the risk of collision is low since the map is relatively sparse in the vicinity of the robot. When a good match is found it dominates the localisation adjustment thus obviating the intrinsic accumulative error aspects of odometry.

Slow changes in the environment are accommodated by incremental readjustments of the occupancy credibility associated with each cell. However, corrections in process can contribute to poor range data matching due to disparities between the current range scan data and the pre-existing environment map component which is within the range scope of the robot. In fact, the range data of the current scan automatically selects the scope since data in the map not corresponding to the current data does not contribute to the matching process.

It is interesting to note that the methodology adopted for these experiments puts emphasis on localising the robot relative to its immediate surroundings (since the range scanner is limited in scope), perhaps at some cost of lower global localisation accuracy. Somehow this does not seem to matter very much. In a sense some 'reaction' based navigation has crept into an essentially 'planning' based strategy in a natural (not particularly intended) way.

*Autonomous Radio Ethernet 'Tetherd' Robot*

In early 1996 our local participatory technology museum (ScienceWorks in Spotswood, Melbourne) approached us (The Intelligent Robotics Research Centre) to build them a robot navigation system [15] which could be used as an exhibit in a new project they were planning. We designed, built and tested a suitable robot system over three months before it was exhibited daily for 10 months (now moved to another location-hopefully still operative). The robot is shown in Figure 12. Even we were surprised at the reliability of the system which used a Denning localiser and a Erwin Sick scanning range finder for sensory data and a Silicon Graphics Indy for computational support. Though this system was extremely reliable the robot could still only be regarded as of the 'hot house' variety because it operated on a physical cable tether with a trickle charger on-board for battery recharging (wheels and sensors operating on battery) and its computational resources sitting on a table where members of the public could operate it and observe its mapping, localisation, path planning and locomotion activities.

The availability of an affordable radio ethernet bridge operating at 2 megabits per second gave us the inspiration to redesign and build a new system with the advantages of a physical tether but with only a radio link. We called the new system 'Etherbot' [31].

Using a rectangular grid system within a DT context, the Denning localiser provided accurate position and orientation fixes, more than the minimum number of bar code beacons being used so that this redundancy could be exploited to improve accuracy and accommodate situations where one or more beacons might be visually obscured or geometrically in ill-posed configurations.

An Erwin Sick range finder, set to scan the front half of the obstacle field with a horizontal sheet scan a few inches above the floor, provided 360° range readings at 1/2 degree intervals every two seconds. Since the range scanner can range to 50 metres, half of the environment viewable from the robot could be ranged to at just above floor level. Since we anticipated that many working environments would not have physical extensions outside the projection of the range scan data on the floor, no attempt to collect range data at other heights was made.

Localisation and range data was transmitted to the Silicon Graphics Workstation via a multiple line RS232 serial line server attached to the on-board radio ethernet bridge component. The other end of the radio ethernet bridge was attached to a hard-wired local area network linking the controlling Silicon Graphics Workstation and a large number of other workstations of various makes. Control of each of the two wheels of the robot (standard wheel chair system, including PID controllers) was provided via a Motorolla M68HC11A microcontroller, also serviced through a serial line on the server attached to the on-board radio ethernet unit. The radio ethernet system can operate over 200 metres with standard stub antennas and up to 6km with specialised directional antennas.

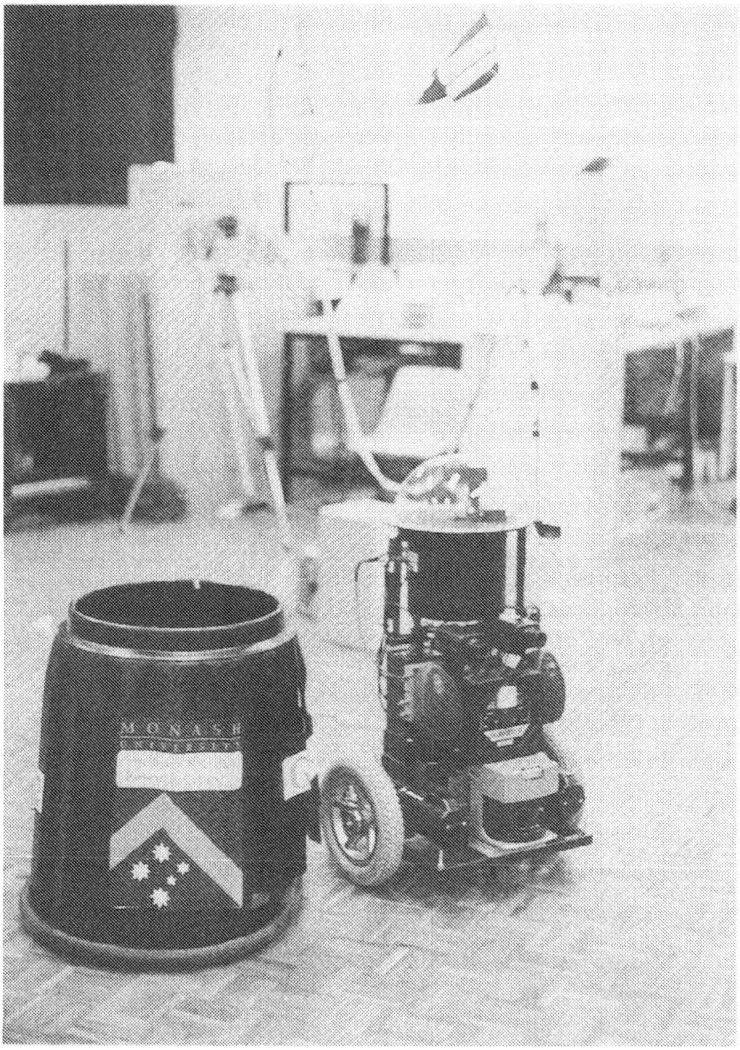


**Figure 12. GO-2 ScienceWorks Museum Robot**

The environmental map is updated after each straight line following of a global path plan by a set distance or until a direction change is needed and a new path calculated. The localiser data corrects the path of the robot to the planned straight line path components. Any new environmental data is accommodated at each planning stage, both additions and deletions taking place at that time. The range finder is of sufficient accuracy and reliability to ignore accumulation of cell occupancy credibility values. Any new obstacle blocking a previously planned path will be gone around in the new plan unless its position requires an entirely new plan to be devised. The computation of new plans is sufficiently simple as to permit regeneration of plans at 1 second intervals if necessary.Aglobal perspective of planning is maintained throughout, despite repeated updates.

Slow changing environments are well accommodated using this simple scheme. One possible weakness is that, during a small straight line movement, range data is not incorporated into the control scheme. However, it is possible to use the range finder as an obstacle proximity detector to stop the vehicle before impending collision.

On the whole this system works very well and has considerable market potential. Using the radio ethernet has not only allowed the fragile and bulky computational support to remain off-board, thus reducing on-board power consumption and space congestion (allowing for a greater pay load), but also makes feasible the possibility of time sharing the robot amongst people working on other Silicon Graphics Workstations on the local area network. This possibility allows the robot to be conveniently used for carrying mail, packages and food/beverages around an office environment. Other applications can be readily considered. Furthermore, many robots could be controlled as a fleet using this approach since control can be centralised for purposes of enforcing precedence protocols or providing efficient work sharing allocations.

The robot itself is shown in Figure 13. Note its skeletal appearance which reflects its real nature.

*All-Terrain Autonomous Tracked Robot* [32]

A tracked vehicle (MD15 see Figure 14) designed for outdoor robotic operations was purchased from Modulaire of Finland following the Intelligent Autonomous Vehicles Conference in Espoo, (just outside Helsinki) in June of 1995, at which meeting it won the outdoor robot navigation competition.

The following instrumentation was mounted on-board.

1. An Erwin Sick range finder attached to a Puma 560 robotic manipulator mounted at the front of the vehicle.

2. A Novatel differential/carrier phase Global Positioning System.

3. A pair of odometry wheels, one on each side of the centre of the vehicle.

**Figure 13. 'Etherbot' Radio Ethernet 'Tethered' Robot**

Figure 14. MD15 Rough Terrain Tracked Robot

4. A TCM digital flux gate compass and pitch/tilt sensor on a wooden vertical rod to reduce ferric metal related electromagnetic interference from the vehicle.

5. A Aironet Arlan 640 radio ethernet bridge unit and a Specialix multichannel RS232 serial line server.

6. A 600 watt petrol electrical/power generator to power all on-board instrumentation and the Puma 560 manipulator.

The base station for the GPS system consisted of a GPS unit and receiver/transmitter at the top of a 30 metre tower in the vicinity of our experimental test grounds. The odometry wheels are large and thin and are mounted on linear bearings so that their contact with the ground is only through their own weight.[see Figure 2]

We attempted to repeat the same kind of experiments as were carried out with the indoor radio ethernet system described earlier by substituting the GPS system for the Denning localiser. However, it turned out not to be quite that simple. It was found that the flux gate compass was subject to electromagnetic variations and that averaging over a number of readings was necessary to get stable results. Also, we had initially hoped to use an optical gyroscope for controlling planned turns on-the-spot by driving the vehicle's tracks in opposite directions, only to find that our optical gyroscope was not properly calibrated for temperature variation; this was in fact why we reverted to using odometry wheels. GPS can return heading data only whilst moving as was not therefore a suitable heading information source for on-the-spot turns. Delays along the serial line server, radio ethernet, Silicon Graphics data acquisition chain made it difficult to stop a turn at precisely the appropriate position as indicated by the odometry wheels. A local (on-board) rotation angle control system would be much better but we have not yet implemented such.

Finally, the Erwin Sick range finder, designed for indoor use, could not handle glint effects in full sunlight, although cloudy dry trials were quite successful. An outdoor version on the range finder is available and may prove more appropriate for our experiments.

However, despite some initial difficulties, the tracked vehicle was in fact capable of outdoor point to point navigation in initially unknown, slowly time-varying obstacle fields.

The capacity of the robot manipulator holding the range finder to wave it around to collect detailed 3D range data has not yet been exploited.

## 7. Discussion

Various interesting problems arise in devising autonomous mobile robot navigation schemes for time-varying environments. The two most salient aspects of such studies are the irreversibility of time as a point of departure in regarding time merely as another dimension in path search space and the uncertainties which arise from unpredictability of environmental change, flawed algorithmic formulations and sensor error.

It was seen in the section on time-varying environments with perfect predictability, that the irreversibility of time is actually an advantage in terms of simplifying the cost propagation algorithms and rendering them parallelisable to some extent. This, at the time of the research, was a surprising outcome which is somewhat counter intuitive.

As far as sensor error is concerned, new faster, more accurate and robust range sensors are becoming available. So far they are quite expensive, but perhaps high volume use will bring their price down. Also, multiple camera stereo based passive range finding is now a mature research area and a number of fast (approaching T.V.. camera frame rate) devices will soon be available on the market. In addition, accurate GPS systems are dropping in price and excellent optical gyro and Inertial Guidance Systems are becoming available at medium to low cost.

Given also that computational power is continuing to be provided in smaller, faster packages at lower cost, the future is bright for autonomous mobile robots to finally find their way into the marketplace as a large scale development.

## 8. Conclusions

This chapter has covered the topic of autonomous mobile robot navigation in time-varying environments, described the types of sensors needed to build such systems and some algorithmic constructions to support path planning, mapping and localisation. A number of experimental case studies were used to illustrate some of the practical aspects of this topic.

## 9. References

1. Kleeman, L. Ultrasonic Autonomous Robot Localisation System, IEEE International Conference - Intelligent Robotics and Systems '89, Tsukuba, Japan, Sept. 1989, pp 212-219.

2. Jarvis, R.A. A Selective Survey of Localisation Methodology for Autonomous Mobile Robot Navigation, Proc. Robots for Competitive Industry Conference, July 14-16, 1993, Brisbane Australia, pp. 310-317.

3. Clark, S. and Durrant-Whyte, H. The Design of a High Performance MMW Radar System for Autonomous Land Vehicle Navigation, Proc. International Conference on Field and Service Robots, Canberra, Australia, 8-10 Dec. 1997, pp. 292-299.

4. Kong, X., Nebot, E., Durrant-Whyte, H. Use of Quaternions in a Low Cost Strapdown INS Unit, Proc. International Conference on Field and Service Robots, Canberra, Australia, 8-10 Dec. 1997, pp. 286-291.

5. Brooks, R.A. Elephants Don't Play Chess, Journal of Robotics and Autonomous Systems, Vol. 6, 1990, pp. 3-15.

6. Chatila, R. and Lacroix, S. Adaptive Navigation for Autonomous Mobile Robots, Proc. The Seventh International Symposium on Robotics Research (edited by G. Giralt and G. Hirzinger), 1996, pp. 450-458.

7. Jarvis, R.A. A Perspective on Range Finding Techniques for Computer Vision, IEEE Transactions on Pattern Analysis and Machine Intelligence. Vol. PAMI-5, No. 2, March 1983, pp. 122-139.

8. Moravec, H.P. Visual Mapping by a Robot Rover, Proc. 6th Int. Joint Conf. Artificial Intell., 1979, pp. 598-620.

9. Faugeras, O.D. How Can Vision Make Mobile Robots Come True, Proc. International Symposium and Exposition on Robots, 6-10 Nov, 1988, Sydney, Australia, pp. 1430-1461.

10. Braunstein, M.L. Depth Perception Through Motion, Academic Press, Inc. 1976.

11. Bolles, R.C. and Woodfill, J. Spatiotemporal Consistency Checking of Passive Range Data, Proc. 6th International Symposium on Robotics Research, Pittsburg, Pensylvania, Oct. 2-5, 1994, pp. 165-183.

12. Prazdny, K. Motion and structure from optical flow. In Proc. 6th Int. Joint Conf. Artificial Intell., Tokyo, Japan, November 8-11, 1979, pp. 702-704.

13. Kanade, T. Very Fast 3-D Sensing Hardware, Proc. The Sixth International Symposium on Robotics Research, Pitsburgh, Pennsylvania, Oct. 2-5, 1993, pp. 185-198.

14. Chong, K.S. and Kleeman, L. Large Scale Sonarray Mapping using Multiple Connected Local Maps, Proc. International Conference on Field and Service Robotics, Canberra, Dec. 8-10, 1997, pp. 538-545.

15. Jarvis, R.A. and Lipton, A.J. GO-2: – An Autonomous Mobile Robot for a Science Museum, Proc. 4th International Conference on Control, Automation, Robotics and Vision, Westin Stamford, Singapore, 3-6 December, 1996, pp. 260-266.

16. Nillson, N.J. Problem-Solving Methods in Artificial Intelligence, McGraw-Hill, 1971.

17. Lozano-Perez, L. and Wesley, M.A. (1979): An Algorithm for Planning Collision-free Paths Among Polyhedral Obstacles, Commun. ACM, 22, 10 Oct 1979, pp. 560-570.

18. Jarvis, R.A. Growing Polyhedral Obstacles for Planning Collision-Free Paths, The Australian Computer Journal, vol. 15, No. 3, August 1983, pp. 103-111.

19. Rosenfeld, A. and Pfaltz, J.L. Sequential Operations in Digital Image Processing, J.A.C.M., Vol. 13, No. 4, Oct. 1966, pp. 471-494.

20. Jarvis, R.A. On Distance Transform Based Collision-Free Path Planning for Robot Navigation in Known, Unknown and Time-Varying Environments, invited chapter for a book entitled 'Advanced Mobile Robots' edited by Professor Yuan F. Zang. World Scientific Publishing Co. Pty. Ltd. 1994, pp. 3-31.

21. Jarvis, R.A. Configuration Space Collision-Free Path Planning for Robotic Manipulators, Proc. 10th Australian Computer Science Conference, Deakin University, Victoria, 4-6 Feb. 1987, pp. 193-204.

22. Tang, K.W. and Jarvis, R.A. Collision-Free Path Finding amongst Polygonal Obstacles using Efficient Free Space Triangulation, Proc. Second International Conference on Automation, Robotics and Computer Vision, 15-18 Sept. 1992, Singapore, pp. RO-11.1.1-RO-11.1.5.

23. Zelinsky. A. Robot Navigation with Learning, The Australian Computer Journal, Vol. 20, No. 2, May 1988, pp. 85-93.

24. Udupa, S.M. (1977): Collision Detection and Avoidance in Computer Controlled Manipulators, Proceedings of 5th International Joint Conference on Artificial Intelligence, August 1977, pp. 737-748.

25. Brooks, R.A. (1983): Solving the Find-Path Problem by a Good Representation of Free Space, IEEE Trans. on Systems, Man and Cybernetics, SMC-13, No. 3, March 1983, pp. 190-197.

26. Chatila, R. (1982): Path Planning and Environment Learning, European Conference on Artificial Intelligence, July 1982, pp. 211-215.

27. Jarvis, R.A. Collision-Free Path Planning in Time Varying Environments, Proc. I.E.E.E./R.S.J. International Workshop on Intelligent Robots and Systems, '89, Tsukuba, Japan, Sept. 4-6 '89, pp. 99-106.

28. Jarvis, R.A. Collision-Free Path Planning in Time-Varying Obstacle Fields Without Perfect Prediction, Proc. '93 ICAR, International Conference on Advanced Robotics, No. 1-2 1993, Tokyo, Japan, pp. 701-706.

29. Jarvis, R.A. Video Plane Robot Swarms, Robotics and Computer Integrated Manufacturing, Vol. 11, No. 4, 1994, pp. 249-258.

30. Jarvis, R.A. Natural Landmark Based Robot Navigation in Distance Transform Space, Proc. AI'95, Canberra, 1995, pp. 323-330.

31. Jarvis, R.A. Etherbot - An Autonomous Mobile Robot on a Local Area Network Radio Tether, Proc. Fifth International Symposium on Experimental Robotics, Barcelona, Catalonia, June 15-18, 1997, pp. 151-163.

32. Jarvis, R.A. An Autonomous Heavy Duty Outdoor Robotic Tracked Vehicle, Proc. International Conference on Intelligent Robots and Systems, Grenoble, France, Sept. 8-12, 1997, pp. 352-359.

# 11

# HIGH INTEGRITY NAVIGATION FOR AUTONOMOUS SYSTEMS

E. M. NEBOT
*Australian Centre for Field Robotics*
*The University of Sydney*
*NSW 2006 Australia*

## 1. Introduction

This chapter addresses the problem of developing ultra-high integrity navigation systems for large outdoor autonomous guided vehicles (AGVs). The term "Navigation" implies the ability to sense, compute and control the position of a vehicle and it is the key to any autonomous system design. To achieve any useful purpose an AGV must know its position in relation to the environment. If this position information is inaccurate, unreliable, or in error, the AGV system will fail, possibly in a catastrophic manner.

A number of research groups around the world have investigated the feasibility of large autonomous vehicle systems. Many of these studies have addressed perceived military needs [1], or the feasibility of using autonomous rovers in the unmanned phases of establishing manned Lunar or Martian research stations [2-6]. Few of these studies and developments have addressed the need for reliable autonomous navigation. None have explored the key issue of system integrity across all facets of operation. This is in part due to the nature of the application; a typical lunar rover is not truly autonomous, an allowance can always be made for some phases of the operation to be tele-operated or placed under the control of a human operator.

Autonomous vehicles have a potentially huge economic importance in mining, stevedoring and agricultural applications [7]. In order to make these applications possible the vehicle navigation system must be ultra-reliable, capable of operating in all environments and weather conditions, capable of operating at speed, and capable of sustaining safe operation during failure or degradation of both navigation and other system components. The key to the development of a high-integrity, and thus useful, navigation system is through the use of a number of physically different sensors whose information is combined in such a way that all possible failures in all possible sensor combinations are detectable. The design of navigation loops based on independent failure mode sensors is also addressed presenting the design of a DGPS/Inertial navigation loop together with a bearing/range/encoder navigation loop. These two examples contemplate linear and non-linear models for the process and observation. They make use of linear and non-linear filtering techniques to implement the data fusion task. The designs presented are very general and can be used as a guide to design a navigation loop based on different combinations of dead reckoning and external sensors. Finally a multi-loop navigation architecture is introduced that makes use of all the available information in the system to generate the best vehicle state estimate. This last topic is of fundamental importance for high integrity systems. It will not only reduce the uncertainty of the state estimation but will increment the information redundancy in the entire frequency spectrum making possible the implementation of low and high frequency fault detection algorithms.

## 2. Navigation Loops

The observations provided by absolute sensors are usually modelled as band-limited measurements corrupted by wide-band noise. State estimators, such as the Kalman filter, which are often used in the main navigation algorithm, implicitly reflect these models and act as low-pass filters for the observation information. As a consequence, true high frequency information, associated with vehicle manoeuvres such as turning, is lost. To overcome this problem, information from encoders or inertial sensors are often fed-forward through the state estimator as a prediction to be corrected by observations from an absolute sensor [8]. The net effect of this is that while absolute observation information is subjected to a low-pass filter, prediction information is subjected to a complementary high-pass filter. Together these two sources of information span the complete frequency spectrum in a complementary manner. As a general rule, absolute sensors provide low-frequency information and rate sensors provide high frequency information. A complete navigation algorithm must exploit this by using low-frequency sensor information to correct low-frequency drift error in high frequency sensors and by using high-frequency sensor information to de-correlate high-frequency noise from high frequency manoeuvres in low-frequency sensor information. Figure 1 shows a block diagram of a navigation loop based on the proposed sensor combination.

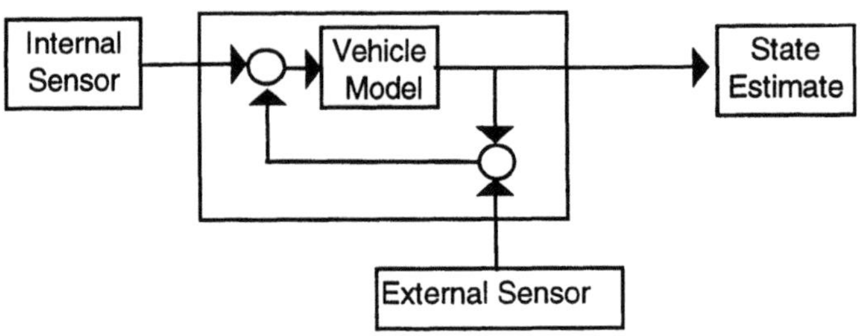


**Figure 1. Single Loop Navigation Architecture**

However, this general rule breaks down when the observation noise is not white. The sensor model associated with these noise sources does not lead to a simple low-pass characteristic for the state estimator. Common examples of such noises are biases and drifts in inertial sensors and noise due to Selected Availability in GPS. In all these cases a shaping filter must be designed whose input is white noise and whose output is the observed colored noise [9]. This filter can be designed by experimentally obtaining a power spectral density of the measurement errors on the assumption that the true states are known.

In the following sections two different navigation loops are presented that make use of inertial sensors and encoders as dead reckoning sensors and GPS and Radar/laser as external sensors. These two examples are very important because they contemplate sensors with independent failure modes and make use of linear and nonlinear Kalman Filtering techniques for their implementation.

### 2.1. INS/GPS Navigation Loop

In this loop a linear Kalman filter approach is used to fuse INS and GPS information. Two approaches are commonly adopted to the problem of fusing information from high and low-frequency sensors. These are the total space formulation and the error state space formulation, frequently referred to as direct and indirect filtering. In the direct approach, Figure 2, the position and velocity are part of the states to be estimated and the indicated INS position and velocity are part of the observation with the external sensors. The drawback of this approach is that, in general, high frequency manoeuvres are not usually contemplated in the linear (or linearised) model. The consequence of this omission is that the filter will unnecessarily attenuate the high frequency information provided by the INS. With this approach the system may not be able to track fast manoeuvres.

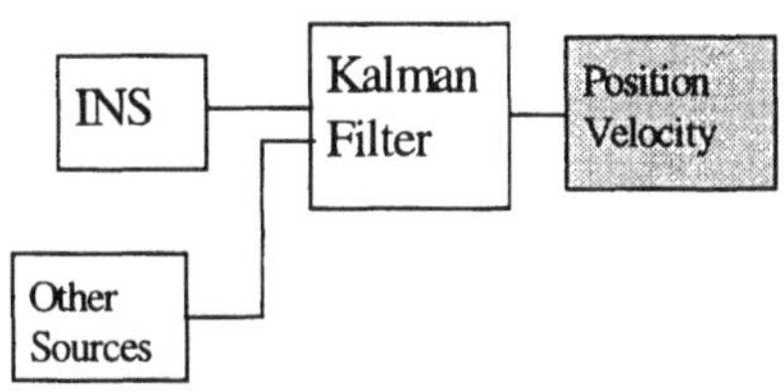


**Figure 2. Direct Kalman Filter**

In the indirect approach, the Kalman filter estimates the errors in position and velocity using the difference between inertial-indicated information and external sensor information. The inertial information can still be obtained even if no additional information is available. This implementation is shown as the indirect feed-back approach in Figure 3. The actual implementation proposed for the GPS/INS integration loop is based on the indirect feedback approach. In this implementation the inertial high frequency information is fed directly to the output without attenuation while the Kalman Filter provides the low frequency correction to the IMU.

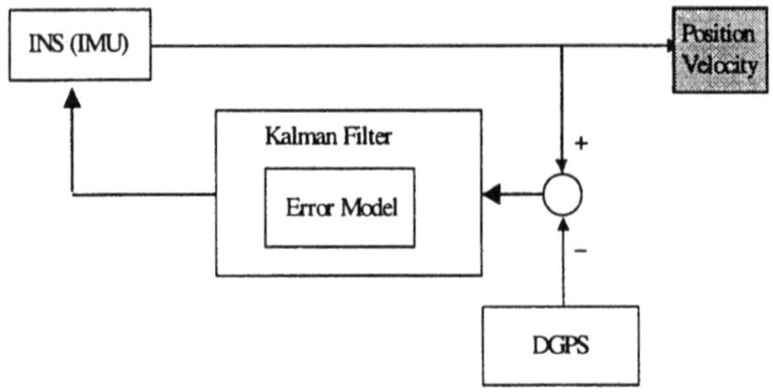


**Figure 3: Indirect Kalman Filter**

The indicated position/velocity information derived from the INS is subtracted from the position/velocity provided by the DGPS system. An error propagation model is used to fuse this observation and estimate errors in position, velocity, attitude and some other parameters to correct the inertial measurement unit, such as biases in accelerometer and gyro and platform misalignment angles. These parameters are fed back to the IMU unit.

### 2.1.1. Coordinate Systems and Transformations.

The navigation algorithm is designed in the geographic frame n, with axes $^{n}\{N, E, D\}$, (North, East and Down). It is necessary to determine the transformation from this frame to other frames since the various sensors provide information in different coordinate frames. Figure 4 shows the various coordinate frames involved in this loop.
The GPS system provides information in the Earth frame e, with axes $^{e}\{X, Y, Z\}$, which is known as the WGS-84 coordinate frame. These axes are fixed to the Earth so that the X and Y axes rotate around the Z axis with the Earth's rotational velocity $\Omega$.

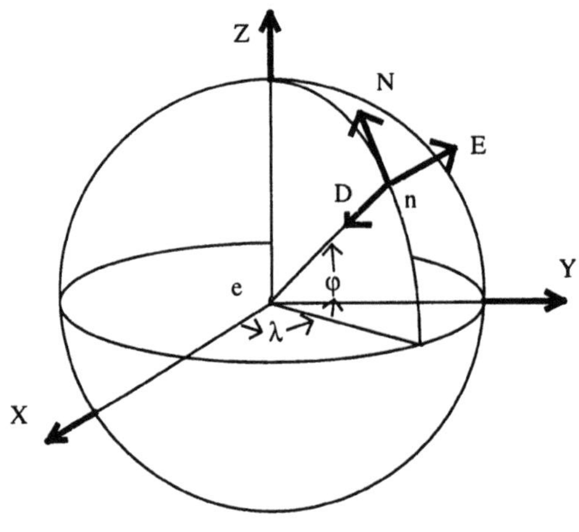


**Figure 4 Coordinate systems**

### 2.1.1.1. Earth centred to latitude longitude and height.

Position information can also be given in Earth frame coordinates but in a polar representation, that is longitude λ, latitude φ and height h, also known as Geodetic coordinate frame. These values can be obtained from x, y and z with the following transformation:

$$f = 1 - \frac{b}{a} \qquad p = \sqrt{x^2 + y^2} \qquad \theta = \tan^{-1}\left(\frac{za}{pb}\right)$$

$$ep = \frac{\left(a^2 - b^2\right)}{b^2} \qquad e = 2f - f^2 \tag{1}$$

$$\lambda = a\tan\left(\frac{y}{x}\right) \qquad N = \frac{a}{\sqrt{1 - e\sin^2(\lambda)}}$$

$$\varphi = \tan^{-1}\left(\frac{2 + ep2 * b\sin^3\theta}{p - e * a\cos^3\theta}\right) \qquad h = \frac{p}{\cos\lambda - N}$$

where a = 6378137 m. and b = 6356752.31m. for WGS-84.

### 2.1.1.2. Earth Reference "e" to Local Reference Frame "n"

The transformation from Earth reference "e" to local reference "n" can be evaluated using:

$$\begin{bmatrix} N \\ E \\ D \end{bmatrix} = \begin{bmatrix} -\sin(\varphi)\cos(\lambda) & -\sin(\varphi)\sin(\lambda) & \cos(\varphi) \\ -\sin(\lambda) & \cos(\lambda) & 0 \\ -\cos(\varphi)\cos(\lambda) & -\cos(\varphi)\sin(\lambda) & -\sin(\varphi) \end{bmatrix} * \begin{bmatrix} x - x_i \\ y - y_i \\ z - z_i \end{bmatrix} \tag{2}$$

In this case the latitude φ and longitude λ correspond to the origin of the local navigation frame 'n', also designated as $x_i$, $y_i$, $z_i$. The GPS information, which is originally in the Earth coordinate frame, is transformed to the local navigation frame with equation 2.

### 2.1.1.3. Transformation from Body "b" to Local "n" reference frame

The inertial measurement unit is mounted on the vehicle constituting a new frame. This is called the body frame "b'', and has axes $^b\{R, P, Y\}$, (Roll , Pitch and Yaw). This frame will be in constant rotation with respect to the "n" frame.
The angular velocities of this rotation are measured by three near-orthogonal gyros. The transformation matrix that relates the "b" and "n" coordinate frames is evaluated solving:

$$\dot{C}_b^n = \Omega_{bn} C_b^n \qquad \Omega_{bn} = \begin{bmatrix} 0 & -\omega_Y & \omega_P \\ \omega_Y & 0 & -\omega_R \\ -\omega_P & \omega_R & 0 \end{bmatrix} \tag{3}$$

Where $\Omega_{bn}$ is the anti-symmetric velocity matrix and $\omega_{Is}$ are the rotation velocities measured by the gyros in the body frame. This integration can be implemented with the following approximation:

$$C_b^n(k+1) = C_b^n(I+\Delta\theta)$$
$$\Delta\theta = \begin{bmatrix} 0 & -\Delta\theta_Y & \Delta\theta_P \\ \Delta\theta_Y & 0 & -\Delta\theta_R \\ -\Delta\theta_P & \Delta\theta_R & 0 \end{bmatrix} \tag{4}$$

where (I+Δθ) is the small angle direction cosine matrix relating the frame at time *k* to the rotating frame at time k+1. This approximation is valid for small angles, which constrains the minimum sampling time of the gyros such that the transformation matrix can be obtained with reasonable accuracy. This sampling time will be a function of the severity of manoeuvres expected from the vehicle. In land vehicle application it is reasonable to assume that the maximum rotation velocity expected will be in the order of 25 deg/sec. Sampling rates of at least 50 hz are necessary to be able to track the orientation of the unit with reasonable accuracy.

### 2.1.2. System Error Model

In order to implement the Kalman Filter proposed, a model of the propagation of the errors is needed. The dynamics of the Earth surface frame navigator can be described by the following set of equations:

$$\dot{R}^n = V^n$$
$$\dot{V} = C_b^n A_{bn}^b + g^n \tag{5}$$
$$\dot{C}_b^n = \Omega_{bn}^n C_b^n$$

For land vehicle applications a simplified system error model can be written in terms of errors in position *R*, velocity *V* and the misalignment angles $\Phi$

$$\begin{bmatrix} \dot{r} \\ \dot{v} \\ \dot{\phi} \end{bmatrix} = \begin{bmatrix} 0 & I & 0 \\ 0 & 0 & a_n \\ 0 & 0 & 0 \end{bmatrix} \begin{bmatrix} r \\ v \\ \phi \end{bmatrix} + \begin{bmatrix} 0 \\ C_b^n e_a \\ C_b^n e_g \end{bmatrix} \quad \dot{x} = F_g x + Gw \tag{6}$$

The majority of the INS platforms commercially available is built with low cost gyros and accelerometers with non-constant biases. In equation 6, $e_g$ and $a_g$ are vectors in body frame that consider the gyros and accelerometers errors. This model can be extended to contemplate for the biases and drift of the accelerometers and gyros. The gyro error model proposed consists of a first order Markov process with correlation time τ plus white noise $\nu$:

$$\dot{\theta} = -(1/\tau)\theta + \nu_g \quad E[\nu_g] = 0, \; E[\nu_g \nu_g^{\ T}] = R_g \tag{7}$$

The $T_g$ matrix contemplating the time constant for the three gyros becomes

$$T_g = \begin{bmatrix} -1/\tau_x & 0 & 0 \\ 0 & -1/\tau_y & 0 \\ 0 & 0 & -1/\tau_z \end{bmatrix} \tag{8}$$

The error model for the accelerometers consists of a random constant component plus white noise:

$$a = \nu_a \qquad E[\nu_a] = 0, \quad E[\nu_a \nu_a^{\ T}] = R_a \tag{9}$$

The $T_a$ matrix for the accelerometer errors is:

$$T_a = \begin{bmatrix} 0 & 0 & 0 \\ 0 & 0 & 0 \\ 0 & 0 & 0 \end{bmatrix} \tag{10}$$

Finally the complete state error model has the following form:

$$\dot{x} = Fx + Gw \tag{11}$$

with

$$F = \begin{bmatrix} & & & 0 & 0 \\ & F_g & & C_b^n & 0 \\ & & & 0 & C_b^n \\ 0 & 0 & 0 & T_a & 0 \\ 0 & 0 & 0 & 0 & T_g \end{bmatrix} \quad G = \begin{bmatrix} 0 & 0 & 0 & 0 & 0 \\ 0 & C_b^n & 0 & 0 & 0 \\ 0 & 0 & C_b^n & 0 & 0 \\ 0 & 0 & 0 & I & 0 \\ 0 & 0 & 0 & 0 & I \end{bmatrix}$$

$$x = \lfloor r_i \quad v_i \quad \Phi_i \quad b_{gi} \quad b_{ai} \rfloor$$

Where $r, v$ and $\phi$ are the errors in position, velocity and angle misalignments, and $b_{gi}$ and $b_{ai}$ are the errors in the gyros and accelerometers in the body frame.

### 2.1.3. Observations

The acceleration measured by the inertial unit is transformed to the navigation coordinate frame and then integrated to obtain velocity and position information as shown in Figure 5.

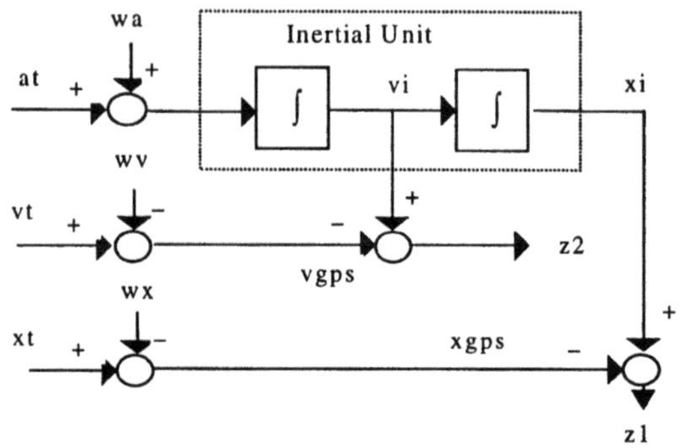


**Figure 5 Filter Observations**

According to the figure the observation $z_i$ can be written in the following form

$$z(k) = \begin{bmatrix} xi(k) - xgps(k) \\ vi(k) - vgps(k) \end{bmatrix} = \qquad (12)$$

$$\begin{bmatrix} [xt(k) + \delta x(k)] - [xt(k) - wx(k)] \\ [vt(k) + \delta v(k)] - [vt(k) - wv(k)] \end{bmatrix} = \begin{bmatrix} \delta x(k) + wx(k) \\ \delta v(k) + wv(k) \end{bmatrix}$$

It can be seen that the observation $z_i$ is composed of the error in indicated position and velocity plus white noise. The observation equation for the filter is then written

$$z = Hx + Dv \qquad (13)$$

with

$$H = \begin{bmatrix} I_{3*3} & 0 & 0 \\ 0 & I_{3*3} & 0 \end{bmatrix} \quad v^T = [1 \quad 1 \quad 1 \quad 1 \quad 1 \quad 1]$$

### 2.1.4. Filter Implementation

This section describes the software distribution of the different tasks involved and the practical aspects of the filter implementation.

#### 2.1.4.1. Partition of Navigation Functions

The navigation functions required for the aided INS strapdown system are partitioned between the fast guidance unit and the navigation processor, as shown in Figure 6.

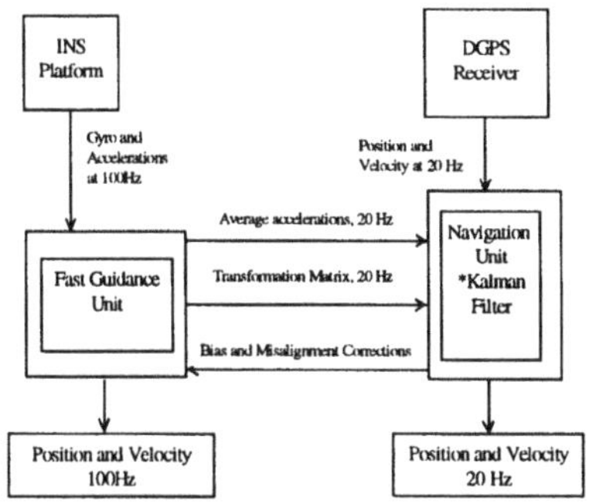


**Figure 6: INS/GPS Navigation architecture**

The system is usually implemented as two independent processes. The first process is implemented in the fast guidance unit. This process communicates with the inertial unit through a very fast link. The inertial measurement unit transmits triaxial gyro and acceleration information at a high frequency rate. The guidance unit integrates the direction cosine matrix using the algorithm detailed in equation 4. It also generates the predictions for velocity and position through single and double integration respectively. The positions, velocities, accelerations and the transformation matrix are transmitted to the navigation unit at a much lower rate. Although the vehicle state information is already transformed into the local navigation coordinate frame, the transformation matrix is still required to implement the data fusion algorithm. The navigation unit receives observed position and velocity information from the DGPS receiver at rates of up to 20 Hz and is responsible for implementing the feedback Kalman Filter fusion algorithm.

#### 2.1.4.2. Discrete Model

The continuous model of the system is given in equations (6) and (13). The discrete model is given by:

$$\begin{aligned} x(k+1) &= F_d x(k) + Q_d(k) w(k) \\ z(k) &= H_d x(k) + D_d v(k) \end{aligned} \tag{14}$$

Although the system is linear the matrices F and G are time variant. They are mainly functions of the actual acceleration and the transformation matrix $C_b^n$. The matrices $F_d$ and $Q_d$ need to be evaluated on line for each sampling time. The exact solution of these matrices is out of the question due to the large computational requirement. The following approximations are used to evaluate these matrices:

$$\begin{aligned} F_d(t_{i+1}, t_i) &= I + F(t_i)\Delta t + \frac{1}{2} F(t_i) * F(t_i) * \Delta t^2 \\ Q_d(t_i) &= G(t_i) Q(t_i) G^T(t_i) \Delta t \end{aligned} \tag{15}$$

With these approximations the state error and state error covariance prediction becomes:

$$\begin{aligned} x(k/k-1) &= F_d(k)\hat{x}(k-1/k-1) \\ P(k/k-1) &= F_d(k)P(k-1/k-1)F_d^T(k) + Q_d(k) \end{aligned} \tag{16}$$

The update equations are only processed when external information is available. In this example the GPS position and velocity are the external observations and the update equations are:

$$\begin{aligned} \tilde{x}(k/k) &= \hat{x}(k/k-1) + W(k)\left[z(k) - H\hat{x}(k/k-1)\right] \\ W(k) &= P(k/k-1)\ H\ S^{-1}(k) \\ S(k) &= H\ P(k/k-1)H^T + R \\ P(k/k) &= P(k/k-1) - W(k)S(k)W^T(k) \end{aligned} \tag{17}$$

### 2.1.5. Experimental Results

Some experimental results are shown to demonstrate the behaviour of the algorithm described. In this example the GPS and INS sensors are mounted on the car and driven along a trajectory approximately 1 km. long. Initially the INS is calibrated and aligned while the car is stationary [10]. Figure 7 shows the trajectory generated by INS alone, (black) and DGPS (grey). It can be seen that the error increased significantly after the first 500 meters. In all the plots DGPS information is drawn in grey while the output of the filter is plotted in black at a much higher frequency. Figure 8 presents the results of DGPS and the filter estimation. In this case the navigation unit is processing the Kalman Filter algorithm to incorporate the DGPS information. It can be seen that low frequency errors are eliminated in this stage. Figure 9 presents an enlargement of the trajectory after the first turn. In this case the GPS is generating a solution affected by multi-path and the filter is incorporating this erroneous information.

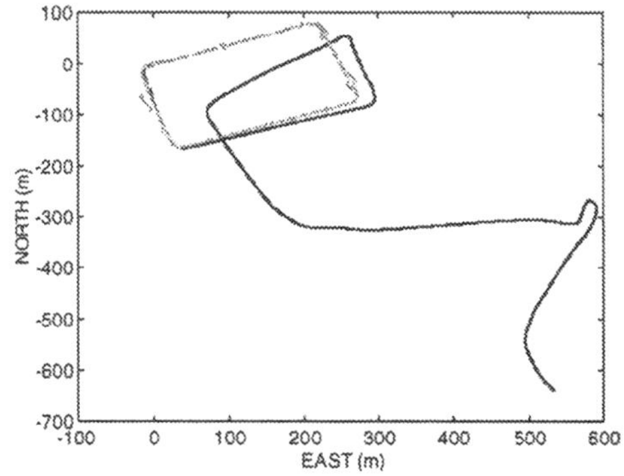


**Figure 7. Trajectory without GPS**

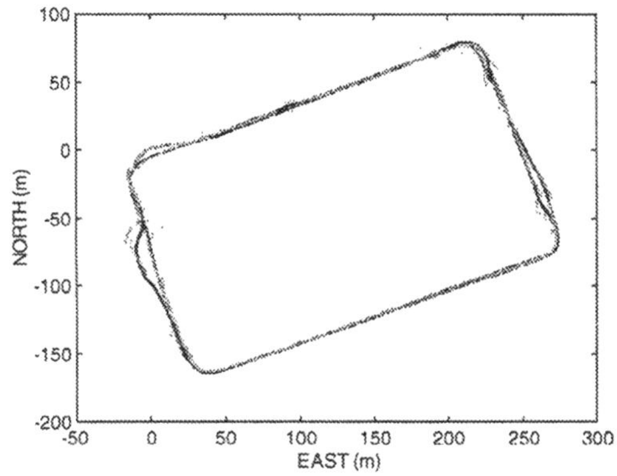


**Figure 8. Trajectory with GPS**

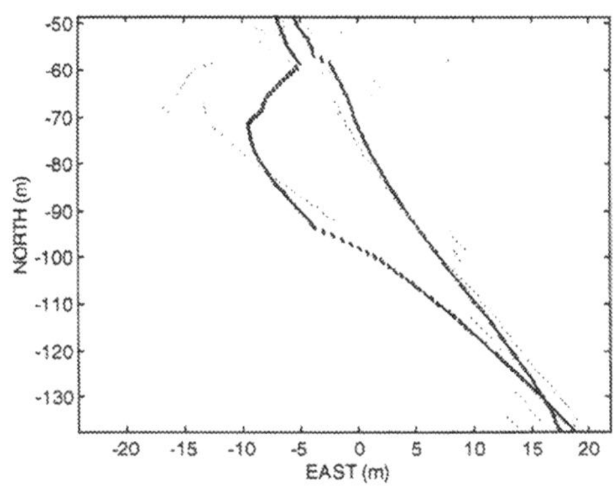


**Figure 9. Enhanced Trajectory.**

Fault detection capabilities are included in the run shown in Figure 10. In Figure 11 it can clearly be seen that the fault is detected and the filter navigates with INS alone during the period that the GPS is in fault. There is no noticeable degrading of performance during the GPS fault. Finally in Figure 12, the results of disconnecting the GPS after the first 500 meters are presented. It can be seen that the attitude of the unit has been updated properly since the filter is able to predict the trajectory for another 500 meters with small errors.

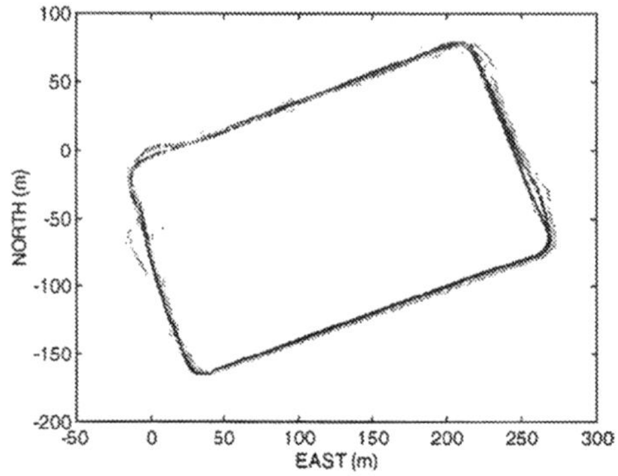


**Figure 10. With GPS and Fault Detection**

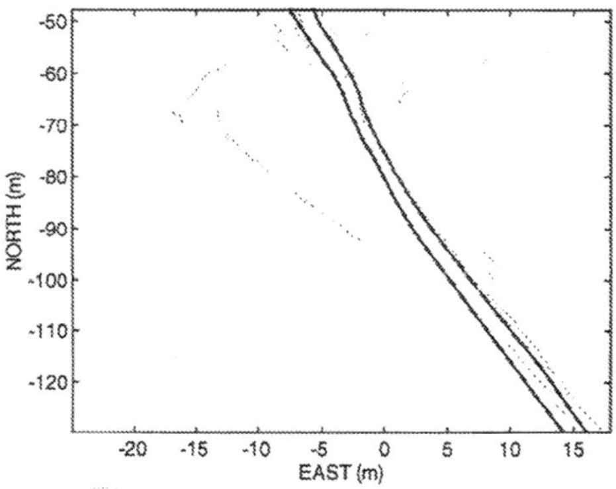


**Figure 11. Enhanced GPS and FD**

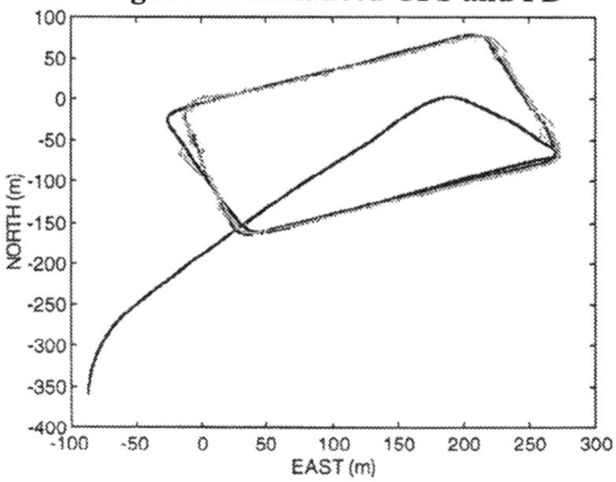


**Figure 12. Disconnecting GPS**

## 2.2. Encoder / Bearing, Range Navigation loop

In this section a navigation loop is presented based on encoders and range/ bearing information. In this example the models for the process and observation are non-linear and are used to estimate position and orientation with the use of Extended Kalman filtering techniques.

The encoders provide velocity and steering angle information that is used with a kinematic model of the vehicle to predict its position and orientation. This prediction is updated with external range and bearing information provided by a laser or radar system.

The external sensor provides information about features of the environment that are located at known positions. These features are usually called beacons and can be detected with radars, lasers and vision systems. The sensor provides information on the location of the beacon with respect to the vehicle position. This information is range or bearing or both.

### 2.2.1. Modeling Aspect

A simple kinematic model is used in this design example. This model can be extended to consider other parameters that can have significant importance in some applications such as wheel radius and slip angle [11],[12].

Consider the example in Figure 13 where the vehicle located at global coordinates $x(k) = (x, y, \phi)$. The steering control $\gamma$ is defined in vehicle coordinate frame. The sensor installed in the vehicle measures range and bearing to the beacons $Bi_{(i=1..n)}$ with respect to the vehicle coordinates, that is $z(k) = (r, \theta)$, where r is the distance from the beacon to the centre of the vehicle and $\theta$ is the sensor bearing measured with respect to the vehicle coordinate frame.

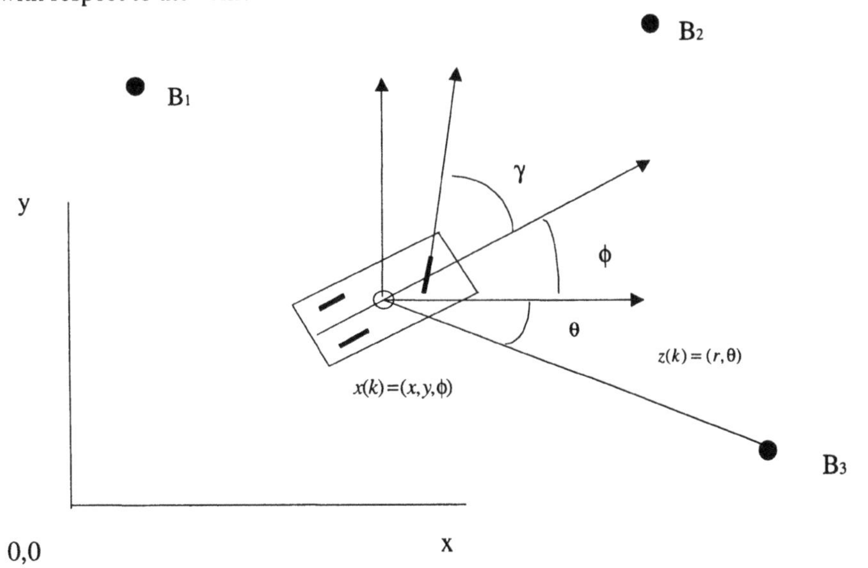


**Figure 13. Vehicle coordinate system**

Considering that the vehicle is controlled through a demanded velocity V and steering angle $\gamma$ the process model can be written as:

$$\begin{bmatrix} x(k) \\ y(k) \\ \phi(k) \end{bmatrix} = \begin{bmatrix} x(k-1) + \Delta T\, V(k-1)\cos(\phi(k-1)+\gamma(k-1) \\ y(k-1) + \Delta T\, V(k-1)\sin(\phi(k-1)+\gamma(k-1) \\ \phi(k-1) + \dfrac{\Delta T\, V(k-1)}{L}\sin(\gamma(k-1)) \end{bmatrix} + \begin{bmatrix} q_x \\ q_y \\ q_\phi \end{bmatrix} \tag{18}$$

where $\Delta T$ is the sampling time, L is the wheel base line and the vector $q$ represents the uncertainty in the model. The process can then be written as a nonlinear equation

$$x(k) = f(x(k-1), u(k-1), k) + v(k-1) \tag{19}$$

where *x(k-1)* and *u(k-1)* are the estimate and input at time *k-1* and $v$ is process white additive noise.
The equation that relates the observation with the states is

$$\begin{bmatrix} z_r^i(k) \\ z_\theta^i(k) \end{bmatrix} = \begin{bmatrix} \sqrt{(X_i - x(k))^2 + (Y_i - y(k))^2} \\ \tan^{-1}\left(\dfrac{Y_i - y(k)}{X_i - x(k)}\right) - \phi(k) \end{bmatrix} + \begin{bmatrix} r_r \\ r_\theta \end{bmatrix} \tag{20}$$

where *z(k)* and *x(k)* are the observation and state values at time *k* and $X_i$ and $Y_i$ are the known positions of the beacons or natural landmarks, and *r* is an additive noise. The observation equation can be expressed in short form as

$$z(k) = h(x(k)) + \omega(k) \tag{21}$$

The noises $\omega(k)$ and $v(k)$ are assumed to be Gaussian, temporally uncorrelated and zero mean, that is

$$E[v(k)] = E[\omega(k)] = 0 \tag{22}$$

with corresponding covariance

$$E[(v(i)\ v^T(j)] = \delta_{ij} Q(i), \qquad E[(\omega(i)\ \omega^T(j)] = \delta_{ij} R(i) \tag{23}$$

### 2.2.2. Prediction Stage

Assuming the state estimate is available at time *k-1*, that is

$$\hat{x}(k-1/k-1) \approx E[x(k-1)/Z^{k-1}] \tag{24}$$

with error covariance $P(k-1/k-1)$.

The prediction $\hat{x}(k/k-1)$ for the state at time $k$ can now be evaluated. The process model is used to predict the value of the state considering the previous states, velocity and steering angle information:

$$\begin{bmatrix} \hat{x}(k/k-1) \\ \hat{y}(k/k-1) \\ \hat{\phi}(k/k-1) \end{bmatrix} = \begin{bmatrix} \tilde{x}(k-1/k-1) + \Delta T\,V(k-1)\cos(\tilde{\phi}(k-1/k-1) + \gamma(k-1) \\ \tilde{y}(k-1/k-1) + \Delta T\,V(k-1)\sin(\tilde{\phi}(k-1/k-1) + \gamma(k-1) \\ \tilde{\phi}(k-1/k-1) + \dfrac{\Delta T\,V(k-1)}{L}\sin(\gamma(k-1)) \end{bmatrix} + \begin{bmatrix} q_x \\ q_y \\ q_\phi \end{bmatrix} \quad (25)$$

The prediction error covariance is evaluated according to:

$$P(k/k-1) = \nabla f_x(k)P(k-1/k-1)\nabla f_x^T(k) + Q \quad (26)$$

where $\nabla f_X(k)$ is the gradient of $f$ evaluated at $x(k)$. For this particular problem the gradient can be expressed in the form:

$$\nabla f_x = \begin{bmatrix} 1 & 0 & -\Delta T\,V(k-1)\cos(\tilde{\phi}(k-1/k-1) + \gamma(k-1) \\ 0 & 1 & \Delta T\,V(k-1)\cos(\tilde{\phi}(k-1/k-1) + \gamma(k-1) \\ 0 & 0 & 1 \end{bmatrix} \quad (27)$$

These are the standard prediction equations that are used to generate and publish the state values based on the information available up to time *k-1*. The state values can then be used by other modules of the system to perform the guidance tasks. Such is the case of the controller or the path planner. The prediction module evaluates the states at regular time intervals. The previous equations assumed that the estimate at time *k-1* and a prediction for the state at time *k* is desired. If an estimate at time *k* is available and a prediction at time *k+1* is required the equation needs to be modified by adding a one to the sub-index k. This will depend on the actual implementation of the algorithm.

### 2.2.3. Estimation stage

Once an external observation becomes available it can be used to update the value of the states. The predicted observation is evaluated to form the Kalman filter innovation sequence:

$$\nu(k) = z(k) - h(\hat{x}(k/k-1)) \quad (28)$$

Finally the states are updated with

$$\tilde{x}(k/k) = \hat{x}(k/k-1) + W(k)\nu(k) \quad (29)$$

where

$$
\begin{aligned}
W(k) &= P(k/k-1)\nabla h_x^T(k)S^{-1}(k) \\
P(k/k) &= P(k/k-1) - W(k)S(k)W^T(k) \\
S(k) &= \nabla h_x(k)P(k/k-1)\nabla h_x^T(k) + R
\end{aligned}
\tag{30}
$$

In the previous equations *W(k)* is the Kalman gain, *P(k/k)* is the covariance of the states and *S(k)* is the covariance of the innovation sequence. Since the relation of the states and the observation is non-linear the matrix $\nabla h_x$ needs to be evaluated on line at every update interval. For this particular example the gradient $\nabla h_x$ is

$$
\begin{aligned}
\nabla h_x &= \frac{1}{d}\begin{bmatrix} \hat{x}(k/k-1) - X_i & \hat{x}(k/k-1) - Y_i & 0 \\ -\hat{x}(k/k-1) - Y_i & \hat{x}(k/k-1) - X_i & -1 \end{bmatrix} \\
d &= (X_i - \hat{x}(k/k-1))^2 + (Y_i - \hat{y}(k/k-1))^2
\end{aligned}
\tag{31}
$$

Before using this information in the update stage two tasks need to performed:

1. Data validation: to determine with a certain level of confidence that the observation obtained comes from a registered beacon.
2. Beacon identification: to determine which beacon has been detected.

These algorithms must be very reliable, since a false matching may produce catastrophic results in the position estimation algorithm. The test in these applications must be very conservative since it is much safer to reject good beacon data than it is to use an observation matched with the wrong beacon.

Another implementation issue that has to be taken into account is the synchronisation of the external information data.

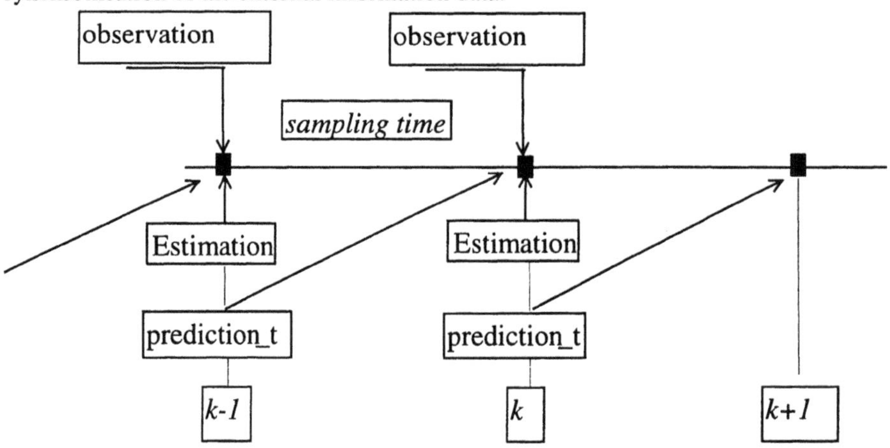


**Figure 14. Synchronous Kalman Filter implementation**

Kalman filters have been widely applied in process control and in many other areas where the time constants of the processes are very large compared to the prediction and observation rates. In most of these applications the observation is assumed to be taken at fixed interval of time, usually the sampling time of the filter. In this case, as shown in Figure 14, the filter always performs a prediction and an update at every sampling time. It can be seen that the observations are synchronised with the prediction intervals. In autonomous vehicles applications it is not usually possible to determine when an external observation will be available. Bearing or range will not in general be synchronised with the prediction intervals. Furthermore, external information may not be available for many prediction intervals. In general the scenario shown in Figure 15 is present.

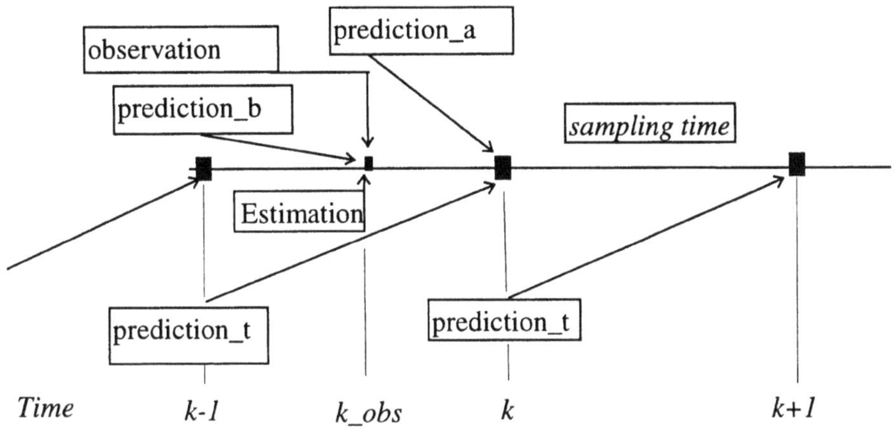


**Figure 15. Asynchronous Kalman Filter implementation.**

At time *k-1* the system will obtain and publish the predicted states at *k*. This information is readily available for all the other tasks in the system. At time *k_obs* an external observation is obtained. Then a new prediction is done from *k-1* to *k_obs*. At the observation time the state is updated and then propagated from *k_obs* to time *k*. Now the new state at time *k* has incorporated all the information available up to time *k*.

### 2.2.4. Final Remarks

This navigation loop has considered the case of non-linear models for the process and observation. The extended Kalman filter was used to address this problem. In this case the gradient of the process and observation models has to be evaluated every time a prediction or update is required. It is important to mention that the non-linear model has been linearised around a nominal point and will be valid if the deviation from the linearization point is not significant. This is usually contemplated by considering a sampling time short enough compared to the time constant of the process under consideration.

Another limitation of the EKF is the computation requirement. Since the gradient is dependent on the actual value of the states, the Kalman gain matrix can not be pre-computed and has to be evaluated on line for every trajectory. Finally EKF implementations are a lot more difficult to tune. Due to the non-linearity of the system many cases need to be contemplated during the testing and tuning stage.
Even with all these limitations the EKF is one of the most widely used tools to fuse data when the prediction or observation model incorporates non-linear terms.

## 3. Multi-loop Navigation Architecture

A high integrity navigation system has to be implemented with multiple navigation loops. Each loop will be based on sensors with different failure modes. Such architecture is shown in Figure 16.

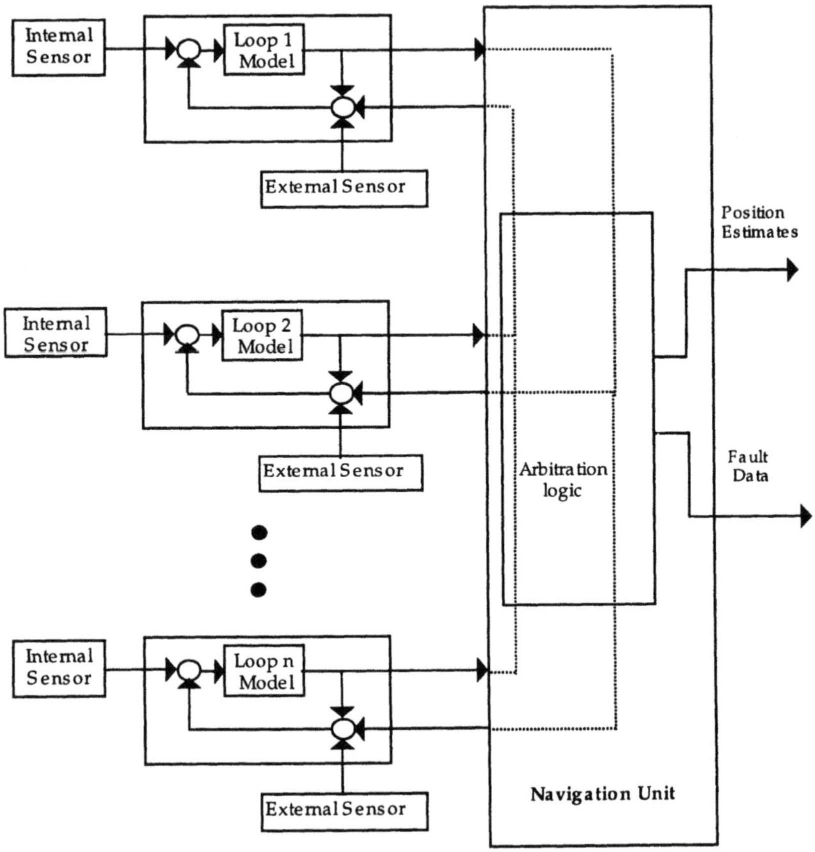


**Figure 16: Multi-Loop Navigation Architecture**

With this approach high frequency faults, such as jumps in sensors, and low frequency faults such as biases, can be detected since it is possible to have redundant

information in the frequency domain with different types of sensors. Readers interested in fault detection techniques can see [11].
The new information obtained from the individual loops can be incorporated into other loops by the navigation unit. This unit can also include fault detection capabilities to detect low frequency faults before using the loop's broadcast information.

### 3.1. Decentralised Systems with Asynchronous Observations

A decentralised estimation system is shown in Figure 17. The n local navigation loops work with their own sensory system (Local Filter and sensor blocks).

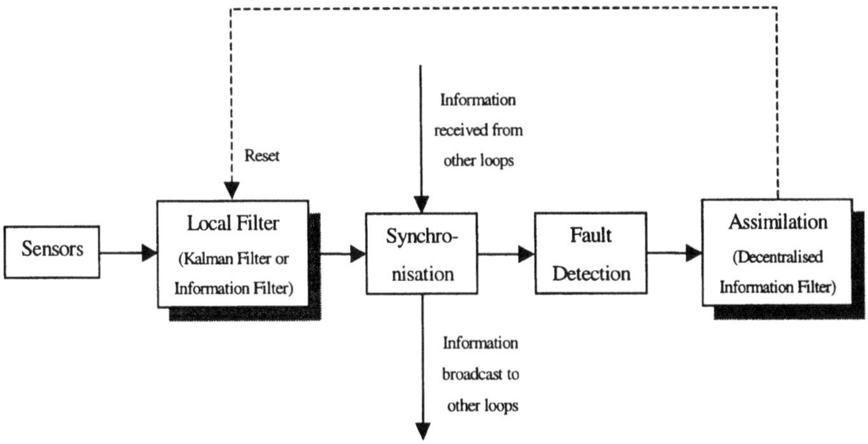


**Figure 17. Flow of information in a local loop**

These local loops may also include high frequency fault detection capabilities to reject sensory faults. Each loop has a local Kalman filter in which prediction, observation and estimation takes place independent of the other loops. Each loop has a prediction time interval and observation rate dictated by their particular sensory system. The Kalman filters implemented in the local loop will make predictions of states and covariances ( $\hat{x}$ and $\hat{P}$ ), at constant interval of times. When new observations become available, the filter will fuse the data to obtain new estimations for the states and covariance matrices ( $\tilde{x}$ and $\tilde{P}$ ). At this point the loop has gathered new information that can be transmitted to the rest of the system.
To fuse the information obtained from the different loops in the assimilation stage, it is required that all of the information corresponds to the same time. Therefore, considering that the sensors observe at different times, it is required to synchronize all the information obtained from the sensors before the assimilation stage.
To obtain the new information at the assimilation time, $\hat{x}$, $\hat{P}$, $\tilde{x}$ and $\tilde{P}$ are propagated forward to the next assimilation time. Denote the values corresponding to the next assimilation time by $\hat{x}_a$, $\hat{P}_a$, $\tilde{x}_a$ and $\tilde{P}_a$. The information form of the above vectors are obtained as:

$$\hat{y}_a = \hat{P}_a^{-1}\hat{x}_a$$
$$\hat{Y}_a = \hat{P}_a^{-1}$$
$$\tilde{y}_a = \tilde{P}_a^{-1}\tilde{x}_a \tag{32}$$
$$\tilde{Y}_a = \tilde{P}_a^{-1}.$$

The new information obtained from the observation is then calculated by

$$i_a = \tilde{y}_a - \hat{y}_a$$
$$I_a = \tilde{Y}_a - \hat{Y}_a. \tag{33}$$

where $i_a$ and $I_a$ are the information state vector and information matrix corresponding to the assimilation time, respectively.

In the assimilation stage, all the information obtained from the various loops (after having been synchronised) are fused with the local loop prediction through the equations:

$$\tilde{y}_l(k \mid k) = \hat{y}_l(k \mid k-1) + \sum_{j=1}^{N} (i_a)_j$$
$$\tilde{Y}_l(k \mid k) = \hat{Y}_l(k \mid k-1) + \sum_{j=1}^{N} (I_a)_j, \tag{34}$$

where the index "*l*" is the loop number, N is the total number of loops and $(i_a)_j$ and $(I_a)_j$ are the propagated values of the new information obtained from the loop number "*j*" to the assimilation time.

Once $\tilde{y}(k/k)$ and $\tilde{Y}(k/k)$ have been updated with the new information provided by all the other loops of the system, the local state vector $\tilde{x}_a$ and covariance $\tilde{P}_a$ are finally updated in the local loop using the relation described in equation 32.

Another aspect worth discussing is the assimilation rate. A system that assimilates all the information present in the application will be optimal with respect to the estimate covariances but will not be fault tolerant [14],[15]. There are low frequency faults that the individual loops will not be able to detect and will be broadcast to the other loops. It is important to include low frequency fault detection capabilities here, since it is the only stage where these faults can be detected. There will be a trade off between the assimilation rate and the lower frequency of the fault that the system will be able to detect. A more comprehensive discussion of synchronous and asynchronous data fusion schemes is presented in [16] and [17].

Figure 18 presents the actual experimental setup used in the ACFR to test these algorithms.

**Figure 18 Experimental setup**

## 4. Conclusion

This chapter presented the designs of two navigation loops based on linear and non-linear Kalman filter implementation. The methodology presented can easily be extended to contemplate for more complicated models and different external sensors. A multi-loop architecture is also presented with an asynchronous data fusion scheme to incorporate all the information available in the system.

## 5. References


1. Nease, A.D. "Air force construction automation/robotics". Proc. National Telesystems Conference, May 1992.
2. Bernold L E. "Experimental studies on mechanics of lunar excavation" J. Aerospace Engineering, 4 (1), Jan 1991, 9-22.
3. Krotkov E., Simmons R., Cozman F., Koenig S., "Safeward Teleoperation for Lunar Rovers: From Human factors to Field trials", Workshop WT1, 1996 IEEE International Conference on Robotics and Automation.
4. Chatila R., Lacroix S., Betge-Brezetz S., Devy M., Simeon T., "Autonomous Mobile Robot navigation for planet Exploration, The EDEN Project", Workshop WT1, 1996 IEEE International Conference on Robotics and Automation, Minneapolis, Minnesota, USA.
5. Giralt G., Laurence B., Luc M., "The Iares Project: Rovers for the human conquest of the Moon and Mars", Workshop WT1, 1996 IEEE International Conference on Robotics and Automation, Minneapolis, Minnesota, USA.
6. Volpe R., Balaram J., Ohm T., Ivlev R., "The Rocky 7 Mars Rover prototype", Workshop WT1, 1996 IEEE International Conference on Robotics and Automation, Minneapolis, Minnesota, USA.

7. Scheding, E. M. Nebot , M. Stevens and H. Durrant-Whyte, J. Roberts and P. Corke, . Cunningham and B. Cook, "Experiments in Autonomous Underground Guidance", IEEE Conference on Robotics and Automation, 1997, pp. 1898-1903.

8. Maybeck P., "Stochastic Models, Estimation and Control", Volume 1, Academic Press, 1979.

9. Nebot E., Scheding S., Durrant-Whyte H., "Frequency domain Modeling of Aided GPS with Applications to High-Speed Vehicle Navigation Systems". Proc. IEEE International Conference on Robotics and Automation, New Mexico, April 1997, pp. 1892-1897.

10. Nebot E. , Durrant-Whyte H., "Initial Calibration and Alignment of an Inertial Navigation Unit", pp. 175-180, Proc. of M2VIP-97, Australia, September 1997.

11. Scheding S., G. Dissanayake, E. Nebot and H. Durrant-Whyte, "Slip Modeling and Aided Inertial Navigation of an LHD", IEEE Conference on Robotics and Automation, 1997, pp. 1904-1909.

12. Durrant Whyte H., "An autonomous guided vehicle for cargo handling applications", Int. Journal of Robotics Research, 15(5), pp. 407-441, 1996.

13. Scheding S., Nebot E., Durrant-Whyte H. "Fault Detection in the Frequency Domain Designing Reliable navigation Systems", FSR97, Canberra Australia.

14. Carlson N. A., "Federated filter for computer-efficient, near-optimal GPS integration," IEEE Position Location and Navigation Symposium (PLANS), pp. 306-314, 1996.

15. Lawrence P. J., and Berarducci M. P., "Navigation sensors, filter, failure mode simulation results using the distributed Kalman filter simulator (DKFSIM)," IEEE Position Location and Navigation Symposium (PLANS), pp. 697-710, 1996.

16. Bozorg M., Nebot E., Durrant-Whyte H., "Decentralised Filter", Submitted to FSR97, Canberra Australia.

17. Grime S. and Durrant-Whyte H. F., "Data fusion in decentralised sensor networks," Control Engineering Practice, vol. 2, pp. 849-863, 1994.

# 12

# AUTONOMOUS ROBOT MOTION PLANNING AND CONTROL IN UNCERTAIN ENVIRONMENTS: OVERVIEW AND A NEW ALGORITHM BASED ON SLIDING-MODE CONTROL

S.G. TZAFESTAS, M.P. TZAMTZI and G.G RIGATOS

Intelligent Robotics and Automation Laboratory
Department of Electrical and Computer Engineering
National Technical University of Athens
Zographou 15773, Athens, Greece

## 1. Introduction

Motion planning and execution for autonomous robots moving inside an environment populated by static and moving obstacles, constitutes one of the most important up to date problems of robotic research. The need for the derivation of safe feasible solutions to this problem increases significantly as the applications of robots extend to complicated, hazardous or even unknown environments. Indoor mobile and space robots are two representative cases of robotic applications where the robot must be able to move safely, avoiding collisions with obstacles, inside an environment where unforeseen changes might take place. Moreover, all these tasks must be performed with a minimum interaction of the human operator.

Although the aforementioned examples are rather extreme with respect to the complexity of the robot's workspace, the need for autonomous motion planning remains, in a more or less complicated form, in all robotic applications. Depending on the specific task performed by the robot, several other factors besides safety, have to be taken into account during motion planning. For example, minimization with respect to the length of the path, the duration of motion or the energy consumed must usually be achieved, in order to consider the robot's mission effectively accomplished.

The method applied for motion planning depends on the type of the robot, the information available about its workspace and the limitations posed by the performed task. Most of the existing research results concern the path planning problem of mobile robots moving inside static precisely known environments. However, the most interesting and demanding applications involve dynamic environments, where other objects besides the robot are moving. Such cases are faced using more complicated methods, that perform an overall design of the robot's motion, that is path planning plus velocity profile design.

The motion planning of manipulators has also been treated, resulting to poorer results, since the existence of many degrees of freedom increases significantly the complexity of the problem. The cases of redundant and mobile manipulators constitute also interesting aspects of the problem.

More recently, research results have appeared for the motion planning of multiple robots moving inside the same workspace. Special care has to be taken when the robots are cooperating to perform a task, since in this case their motions cannot be designed independently.

The complexity of the motion planning problem increases considerably, when the dynamic and kinematic constraints of the robot are taken into account. The difficulty in this case is caused by the fact that a motion design, that avoids collisions with obstacles, might not be feasible with respect to the posed constraints.

In addition to the motion planning problem, a control law, whose application accomplishes the desirable motions execution, has to be designed. Moreover, the treatment of real situation problems demands taking into account several factors, like the model uncertainty, the declination between the executed and the desirable motion or the computational efficiency of the proposed algorithm.

A short overview of the existing results, concerning the aforementioned aspects of the motion planning problem, will be presented in the following. A detailed discussion of the existing methods is presented in [1].

**Mobile Robots**: Most of the research about motion planning of mobile robots concerns the path planning problem, that considers only the geometry of the robot and the obstacles in the workspace, ignoring dynamic constraints. This aspect of the problem is known as the *geometric motion planning*, the *piano mover's* or the *FindPath* problem [2].

Many of the methods proposed for the solution of the FindPath problem are using graph representations of the robot's configuration space. The edges in the graph correspond to subpaths between feasible configurations. The determination of the path between a start and a goal point is performed using graph search methods, such as the $A^*$-search method [3,4]. Alternative approaches are the Voronoi diagrams [5-9], the visibility graph [10] and the approximate cell decomposition method [11,12].

An interesting approach is proposed by the artificial potential field method. According to this method, a potential field is considered in such a way that the forces of the field push the robot towards the goal and away from the obstacles [13-16].

The aforementioned methods are divided into two classes: global and local [17]. The global methods, such as the cell decomposition, assume that a model of the whole robot's workspace is available, while the local ones, such as the potential field method, study at each instant only a small part of the workspace around the present position of the robot. The global approaches have the advantage that they provide a complete path from the start to the goal point. However they are rather slow, and become inapplicable in cases where the environment is inaccurate or changes rapidly. On the other hand, the local methods are fast, with low computational complexity, but suffer from local minima, where the robot may be trapped. Approaches that cope with this problem for the case of artificial potential field are given in [18,19].

Most of the motion planning methods try to avoid contact with obstacles. A different approach is presented in [20,21], where contact-based motions are designed, either due to the nature of the performed task or for reducing the position uncertainty.

When there are moving obstacles in the robot's workspace, more demanding motion planning methods have to be applied. The decomposition of the motion planning problem inside an environment with stationary and moving obstacles to the problems of path planning and design of velocity profile along this path, has been proposed in [22-24] by Kant and Zucker, as a method to simplify the solution of the collision avoidance problem. The same method has also been followed in subsequent works [25,26]. Other approaches, where both the path and the velocity are designed in order to avoid moving obstacles are presented in [27].

In many robotic applications, the motion of the robot must be designed taking into account several cost factors, such as time, energy, etc. A path planning method that selects optimal paths with respect to multiple objectives is given in [28].

**Manipulators**. The existing results concerning motion planning of robotic manipulators are rather poor compared to those for mobile robots. This is due to the following two facts [29]: i) the representation of the configuration space is rather difficult, and ii) the dimension of the configuration space is high for most practical manipulators. In order to face the first problem, discrete representations of the configuration space have been proposed in [30-32]. However, the second fact makes practically impossible the treatment of manipulators with more than six degrees of freedom [32-34]. In order to face this problem a solution is proposed in [35] and [36] where the motion planning is performed sequentially for the links, starting from the base link. In [29] a backtracking mechanism is incorporated into this motion planner in order to increase its effectiveness. Other approaches to this problem involve two-level local-global planners [37] or random search techniques [19].

However, when the manipulator moves inside a cluttered environment, the previous methods that avoid contact with obstacles may prove to be impractical. A methodology called *obstacle accommodation* that controls contact, instead of avoiding it, in order to prevent damages to the robot, has been presented in [38]. The contact points can be at any position on any link.

**Redundant Manipulators.** An interesting case of motion planning for manipulators is the case of redundant manipulators. A manipulator is called *redundant* when it has more degrees of freedom that required for covering the workspace [39]. This property provides the capability of reconfiguring the manipulator without affecting the end-effector. The kinematic redundancy has the result that, for each task coordinates, the equation that describes the forward kinematics of the manipulator, has an infinite number of joint coordinates solutions. The motion planning is performed by selecting specific solutions following the *task augmentation* or the *optimization* approach [40]. The first method adds degrees of freedom to the performed task, eliminating this way the redundancy. The *Jacobian extension* technique [41-43] follows this approach.

The optimization methods exploit the extra degrees of freedom in order to optimize certain cost functions. Methods using *Jacobian pseudoinverses* [44,45] follow this approach. Furthermore, methods using potential functions designed in such a way that obstacle avoidance or other performance criterions are satisfied have been presented in [16,46-50]. However, these methods have the disadvantage of stacking in local minima.

The aforementioned methods deal with local motion planning, so they cannot give solutions when the range of motion is large and the workspace is cluttered with obstacles. Then global optimization techniques have to be applied. Such methods using global optimization are presented in [51,52]. A methodology that results in feasible but not optimal paths, with lower computational complexity is given in [40]. Finally, a global path planning approach using discretization of the configuration space and achieving design at a minimum overall cost is presented in [39].

**Mobile Manipulators**. Mobile manipulators are manipulators attached to a mobile platform. The mobile platform provides some form of redundancy, different from the one due to extra joints. The difference consist in the fact that extra joints increase the flexibility within the workspace, while the mobility of the platform increase both the flexibility and the effective workspace [53].

The motion planning problem of mobile manipulators can be treated following methods developed for redundant manipulators [54]. Moreover, motion planning methods have been developed specifically for mobile manipulators [55,56]. In [53] a motion planning method for manipulators that perform a sequence of tasks, described by a prespecified trajectory of the end-effector, is developed. Finally, a planning and control algorithm for the platform, that takes the measured joint displacement of the manipulator as the input for motion planning, and controls the platform in order to bring the manipulator into a preferred operating region, is presented in [57].

**Nonholonomic Mobile Robots**. In the geometric formulation of the motion planning problem for mobile robots, the robotic system is considered to be subject to holonomic constraints, such as the linking between the joints of a manipulator. This type of constraints can be described by restricting the configuration space. On the contrary a nonholonomic constraint is expressed as a non-integrable equation involving the derivatives of the configuration parameters. As a result, these constraints cannot be eliminated by restricting the configuration space. So, there is the possibility that a path in the configuration space does not correspond to a feasible trajectory.

Nonholonomic motion planning has been approached with mainly two different ways: i) search based methods, that use some type of decomposition of the configuration space [58-60] and ii) control theoretic methods that use tools of geometric control theory [61,62]. Most of the methods of the second class treat the motion planning problem ignoring the presence of obstacles. A nonholonomic motion planner that treats collision avoidance, has been presented in [63]. A collision free path that ignores nonholonomic constraints is designed first. Then, this path is transformed to a feasible one by using an exact characterization of the minimal-length constrained path connecting any two configurations in the absence of obstacles. The presence of obstacles is taken into account by computing subgoals along the initial path. A similar approach is presented in [64].

**Multiple robots**. As the number and kinds of robotic applications increase, there appear more cases where more than one robots are moving within the same workspace. Then, the motion planning problem must be performed taking into account the interaction between the robots. Many methods treat the other robots as moving obstacles. However, this might prove impractical, especially when the robots cooperate to perform a task.

The problem of simultaneous motion planning for manipulators moving in the same workspace is treated in [65,66], using configuration space methods. More specifically, in [66] the problem of two planar robot arms moving a common object is considered.

The motion planning of multiple robots is studied in [67]. The approach proposed is based on dynamically grouping the robots. Each group consists of individual robots that have the same goal position and keep relationship with one another to reach the goal. The algorithm proposed suggests that the members of the group follow a member characterised as the *leader*. Similar approaches have been proposed in [68,69].

It is worth noting, before ending this section, that methods developed for motion planning of robots can find applications in other scientific areas. A characteristic example is the work in [70], where an experimental motion planner is developed for the beam of radiation used to destroy brain tumors, during stereotaxic radiosurgery.

The following sections study the problem of control and motion planning in the presence of uncertainty, for a mobile robot subject to state and actuator constraints, that moves along a predetermined path inside a terrain with moving obstacles. The basic idea followed consists in the combination and mutual interaction of a known motion planning method, called *minimum interference strategy (MIS)*, and a sliding mode controller.

## 2. Motivation and Background Information for the Proposed Method.

The problem of motion along predefined paths appears in many robotic applications [27], [71-77]. In some cases the path on which the robot moves is determined by the nature of the job performed by the robot, e.g. arc welding or laser cutting. In other cases the robot cannot move freely because of limitations concerning its workspace. For example, there may be a specific network of roads that have to be followed.

In any of the previous cases, the collision avoidance with other objects moving in the workspace of the robot can be achieved by adjusting the motion along the predefined path. The problem of designing the robot's velocity profile along this path in a such a way that guarantees collision avoidance and/or minimum time optimisation has been treated in several papers, e.g. [27], [71-77] .

Following the idea of decomposition of the motion planning problem to the problems of path planning and design of velocity profile introduced by Kant and Zucker in [22-24], Kyriakopoulos and Saridis [71] proposed a method for motion planning along a predefined path, that guarantees avoiding collisions with moving obstacles. The basic idea followed is to alter the velocity along the path, but at the same time satisfy the dynamic constraints of the robot and stay as close as possible to a nominal plan. The proposed method, called *Minimum Interference Strategy* gives suboptimal with respect to the time of the motion, but very computationally efficient solutions, a fact that makes it attractive for cases where on-line motion planning is required.

The motion planning method proposed in [71] is performed taking into account velocity and actuator constraints. The solution that comes from the MIS, in some intervals requires that the robot moves with the minimum or maximum velocity, while in other intervals requires the implementation of the minimum or maximum actuator force. However, driving the system to its limits provides no possibility to face disturbances or uncertainty. Any deviation from the nominal design could cause undesirable effects or demand the application of control laws outside the actuator's range.

The presence of uncertainty in the system, as well as in the workspace of the robot, has been treated in several works concerning motion planning [20,21,76-78]. A very interesting approach to the problem of facing uncertainties during the path-constrained robot control with limited torques, aiming to minimum time optimisation is presented in [76] and [77]. There, a path velocity controller is designed that modifies on-line the nominal velocity profile.

In the present chapter we propose a method of motion planning and control law design for the problem treated in [71], under the additional assumption of parametric uncertainty in the model that describes the robot's motion. The basic idea followed consists in a combination of the MIS with a robust control law design method and suggests that the desired velocity profile should be designed using smaller velocity and actuator ranges, in such a way that, despite the errors due to uncertainty and the deviation between the nominal and the robust control law, the actual velocity and actuator limits are not violated. However, the aforementioned makes obvious the need for a robust control law that provides a simple and accurate estimation of the errors between the desired and the performed motion, as well as the declination from the nominal control law. These requirements have led to the selection of sliding mode control [79].

Sliding mode control has been widely used in robotic applications [80-83], since it provides an easily designed and implemented method to solve the trajectory following problem for nonlinear systems in the presence of parametric uncertainty. In the present work, the sliding mode control is not used for the solution of the tracking problem in its classical sense, but for the control of the robot's velocity along the predefined path.

## 3. MIS Method for Perfectly Known Models

The problem studied in [71], concerns the motion of a vehicle between two points A and B, on a surface $g(x,y,z)=0$ along a path $\boldsymbol{r}(s)=[x(s),y(s),z(s)]^{\mathrm{T}}$, chosen in such a way that collisions with stationary obstacles are avoided (A corresponds to $s=s_0$, and B to $s=s_f$). The path $\boldsymbol{r}(s)$ describes the motion of the center of gravity $G$ of the robot with respect to a world coordinate frame $W$. It has been proved in [71] that under the assumption that neither sliding nor skidding occurs (and some additional simplifying assumptions), the motion of the robot along $\boldsymbol{r}(s)$ is described by the model equation

$$(m + I_R(s))\,\ddot{s} + I_R'(s)\,\dot{s}^2 + mgz'(s) = u \tag{1}$$

with initial conditions

$$s(0) = s_0\,,\ \dot{s}(0) = v_0 \tag{2}$$

final conditions

$$s(t_f) = s_f\,,\ \dot{s}(t_f) = v_f\,,\ t_f = \text{free} \tag{3}$$

and state restrictions

$$0 < \varepsilon \le \dot{s}(t) \le v_{\max}\big(s(t)\big) \tag{4}$$

where

- $m$ denotes the mass of the robot,
- $I_R(s) := < \Omega(s), \boldsymbol{I}\Omega(s) >$,
- $\boldsymbol{I}$ is the moment of inertia of the robot around the center of gravity $G$,
- $\Omega(s)$ is determined by the equation $\omega = \Omega(s)\,\dot{s}$,
- $\omega$ is the vector of angular velocity,
- $g$ is the gravity constant,
- $z(s)$ is the height of $G$ with respect to the world coordinate frame,
- $u$ is the sum of the actuator forces on the tangent of $\boldsymbol{r}(s)$,
- $\varepsilon$ is a positive constant,
- the dot and the prime denote derivatives with respect to $t$ and $s$, respectively.

Furthermore, the input $u$ applied to the system (1) is restricted according to the condition

$$-U_2 \le u \le U_1 \tag{5}$$

where $U_1, U_2$ express the actuator constraints.

Consider now, that there are other moving objects in the robots workspace, which are assumed not to permanently stay on or move parallel to $\boldsymbol{r}(s)$. Then, the probable collision sets

$$D_s = \{s \mid \exists\, t : \text{collision occurs if } G \text{ is on } s \text{ at time } t\}$$
$$D_t = \{t \mid \exists\, s : \text{collision occurs if } G \text{ is on } s \text{ at time } t\}$$

can be derived with the use of information provided by sensors.

The MIS is based on the fact that collision with the moving objects can be avoided when either one of the two conditions

$$s(t) \ge s_2 = \max D_s,\ \forall\, t \ge t_1 = \min D_t \tag{6}$$
$$s(t) \le s_1 = \min D_s,\ \forall\, t \le t_2 = \max D_t \tag{7}$$

is satisfied. When condition (6) is satisfied the method is called **Accelerating MIS (AMIS)**, while in the second case when (7) is satisfied, it is called **Decelerating MIS (DMIS)**. Furthermore, the AMIS and DMIS methods propose velocity profiles, such that the overall motion time $t_f$ is as close as possible to $T$, that is the total time of motion according to a nominal velocity profile $v_n(s)$, chosen with the use of an optimality criterion. In this sense, the solutions are suboptimal with respect to time.

Let $I_i = [v_{\min}^i, v_{\max}^i] \subseteq [\varepsilon, v_{\max}(s_i)]$, $i = 1, 2$ be an interval of feasible velocities at $s_i$, such that the following property holds: for all $v \in I_i$, there exists a velocity profile for the robot that i) satisfies the restrictions (2)-(4), ii) can be achieved with a control law satisfying (5), iii) the robot reaches $s_i$ at $t_{2-i+1}$ with velocity $v$ and iv) the time of motion $t_f - t_{2-i+1}$ from $s_i$ to $s_f$ is as close as possible to $T - t_{2-i+1}$. Then, the velocity profile design is decomposed to the following two problems:

1) find $v(s), s \in [s_0, s_i]$, $i = 1, 2$, such that $v(s_0) = v_0$, $v(s_i) = v \in I_i$ and $\int_{s_0}^{s_i} \frac{1}{v(s)} ds = t_{2-i+1}$,

2) find $v(s), s \in [s_i, s_f]$, $i = 1, 2$, such that $v(s_i) = v$, $v(s_f) = v_f$ and $\int_{s_i}^{s_f} \frac{1}{v(s)} ds = t_f - t_{2-i+1}$.

where $v(s)$ must satisfy (4) and be achievable with a control law satisfying (5). Note that $i = 1$ corresponds to DMIS and $i = 2$ to AMIS. The velocity profiles proposed in [71] are as follows:

**Accelerating Minimum Interference Strategy (AMIS)**

1. A part with velocity $v(s) = v_n(s)$, for $s \in [s_0, s_c]$(nominal velocity function).
2. A part with input $u(s) = U_1$, for $s \in [s_c, s_v]$.
3. A part with velocity $v(s) = v_{\max}(s)$, for $s \in [s_v, s_b]$.
4. A part with input $u(s) = -U_2$, for $s \in [s_b, s_2]$.
5. A part with input $u(s) = -U_2$, for $s \in [s_2, s_g]$.
6. A part with velocity $v(s) = v_{\min}(s)$, for $s \in [s_g, s_u]$.
7. A part with input $u(s) = U_1$, for $s \in [s_u, s_r]$.
8. A part with velocity $v(s) = v_n(s)$, for $s \in [s_r, s_f]$ (nominal velocity function).

**Decelerating Minimum Interference Strategy (DMIS)**

1. A part with velocity $v(s) = v_n(s)$, for $s \in [s_0, s_c]$(nominal velocity function).
2. A part with input $u(s) = -U_2$, for $s \in [s_c, s_v]$.
3. A part with velocity $v(s) = v_{\min}(s)$, for $s \in [s_v, s_b]$.
4. A part with input $u(s) = U_1$, for $s \in [s_b, s_1]$.
5. A part with input $u(s) = U_1$, for $s \in [s_1, s_g]$.
6. A part with velocity $v(s) = v_{\max}(s)$, for $s \in [s_g, s_u]$.
7. A part with input $u(s) = -U_2$, for $s \in [s_u, s_r]$.
8. A part with velocity $v(s) = v_n(s)$, for $s \in [s_r, s_f]$ (nominal velocity function).

The points $s_c, s_v, s_b, s_g, s_u$ and $s_r$ can be computed iteratively using arithmetic analysis methods, following the proceeding rules:

1) For AMIS, velocity $v(s_2) \in I_2$ is selected as big as possible. For DMIS velocity $v(s_1) \in I_1$ is selected as small as possible.

2) The points $s_c$ and $s_r$ must be as close as possible to $s_2$ for AMIS ($s_1$ for DMIS), in order for the robot to move with its nominal velocity for as long as possible.

3) The values of these six points must be selected so that the requirements regarding the duration of motion from $s_0$ to $s_i$ and from $s_i$ to $s_f$ are satisfied.

Before completing this Section, we have to note that in some cases one of the two methods, or even both of them, may not be feasible under the state and actuator constraints posed. For further details regarding the design of MIS velocity profiles, the reader is referred to [71].

**4. Robust Control and Motion Planning**

When the parameters that describe the dynamic behaviour of the system (1) are perfectly known, then it is easy to compute a control law $u$, so that the robot executes the motion proposed by MIS. However, when the system suffers from uncertainty, then a robust control law has to be implemented.

As can be seen from the previous section, the motion proposed by MIS drives the system to its limits, as in some parts the system moves with the maximum or minimum allowed velocity, while in other parts the maximum acceleration or deceleration has to be implemented. Although, this fact can be accepted when the model of the robot is perfectly known, it can cause problems in the presence of uncertainty. For example, if the robot moves with velocity greater than the maximum allowed, then skidding could occur, or if the executed motion deviates significantly from the desirable one, collisions might not be avoided. Even when a robust control law is designed, that would possibly guarantee the exact execution of the desirable motion, the problem of keeping this control law inside the range defined by the actuator constraints has to be taken under consideration, because, otherwise, the actuators will not be able to implement the control law.

These problems can be faced, so that undesirable behaviour is avoided, by taking under consideration the expected extent of errors, during the motion planning with the use of MIS. For these reasons, it is necessary to select a robust control law that provides an easy way to compute the bounds of the expected errors. In addition, this control law must have a simple structure and be easily implemented, in order to avoid the increase of the computational burden. Sliding mode control appears to be a suitable choice, satisfying the previous requirements.

Sliding mode control is a robust control law for trajectory following [79], that gives the possibility to easily compute the bounds of the error between the desired and the executed trajectory. More specifically, when noncontinuous sliding mode control law is applied, then exact trajectory following can be achieved, under the assumption of zero initial condition error. This assumption seems rather reasonable for the problem under consideration, since the position and velocity of the robot can be considered to be accurately measured. The application of a continuous sliding mode control law, under the same assumption, can lead to a bounded error with a bound known in advance. Furthermore, the difference between the nominal control law that would achieve the desired performance in the ideal case of exact knowledge of the model, and the sliding mode control law is bounded with a bound that depends on the extent of uncertainty in the model.

In the following, the mathematical formulation of the proposed method is presented. Consider, that the uncertainty of the parameters that appear in (1) is described by the following inequalities

$$m_l \le m \le m_u, \quad \boldsymbol{I}_l \le \boldsymbol{I} \le \boldsymbol{I}_u, \quad \boldsymbol{\Omega}_l(s) \le \boldsymbol{\Omega}(s) \le \boldsymbol{\Omega}_u(s) \tag{8a}$$

$$\boldsymbol{V}_l(s) \le \boldsymbol{\Omega}'(s) \le \boldsymbol{V}_u(s), \quad z_l'(s) \le z'(s) \le z_u'(s) \tag{8b}$$

where $m_l$, $m_u$ are known constants, $\boldsymbol{I}_l, \boldsymbol{I}_u$ are matrices of known constants, $\boldsymbol{\Omega}_l(s), \boldsymbol{\Omega}_u(s)$, $\boldsymbol{V}_l(s), \boldsymbol{V}_u(s)$ are vectors of known continuous functions and $z_l'(s)$, $z_u'(s)$ are known continuous functions of $s$. In addition, some estimates $\hat{m}, \hat{\boldsymbol{I}}, \boldsymbol{\Omega}(s), \hat{\boldsymbol{\Omega}}'(s)$ and $\hat{z}'(s)$

of these parameters are available. The dynamic model that comes from (1) when these estimates are used, will be referred to as *nominal model.*

The uncertainty in the model expresses lack of knowledge or time variation and can be caused by several reasons. For example, the mass and the moment of inertia could vary corresponding to the burden that the robot carries. The moment of inertia could also vary, when other parts, like a robot arm whose position might change with time, are attached to the robot. Moreover, the functions $\Omega(s)$ and $z'(s)$ depend on the path $\boldsymbol{r}(s)$ and the surface $g(x,y,z)=0$ on which the robot is moving. When the functions $\boldsymbol{r}(s)$ and $g(x,y,z)$ cannot be accurately defined, then $\Omega(s)$ and $z'(s)$ are not perfectly known.

Writing (1) in the form

$$\ddot{s} = f(s,\dot{s}) + b(s) \tag{9}$$

where

$$f(s,\dot{s}) := \frac{-I_R'(s)\dot{s}^2 - mgz'(s)}{m + I_R(s)} \tag{10}$$

$$b(s) := \frac{1}{m + I_R(s)} \tag{11}$$

and using (8), there can be found a continuous function $F(s,\dot{s})$ and two positive continuous functions $b_{\max}(s)$ and $b_{\min}(s)$, such that

$$\left| f(s,\dot{s}) - \hat{f}(s,\dot{s}) \right| \le F(s,\dot{s}) \tag{12}$$

$$b_{\min}(s) \le b(s) \le b_{\max}(s) \tag{13}$$

where $\hat{f}(s,\dot{s})$ is the function defined by (10) that corresponds to the nominal model. Consider now, that a discontinuous sliding mode control law

$$u = [u_1 - k\,\mathrm{sgn}(\sigma(s,\dot{s}))]/\hat{b}(s) \tag{14}$$

is designed, with

$$k \ge \beta(s)(F(s,\dot{s}) + \eta) + (\beta(s) - 1)\,|\,u_1| \tag{15}$$

where

$$u_1 = -\hat{f}(s,\dot{s}) + \ddot{s}_d - \lambda\dot{e} \tag{16}$$

$$\beta(s) = \max\left\{ \left(\frac{\hat{b}(s)}{b(s)}\right)_{\max}, \left(\left(\frac{\hat{b}(s)}{b(s)}\right)_{\min}\right)^{-1} \right\} \tag{17}$$

$$\sigma(s,\dot{s};t) = \dot{e} + \lambda e = 0,\quad \boldsymbol{e} = [e,\dot{e}] = \boldsymbol{s} - \boldsymbol{s}_d,\ \boldsymbol{s} = [s,\dot{s}],\ \boldsymbol{s}_d = [s_d,\dot{s}_d]$$

$\lambda$ and $\eta$ are strictly positive constants, $\hat{b}(s)$ is the function defined by (11) that corresponds to the nominal model, and $s_d(t)$, $\dot{s}_d(t)$ are the desired position and velocity time functions proposed by the implementation of MIS for the nominal model.

The sliding mode control law (14) can be written in the form

$$u = \hat{u}_n - [\,\lambda\dot{e} + k\,\mathrm{sgn}(\sigma(s,\dot{s}))]/\hat{b}(s) \tag{18}$$

where

$$\hat{u}_n = [\ddot{s}_d - \hat{f}(s,\dot{s})]/\hat{b}(s) \tag{19}$$

is the control law that would achieve perfect trajectory following for the nominal model. According to [79], the implementation of the control law (14) to the system (9), results in perfect trajectory following, under the assumption of zero initial condition error. Setting $|\dot{e}| = 0$ in (18), it follows that the deviation between the nominal and the sliding mode control law is absolutely bounded, according to the relation

$$|\,u - \hat{u}_n\,| \le k/\hat{b}(s) \tag{20}$$

This deviation can cause problems in those parts of the desired motion, where the nominal control law takes values close to the actuator limits, since in that case the actuators used might not be able to implement the sliding mode control law. This problem can be solved by implementing the MIS for the nominal model using a fictitious input constraint given by the relation

$$-U_2 + k\max\{1/\hat{b}(s) : s \in [s_0, s_f]\} \le u \le U_1 - k\max\{1/\hat{b}(s) : s \in [s_0, s_f]\} \qquad (21)$$

Using (16) and (19), relation (15) takes the form

$$k \ge \beta(s)(F(s,\dot{s}) + \eta) + (\beta(s) - 1)\left|\hat{b}(s)\hat{u}_n + \lambda\dot{e}\right| \qquad (22)$$

Since $\beta(s), \hat{b}(s)$ and $F(s,\dot{s})$ are continuous functions, while $s, \dot{s}$ and $\hat{u}_n$ are bounded with bounds determined by (2)-(5) and $|\dot{e}| = 0$, it follows that the valid values of $k$ can be estimated before the motion planning, and then be used for the determination of the fictitious actuator limits from (21).

However, in practice the application of discontinuous sliding mode control leads to control chattering. In order to smooth out the control discontinuity, (14) can be replaced by the control law

$$u = [\,u_1 - k\,\mathrm{sat}(s/\Phi)]/\hat{b}(s) \qquad (23)$$

where $\mathrm{sat}(\cdot)$ is the saturation function

$$\mathrm{sat}(y) = \begin{cases} -1, & y < -1 \\ y, & |y| \le 1 \\ 1, & y > 1 \end{cases}$$

and $\Phi$ is a positive constant.

Smoothing the control law according to (23), results to tracking within a precision $\varepsilon = \Phi/\lambda$ [79]. More specifically, under the assumption of zero initial condition error, it holds that

$$|e(t)| \le \varepsilon,\ |\dot{e}(t)| \le 2\Phi, \forall t \ge 0 \qquad (24)$$

From the aforementioned it follows that the term that contains the derivative of the error in (18) can no longer be ignored, so (20) takes the form

$$|u - \hat{u}_n| \le (k + \lambda|\dot{e}|)/\hat{b}(s) \le (k + 2\lambda\Phi)\max\{1/\hat{b}(s) : s \in [s_0, s_f]\}$$

where $k$ is determined from (22) taking into account (24). Hence, when continuous sliding mode control is applied, (21) has to be replaced by the condition

$$-U_2 + (k + 2\lambda\Phi)\max\{1/\hat{b}(s) : s \in [s_0, s_f]\} \le u$$
$$U_1 - (k + 2\lambda\Phi)\max\{1/\hat{b}(s) : s \in [s_0, s_f]\} \qquad (25)$$

In the case of discontinuous sliding mode control, there is no need to change the velocity limits of the desired motion, as perfect tracking is achieved. However, when continuous sliding mode control is implemented, the expected errors given by (24) have to be taken under consideration during the motion design, in order for the robot not to go beyond the velocity limits or get inside the interval $[s_1, s_2]$ for time instants inside $[t_1, t_2]$. This can be accomplished by designing the MIS velocity profile with velocity limits given by the relation

$$v_{\min}(s) + 2\,\Phi \le v(s) \le v_{\max}(s) - 2\,\Phi \qquad (26)$$

Furthermore, the MIS is applied for $s'_1, s'_2$, instead of $s_1, s_2$, where

$$s'_1 = s_1 - \varepsilon = s_1 - \Phi/\lambda\,,\ s'_2 = s_2 + \varepsilon = s_2 + \Phi/\lambda \qquad (27)$$

**5. Notes on the Practical Implementation of the Method**

The robust control and motion planning method described in the previous Section is suitable for on-line implementation, since its requirements for on-line computations are almost the same with those of the MIS for perfectly known models. This is due to the fact that facing the uncertainty is accomplished with a proper choice of the control law parameters and the input and state constraints, which is performed off-line. Then, the MIS velocity profiles are designed on-line, as if the model of the robot was perfectly known. Moreover, the sliding mode control has a simple structure, so its application does not demand complex computations.

However, several factors must be taken into account before the application of the method, in order to guarantee its effectiveness. Relations (21),(22) and (25)-(27) show that, in order to face the uncertainty, it is necessary to design the desired motion under constraints stricter than the initially posed. Furthermore, the sliding mode control parameters $k, \eta, \Phi$ and $\lambda$ have to be chosen in such a way that the new constraints do not become too restrictive. This interaction between MIS and sliding mode control poses constraints on the extent of uncertainty that can be faced by the specific method. The previous discussion makes obvious the need for an accurate estimation of the bounds of uncertainty described in (12) and (13). Note, that the functions $F(s,\dot{s}), b_{\min}(s)$ and $b_{\max}(s)$ can be determined taking into account that $s,\dot{s}$ remain inside intervals defined by (2)-(4).

Before the end of this section, the following point must be clarified. As can be seen from Section 3, the velocity profiles of MIS do not provide directly the functions $s_d(t), \dot{s}_d(t)$ and $\ddot{s}_d(t)$ which are needed for the design of the control law. In order to determine these functions, the following tasks have to be performed: Let $v_d(s)$ denote the velocity profile proposed by MIS.

i) Using the relation [71]

$$\tfrac{1}{2}(m+I_R(s))v_d^2(s) = \tfrac{1}{2}(m+I_R(s_k))v_d^2(s_k) + mg(z(s)-z(s_k)) + \int_{s_k}^{s} u(r)dr \quad \forall s, s_k \in [s_0, s_f]$$

we determine the values of $v_d(s)$ for $s \in [s_c, s_v], s \in [s_b, s_g]$ and $s \in [s_u, s_r]$, setting $u = U_1$ or $u = -U_2$.

ii) We determine the function $t_d(s) = \int_{s_0}^{s} \frac{1}{v_d(r)} dr$ , $\forall s \in [s_0, s_f]$. Since $t_d(s)$ is a strictly increasing function, the inverse function exists and constitutes the desirable position function $s_d(t)$.

iii) The desirable velocity and acceleration functions are determined by the relations

$$\dot{s}_d(t) = \frac{ds_d}{dt}(t) = \frac{1}{\dfrac{dt_d}{ds}(s_d(t))}, \qquad \ddot{s}_d(t) = \frac{d^2 s_d}{dt^2} = \frac{-\dfrac{d^2 t_d}{ds^2}(s_d(t))}{\left(\dfrac{dt_d}{ds}(s_d(t))\right)^3}$$

using analytic or approximating methods.

**6. Example**

Consider that the robot is moving on a flat horizontal surface ($z = 0$), along a curved path with radius $\rho$ greater or equal to 6m, while $s$ belongs in the interval [0,15], that is $s_0 = 0$ and $s_f = 15$. The mass $m$ of the robot varies between 50kg and 70kg, while the moment of inertia $I$ is considered equal to 32kgm$^2$. The limit values of the actuator forces are $U_1$=140N and $-U_2 = -60$N. The minimum velocity of the robot is taken equal to 0.1m/sec and the maximum one equal to 4m/sec. The nominal velocity function is chosen equal to 1m/sec, so the total time of the nominal motion is 15sec and the initial and final desired velocities are equal to 1m/sec. The robot's workspace is given in Fig. 1. The robot is moving from A to B, while another vehicle is moving from Γ to Δ. The values of $s$ at which collision can happen belong in the interval $[s_1, s_2] = [7m, 8m]$, while the corresponding instants of time belong in the interval $[t_1, t_2]$=[6sec,9sec]. The nominal trajectory which is used for the design of the desired motion is considered to be a straight line, while the mass of the nominal model is chosen equal to 60kg.

From the aforementioned requirements and relations (9)-(13), it follows that for this application $f(s,\dot{s}) = F(s,\dot{s}) \equiv 0, \forall s$ and $b(s) = 1/(m + I/\rho^2)$. The function $b(s)$ is bounded with $b_{\min}(s) = 0.014, b_{\max}(s) = 0.02$, and the corresponding nominal function is $\hat{b}(s) = 0.017$. Moreover, from (17) it follows that $\beta = 1.2$.

According to (22), for the case of discontinuous control law, the constant $k$ must be selected so that $k \geq \beta\eta + (\beta - 1)\left|\hat{b}\hat{u}_n\right|$, where $\hat{u}_n$ is the nominal control law that satisfies $-60 \leq \hat{u}_n \leq 140$. Choosing $\eta = 0.1$, it follows that $k \geq 0.586$. Let $k = 0.6$. Then, (21) implies that the fictitious values of the actuator limits that are used for the implementation of MIS become $U_1 =$ 104N and $-U_2 = -24$N. The discontinuous sliding mode is designed with $\lambda = 0.1$.

The simulation results from the application of the proposed method, with DMIS velocity design, on a model described by a curve with radius $\rho$=8m, and mass 55kg are presented in Figs. 2-4. As can be seen from Fig. 2, the robot's motion satisfies the condition $s(t) \leq 7$ for all $t \leq 9$, so collisions are avoided. In addition the total time of motion is almost equal to the nominal $T = 15\,\text{sec}$. Furthermore, Fig. 3 shows that the robot's motion satisfies the velocity limits, despite the presence of uncertainty. Finally, from Fig.4 it appears that the applied sliding mode control law remains inside the actual actuator limits. The chattering that appears in the simulation results is due to the use of a discontinuous control law and to the simulation errors.

In order to eliminate chattering, a continuous sliding mode control can be applied. The constant Φ is chosen equal to 0.01. Then, working as previously, taking also into account the tracking errors given from (24), we find that $k$ must be selected greater than 0.5864. Choosing again $k = 0.6, \lambda = 0.1$, it follows from (25) that the fictitious values of the actuator limits that are used for the implementation of MIS become $U_1 = 103.8$ N and $-U_2 = -23.88$ N. Furthermore, the velocity limits and the values of $s_1, s_2$ have to be changed. Specifically, according to (26) and (27), the MIS is applied with $v_{\min} = 0.12, v_{\max} = 3.9$, $s_1' = 6.9$ and $s_2' = 8.1$.

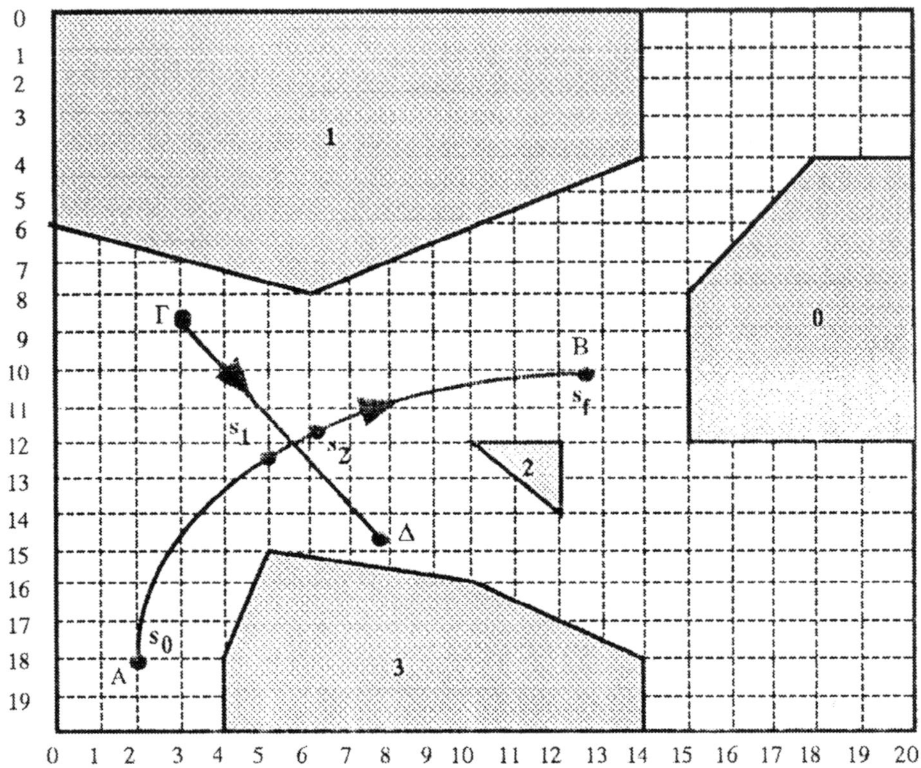


**Fig. 1 The workspace of the robot**

The simulation results appear in Figs. 5-7. From Fig. 7 it follows that the chattering has decreased significantly comparing to the case of discontinuous sliding mode control. However, the selected value of $\Phi$ is small, so the chattering has not been completely eliminated. Moreover, Figs. 5-7 show that collisions with moving objects are avoided, the total time of motion is almost equal to the nominal, and, in addition, the velocity and the control law applied remain always inside the actual limits.

## 7. Conclusions

The robust control and motion planning method proposed in the present chapter, provides an effective way of facing on-line the problem of collision avoidance with moving obstacles during the motion of a robot along a predefined path, without violating the state and input constraints, even under the presence of uncertainty. Furthermore, the motion executed remains as close as possible, with respect to the time of motion, to a nominal velocity profile. The method is suitable for on-line implementation. The presence of uncertainty does not increase the computational burden, because the new input and velocity constraints used for the implementation of MIS, as well as the sliding mode control law parameters can be determined off-line.

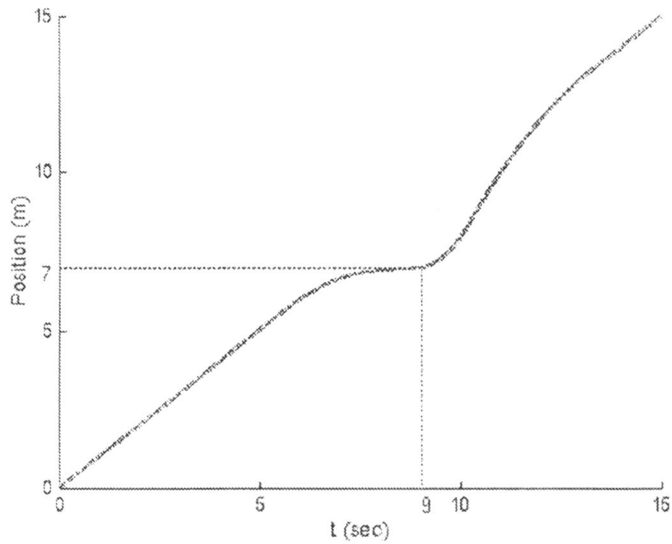


**Fig. 2 Position of the robot for discontinuous sliding mode control**

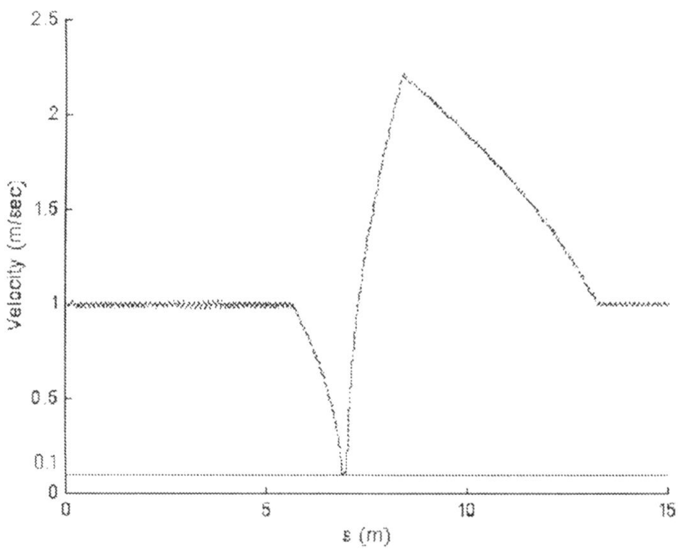


**Fig. 3 Velocity of the robot for discontinuous sliding mode control**

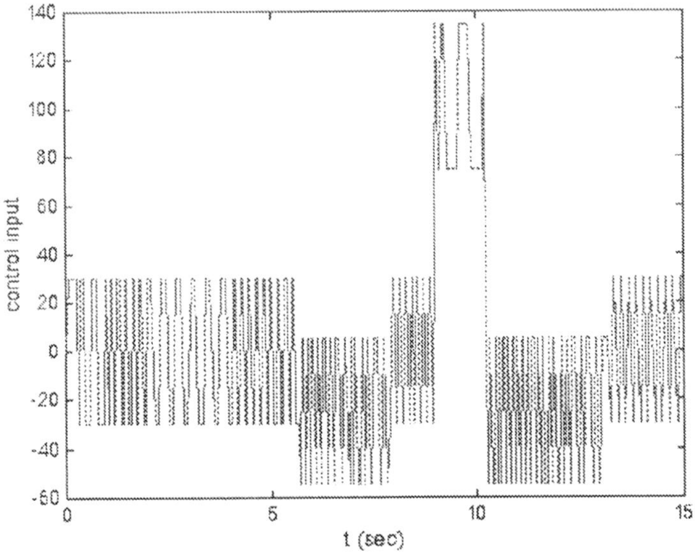


**Fig. 4 Discontinuous sliding mode control law**

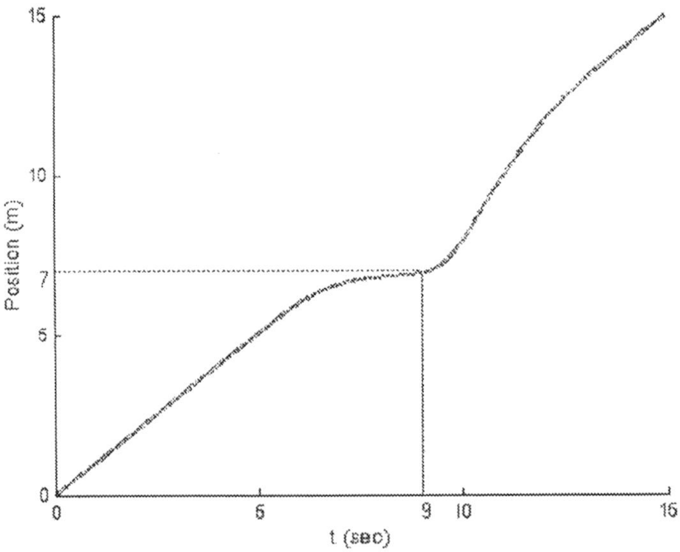


**Fig. 5 Position of the robot for continuous sliding mode control**

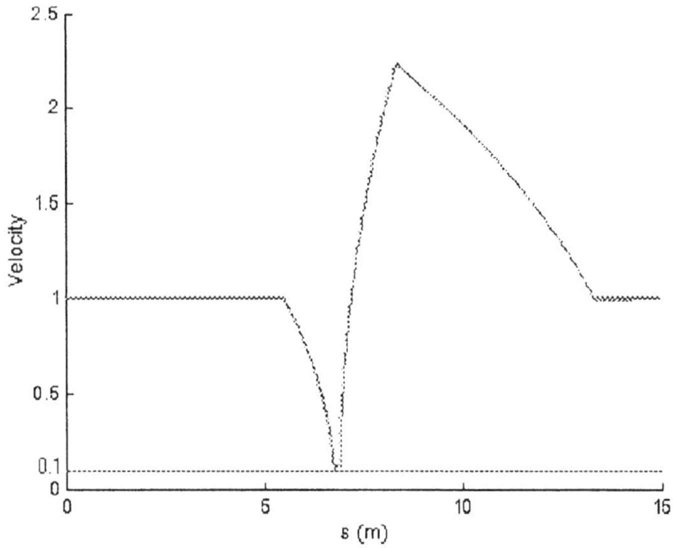


**Fig. 6 Velocity of the robot for continuous sliding mode control**

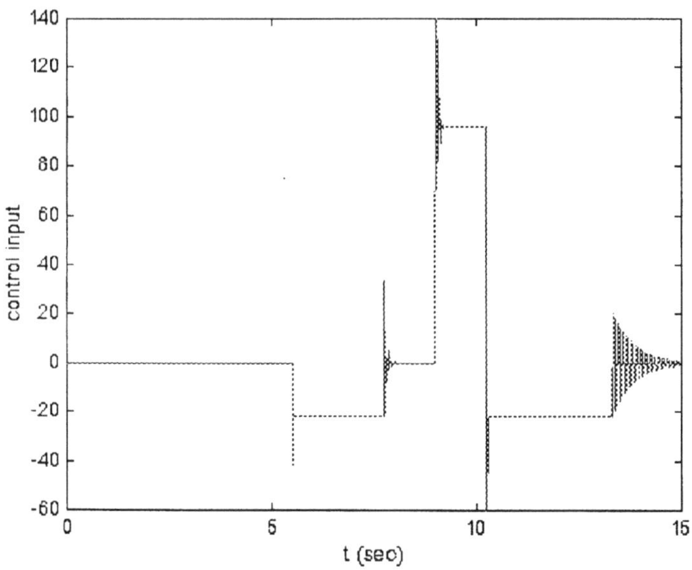


**Fig. 7 Continuous sliding mode control law**

**Acknowledgement.**

The work described in the present chapter was partly supported by the Greek Secretariat of Research and Technology (PENED Programme). M.P. Tzamtzi is a scholar of the Greek Scholarships Foundation.

## 8. References

1. Latombe, J.C.: *Robot Motion Planning*, Kluwer Academic Publishers, 1991.
2. Barbehenn, M., and Hutchinson, S.: Efficient Search and Hierarchical Motion Planning by Dynamically Maintaining Single-Source Shortest Paths Trees, *IEEE Trans. Robotics Automat.*, **11** (1995), 198-214.
3. Brooks, R., and Lozano-Perez, T.: *A Subdivision Algorithm in Gonfiguration Space for Findpath with Rotation*, Technical Report AIM 684, MIT AI Laboratory, 1982.
4. Brooks, R., and Lozano-Perez, T.: A Subdivision Algorithm in Gonfiguration Space for Findpath with Rotation, In: *Proc. Int. Joint Conf. on Art. Intell.*, (1983), 799-806.
5. Tzionas, P.G., Thanailakis, A., and Tsalides, P.G.: Collision-Free Path Planning for a Diamond-Shaped Robot Using Two-Dimensional Cellular Automata, *IEEE Trans. Robotics Automat.*, **13** (1997), 237-250.
6. Rao, N.S.V.: An Algorithmic Framework for Navigation in Unknown Terrains, *IEEE Trans. Comput.*, (1989), 37-43.
7. Rao, N.S.V., Stoltzfus, N.,and Iyengar, S.S.: A 'Retraction' Method for Learned Navigation in Unknown Terrains for a Circular robot, *IEEE Trans. Robotics Automat.*, **7** (1991), 699-707.
8. Sugihara, K.: Approximation of Generalised Voronoi Diagrams by Ordinary Voronoi Diagrams, *Graphical Models and Image Processing*, **55** (1993), 522-531.
9. O'Dunlaing, C., and Yap, C.K.: A Retraction Method for Planning the Motion of a Disc, *J. Algorithms*, **6** (1983), 187-192.
10. Asano, T., Guibas, L., Hershberger, J., and Imai, H.: Visibility-Polygon Search and Euclidean Shortest Path, In: *26th Symp. Foundations Comput. Sci.*, Portland, OR, (1985), 155-164.
11. Schwartz, J.T., and Sharir, M.: On the Piano Mover's Problem: I. The Case of a Two-Dimensional Rigid Polygonal Body Moving Amidst Polygonal Barriers, *Commun. Pure Appl. Math.*, 36 (1983), 345-398.
12. Lozano-Perez, T.: Spatial Planning: A Configuration Space Approach, *IEEE Trans. Comput.*, **32** (1983), 108-120.
13. Koditschek, D.E.: Robot Planning and Control via Potential Functions, In: *Robotics Review*, (Khatib, O., Graig, J., and Lozano-Perez, T., Eds.) Cambridge, MA: MIT Press, **1**, (1989).
14. Barraquand, J., and Latombe, J.C.: Nonholonomic Multi-Body Mobile Robots: Controllability and Motion Planning in the Presence of Obstacles, In: *Proc. IEEE Intl. Conf. Robotics Automat.*, (1991), 2328-2335.
15. Donald, B.R.: *Error Detection and Recovery for Robot Motion Planning with Uncertainty*, Ph.D. Dissertation, Dept. EE&CS, MIT, 1987.
16. Khatib, O.: Real-Time Obstacle Avoidance for Manipulators and Mobile Robots,

*Int. J. Robotics Res.*, **5** (1986), 90-98.
17. Fox, D., Burgard, W. and Thrun, S.: The Dynamic Window Approach to Collision Avoidance, *IEEE Robotics and Autom. Mag.*, **4** (1997), 23-33.
18. De Luca, A., Oriolo, G.: Local Incremental Planning for Nonholonomic Mobile Robots, *IEEE Int. Conf. on Robotics and Automation*, San Diego, (1984).
19. Barraquand, J., and Latombe, J.C.: Robot Motion Planning: A Distributed Representation Approach, IJJR, **10** (1991).
20. De La Rosa, F., Najera, J., and Laugier, C.: Dealing with Uncertainty Constraints in Motion Planning, In: *Intelligent Autonomous Systems*, (Rembold, U., et.al., Eds.) IOS Press, (1995), 497-504.
21. De La Rosa, Laugier, C.F., and Najera, J.: Robust Path Planning in the Plane, *IEEE Trans. Robotics Autom.*, **12** (1996), 347-353.
22. Kant, K., and Zucker, S.: Trajectory Planning Problems, i: Determining Velocity Along a Fixed Path, In: *Proc. of the IEEE 8th Int. Conf. Pattern Recogn.*, (1984), 196-198.
23. Kant, K., and Zucker, S.: Toward Efficient Trajectory Planning: The Path-Velocity Decomposition, *International Journal of Robotics Research*, (1986), 72-89.
24. Kant, K., and Zucker, S.: Planning Collision-Free Trajectories in Time-Varying Environments: A Two-Level Hierarchy, In: *Proc. of the 1988 IEEE International Conference on Robotics and Automation*, (1988), 1644-1649.
25. Griswold, N., and Eem, J.: Control of Mobile Robots in the Presence of Moving Objects, *IEEE Trans. Robotics Automat.*, **6** (1990), 263-268.
26. Wu, C., and Jou, C.: Design of a Controlled Spatial Curve Trajectory for Robot Manipulators, *Proc. 27th Conference on Decision and Control*, (1988), 161-166.
27. Fujimura, K.: Motion Planning in Time-Varying Environments, in: *Recent Trends in Mobile Robots*, (Zheng, Y.F., Ed.), World Scientific, Singapore, (1993), 33-54.
28. Fujimura, K.: Path Planning with Multiple Objectives, *IEEE Robotics and Autom. Mag.*, **3** (1996), 33-38.
29. Gupta, K.K., and Guo, Z.: Motion Planning for Many Degrees of Freedom: Sequential Search with Backtracking, *IEEE Trans. Robotics Automat.*, **11** (1995), 897-906.
30. Faverjon, B.: Obstacle Avoidance Using an Octree in the Configuration Space of a Manipulator, In: *IEEE Conf. Robotics and Automation*, (1984), 504-512.
31. Gouzenes, L.: Strategies for Solving Collision-Free Trajectories Problems for Mobile and Manipulator Robots, *Int. J. Robotics Res.*, **3** (1984), 51-65.
32. Lozano-Perez,T.: A Simple Motion Planning Algorithm for General Robot Manipulators, *IEEE Trans. Robotics Automat.*, **3** (1987), 224-238.
33. Hasegawa, T., and Terasaki, H.: Collision Avoidance: Divide and Conquer Approach by Space Characterization and Intermediate Goals, *IEEE Trans. Syst, Man, Cyber.*, **18** (1988), 337-437.
34. Kondo, K.: Motion Planning with Six Degrees of Freedom by Multi-Strategic Bidirectional Heuristic Free-Space Enumeration, *IEEE Trans. Robotics Automat.*, **7** (1991), 278-288.
35. Gupta, K.K.: Fact Collision Avoidance for Manipulator Arms: A Sequential Search

Strategy, *IEEE Trans. Robotics Automat.*, **6** (1990), 522-532.
36. Gupta, K.K and Guo, Z.: Toward Practical Motion Planners: Experiments with a Sequential Search Strategy, In: *Int. Conf. Robotics and Automation*, (1991), 1006-1011.
37. Faverjon, B, and Tournassoud, P.: A Local Based Approach for Path Planning of Manipulators with a High Number of Degrees of Freedom, In: *IEEE Conf. Robotics and Automation*, (1987), 1152-1159.
38. Shan, Y., and Koren, Y.: Obstacle Accommodation Motion Planning, *IEEE Trans. Robotics Automat.*, **11** (1995), 36-49.
39. Lück, C.L.: Self-Motion Representation and Global Path Planning Optimization for Redundant Manipulators through Topology-Based Discretization, *J. Int. Robotic Syst.*, **19** (1997), 23-38.
40. Seereeram, S, and Wen, J.T.: A Global Approach to Path Planning for Redundant Manipulators, *IEEE Trans. Robotics Automat.*, **11** (1995), 152-160.
41. Baillieul, J.: Kinematic Programming Alternatives for Redundant Manipulators, In: *Proc. 1985 IEEE Robotics and Autom. Conf.*, (1985), 722-728.
42. Chang, P.H.: A Closed-Form Solution for Inverse Kinematics of Robot Manipulators with Redundancy, *IEEE Trans. Robotics Automat.*, 3 (1987), 393-403.
43. Seraji, H.: Configuration Control of Redundant Manipulators: Theory and Implementation, *IEEE Trans. Robotics Automat.*, 5 (1989), 472-490.
44. Klein, C.A. and Huang, C.H.: Review of Pseudoinverse Control for Use with Kinematically Redundant Manipulators, *IEEE Trans. Syst., Man, Cybernet.*, **13** (1983), 245-50.
45. Wampler, C.W.: Inverse Kinematic Functions for Redundant Manipulators, In: Proc. *IEEE Int. Conf. Robotics Automat.*, (1987), 610-617.
46. Kircanski, M., and Vukibratovic, M.: Trajectory Planning for Redundant Manipulators in the Presence of Obstacles, In: *Preprints 5th CISM-IFToMM Symp. Theory and Practice of Robots and Manipulators*, (1984), 43-50.
47. Hollerbach, J.M., and Suh, K.C.: Redundancy Resolution of Manipulators through Torque Optimization, In: *Proc. 1985 IEEE Robotics and Autom. Conf.*, (1985), 308-315.
48. Yoshikawa, T: Dynamic Manipulability of Robot Manipulators, *J. Robotic Syst.*, **2** (1985), 113-124.
49. Klein, C.A., and Blaho, B.E.: Dexterity Measures for the Design and Control of Kinematically Redundant Manipulators, Int. J. Robotics Res., **6** (1987), 72-83.
50. Chiu, S.L.: Control of Redundant Manipulators for Task Compatibility, In: *Proc. 1987 IEEE Robotics and Autom. Conf.*, (1987), 1718-1724.
51. Kazerounian, K., and Wang, Z.: Global versus Local Optimization in Redundancy Resolution of Robotic Manipulators, *Int. J. Robotics Res.*, **7** (1988), 3-12.
52. Baillieul, J., Martin, D.P., and Hollerbach, J.M.: Resolution of Kinematic Redundancy Using Optimization Techniques, *IEEE Trans. Automat. Cont.*, **5** (1989), 529-533.
53. Lee, J.K., and Cho, H.S.: Mobile Manipulator Motion Planning for Multiple Tasks

Using Global Optimization Approach, *J. Int. Robotics Res.*, **18** (1997), 169-190.

54. Seraji, H.: Configuration Control of Rover-Mounted Manipulators, In: *Proc. IEEE Int. Conf. on Robotics and Automation*, (1995), 2261-2266.
55. Pin, F.G., and Culioli, J.C.: Optimal Positioning of Combined Mobile Platform Manipulator Systems for Material Handling Tasks, *J. Intelligent and Robotic Systems*, **6** (1992), 165-182.
56. Zhao, M., Ansari, N., and Hou, E.S.: Mobile Manipulator Path Planning by a Geretic Algorithm, *J. Robotic Systems*, **11** (1994), 143-153.
57. Yamamoto, Y., and Yun, X.: Coordinating Locomotion and Manipulation of a Mobile Manipulator, In: *Recent Trends in Mobile Robots*, (Zheng, Y.F.,Ed.), World Scientific, 1993.
58. Jacobs, P., Laumond, J.-P., and Taix, M.: A Complete Iterative Motion Planner for a Car-Like Robot, *J. Geom. Algorithm.*, (1990).
59. Hague, T., and Cameron, S.: Motion Planning for Nonholonomic Industrial Robot Vehicles, In: IEEE/RSJ Int. Workshop Intell. Robots Syst., Japan, (1991), 1275-1280.
60. Cherif, M., and Laugier, C.: On Physically-Based Modelling and Off-Road Vehicle Motion Planning, In: *Intelligent Autonomous Systems*, (Rembold, U., et al., Eds.), IOS Press, 1995.
61. Li, Z., and Canny, J.: Motion of Two Rigid Bodies with Rolling Constraints, *IEEE Trans. Robotics Automat.*, **6** (1990), 62-72.
62. Barraquand, J., and Latombe, J.C.: On Non-holonomic Mobile Robots and Optimal Maneuvering, *Revue d'Intelligence Artificielle*, **3** (1989), 77-103.
63. Laumond, J.-P., Jacobs, P., Taix, M., and Murray, R.M.: A Motion Planner for Nonholonomic Mobile Robots, *IEEE Trans. Robotics Automat.*, **10** (1994), 577-593.
64. Divelbiss, A.W., and Wen, J.T.: A Path Space Approach to Nonholonomic Motion Planning in the Presence of Obstacles, *IEEE Trans. Robotics Automat.*, **13** (1997), 443-451.
65. Conte, G., and Zulli, R.: Hierarchical Path Planning in a Multi-Robot Environment with a Simple Navigation Function, *IEEE Trans. Syst., Man, Cybernet.*, **25** (1995) 651-655.
66. Chien, Y.P., and Xue, Q.: Path Planning for Two Planar Robots Moving in Unknown Environment, *IEEE Trans. Syst., Man, Cybernet.*, **22** (1992), 307-317.
67. Arai, T., and Ota, J.: Motion Planning of Multiple Mobile Robots Using Dynamic Groups, In: *Recent Trends in Mobile Robots*, (Zheng,Y.F., Ed.), World Scientific, 1993.
68. Wang, P.K.C.: Interaction Dynamics of Multiple Mobile Robots with Simple Navigation Strategies, *J. Robotics Systems*, **6** (1989), 77-101.
69. Arai, T., Ogata, H., and Suzuki, T.: Collision Avoidance among Multiple Robots Using Virtual Impedance, In: *Proc. IROS'89*, (1989), 479-485.
70. Schweikard, A., Adler, J.R., and Latombe, J.C.: Motion Planning in Stereotaxic Radiosurgery, *IEEE Trans. Robotics Autom.* **9** (1993), 764-774.
71. Kyriakopoulos, K.J., and Saridis, G.N.: Optimal and Suboptimal Motion Planning

for Collision Avoidance of Mobile Robots in Non-stationary Environments, *J. Intelligent and Robotics Systems*, **11** (1995), 223-267.

72. Kyriakopoulos, K.J., and Saridis, G.N.: Minimum Distance Estimation and Collision Prediction under Uncertainty for On-Line Robotic Motion Planning, In: *Proc. of the 1990 IFAC World Congress*, (1990).
73. Kyriakopoulos, K.J., and Saridis, G.N.: Collision Avoidance of Mobile Robots in Non-Stationary Environments, In: *Proc. of the 1991 International Conference on Robotics and Automation*, (1991).
74. Kyriakopoulos, K.J., and Saridis, G.N.: Distance Estimation and Collision Prediction for On-Line Robotic Motion Planning, *Automatica*, **28** (1992), 389-394.
75. Cho, H.K., Lee, B.H, and Ko, M.S.: Time-Optimal Motion of a Cooperating Multiple Robot System Along a Prescribed Path, *IEEE Trans. Syst. Man Cyber.*, **25** (1995), 1648-1657.
76. Dahl, O., and Nielsen, L.: Torque Limited Path Following by On-Line Trajectory Time Scaling, *IEEE Trans. Robotics Automat.*, **6** (1991).
77. Dahl, O.: Path-Constrained Robot Control with Limited Torques-Experimental Evaluation, *IEEE Trans. Robotics Automat.*, **10** (1994), 658-669.
78. Hu, T.C., Kahng, A.B., and Robins, G.: Optimal Robust Path Planning in General Environments, *IEEE Trans. Robotics Automat.*, **9** (1993), 775-784.
79. Slotine, J.J.E., and Li, W.: *Applied Nonlinear Control*, Prentice Hall, Englewood Cliffs, New Jersey, 1991.
80. Su, C.Y., and Leung, T.P.: A Sliding Mode Controller with Bound Estimation for Robot Manipulators, *IEEE Trans. Robotics Automat.*, **9** (1993), 208-214.
81. Guldner, J., and Utkin, V.I.: Sliding Mode Control for Gradient Tracking and Robot Navigation Using Artificial Potential Fields, *IEEE Trans. Robotics Automat.*, **11** (1995), 247-254.
82. Tzafestas, S.G., Raibert, M., and Tzafestas, C.S.: Robust Sliding Mode Control Applied to a 5-Link Biped Robot, *J. Intelligent and Robotic Systems*, **15** (1996), 67-133.
83. Kung, C.C, and Lin, S.C.: Fuzzy Controller Design: A Sliding Mode Approach, In: *Fuzzy Reasoning in Information, Decision and Control Systems*, (Tzafestas, S.G., and Venetsanopoulos, A.N., Eds.), Kluwer, Dordrecht-Boston, 1994.

# 13

# AUTONOMOUS CONTROL FOR AN OMNIDIRECTIONAL MOBILE ROBOT WITH FEEDBACK CONTROL SYSTEM

KEIGO WATANABE AND YAMATÕ SHIRAISHI
*Department of Mechanical Engineering, Saga University*
*1-Honjomachi, Saga 840, Japan*

JUN TANG
*Department of Mechanical Engineering, Yamaguchi University*
*2557 Tokiwadai, Ube 755, Japan*

TOSHIO FUKUDA
*Department of Micro Systems Engineering, Nagoya University*
*Furo-cho, Chikusa-ku, Nagoya 464, Japan*

AND

SPYROS G. TZAFESTAS
*Computer Science Division, National Technical University of Athens*
*Zografou Campus, 15773 Athens, Greece*

## 1. Introduction

Recently, an industrial mobile robot has been required to have some abilities that it can freely travels within a confused factory and achieve an accurate positioning in a work station. For this purpose, an active investigation is now focused on the study of omnidirectional mobile robots, instead of a conventional mobile robot with two independent driving wheels, or with the front-wheel handling and rear-wheel driving. The omnidirectional mobile robot has been studied by using a variety of mechanisms. In particular, a holonomic vehicle has a full omnidirectionality with simultaneous and independently controlled rotational and translational motion capabilities. Several omnidirectional platforms have been known to be realized by driving wheel with steering (or offset driving wheel) [1], [2], [3], universal wheels [4], [5], spherical tires [6], or crawler mechanisms [7], [8].

A new family of holonomic wheeled platforms has also been proposed by Pin and Killough [9], in which an orthogonal-wheel concept has been introduced. It should be noted, however, that the conventional studies on the omnidirectional vehicles have been focused on the development of mechanisms or only on the analysis of kinematics, and hence there are little studies on the development of dynamic models and highly accurate control systems.

In this chapter, we introduce one dynamic model for an omnidirectional mobile robot, in which it is assumed that the platform is based on three orthogonal-wheel assemblies. Then, to develop a control system with high performance for such a mobile robot, a resolved acceleration control system can be shown to be designed using such a dynamic model. The practical experiments are presented to illustrate the full omnidirectionality of the platform with decoupled rotational and translational motions.

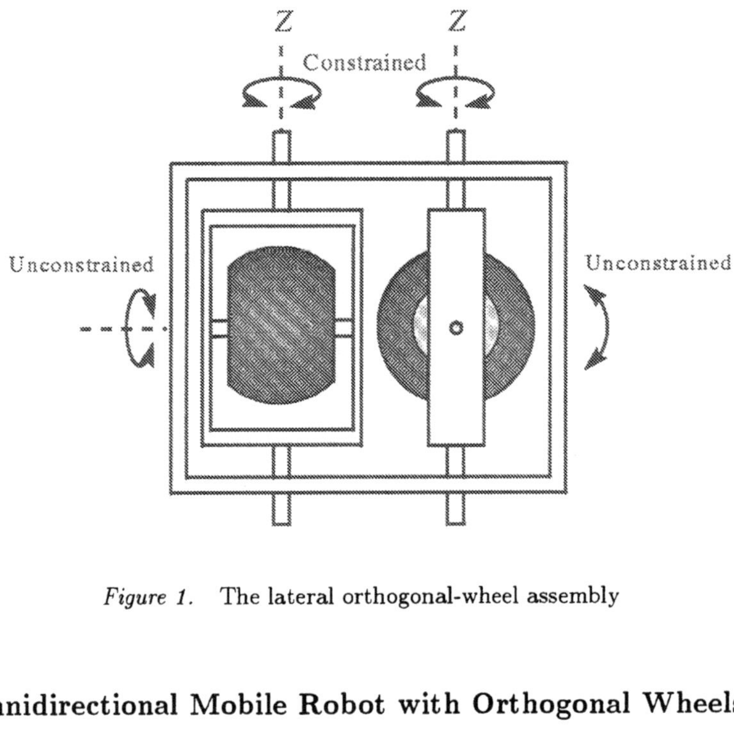


*Figure 1.* The lateral orthogonal-wheel assembly

## 2. Omnidirectional Mobile Robot with Orthogonal Wheels

In this section, we assume that the omnidirectional mobile robot consists of the orthogonal-wheel assembly mechanism proposed by Pin and Killough [9].

The basic lateral orthogonal-wheel assembly is illustrated in Fig. 1. The major components are two spheres of equal diameter which have been sliced to resemble wide, i.e., rounded-tire wheels. The axle of each wheel is perpendicular to the sliced surfaces and is mounted using ball bearing so that the wheel is freewheeling around its axle. Through a bracket which is holding the extremities of the wheel axle, each wheel can be driven to roll on its portion of spherical surface, rotating around an axis $Z$, perpendicular to the wheel axle. When these axes $Z_1$ and $Z_2$ are maintained parallel and at a constant distance from each other, and when the wheel rotations around these axes are synchronized, contact with the ground can be assured by at least one wheel, while allowing enough space for the brackets holding the wheel axles to clear the ground. These assemblies can be constructed by using more than two wheels, in which each wheel axle should be allocated with less than 90° offset.

In the following, it was assumed that a platform is based on three assemblies allocated at an equal distance from the center of gravity (c.g.) for the robot, in which one assembly consists of two wheels. The actual experimental mobile robot is appeared in Fig. 2.

*Figure 2.* The appearance of an omnidirectional mobile robot

## 3. Derivation of Dynamic Model for the Omnidirectional Mobile Robot

Let the mobile robot be rigid moving on the work space. It is assumed that the absolute coordinate system $O_w - X_wY_w$ is fixed on the plane and the moving coordinate system $O_m - X_mY_m$ is fixed on the c.g. for the mobile robot as shown in Fig. 3.

When defining the position vector of the c.g. for the mobile robot such as $\boldsymbol{S}_w = [x_w \; y_w]^T$, we have

$$\boldsymbol{M}\ddot{\boldsymbol{S}}_w = \boldsymbol{F}_w \tag{1}$$

where $\boldsymbol{F}_w = [F_x \; F_y]^T$ is the force vector in the absolute coordinate system applied to the center of gravity for the mobile robot and $\boldsymbol{M}$ is a symmetric positive-definite matrix as $\boldsymbol{M} = \mathrm{diag}(M, M)$ with the mass $M$.

Let $\phi$ denote the angle between $X_w$- and $X_m$-coordinates, i.e., the rotational angle of the moving coordinate system with respect to the absolute coordinate system. When introducing the coordinate transformation matrix from the absolute coordinate system to the moving coordinate system such as

$${}^wR_m = \begin{bmatrix} \cos\phi & -\sin\phi \\ \sin\phi & \cos\phi \end{bmatrix} \tag{2}$$

it follows that

$$\dot{\boldsymbol{S}}_w = {}^wR_m \dot{\boldsymbol{s}}_m \tag{3}$$

$$\boldsymbol{F}_w = {}^wR_m \boldsymbol{f}_m \tag{4}$$

where $\boldsymbol{s}_m = [x_m \; y_m]^T$ and $\boldsymbol{f}_m = [f_x \; f_y]^T$ are the position vector of the c.g. and the force vector applied to the c.g. in the moving coordinate system, respectively.

Therefore, transforming equation (1) to the moving coordinate system gives

$$\boldsymbol{M}({}^wR_m^T \, {}^w\dot{R}_m \dot{\boldsymbol{s}}_m + \ddot{\boldsymbol{s}}_m) = \boldsymbol{f}_m \tag{5}$$

Then, the dynamic properties for the mobile robot can be described as [10]

$$M(\ddot{x}_m - \dot{y}_m\dot{\phi}) = f_x \tag{6}$$

$$M(\ddot{y}_m + \dot{x}_m\dot{\phi}) = f_y \tag{7}$$

$$I_v\ddot{\phi} = M_I \tag{8}$$

where $I_v$ is the moment of inertia for the robot, $M_I$ is the moment around the c.g. for the robot, and $f_x, f_y, M_I$ are given by

$$f_x = -\frac{1}{2}D_1 - \frac{1}{2}D_2 + D_3 \tag{9}$$

$$f_y = \frac{\sqrt{3}}{2}D_1 - \frac{\sqrt{3}}{2}D_2 \tag{10}$$

$$M_I = (D_1 + D_2 + D_3)L \tag{11}$$

In addition, the driving system property [11] for each assembly is assumed to be given by

$$I_w\dot{\omega}_i + c\omega_i = ku_i - rD_i \quad (i = 1, 2, 3) \tag{12}$$

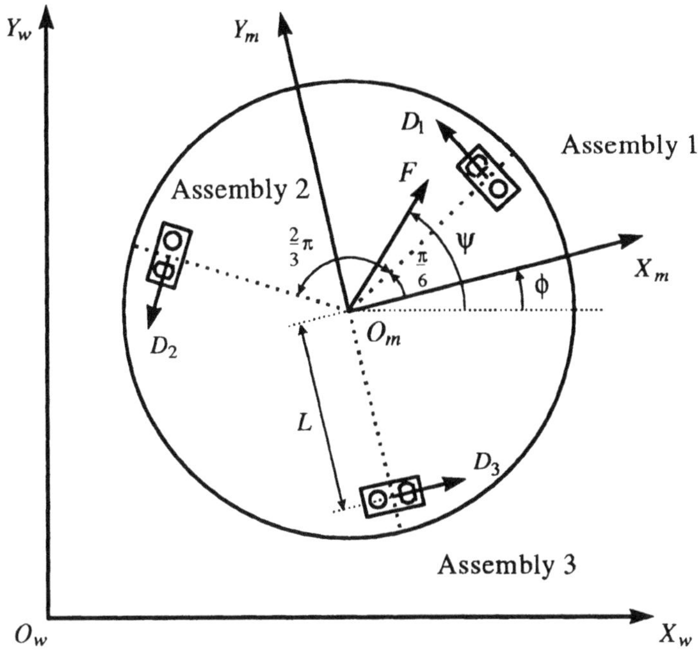


*Figure 3.* Model of an omnidirectional mobile robot

where $L$ is the distance between any assembly and the c.g. of the robot; $c$ is the viscous friction factor for the wheel; $D_i$ is the driving force for each assembly; $r$ is the radius of the wheel; $I_w$ is the moment of inertia of the wheel around the driving shaft; $\omega_i$ is the rotational rate of the wheel; $k$ is the driving gain factor; and $u_i$ is the driving input torque.

On the other hand, the geometrical relationships among variables $\dot{\phi}$, $\dot{x}_m$, $\dot{y}_m$, and $\omega_i$, i.e., the inverse kinematics are given by

$$r\omega_1 = -\frac{1}{2}\dot{x}_m + \frac{\sqrt{3}}{2}\dot{y}_m + L\dot{\phi} \tag{13}$$

$$r\omega_2 = -\frac{1}{2}\dot{x}_m - \frac{\sqrt{3}}{2}\dot{y}_m + L\dot{\phi} \tag{14}$$

$$r\omega_3 = \dot{x}_m + L\dot{\phi} \tag{15}$$

Therefore, using equations (6)~(15) gives

$$\ddot{x}_m = a_1\dot{x}_m + a_2'\dot{y}_m\dot{\phi} - b_1(u_1 + u_2 - 2u_3) \tag{16}$$

$$\ddot{y}_m = a_1\dot{y}_m - a_2'\dot{x}_m\dot{\phi} + \sqrt{3}b_1(u_1 - u_2) \tag{17}$$

$$\ddot{\phi} = a_3\dot{\phi} + b_2(u_1 + u_2 + u_3) \tag{18}$$

where

$$
\begin{aligned}
a_1 &= -3c/(3I_w + 2Mr^2), \quad a_2' = 2Mr^2/(3I_w + 2Mr^2) \\
a_3 &= -3cL^2/(3I_wL^2 + I_vr^2) \\
b_1 &= kr/(3I_w + 2Mr^2), \quad b_2 = krL/(3I_wL^2 + I_vr^2)
\end{aligned}
$$

It is easy to find that combining equations (2), (3), (16) and (17) yields the appropriate dynamic equations in the absolute coordinate system for the robot. Thus, defining the state variable for the robot as $\boldsymbol{x} = [x_w\ y_w\ \phi\ \dot{x}_w\ \dot{y}_w\ \dot{\phi}]^T$, the manipulated variable as $\boldsymbol{u} = [u_1\ u_2\ u_3]^T$, and the output variable as $\boldsymbol{y} = [\dot{x}_w\ \dot{y}_w\ \phi]^T$ yields the following state equation:

$$
\begin{aligned}
\dot{\boldsymbol{x}} &= A(\boldsymbol{x})\boldsymbol{x} + B(\boldsymbol{x})\boldsymbol{u} && (19) \\
\boldsymbol{y} &= C\boldsymbol{x} && (20)
\end{aligned}
$$

where

$$
\begin{aligned}
A(\boldsymbol{x}) &= \begin{bmatrix} 0 & 0 & 0 & 1 & 0 & 0 \\ 0 & 0 & 0 & 0 & 1 & 0 \\ 0 & 0 & 0 & 0 & 0 & 1 \\ 0 & 0 & 0 & a_1 & -a_2\dot{\phi} & 0 \\ 0 & 0 & 0 & a_2\dot{\phi} & a_1 & 0 \\ 0 & 0 & 0 & 0 & 0 & a_3 \end{bmatrix} \\
B(\boldsymbol{x}) &= \begin{bmatrix} 0 & 0 & 0 \\ 0 & 0 & 0 \\ 0 & 0 & 0 \\ b_1\beta_1 & b_1\beta_2 & 2b_1\cos\phi \\ b_1\beta_3 & b_1\beta_4 & 2b_1\sin\phi \\ b_2 & b_2 & b_2 \end{bmatrix} \\
C &= \begin{bmatrix} 0 & 0 & 0 & 1 & 0 & 0 \\ 0 & 0 & 0 & 0 & 1 & 0 \\ 0 & 0 & 1 & 0 & 0 & 0 \end{bmatrix}
\end{aligned}
$$

$$
\begin{aligned}
a_2 &= 1 - a_2' = 3I_w/(3I_w + 2Mr^2) \\
\beta_1 &= -\sqrt{3}\sin\phi - \cos\phi, \quad \beta_2 = \sqrt{3}\sin\phi - \cos\phi \\
\beta_3 &= \sqrt{3}\cos\phi - \sin\phi, \quad \beta_4 = -\sqrt{3}\cos\phi - \sin\phi
\end{aligned}
$$

Let the translational velocity of the robot in the absolute coordinate system be $V = (\dot{x}_w^2 + \dot{y}_w^2)^{\frac{1}{2}}$ and the azimuth of the robot in the absolute coordinate system be $\psi = \theta + \phi$. Here, $\theta$ denotes the angle between $X_m$-coordinate and $\boldsymbol{f}_m$, i.e., the azimuth of the robot in the moving coordinate system. Then, it is found that

$$\dot{x}_w = V \cos \psi \tag{21}$$

$$\dot{y}_w = V \sin \psi \tag{22}$$

$$\psi = \arctan \frac{\dot{y}_w}{\dot{x}_w} \tag{23}$$

where note that a counterclockwise rotation denotes the positive direction for the rotational motion of the robot.

It should be noted that $X_w$- and $Y_w$-directional motions in (19) are coupled, because the equation of motions is derived in the absolute coordinate system. However, since the rotational angle of the robot can be assured as $\phi = \psi - \theta$ even if the azimuth of the robot is changed arbitrarily, the mobile robot can realize a translational motion without changing the pose.

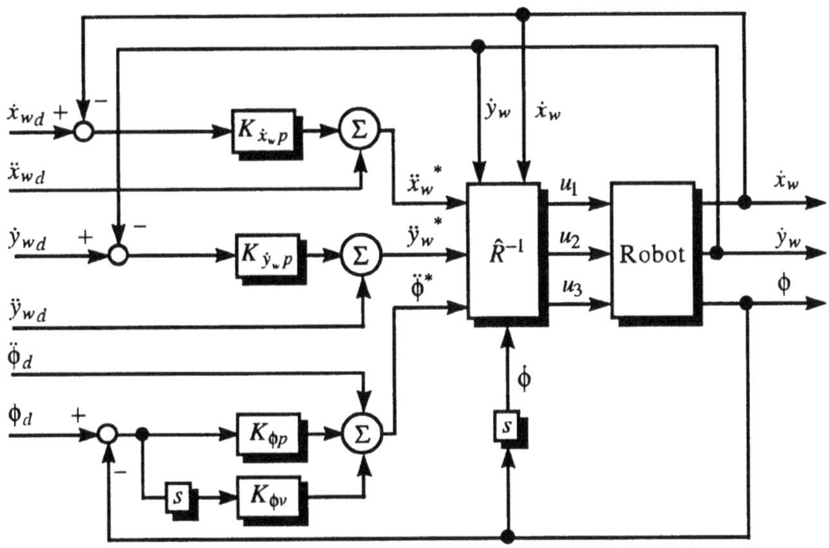


*Figure 4.* A resolved acceleration control system

## 4. Resolved Acceleration Control

Since the dynamic equation is finally considered in the absolute coordinate system, solving equation (19) with respect to $u_i, i = 1, 2, 3$ gives the following resolved acceleration control:

$$u_1 = \frac{\beta_1(\ddot{x}_w^* - a_1\dot{x}_w + a_2\dot{\phi}\dot{y}_w)}{6b_1}$$

$$+ \quad \frac{\beta_3(\ddot{y}_w^* - a_1\dot{y}_w - a_2\dot{\phi}\dot{x}_w)}{6b_1} + \frac{\ddot{\phi}^* - a_3\dot{\phi}}{3b_2} \tag{24}$$

$$u_2 = \frac{\beta_2(\ddot{x}_w^* - a_1\dot{x}_w + a_2\dot{\phi}\dot{y}_w)}{6b_1}$$

$$+ \quad \frac{\beta_4(\ddot{y}_w^* - a_1\dot{y}_w - a_2\dot{\phi}\dot{x}_w)}{6b_1} + \frac{\ddot{\phi}^* - a_3\dot{\phi}}{3b_2} \tag{25}$$

$$u_3 = \frac{\cos\phi(\ddot{x}_w^* - a_1\dot{x}_w + a_2\dot{\phi}\dot{y}_w)}{3b_1}$$

$$+ \quad \frac{\sin\phi(\ddot{y}_w^* - a_1\dot{y}_w - a_2\dot{\phi}\dot{x}_w)}{3b_1} + \frac{\ddot{\phi}^* - a_3\dot{\phi}}{3b_2} \tag{26}$$

where $\ddot{x}_w^*, \ddot{y}_w^*$ and $\ddot{\phi}^*$ are given by adding PI-servo or PD-servo to $\ddot{x}_{wd}, \ddot{y}_{wd}$ and $\ddot{\phi}_d$ such that

$$\ddot{x}_w^* = \ddot{x}_{wd} + K_{\dot{x}p}e_{\dot{x}} + K_{\dot{x}i}\int_0^t e_{\dot{x}}dt \tag{27}$$

$$\ddot{y}_w^* = \ddot{y}_{wd} + K_{\dot{y}p}e_{\dot{y}} + K_{\dot{y}i}\int_0^t e_{\dot{y}}dt \tag{28}$$

$$\ddot{\phi}^* = \ddot{\phi}_d + K_{\phi v}\dot{e}_\phi + K_{\phi p}e_\phi \tag{29}$$

Here, $K_{\cdot p}$ is the proportional gain, $K_{\cdot i}$ is the integral gain, $K_{\cdot v}$ is the derivative gain, and each error is defined by

$$e_{\dot{x}} = \dot{x}_{wd} - \dot{x}_w \tag{30}$$

$$e_{\dot{y}} = \dot{y}_{wd} - \dot{y}_w \tag{31}$$

$$e_\phi = \phi_d - \phi \tag{32}$$

where $\dot{x}_{wd}, \dot{y}_{wd}$ and $\phi_d$ denote the references for $\dot{x}_w, \dot{y}_w$ and $\phi$, respectively. It should be noted that the solutions $u_i, i = 1, 2, 3$ exit for all $t$, because the input distribution factors $b_1$ and $b_2$ are always non-zero constants. The block diagram of a resolved acceleration control system for the mobile robot is shown in Fig. 4, in which $\hat{R}^{-1}$ in the figure denotes the estimated inverse dynamic model.

## 5. Experimental Results

We have applied the above control system to an actual mobile robot. It was assumed that the control sampling period is 50 [ms] and the total experimental time is 20 [s]. The physical parameters of the mobile robot are as follows:

$$I_v = 11.25\ [\mathrm{kgm}^2], I_w = 0.02108\ [\mathrm{kgm}^2], c = 5.983 \times 10^{-6}\ [\mathrm{kgm}^2/\mathrm{s}],$$

$$M = 9.4\ [\mathrm{kg}], L = 0.178\ [\mathrm{m}], r = 0.0245\ [\mathrm{m}], k = 1.0$$

It was also assumed that the initial moving coordinate system is equal to the absolute coordinate system, and the initial value of state variable is given as $\boldsymbol{x} = [0\ 0\ 0\ 0\ 0\ 0]^T$.

### 5.1. SIMULTANEOUSLY TRANSLATIONAL AND ROTATIONAL MOTIONS

The omnidirectional mobile robot can independently achieve the translational motion and the rotational motion around the c.g. in the two-dimensional plane. To check this property, the holonomic property, it was assumed that the robot must travel with a fixed azimuth angle $\psi_d = \pi/4$ [rad] for 20 [s], but with zero rotational angle for 0~10 [s] and with a uniformly varying rotational angle from $\phi_d = 0$ [rad] to $\phi_d = \pi/2$ [rad] for 10~20 [s]. Although the moving velocity was to be $V_d = 0.05$ [m/s] in the steady-state, a sinusoidal reference velocity was set for each 2 seconds at the starting and ending duration.

Using these $V_d$ and $\psi_d$, the corresponding $\dot{x}_{wd}$ and $\dot{y}_{wd}$ were derived from (21) and (22), and $\ddot{x}_{wd}$ and $\ddot{y}_{wd}$ were also derived by their derivatives.

#### 5.1.1. *Case 1*

Figures 5-7 show the results for the case when the servo gains are given as in Table 1. In this case, it is found that the control of $\phi$ is achieved very well, while the velocity control results have an overshoot.

TABLE 1. Gains for Case 1

| $K_{\phi p}$ | $K_{\phi v}$ | $K_{\dot{x}i}$ | $K_{\dot{x}p}$ | $K_{\dot{y}i}$ | $K_{\dot{y}p}$ |
|---|---|---|---|---|---|
| 2.25 | 3.0 | 1.0 | 2.0 | 1.0 | 2.0 |

#### 5.1.2. *Case 2*

To improve the responses on $\dot{x}_w$ and $\dot{y}_w$, adjusted servo gains are tabulated in Table 2. The corresponding control results are shown in Figs. 8-13. It is found from these figures that the good trajectory control result is obtained, though there exit successive oscillations in the velocity responses. This seems to be attributed to the reason that the dc servo driver has a restricted resolution ability.

### 5.2. CIRCULAR MOTION WITH A FIXED POSE

As a motion that a conventional mobile robot cannot achieve, we can consider a circular motion with a radius of curvature less than the half of

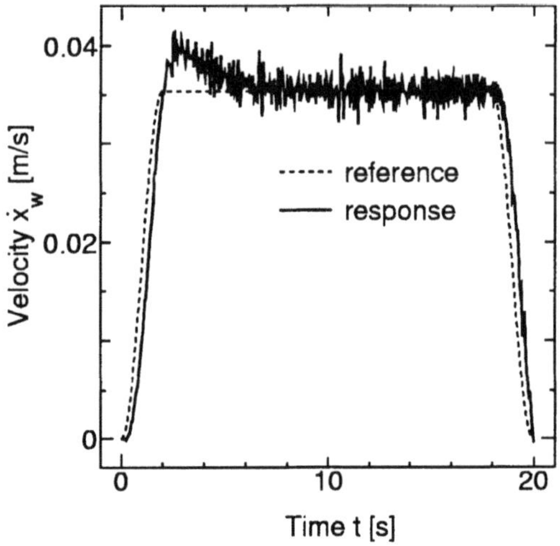


*Figure 5.* Velocity $\dot{x}_w$ for Case 1

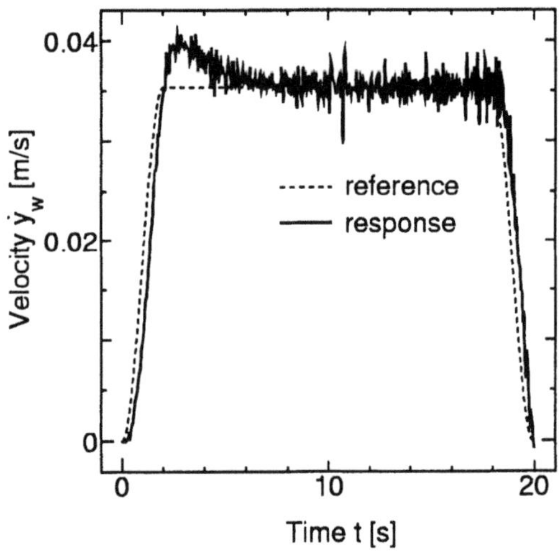


*Figure 6.* Velocity $\dot{y}_w$ for Case 1

distance between wheels, that is, its motion has a center of rotation on the distance between wheels. The omnidirectional mobile robot, of course, can easily achieve such a motion. In the following experiment, it was assumed that a circular trajectory has a radius of curvature 0.1 [m], while the dis-

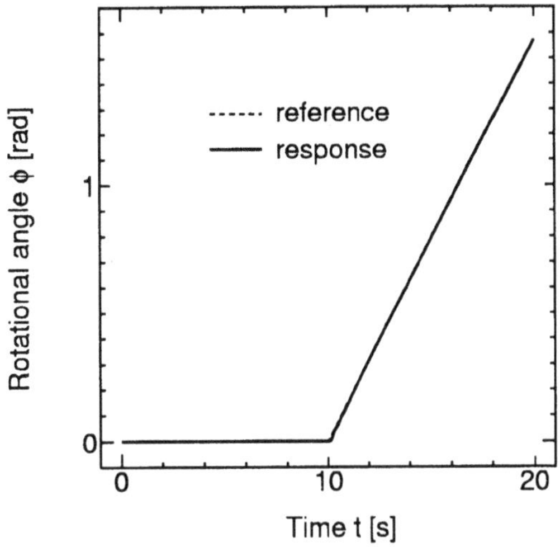


*Figure 7.* Rotational angle $\phi$ for Case 1

TABLE 2. Gains for Case 2

| $K_{\phi p}$ | $K_{\phi v}$ | $K_{\dot{x}i}$ | $K_{\dot{x}p}$ | $K_{\dot{y}i}$ | $K_{\dot{y}p}$ |
|---|---|---|---|---|---|
| 2.25 | 3.0 | 25.0 | 10.0 | 25.0 | 10.0 |

tance between wheels is 0.356 [m]. The resultant moving velocity was given by $V_d = 0.04488$ [m/s], but a sinusoidal reference velocity was set for each 2 seconds at the starting and ending duration. The moving azimuth was $\psi_d = 0.0$ [rad] for each 3 seconds at the starting and ending duration, while it had a uniformly successive increment up to $2\pi$ [rad] for the remaining 14 seconds, where the robot pose, the rotational angle was fixed as $\phi_d = 0.0$ [rad] for the whole duration.

The corresponding control results are shown in Figs. 14-19, where the servo gains are the same as those tabulated in Table 2. From these figures, it is found that the satisfactory control results are obtained without changing the servo gains.

### 5.3. 8 CHARACTER MOTION WITH A SIMULTANEOUS ROTATION

In this section, the following 8 character path was considered to check the tracking performance of the robot for the case when the references $\dot{x}_{wd}$ and

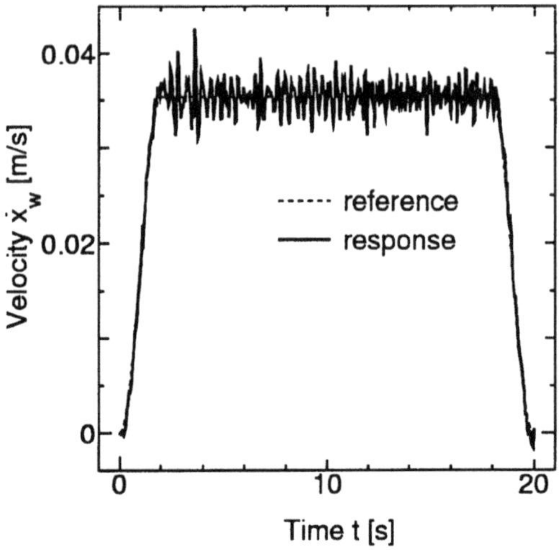


*Figure 8.* Velocity $\dot{x}_w$ for Case 2

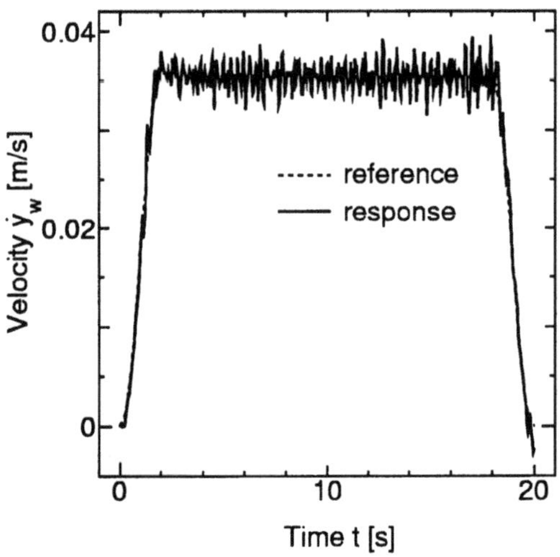


*Figure 9.* Velocity $\dot{y}_w$ for Case 2

$\dot{y}_{wd}$ are changed frequently and to examine the effect of a time-varying $\phi$ on the responses.

It was assumed that $V_d = 0.05$ [m/s], but with an acceleration and deceleration for each 2 seconds at the starting and ending duration. The

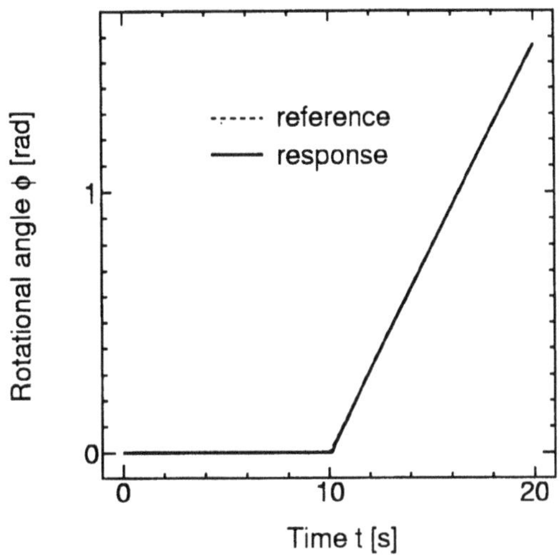


*Figure 10.* Rotational angle $\phi$ for Case 2

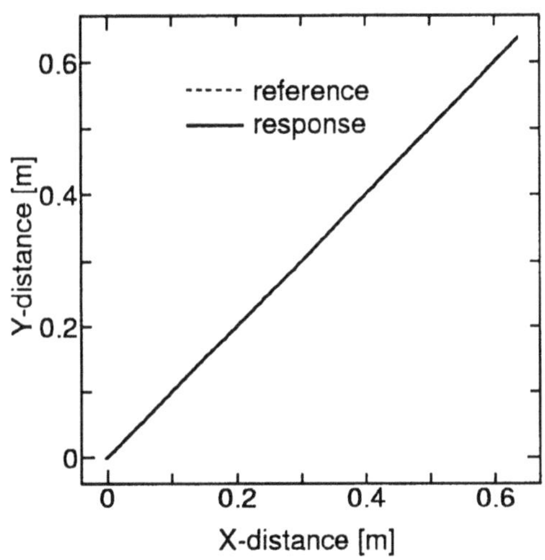


*Figure 11.* Trajectory for Case 2

8 character was composed of two squared path, where there existed an acceleration and deceleration for 1 second at each corner. The resultant path was symmetry with respect to the $x$-coordinate. It was also assumed that the reference rotational angle was zero for 0~10 [s], while it was uniformly

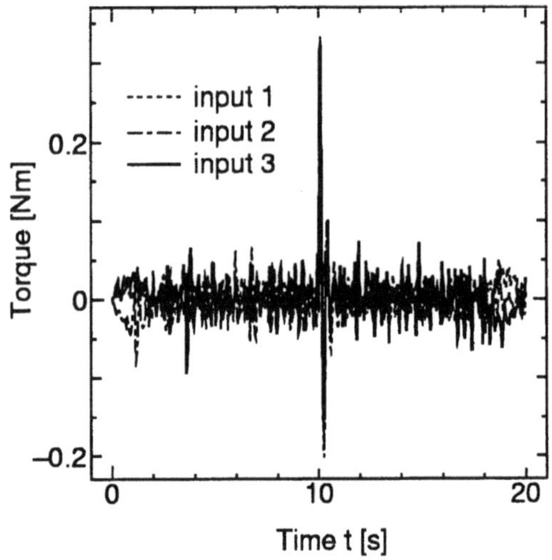


*Figure 12.* Input torque for Case 2

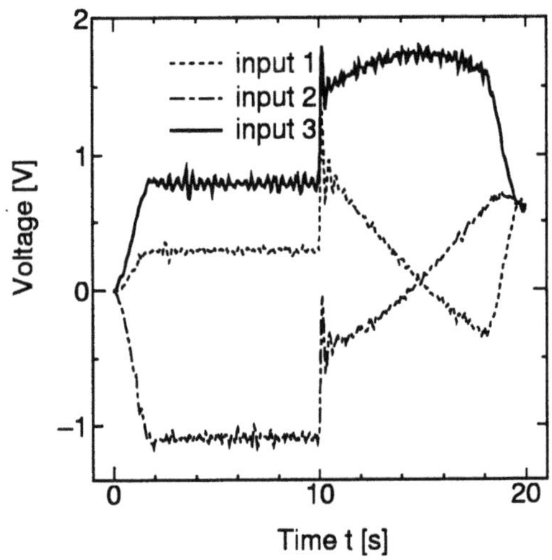


*Figure 13.* Input voltage for Case 2

time-increasing from $\phi_d = 0$ [rad] to $\phi_d = \pi/2$ [rad] for 10~20 [s].

The control results are shown in Figs. 20-22, where the servo gains given in Table 2 were used. It is understood from these figures that some overshoots appear in the responses of $\dot{x}_w$ and $\dot{y}_w$, though the response of

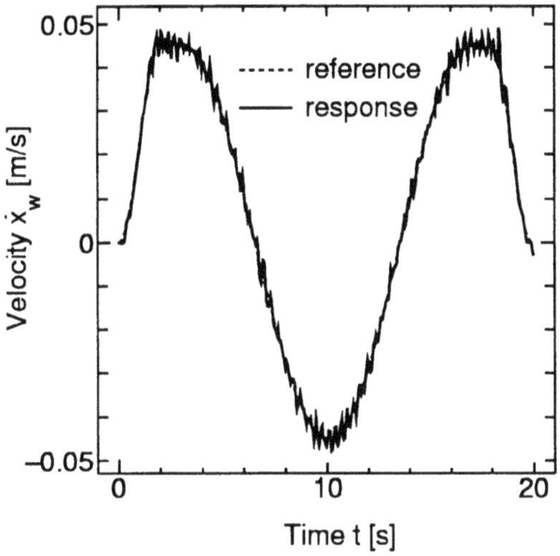


*Figure 14.* Velocity $\dot{x}_w$ for a circular path

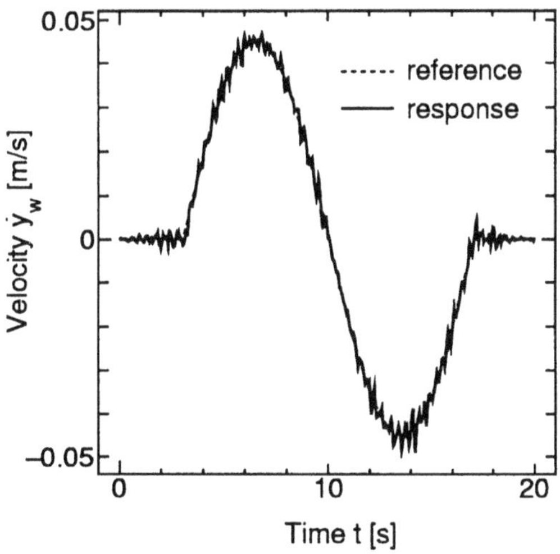


*Figure 15.* Velocity $\dot{y}_w$ for a circular path

rotational angle tracks the reference very well. This seems to be attributed to the fact that the time required for the acceleration and deceleration is too short, and in fact it was not improved even if the gains for $\dot{x}_{wd}$ and $\dot{y}_{wd}$ were increased.

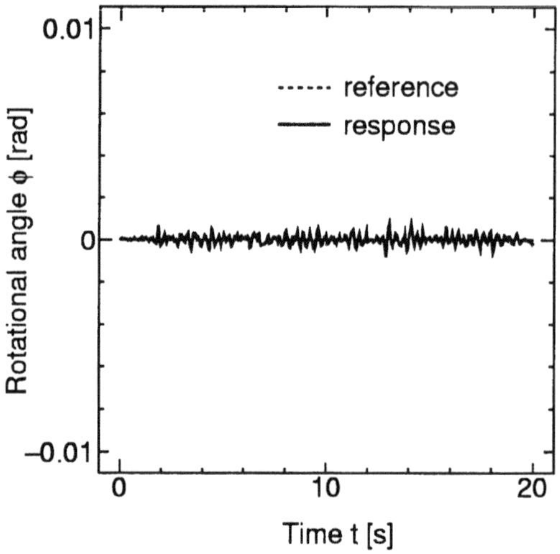


*Figure 16.* Rotational angle $\phi$ for a circular path

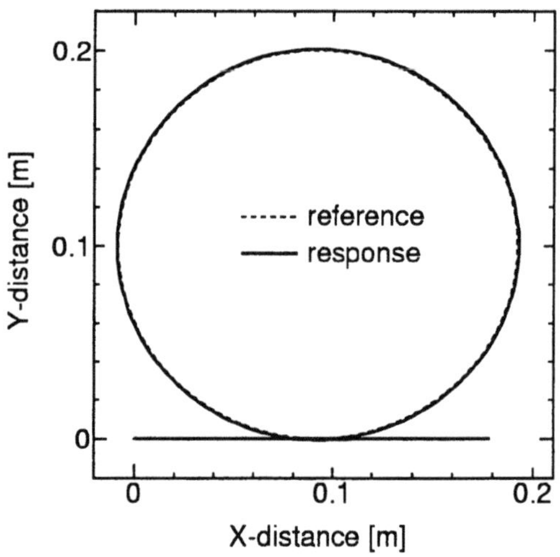


*Figure 17.* Trajectory for a circular path

Furthermore, it should be noted from Fig. 22 that the tracking errors appear at the corners of squared path. It is worth, however, to note that the time-varying or time-invariant property of rotational angle $\phi$ does not affect on the trajectory response, because the responses on the upper and

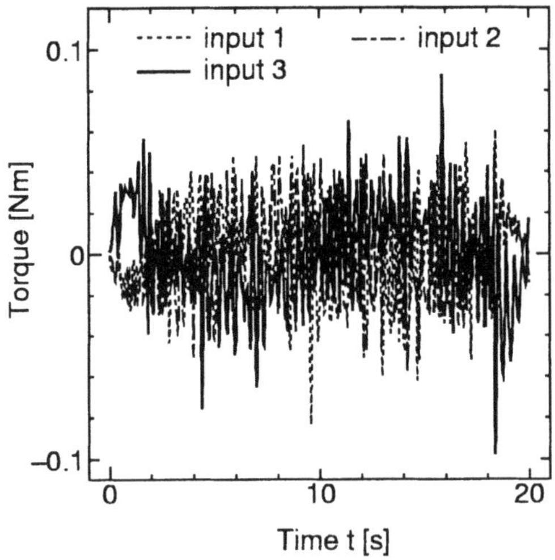


*Figure 18.* Input torque for a circular path

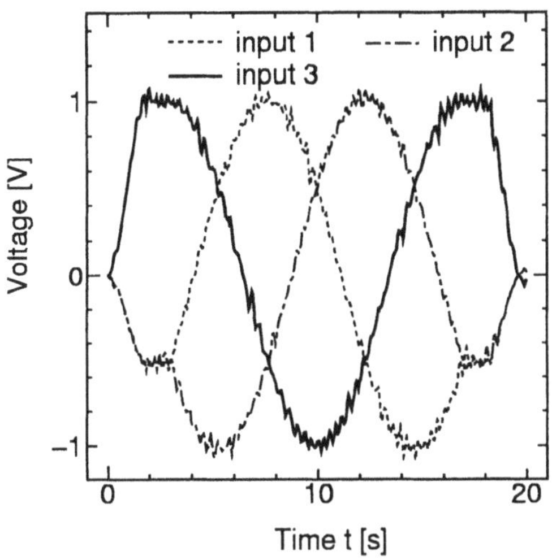


*Figure 19.* Input voltage for a circular path

bottom squared path are approximately symmetry with respect to the $x$-coordinate. It may need a compensate approach such as a sliding mode control for the resolved acceleration control system with a modeling error to improve the control response.

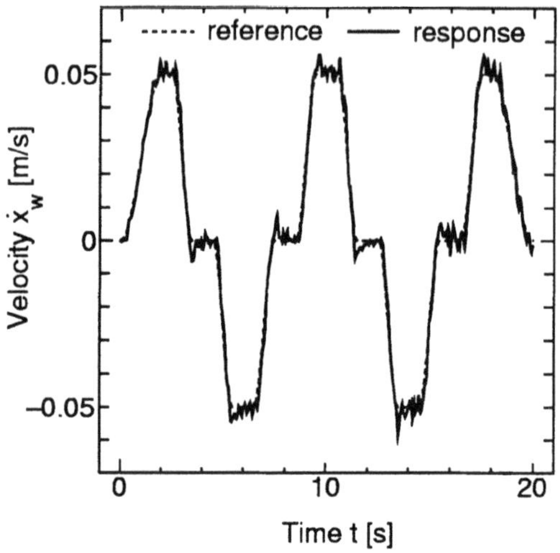


*Figure 20.* Velocity $\dot{x}_w$ for an 8 character path

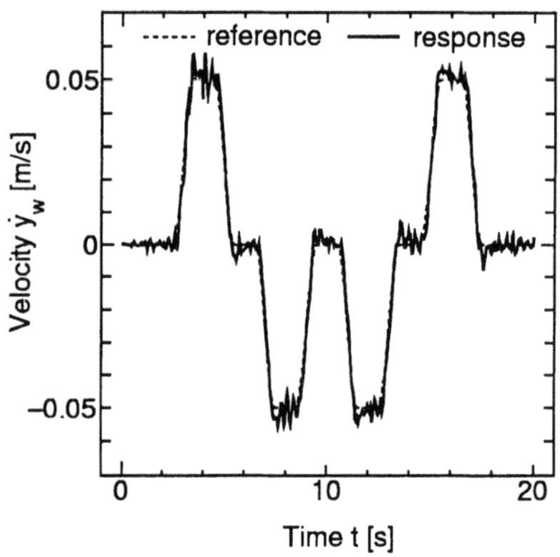


*Figure 21.* Velocity $\dot{y}_w$ for an 8 character path

## 6. Conclusions

In this chapter, we have derived a dynamic model for an omnidirectional holonomic mobile robot, whose platform was assumed to be based on three lateral orthogonal-wheel assemblies. Then, the resolved acceleration control

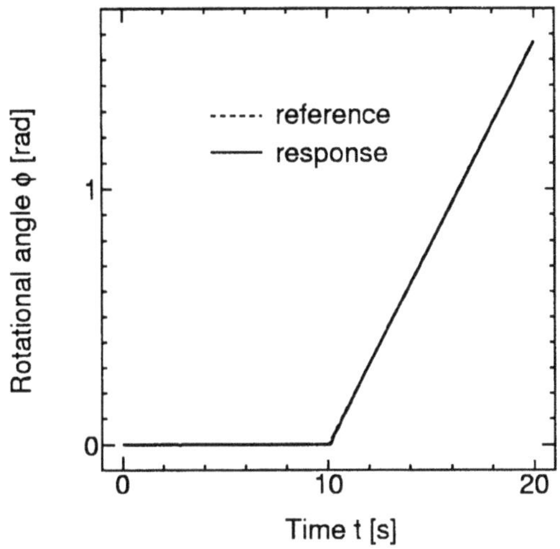


*Figure 22.* Rotational angle $\phi$ for an 8 character path

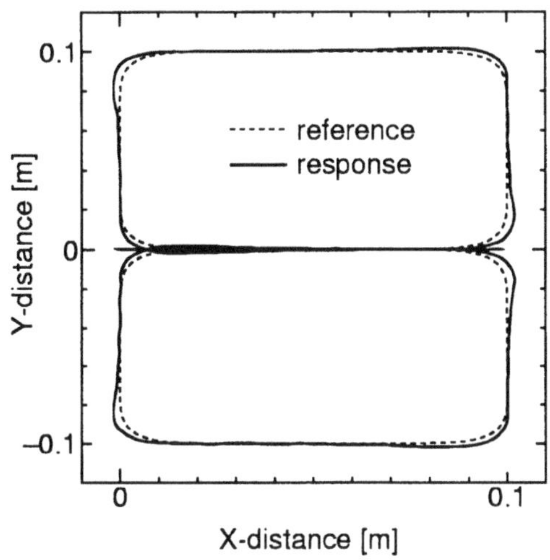


*Figure 23.* Trajectory for an 8 character path

system was derived using such a model, and several experimental results showed that the present control system was effective for controlling such a mobile robot.

We refer to [12] for the control results due to the stochastic fuzzy control approach. See also [13] for the detailed construction problem of the omnidirectional mobile robot using some orthogonal wheel mechanisms.

## References


1. Nakano, E. and Koyachi, N.: An Advanced Mechanism of the Omni-Directional Vehicle (ODV) and Its Application to the Working Wheelchair for the Disabled, *Proc. of '83 Int. Conf. Advanced Robotics* (1983), 277-284.
2. Nakano, E., Mori, Y., and Takahashi, T.: Study of the Mechanism and Control of Omni-Directional Vehicle, *Proc. of the 12th Annual Conference of RSJ* **1** (1994), 369-379. (in Japanese).
3. Wada, M., Tominaga, Y., and Mori, S.: Design of an Omnidirectional Holonomic Vehicle, *Proc. of the 13th Annual Conference of RSJ* **1** (1995), 145-146. (in Japanese).
4. Muir, P.F. and Neuman, C.P.: Kinematic Modeling of Wheeled Mobile Robots, *J. Robotic Systems* **4** (1987), 281-340.
5. Asama, H., Bogoni, L., Sato, M., Kaetsu, H., and Endo, I.: Kinematics of and Omni-Directional Mobile Robot with 3DOF Decoupling Drive Mechanism, *Proc. of the 12th Annual Conference of RSJ* **1** (1994), 367-368. (in Japanese).
6. Nishikawa, A., West, M., and Asada, H.: Development of a Holonomic Omnidirectional Vehicle and an Accurate Guidance Method of the Vehicles, *J. of the Robotics Society of Japan* **13**-2 (1995), 249-256. (in Japanese).
7. Hirose, S. and Amano, S.: The VUTON: High Payload, High Efficiency Holonomic Omni-Directional Vehicle, *Proc. of JSME Annual Conference on Robotics and Mechatronics* (1993), 350-355. (in Japanese).
8. Hirano, T., Chen, P., and Toyota, T.: A Fundamental Study on Omni-Directional-Vehicle (1) – Omni-Directional Running Mechanism for Offroad –, *Proc. of the 14th SICE Kyushu Branch Annual Conference* (1995), 273-274. (in Japanese).
9. Pin, F.G. and Killough, S.M.: A New Family of Omnidirectional and Holonomic Wheeled Platforms for Mobile Robots, *IEEE Transactions on Robotics and Automation* **10**-4 (1994), 480-489.
10. Iwatsuki, M. Nakano, K., and Ohuchi, T.: Target Point Tracking Control of Robot Vehicle by Fuzzy Reasoning, *Trans. of the Society of Instrument and Control Engineers (SICE)* **27**-1 (1991), 70-76. (in Japanese)
11. Saito, M. and Tsumura, T.: Collision Avoidance Among Multiple Mobile Robots – A Local Approach Based on Non-linear Programming –, *Trans. of the Institute of Systems, Control and Information Engineers* **3**-8 (1990), 252-260. (in Japanese).
12. Tang, J., Nomiyama, A., Watanabe, K., and Yubazaki, N.: Stochastic Fuzzy Control for an Autonomous Mobile Robot, *Proc. of 1996 IEEE Int. Conf. on Systems, Man and Cybernetics*, Beijing, China, **1** (1996), 316-321.
13. Tang, J., Watanabe, K., and Shiraishi, Y: Design of Traveling Experiment of an Omnidirectional Holonomic Mobile Robot, *Proc. of 1996 IEEE/RSJ Int. Conf. on Intelligent Robotics and Systems (IROS96)*, Osaka, Japan, **1** (1996), 66-73.

# 14
# ON THE DESIGN OF A CONTROL ARCHITECTURE FOR AN AUTONOMOUS MOBILE ROBOT

E. PEREIRA DA SILVA
*Instituto Superior de Engenharia do Porto*
*and Instituto de Sistemas e Robótica-Porto*
*Rua de S. Tomé, 4200 Porto, Portugal*

F. LOBO PEREIRA and J. BORGES SOUSA
*Faculdade de Engenharia da Universidade do Porto*
*and Instituto de Sistemas e Robótica-Porto*
*Rua dos Bragas, 4099 Porto Codex, Portugal*

## 1. Introduction

In this chapter we present a model of a hierarchic control architecture for Autonomous Mobile Robots (AMR) as well as its use in the design of an autonomous wheeled mobile platform for transportation in an industrial environment. This model results from successive incremental refinements consolidated over the years with the experience accumulated in a number of R&D projects (e.g., [1], [2]) with special emphasis for the NATO funded PO-Robot project [3].

The inexistence of well established systematic methodologies to design control structures for complex systems, such as those arising in autonomous mobile robotics, leads to a multiplicity of approaches reported in the literature (e.g, [4]). Although frequently related to the natural organisation of the system and, usually, poorly linked to the desired performance, architectural options play an important role in handling the inherent design complexity and in enabling the behaviour verification. Furthermore, the design activity suffers from the difficulty inherent to the integration of a wide range of concepts, methods and tools from a variety of disciplines such as Artificial Intelligence, Control Systems Theory, Computer Science and Operations Research.

Additional difficulties arise when the envisaged control architectures should enable autonomous vehicles to operate in semi-structured, partially unknown, and possibly hostile environments. For these critical systems, safety, apriori systems verifiability, mission reliability and survivability are among the requirements to be carefully taken into account in the design. In order to endow the AMR with the ability to adapt to its environment, a complex set of perceivable situations has to be considered. In order to make use of the wealth of the gathered data, the AMR control architecture must take into account the unique set of circumstances arising during mission execution in order to produce the needed correct perceptions and synthesise operational behaviours that efficiently contribute to the accomplishment of the goals embodied in the mission plan.

For a given set of applicational scenarios, the design and development of autonomous vehicles require the consideration of systematic procedures ranging from analysis, verification and validation of users requirements up to the final stages of the development process (see [5]). The adoption of proper architectural options not only allows the incorporation of apriori assumptions on the natural organization of the system, but also strongly contributes to the simplification of behaviour verification and to the modularity of the design.

We consider a model of architecture satisfying a number of general common requirements, such as, autonomy, mission reliability, safety, testability, expandability, error handling and error recovery capabilities, fault-tolerance and redundancy, and sensor guided motion which can be synthesised within a framework characterized by the following general design features [1], [2], [3]:

- Hierarchic control architecture.
- Bottom-up design and development.
- Timed transition modelling.
- Linguistic based approach to decision.
- Task-organized behaviour.
- Embedded error handling and recovery.

The existence of a hierarchy corresponds to the need of well defined planning and control mechanisms [6] which require the consideration of a world model. This underlying knowledge together with mission specification data are the ingredients needed to build a mission plan whose execution produces a real-time response to uncontrolled events.

In this model, the hierarchic architecture relies on the definition of independent linguistic units at each component [7] which support the mapping of mission specification onto the mission execution. Furthermore, mission composition mechanisms are embodied in the mission specification language. The adoption of a bottom-up approach allows to build a basic kernel of mission primitives which may be incrementally enriched with additional functionalities. By considering a list of contingency modes of operation and by using some redundancy of computational resources and different sensors and actuators, some operating capabilities may be preserved in the presence of malfunctions or failures.

The framework considered in this chapter considers three directions: structural, functional and operational or behavioural.

From the structural point of view, the model of the architecture is based on considering components as a hierarchic network of automata, and activities as transitions in a single framework encompassing both clock and discrete events.

The functional perspective of the model is characterized by a:

Hierarchic organization of the components where, on the one hand, more abstract functions, at the higher levels, perform decision activities, and one the other hand, more concrete functions are at the lower levels.

Linguistic decision scheme defining the scope of the executed control actions so that the associated state trajectories correspond to the specified mission goals.

On the operational side, the model of the architecture is based on the specification of task oriented behaviours, on the definition of tasks via the composition of primitive functions and on the hierarchic synthesis of behaviours.

This chapter is organized as follows:

In section 2, some design issues are addressed in order to situate the control architecture design within the proper engineering context. A brief survey of some of the most representative approaches that bear a more relevant impact on our work is presented in section 3. The framework within which the Po-Robot System (partially described in the last section) was developed is the object of detailed consideration in section 4.

## 2. Some System Design Issues

Clearly, the control architecture is intimately connected to the performance of the system whose design attempts to meet the enduser requirements. Therefore, implementation issues may dictate some of the design options underlying the control architecture. This discussion is taken from [4 ].

### 2.1 ISSUES FROM SYSTEMS ENGINEERING

Software engineering is a field where the management of diversity and complexity via the adoption of a methodological basis has been of concern for some time (see [8], [9]), being the IEEE trial-and-use standard, [10], a recent illustration. This standard offers an integrated approach to engineer a system and involves a number of steps which, embedded in an iterative spiral scheme, allows the gradual improvement of the design. The main steps are:

- Requirement analysis and validation.
- Functional analysis and verification.
- Synthesis.

Scott Harmon, [5], discusses the differences between the main approaches that he classified as implementors and developers. Given their strong research orientation, they have focused their efforts, respectively, on the decomposition of the functional requirements into implementable pieces, and in the definition of the hardware and the functional processing which will meet the design requirements. The major limitations in the design and implementation of such systems lie essentially in the definition of the design requirements, in the mapping of the functional processing into the hardware and in the overall verification of the designed system.

The implementation of a controller for the class of complex systems encompassing AMRs is not yet a systematic procedure and it involves the simultaneous consideration of a large number of methodologies whose complete integration in a unique design framework is not currently available. An analog to concurrent engineering is suggested by the relations between design and implementation steps. Methods and tools are needed to support the parallelism of these design steps. The lack of experience in performing all the design steps concurrently is a major limitation of the available technology. The implementation of practical applications will certainly contribute to the development of this technology.

## 2.2 ISSUES FROM CONTROL AND COMPUTER ENGINEERING

Since AMRs are designed to accomplish missions composed of sets of distinct tasks requiring the execution of different activities, there is the need of controlling a set of dynamic systems representing the involved physical components, and, simultaneously, guaranteeing the apriori correctness of corresponding programs, [11], [12].

Concepts and methods of the various cognitive disciplines of classic Artificial Intelligence, such as knowledge base systems, and planning and reasoning about actions, have encountered difficulties in meeting the requirements inherent to real time sensing and actuation which have been dealt with by control researchers. As a result, reactive planning and any-time algorithms emerged. While, in the first class of approaches, reactive behaviours are generated by tightly coupling planning and execution, in the second one, the best compromise between accuracy and available computational time is sought.

Furthermore, from this cross fertilization and from the undertaken work on reactive systems, a variety of notions of "intelligence" became apparent. Instead of rigid, deliberative decision processes, the behavioural approaches consider the real-time interaction with the environment as the essence of intelligence. Intelligent real-time systems should make correct and opportune decisions in various situations, and accurately perform complex tasks in highly variable, unstructured and unpredictable environments.

Recent developments show a clear trend towards new concepts and methods conjugating techniques of modelling, synthesis and analysis developed in the framework of Control Engineering, with procedures of specification, implementation and verification traditionally addressed in the field of Computer Engineering, [13]. A general approach to model hybrid systems, natural to Control Engineering, considers the overall system organized as a collection of dynamic control systems interacting concurrently in real-time within the given environment. Each dynamic control system may correspond to some subsystem or component of a subsystem with one or more operating points. There is a wide acceptance that the design of a hybrid controller [12], [14], [15], [16], [17], should involve several phases including the:

- Design of the usual (in the sense of Control Theory) controllers for dynamic control systems closely describing the underlying physical system, [18].
- Design of the mechanisms for handling typed events, such as finite state machines, automata or Petri nets, that will intervene by adequately reconfiguring activities.
- Build interfaces between continuous and discrete levels enabling the identification of relevant discrete events based on the evolution of the physical state, and the generation of appropriate reference signals to the dynamic controllers at the time-continuous physical level.

Given the high cost to be paid for any exceptional repairing or recovery of an AMR in the course of operation, especially for a long range one, it is crucial to ensure its behaviour in spite of the high variability and unpredictability of its environment. This verification includes not only the correctness of logical aspects but also the satisfaction of real-time constraints and the agreement between programming and specifications [12]. Thus, an adequate design and control framework for the class of dynamic systems encompassing AMRs will have to consider concepts and associated techniques for requirements specification, hybrid systems modelling, control synthesis, and behaviour verification. There are a number of approaches and tools supporting verification and validation tools, such as [11], [12], [13], [16], [19], [20] to name just a few.

## 3. Related Work

It is not in the scope of this chapter to cover substantially the state-of-the-art in control architectures for autonomous mobile robots. We simply, find it useful to briefly visit diverse approaches that in one sense or another relate to our work. This permits not only to situate our approach to the design of control architectures, but also helps to identify the direction of progress of the current work.

### 3.1 FUNCTIONAL MODEL ARCHITECTURES

#### 3.1.1 The Reference Model Architecture

The Reference Model Architecture addresses control systems for Intelligent Machines which includes automated, embedded and robotic systems. Intelligent Machines are closed loop systems in the sense that they utilize feedback from the environment to manifest "Intelligent Behaviour" in real-time. This paradigm, depicted in Figure 1, is best represented by hierarchical architectures like NASREM [21], [22], [23] which encompass the purpose of enforcing modularity and a software methodology. In this approach, the control problem is partitioned into:

- Behaviour Generation (Task Decomposition)
- World Modelling
- Sensory Processing
- Value Judgement

The various elements of the control structure are clustered into computational nodes arranged in hierarchical layers, being each layer provided with a characteristic functionality and timing. The Real-time Control System (RCS) design paradigm begins with the analysis of tasks and goals, then develops control algorithms, data structures, and sensory information necessary to perform the tasks and reach the goals.

This paradigm has been refined over the years. Starting with the cerebellar model arithmetic computer in the 70's, it evolved to the RCS-4 in the 90's, having architectures of intermediate complexity been applied to Automated Manufacturing Facilities, an Army Field Material Handling, Multiple Underwater Vehicles and a Flight Telerobotic Servicer.

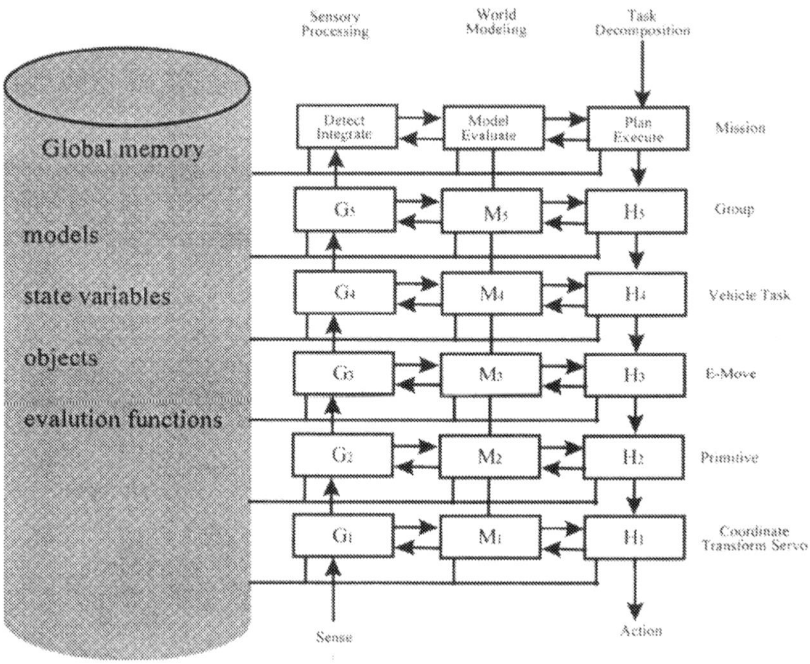


*Figure 1. The NIST hierarchical architecture*

The methodology associated with the RCS architecture explicitly addresses the design of the following critical components:

- Real-time task behaviour and software execution modes.
- Interfaces and communication methods between software modules hardware resources and human operators.
- Information models and knowledge base management.
- Allocation of resources.

- Rules for decomposing the design spatially and temporally.

3.1.2 The Hierarchic Intelligent Control Architecture

Saridis and collaborators have been developing a consistent theory for intelligent machines based on the theory of hierarchical intelligent control ([7], [24], [25]) formally supported by a Petri Net language. The proposed framework involves three major levels, organization, coordination and execution as described in Figure 2. One distinctive characteristic of this approach arises from the notion of task composition of primitive functions.

From the mission plan, the organization level (the task organizer) generates high level tasks for the Autonomous Intelligent Mobile Robot System to accomplish.

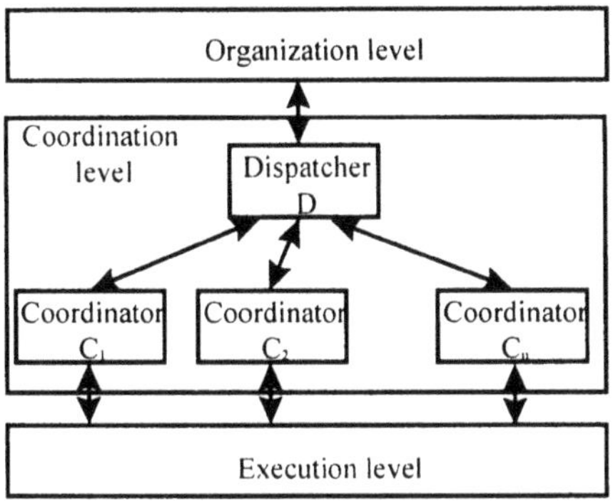


*Figure 2. The hierarchical structure of an Intelligent Machine.*

The coordination level, [26], acts as an interface between the organization and execution levels. This level consists of one dispatcher and several task coordinators. The dispatcher decomposes the tasks of the mission plan generated by the organizer into coordinator-oriented control actions and then dispatches them to the corresponding coordinators at the most appropriate times. The coordinators translate control commands into operation instructions to be loaded into the execution level and made available at the corresponding execution mechanisms.

In the proposed coordination structure, places represent processes and transitions represent algorithms. Since Petri net models are incapable of describing language translation as required for the implementation of the linguistic decision schemata [7], Petri Net Transducers (PNT) are introduced [27]. These are language translators transforming a given input task plan into an output task plan.

The execution level operates in real-time. It executes the instructions coming from the coordination level and provides information feedback to it.

The analytic model underlying the hierarchic control structure based on the principle of Increasing Precision with Decreasing Intelligence has been formulated in a probabilistic framework where uncertainty and imprecision are expressed in terms of entropies. The optimal strategy for the decision planning and task execution of the intelligent machine is to minimize the total entropy of the system. Thus, optimal control problems may be formulated in order to address intelligent machine design issues. Recent developments of this analytical framework involve the development of a methodology supported on Information-Based Complexity for joint measure of algorithm cost and reliability, [28]. This approach further improves the basis for the analytical design of intelligent machines.

### 3.1.3 The LAAS Architecture

Lot of work has been developed at LAAS in the design of autonomous mobile platforms [6], [29], [30], [31], [32], [33]. Although several robots have been built, we will concentrate our attention on the conceptual framework. General criteria for the design of a control structure of an experimental prototype of an autonomous machine include reactivity, flexibility, reconfigurability, autonomy, and programmability.

The proposed architecture is organized into three levels corresponding to two decisional levels above the functional level. The two higher levels are built according to the paradigm of the supervisor/planner. While the higher level uses a temporal planner, the second level receives tasks and transforms them into procedures composed by elementary actions and supervises the execution of these procedures while reacting to asynchronous events. At this level, planning consists in the refinement using specific knowledge concerning the task or the domain.

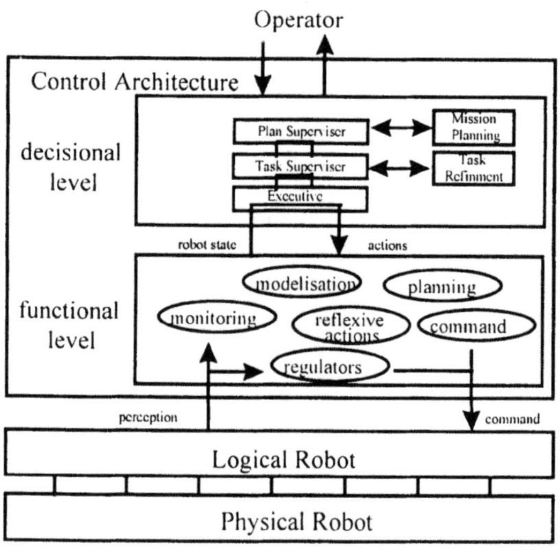


*Figure 3. The LAAS hierarchic Architecture*

In what concerns the mission planning level, a temporal planner IxTeT was developed to reason about the numerical or symbolic temporal relations. The plan produced by

IxTeT consists in a partially ordered set of tasks associated with the temporal constraints in such a way that maximal and minimal bounds on the durations of execution are defined and taken into account in the synchronization with external events.

Supervision and refinement of tasks are integrated via a Procedural Reasoning System which offers a context well adapted to support the interaction between decision and reaction.

The nature of the execution level role is purely reactive without any planning capabilities. It controls the execution of actions according to pre-defined schemes selected according to the context. Its implementation is based on a set of rules that permits to compile off-line a set of rules producing a program to search a decision tree in bounded time.

A good description of the developed work with some further conceptualization of the control architecture (see Figure 3) is presented in [32], [33]. In [29], [34] a generic architecture for intervention robots is described with particular emphasis to the deliberative level. Here, the supervision is implemented as a procedural reasoning module adapted to the "real-time" requirements associated with the execution of the functionalities which are endowed with some reactivity.

### 3.1.4 The Subsumption Architecture

As an alternative to the hierarchic architectures the subsumption approach Brooks (in [35], [36]) avoids the use of a explicit world model and instead implements both primitive and complex behaviour by a more direct coupling between perception and action. This structure was inspired by studies from the field of ethology on animal behaviour.

It appears to be better suited for reaction to unforeseen events, although less predictable in more routine circumstances. Brooks argues that the "world is its own best model" and that complex and apparently purposeful action can arise from the competition of layered behaviours according to a predetermined priority scheme without recourse to an internal world model. In this setting there are no planning or explicit control actions.

In the Brooks approach the reactions are organized in a hierarchy of levels where each level corresponds to a set of possible behaviours. Under the effect of an internal or external stimulation a particular behaviour is required. Then it emits a influx towards the inferior level. At this level, another behaviour arises as the result of simultaneous action of the influx and other stimulae. The process continues until terminal behaviours are activated. In this approach each level implements a unique behaviour which may inhibit the inputs and outputs the immediately bellow level. Behaviours at a higher level of abstraction are said to subsume those at a lower level. Brooks claims that the interactions between those behaviours permit to achieve more complex behaviours

### 3.1.5 The Fate and KAMARA Architecture

KAMARA is a distributed control architecture that evolved from FATE [37], [38]. The FATE architecture consists of a blackboard planning level to generates situation dependent manipulator-specific elementary operations that are executed under a real-time robot controller RT-Cs. The real-time controller is able to control the manipulators either independently or integrated in a closed kinematics chain.

The new control architecture KAMARA uses a recursive algorithm to exchange tasks at the cell-level (cooperation among robots) and at the robot level (cooperation among components of a robot) allows the distributed coordination and scheduling of activities. This architecture can be properly formalized in a multi-agent framework where each agent is composed by the following components: Communicator, Planner/Action selector and Executor. It is worth to note that communication among the various agents may take place either synchronously or asynchronously. Since it uses several components of the former centralized FATE architecture, the execution of an explicit elementary operation is in both systems performed by the real-time robot control systems RT-RCS.

## 3.2 INTEGRATION CONTEXTS AND REAL-TIME LANGUAGES

### 3.2.1 ControlShell

ControlShell [39]is a very light and reconfigurable system developed to define a real-time integration context having in mind system's validation via experimentation rather than the formal verification of the global behaviour of the system. It offers:

- Set of tools to build shareable functional components and to specify their interactions via data their data exchange requirements
- Finite state machine implementing a programmable executive system that supervises the execution activities of the various components.

The basic elements are organized in a library of expandable, shareable and reusable software modules which might be articulated via a graphic editor which enables to specify the data flow at the component level. Strategies for module activation/ deactivation can be programmed in the form of a state-graph which will be implemented by a state-transition model.

### 3.2.2 Chimera

Chimera also defines a modular real-time integration environment [40]. Components are defined as instances of a generic object that includes an internal state, an interface constituted by input-output ports, a method with encapsulated data and resources. Ports enable the interconnection among the various objects.

Each object has four states: NOT-CREATED, ON, OFF and ERROR. When activated, each object gives rise to an activity which is executed independently of those of other objects. The execution of an activity generates or requires data to or from a local memory area which is exported to or imported from a common memory.

### 3.2.3 Orccad

General methods for real-time control and for consistency verification have been developed at several institutions (e.g., see [15], [16], [17], [19], [20],[41]and references therein), to name just a few). In particular, at INRIA, some work has been focused in the development of a general framework to support architectural design options explicitly for AMR control systems. Special attention was paid to specific issues arising in the underwater environment where real-time adaptivity like self-replanning and sensor-based control play an important role. An implementation has taken place in Vortex at IFREMER and applications to underwater robotics have been published in [11], [12]. In [20], methods of formal verification are applied to various problems in robotics within ORCCAD, a development environment for robotic controllers designed at INRIA. It provides a set of tools to check before the mission starts that the vehicle will perform the intended tasks correctly. That is, to verify logical consistency and the satisfaction of temporal constraints, thus allowing to conclude whether:

- specifications are correct,
- programming conforms to specifications, and
- real-time and implementation related constraints do not perturb the behaviour.

In this environment [11], the Robot Task, RT, is the main entity which is associated with a local behaviour controlled by a set of observers and modelled by a finite state automaton. It is composed by a set of real-time communicating tasks, the module tasks whose specification require explicit temporal and synchronization constraints. The RT is composed of an elementary servo-control task and a logical module instantiated in a automaton which manages the overall behaviour. The module task, MT, is the basic entity which implements each elementary process of the control law and therefore handles all temporal issues. With respect to dialogue and verification tools, the ORCCAD environment includes a friendly graphical human-system interface and simulation software.

## 4. The Model for The Control Architecture

### 4.1 DESIGN REQUIREMENTS

The definition of system requirements is the first of a number of steps towards the design and implementation [42]. The approach presented in this chapter was motivated by the analysis underlying the design of the control architecture for the AMR PO-ROBOT being developed under the NATO Science for Stability Program (see [43]).

The objective of the PO-ROBOT project consists in developing a mobile platform for industrial environments able to: move to a number of specified locations while avoiding obstacles, perform a set of operations such as load and download of materials and to activate load and download devices or tools without the human intervention. The system should also be endowed with motion autonomy in semi-structured and variable environments, as well as with some ability to establish communication with other systems [43]. The following general requirements were extracted:

- autonomy,
- mission reliability,
- safety,
- testability,
- expandability,
- error handling and error recovery capabilities,
- fault-tolerance,
- friendly user interface amenable to task definition and mission specification, and
- easy user operation.

## 4.2 DESIGN CONSIDERATIONS

A careful analysis of the stated requirements and of the consequences to system's implementation leads to the following considerations:

The emerging functional multiplicity of the system inherent to

- multiplicity of "input focus",
- simultaneous interaction with external environment,
- mission with multiple goals,
- subsystems interaction,

is only viable if composed by a set of entities within a framework of enabling parallel and concurrent execution.

The experience accumulated in the execution of the Po-Robot project, strongly advocates the modularity of modelling and implementation. This is due, not only to the increased execution efficiency, but also to the improvement of analysis and refinement in the design process. On the other hand, the need of systematization led to establishment of a generalized model for the various entities, i.e., a common structure invariant with respect to the function characterization. Complexity is the main draw back associated to the modular modelling as the sate of the system results as the composition of the state of each subsystem which promotes a rapid increase of the number of viable trajectories.

Additionally, a property of composability of behaviours of the model of the architecture should ensure that the admissibility of behaviours is preserved under the allowed operations of behaviour composition (sequences, parallelization or combination). This means that the defined event language is closed and therefore the system controllability is ensured. By properly restricting the domain, composability guarantees the decomposability (composition and decomposition behaviours) which results in the independence of the basic behaviours. It follows the ability to interpret and to isolate failures and the consequent viability of their recovery, as well as the possibility of modelling safe behaviours.

The above mentioned requirements and the consequent analysis led us to consider a development framework characterized by the following design options [3], and [44]:

Hierarchic control architecture

From the above considerations, it is clear the need to coordinate the activity of the various entities since the overall system's behaviour depends on the various subsystems.

The attempt to encompass within a single framework, the requirements underlying the AMR, promote the emergence of two conflicting vectors :

- Structural and functional parallelism and concurrency .
- Behaviour composition.

A two dimensional architecture is adopted to conjugate these two vectors: A temporal one required for activities scheduling and a functional one, to their parallel coordination.

Bottom-up design and development

The adoption of a bottom-up approach allows the construction of a basic kernel of functionalities primitives which, as knowledge increases with operational experience, may be incrementally enlarged with additional functionalities. This same idea underlies the safety of mission specification as it avoids the definition of tasks and functions whose implementation is not semantically viable.

Timed transition modelling of activities

In spite of the "continuous" nature of the many subsystems of the AMR, the complexity due the multiplicity of interactions suggested the adoption of a discrete event dynamic model. Temporal constraints in addition to the subsystems interaction, led to the adoption of a framework of timed transitions developed by Ostroff [45] and [46]. Furthermore, this model allows to encompass modelling, analysis and synthesis in a single tool. However, since this framework addressed purely discrete event systems, some modifications were needed in order to handle the inherently hybrid nature of the AMR.

Decision Linguistic approach

An autonomous system usually involves handling information at various abstraction levels and, in particular, the one interfacing the human operator which has, for example, to define the mission or supervise activities.

Therefore, the decision process involves several vistas:

- Human operator which has to be liable for the mission specification at a high abstraction level. With this respect, careful grammar design should facilitate plan generation.
- System or component since sets of output events are generated in response to other set of input events.

Since activities in each component are represented by transition models, it can be seen as a processor of a set of events whose output may be represented by another set of events. These may represent commands of the higher hierarchic levels, state feedback from the lower levels, or even contextual information from the same level. Therefore, a decision scheme can be seen as a distributed mechanism that controls the devices operating the input-output translation processes and therefore the overall coordination is achieved through language translation.

Furthermore, the atomic nature of the linguistic structure contribute to ensure decomposability and composability with the consequent protection as the scope of errors can only be controlled only if modules are syntactically delimited.

Task oriented behaviours

In order to ensure that a mission specification corresponds to an admissible set of behaviours, a framework encompassing a given set of admissible elementary behaviours as well as a composition language is considered.

The behaviour of the robotic system results from the execution of a set of tasks (mission) involving one or more resources of the system. A task is carried out by executing a combination of some elementary activities.

Although it is possible to increase the number of tasks by composing elementary tasks, the number of different elementary tasks can not increase without deep modifications of the computational or physical systems.

The hierarchic synthesis of the behaviour is defined via a time (scheduling) and a spatial (parallelism) decomposition of tasks. While time resolution is expressed in terms of bandwidth [21], spatial resolution is defined by the actuation domains of subtasks.

In parallel to the task synthesis, occurs the hierarchic sensory processing, via a temporal composition and a spatial integration. The resolution of the behaviour synthesis process at the various hierarchic levels has to be compatible with the required levels of the sensory system.

Error handling and recovery procedures embedded

The hostile nature of the operational environment and the relative inaccessibility to the vehicle imposes strong requirements concerning safety and mission reliability. These imply the consideration of stringent error handling and recovery procedures. By considering a list of contingent modes of operation and by using redundant sensors, actuators and computational resources, some operational capabilities may be preserved in spite of the occurrence of malfunctions or failures. Since error handling and recovery procedures are embedded in overall system, error detection and recovery is attempted as locally as possible in each component.

## 4.3 THE CONTROL ARCHITECTURE MODEL

In this section, we propose a model specified along the three vectors referred to in the introduction: structural, functional and operational.

Five hierarchic stages are considered: supervision, coordination, functional, abstraction and physical. While the physical level represents the vehicle, the abstraction level extracts a representation so that the control architecture becomes independent of the specific hardware platform.

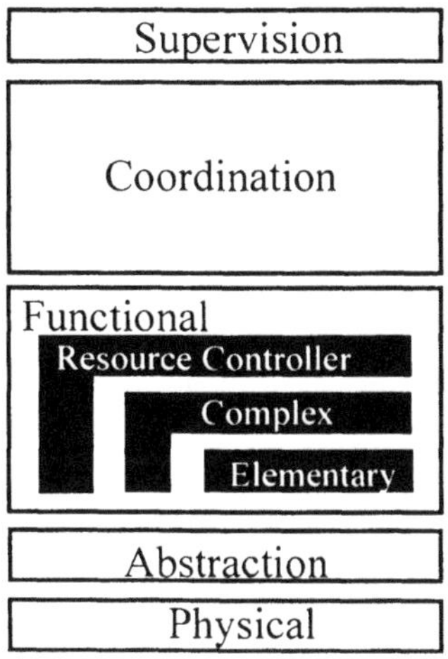


*Figure 4. Model for a control architecture of autonomous system.*

The framework to model the architecture is that of a hierarchic network of dynamic automata, where each component is defined according to its position in the hierarchy.

The functional layer consists in a set of organized entities composing the group of functionalities that either materialize the interface between the decision system and the physical world, or produce the liaison between the decision system and the maintained representation of the world. Most of the components contribute to the implementation of a set of control laws parametrized by the coordination layer which, based on sensory information (physical system), produces references to the actuators (physical system).

The coordination level is essentially composed by a set of supervisory automata (tasks) which implement task coordinators. In parallel, there are entities that, in order to control the mission execution, command the composition of supervisors. In order to adapt the clearly discrete and abstract nature of the coordination and the mostly continuous character of the control and sensing activities of the functional layer, a set of interface entities was introduced in the later. These entities not only promote the independence of the coordination level with respect to the functionalities but also provide to the former an interpretation of the functional layer state. This entity provides the Coordination/Functional layer interface and is designated by resource controller, thus assuming a structuring character.

The supervision layer supports the operator at the top of the hierarchy of the decision system. This layer is composed by an automaton that receives information via the synchronization of coordination events as well as commands from the operator.

### 4.3.1 Physical Layer

This stage represents the vehicle and includes sensors, actuators, energy support system and computational devices. Sensors follow the evolution of both the environment and the state of the vehicle. Actuators update the world state. The energy support system supplies energy according to the needs of the various subsystems. Computational devices process data, memorize the state as well as the inter-components communication channels. By taking into account gathered data, the control system acts on the physical system via a set of modules modelling the physical structure based on automata.

For the designer, the physical layer is a list of sets of variables, being each physical component characterized by the following attributes:

- in (abstract level to physical system) or out (physical system to abstract level)
- domain and units, and
- mode (synchronous or asynchronous) and maximum rate

### 4.3.2 Abstraction Layer

This layer has the function of filtering and formatting data flowing from or to the physical components so that the requirements of the physical components and of the control architecture are both satisfied. In addition, there situations where we might have more than one device offering the same functionality or generating equivalent data. The abstraction layer encapsulates the physical system in order to allow the independence of the control architecture with respect to the specific implementation hardware. In this way, an abstract layer is provided to the control system.

The Abstract layer may be seen as a set of components being each one characterized by:

An interface to the functional level:

a set of variables with the following attributes:

- in ( functional to abstract level) or out (abstract to functional level)
- domain and units
- mode (synchronous or asynchronous) and maximum rate

a set of messages characterizing the discrete behaviour of the component with following attributes:

- in ( functional to abstract level) or out (abstract to functional level)

an interface to physical level:

- a set of variables already defined in the physical layer.

data pre- and post- processing for each entity

Typically, this is an easy processing schema like averaging or selecting. Here, the lowest level of error detection and recovery is included.

In the definition of the abstract components we follow the methodology presented by Sousa. [47]

4.3.3 Functional Layer

The functional is the operational structure in charge of the development of activities. From the point of view of the higher levels of the architecture, these functionalities are organized into groups which are essentially associated with sensing, actuating and computing resources. Reactivity is implemented within this layer, consisting its reconfigurability in the ability to define diverse feedback loops involving the various components. The coordination has in charge the selection of any particular reconfiguration which depends on the active tasks.

The project options pointed out to a hierarchic structural and behavioural scheme represented by standardized modular components. These constitute a dynamic network of automata divided into three categories: resource controllers; complex funcionalities; and elementary funcionalities. The main differences among them lie in the semantic class associated with the implemented function. Each component of the functional layer is characterized by:

- Interface to the components at the level above (coordination and/or functional):
- Interface to the components at the level below (functional and/or abstraction)
- Timed transition model to manage the activity implementing the functionality, and synchronize with other activities via shared events.

The interface with components at the level above is defined in the component being modelled by a:

set of variables to share data (in - from or out - to).

set of messages characterizing the discrete behaviour of the component and having a correspondence with transition events in the TTM (Timed Transition Model) describing the activity of the module (in - from or out - to).

The interface with components at the level below is defined by the ones with which the interaction is established being modelled by a:

set of variables to share data (in - from or out - to).

set of messages characterizing the discrete behaviour of the component and having a correspondence with transition events in the TTM describing the activity of the module (in - from or out - to).

When an activity is modelled, special attention has to be paid to the events shared by the TTM and other modules.

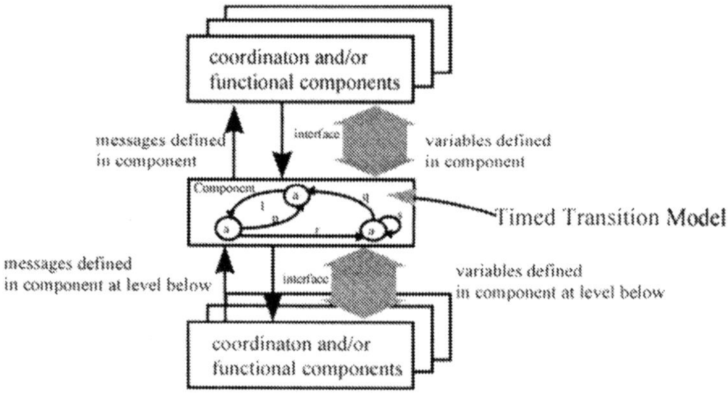


*Figure 5. Component of the function layer*

4.3.3.1 Elementary Functionalities

These are responsible for direct sensing and actuation, and realize the interface between the abstraction level and the control architecture. These modules implement periodic activities and are structurally invariant, that is, their execution does not depend on contextual conditions. Typically, they do not have the ability to terminate their own activities but wait indefinitely for the required external event. Execution of these components is synchronized with a clock transition (fixed frequency) or with an external event (variable frequency). Reactivity, in the most immediate sense, arises from the connections among these modules.

We consider three types of elementary components: filters, regulators and memories.

Filters are data integrators (e.g., an odometric process, or, more generally, a simple position sensor process) receiving data from several sources and providing the adequate format and abstraction character representation.

Regulators are the basic control processes of physical devices (e.g., PID for the motion of a motor)

Memories are components that temporarily organize and store information in abstract form and are fundamental for complex funcionalities.

4.3.3.2 Complex Functionalities.

Complex functionalities consist of activities that combine, aggregate or produce information to or from components of distinct nature and may be organized into successive hierarchic layers. Complex components are responsible by, on the one hand, the system's perception abilities and, on the other hand, by its causality structure. The execution of the activities performed by these components are asynchronous and variable with the system's state.

We consider two main types of the complex functionalities: virtual sensors and servers.

Virtual sensors are activities that observe the logic state of platform. They are organized into three subtypes: Discrete filters or classifiers (e.g., topological localization), alarms (e.g., obstacle detection) which fire events and continuous filters (e.g., geometric localization).

Servers are asynchronous control activities that affect both the world's variables and the data stored in the system. These activities are divided into two subtypes: Control laws (e.g., straight line, curve or l-shaped curve motion controllers) and the calculators (e.g., geometric planning or task planning).

4.3.3.3 Resource Controllers

A resource is a hierarchic structure dynamically reconfigurable in order to allow the emergence of the functionality required by a given task. It is externally characterized by the associated activities.

The role of the resource controller is to ensure the effectiveness of the activity including the interface with the coordination layer. While elementary components depend totally on the envisaged project application, these components depend on the instance of the architecture model.

They can be organized into three large classes: power; computation; and mechanic and sensory.

4.3.3.3.1 Power Controller

The controller of this resource has the goal to control the distribution of power to the various subsystems, monitor their power consumption and also detect related operational or functional failures.

Of course the associated computational system and power supply should be independent of those for all other subsystems. Watchdogs (one line for each subsystem) are used to detect operational failures. By monitoring the power consumption and the respective rate for each subsystem, the controller of this resource implements a basic safety function.

#### 4.3.3.3.2 Computation Controller

This is a distributed resource reaching all the components of the system. It is mostly constituted by a set of elements dedicated to specific functions. Normally, the computational resource to be managed refers to the set of CPU units to which it is possible to dynamically allocate a varied and generalized set of processes. These processes are the logic part the system components.

The presented approach to manage this resource consists in defining the base of the scheduling policy at the system's design phase. During the execution of the system's activities, the controller of the computational resource just has to use the average free time of each CPU in order to allocate computing tasks free time. We reckon that, in more hostile environments, a more elaborated management structure is required.

#### 4.3.3.3.3 Mechanical or Sensorial Resource Controller

These resources can be organized into two classes:

One client-one server-several activities (denoted by 1C1SnA):

This type of resources can perform several distinct activities, but only one at a time in which only one task can be client of the resource. The design of the controller for this type of resources has to take into account the possibility of implementation of multiple activities (motion control laws for the mechanical devices, sensor fusion schemes) while keeping invariant the basic controller [48].

Several clients-on server-one activity (denoted by nC1S1A):

This type of resources continuously execute only one activity that can be reparametrized in order to serve several clients. In the project of this type of controllers, it has to be considered that the reference to the controller depends on a parameter to adjust the intended behaviour. This requirement can be met or not depending on the its state and on the complex functionalities it is serving. Sonar or infra-red sensors are examples of this type of resources outputting data to funcionalities such as obstacle detection and geometric localization [48].

### 4.3.4 Coordination Layer

The coordination system associated with the resource controllers ensures that, in each instant, the overall system functions correctly. The specification of a given task coordination is based on the structure of the respective task.

#### 4.3.4.1 Task

A task may be described by a goal (in case the system has the capability of planning and synthesizing the behaviour of the considered task), or by a composition of tasks or elementary tasks enabling to reach the given goal. A task may globally be characterized by:

- task name or command,

- Goal, state or families of states to be attained,
- Parameters, and information required for task execution,
- state (idle, active, run, rip),
- requirements or families of admissible states to initiate a task,
- conditions to be met in order to suspend or stop a task,
- execution temporal conditions,
- power constraints, and
- procedural data for execution:

  generic task: state automaton (tree or graph) describing the task behaviour. Each node of the automaton represents either a task or a task composition operator.

  elementary task: it may be an instance of an automaton for each type of resource controller.

Control and the respective task execution is a composition of the automaton describing the task with the generic one describing the state of the task.

#### 4.3.4.2 Coordination Skills

In order to the system be able to exhibit an autonomously generated behaviour, the coordination requires the following functions:

Task interpretation. This means that the coordination is able to build an automaton enabling the execution of the task.

Modification of the decision tree. When a task becomes active, a new supervisory function was created by composing in parallel the previous one with the new supervisory automaton. The inverse action takes place when a task is deactivated.

Task execution. The system is able to generate events that ensure the admissibility of trajectories. On the one hand, these events reconfigure the automata network at the functional level and, on the other hand, they activate control laws, alarms, classifiers, etc., required to the mission development.

Transmission of the coordination state trajectory to the supervision.

Continuous acceptance of missions, commands (events) from supervision.

#### 4.3.4.3 Decision strategies

Decision strategies emerge from the network of active tasks. The coordination activity takes place under the interaction of the hierarchy of supervisors over the functional layer. The state "explosion" that could arise from the increase of the number of entities in the functional layer is avoided by the fact that the coordination activity only needs to "track" the state trajectory associated with the composition of automata that interface with the coordination layer.

This abstract character of the coordination layer plays an essential role in the decision system since it allows behaviours that are more complex and adaptable than those permitted by reactive structures.

## 5. The Po-Robot System

Our vehicle is based on the Robuter, an indoors commercial platform. The propulsion system uses two individually controlled drive wheels on either side of the base and two with caster wheels in the front. To the initial configuration, we added a conveyor belt on the top in order to handle materials and a new set of the sensors which includes a compass, IR sensors, and a mobile ultrasonic TOF (Time-Of-Flight) sensors. The computational system was upgraded by replacing the old CPU board by a new one, PEP VM62 with a MC68060, and a new real-time sensor net (ExMIO- extended multi-Input/Output Bus) was developed. The original operating system, Albatroz, was replaced by the VxWorks with real-time, multi-tasking capabilities. The new hardware configuration is depicted in the next figure.

### 5.1 COMPONENTS OF PHYSICAL SYSTEM

The actuating systems of the vehicle are:

Two propulsion motors responsible for the movement of vehicle which is endowed with a differential steering;

a electromagnetic brake associated with each motor;

a motor of conveyor belt used to load and download material in the docking station;

four step motors, each used to rotate a small tower equipped with TOF ultrasonic sensors and IR sensors; and

three power switches to turn on/off some subsystems.

In what concerns the sensors, the platform is equipped with:

Three optical encoders attached to the armature shaft measuring the velocity of the motors;

ultrasonic time-of-flight sensors used to determine the distances to the objects near the platform;

twenty eight IR sensors playing the role of optical bumpers and used to communicate with other vehicles or systems such as a docking station or a operational interface;

two compass sensors for heading readings: one magnetic for absolute and an optical one for incremental readings (the former is due to the fact that the magnetic headings is not always available);

five accelerometers for monitoring the vehicle movement and support localization algorithms;

a front and rear bumpers to sense world objects which allow collisions;

five on/off pressure button for emergency stops and selection of docking station;

one joystick to directly drive the vehicle in emergency malfunction situations; and

one telecontrol station composed by a steering wheel, accelerator, brake and system on/off switches.

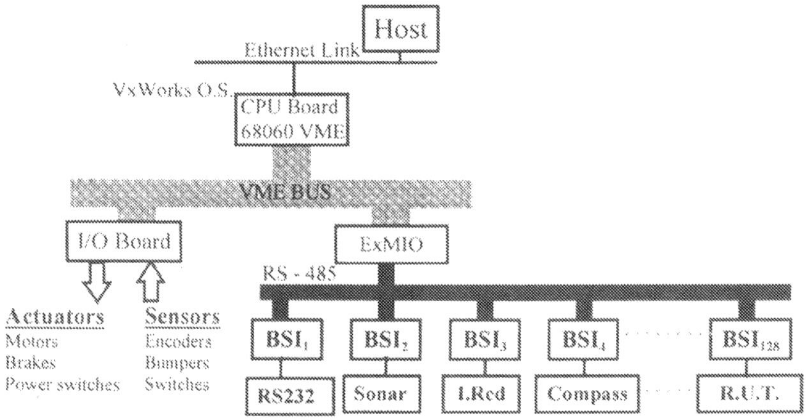


*Figure 6. Po-Robot basic hardware configuration.*

The computational system is distributed in the platform reaching every device. All sensors and actuators are encapsulated with logical processes which run locally in the sensor or actuator interface. In this way, we spare the main computational system to run high level processes. Currently, we have:

One CPU PEP VM62 with a MC68060 supporting the VxWorks Operating System and running all functional and coordination logical process modules. We envisage the need of another CPU in order to incorporate additional complex funcionalities.

One Input/Output card supporting some interface activities required by hardware processes associated with some elementary functional activities.

One ExMIO- extended multi-Input/Output Bus needed to map all sensors and actuators plugged in the net into the memory of the main CPU. Devices are plugged in via a BSI (Bus Sensor Interface). Sensor and actuator devices are encapsulated by computational processes supplied by a local micro processor.

Power system

The power system is based in four traction sealed lead batteries, each battery providing 12 Volts 60Ah. Depending on the characteristics the mission, this power system allows 2 a 6 hours of autonomous operation.

## 5.2 PO-ROBOT ARCHITECTURE

### 5.2.1 Physical and Abstraction Layers

As seen in sections 4.3.1 and 4.3.2, the Physical and Abstract Layers involve several hundreds of variables and messages describing the behaviour of actuators, sensors, computational devices, and power system. For a given variable, the main differences of the two layers reside not only in the interpretation of its value but also in the way it is quantified. For practical reasons is not possible to list here all the variables and we suggest the Po-robot reports [48].

### 5.2.2 Functional Layer

Although not all the designed components of this layer are fully implemented, the main kernel is already in place and running according to the requirements. The design process was guided by the model presented in section 4.3.3. Since TTM diagrams provide an implementable framework, and the real-time attributes of VxWorks operating system (such as queuing mechanisms for messages and shared variables, and dynamic scheduling process policy) are adequate, the implementation of the logic processes follow closely the designed components[48].

Table 1 shows a sample of the functional layer organization where the designed and implemented components are explicit.

### 5.2.3 Coordination Layer

At the current stage of development, we have already implemented the tasks addressing the movement and computational resources controllers. The task "manoeuvre" involves the composition of basic movement activities (i.e., lines, curves, L-shaped curves and docking).

### 5.2.4 Supervision Layer

At this stage, the supervision layer is still very simple and just allows the human operator supervision for system's development purposes. It is based on the management of graphical interface and allows the operator interact with system and perform a limited set of commands such as plan, start, abort, suspend or continue mission.

*Table 1. Architecture Schemata*

| Architecture Level | Type automata | Classes | Sub-classes | Components of architecture |
|---|---|---|---|---|
| Supervision | Supervisor | | | Interface with Operator |
| Coordination | Supervisor | Tasks | | Linecurve |
| | | | | Teleoperation |
| | | | | Dock |
| | | | | Manoeuvre |
| | | | | Planning |
| | | | | Load |
| Functional | | | | |
| | Resource Controllers | | | |
| | | 1C1SnA | | Locomotion Controller |
| | | | | Navigational Controller |
| | | | | Handling Controller |
| | | nC1S1A | | Virtual Sensor System |
| | | | | Sonar Management System |
| | | | | Infrared Management System |
| | | | | World map management |
| | | Computational | | |
| | | Power | | |
| | Complex | | | |
| | | Virtual sensors | | |
| | | | Discrete Filters | Topological localization |
| | | | | Obstacle Avoidance |
| | | | Continuous Filter | Geometrical global localization |
| | | | | Relative Localization |
| | | | Alarms | Obstacle Detection |
| | | Servers | | |
| | | | Control laws | LineCurve |
| | | | | Teleoperation |
| | | | | L-Shape |
| | | | | Handling material control |
| | | | Calculators | Geometrical Trajectory planning |
| | | | | Topological Planning |
| | Elementary | | | |
| | | | | PID- controllers |
| | | Regulators | | |
| | | Filters | | Dead reckoning |
| | | | | Conveyor position |
| | | | | Joystick or teleoperation input |

**References**


1. Pascoal, A. Bjerrun, A. and Coudeville, M,: "MARIUS: An Autonomous Underwater Vehicle for Environmental Surveying", Procs. of the MAST and Euromar Market, CEC, Brussels, Belgium, 1993, pp. 764-758.
2. Sousa, J. Borges, Pereira, F. Lobo and Silva, E. Pereira: "A Dynamically Configurable Architecture for the Control of an AUV", Procs. of the OCEANS 94 Conf., Brest, France, Sept., 1994, pp 131-136.
3. Silva, E. Pereira , Sousa, J. Borges, Pereira, F. Lobo, Sequeira, J. and Ribeiro, I.: "On the Design of the Po-Robot System", Procs. of the IEEE Intelligent Vehicles Symposium 94, Paris, France, 1994 .
4. Sousa, J. Borges, Pereira, F. Lobo and Silva, E. Pereira: "Software Architectures for Autonomous Vehicles: Survey", Workshop AUVs, Porto, Portugal, Sept. 93.
5. Harmon, S. Y.: "Architectures: Designers versus Implementors", in Procs. of the Workshop On Architectures for Intelligent Control Systems , IEEE Int. Conf. on Robotics and Automation, Nice, France, May 1992, pp. 1-6.
6. Chatila, R. Alami, Degallaix and H. Haruelle: "Integrated Planning and Execution Control of Autonomous Robot Actions", Procs. IEEE Int. Conf. on Rob. & Autom., Nice, France, 1992.
7. Saridis, G., Graham, J.: "Linguistic Decision Schemata for Intelligent Robots," Automatica, vol. 20, n° 1, 1984, pp. 12-126.
8. Heninger, K.: "Specifying Software Requirements for Complex Systems: New techniques and their applications", IEEE Trans. Soft. Eng., vol. SE-6, n° 1, Jan. 1980, pp. 2-12.
9. Pressman, R.: Software Engineering: A Practitioner's Approach, Macgraw-Hill Int. Editions, 1992.
10. IEEE P1220 Trial-and-Use Standard for Systems Engineering, IEEE Standards Dept., NY, 1994.
11. Simon, D., Espiau, B., Castillo, C., Kapellos, K.: "Computer-Aided Design of a Generic Robot Controller Handling Reactivity and Real-Time Control Issues", IEEE Trans. on Control Systems Technology, vol. 1, n° 4, Dec. 1993, pp. 213-229.
12. Simon, D., Kapellos, K., Espiau, B.: "Formal Verification of Mission and Tasks Application to Underwater Robotics", pre-print, Workshop on Hybrid Systems, Grenoble, France, Sept. 1995.
13. Ying, Z.: "A Foundation for the Design and Analysis of Robotic Systems and Behaviours", PhD thesis, British Columbia University, Sept. 1994.
14. Antsaklis, P.: "Final Report of the Task Force on Intelligent Control, Technical Committee on Intelligent Control", IEEE Control Systems Society, December 1993.
15. Henzinger, T., Manna, Z., and Pnueli, A.: "Towards Refining Temporal Specifications into Hybrid Systems", in: Hybrid Systems, Grossman, R., Nerode, A., Ravn, A., and Rischel, H., eds., Lect. Notes in Comp. Sci. n° 736, Springer Verlag, 1993, pp. 60-76.
16. Henzinger, T., Ho, P.: "Model Checking Strategies for Linear Hybrid Systems", 7th Int. Conf. Indust. & Eng. Applic. of A.I. & Expert Syst., Austin, TX, May 1994.

17. Ravn, A., Rischel, H., Hansen, K.: "Specifying and Verifying Requirements for Real-Time Systems", IEEE Trans. Soft. Eng., vol.18, N°1, Jan. 1993, pp. 41-55.
18. Samson, C, Espiau, B., Le Borge, M.: "Robot Control: The Task Function Approach", Oxford University Press, 1990.
19. Alur, R., Courcoubetis, C., Henzinger, T. and Ho, P.: "Hybrid Automata: An Algorithm Approach to the Specification and Verification of Hybrid Systems", Hybrid Systems, R. Grossman, A. Nerode, R. Ravn and H. Rischel eds., Lect. Notes in Comp. Sci. n° 736, Springer Verlag, 1993, pp. 209-229.
20. Espiau, B., Simon, D., Kapellos, K.: "Formal Verification of Missions and Tasks", pre-print , INRIA, 1995.
21. Albus J.: "Outline for a Theory of Intelligence," IEEE Transactions on Systems, Man, and Cybernetics, Vol. 21, n°. 3, pp. 473-509, May/June 1991.
22. Albus, J.: "System Description and Design Architecture for Multiple Autonomous Undersea Vehicles", NIST Tech. Note 1251, Washington, DC, Sept., 1988.
23. Albus, J., Quintero R.: "Towards a Reference Model Architecture for Real-Time Intelligent Control Systems (ARTICS)", Robotics and Manufacturing, New York, ASME, vol.3, 1990.
24. Saridis, G.: "Foundation of the Theory on Intelligent Control," Procs. IEEE Workshop on Intelligent Control, Rensselaer Polytechnique Institute, Troy, N.Y. 1985, pp. 23-28.
25. Saridis, G.: "Analytic Formulation of the Principle of Increasing Intelligent with Decreasing Precision for Intelligent Machines," Automatica, Vol. 25, n°. 3, pp.461-467, 1989.
26. Wang, F., Saridis G.: "A Coordination Theory for Intelligent Machines", Automatica, Vol. 20, N° 5, 1990, pp. 833-844.
27. Wang, F., Kyriakopoulos, K., A. Tsolkas and G. Saridis.: "A Petri-net for an Intelligent Mobile Robot", IEEE Trans. on Systems, Man and Cybernetics, Vol. 21, No. 4, July/August 1991, pp. 777-789.
28. Lima, P.: "Feedback-Based Performance Improvement of Intelligent Control Systems", Procs. Int. URIC Program Development, US/Portugal Workshop, March, 2-3, 1995, pp. 105-110.
29. Chatila, R., Ingrand, F. and Alami, R.: "Mission Planning and Execution Control for Intervention Robots", Procs. of the International Program Development in Undersea Robotics & Intelligent Control - a joint US/Portugal Workshop, March, 2-3, 1995, Lisboa, Portugal, pp. 38-43.
30. Giralt, G., Sobek, R. P., Chatila, R.: " A Multi-Level Planning and Navigation System for a Mobile Robot: a First Approach to HILARE", Procs. 6th IJCAI, Tokyo, Japan, August 1979.
31. Giralt, G., Chatila, R., Vaisset, M., "An Integrated Navigation and Motion Control System for Autonomous Multisensory Mobile Robots", Procs 1st Int. Symp. on Robotics Research, M.I.T., Michael Brady, Richard Paul, Eds, 1983.
32. Noreils, F., Chatila, R.: "Plan Execution Monitoring and Control Architecture for Mobile Robots", IEEE Trans. Rob. & Autom., Vol.11 No.2, April 1995, pp. 255-266.

33. Fleury, Sara: "Architecture de Contrôle Distribuée Pour Robots Mobiles Autonomes: Principes, Conception et Applications", PhD thesis, Université Paul Sabatier de Toulouse, Rapport LAAS N° 96156, 1996.
34. Alami, R., Chatila, R. and Freeman, P.: "Task Level Teleprogramming for Intervention Robots", Procs of Mobile Robots for Subsea Environments; IARP, Monterey, CA, 1991, pp 119-136.
35. Brooks R., "Achieving Artificial Intelligence Through Building Robots," M.I.T. Artificial Intelligence Laboratory, Memo 899, May 1986.
36. Brooks , R., "A Robust Layered Control System for a Mobile Robot", IEEE Journal of Robotics and Automation, Vol. RA-2, 1, March 1986.
37- Laengle, TH. And Lueth, T.C.: "Decentralized Control of Distributed Intelligent Robots and Subsystems"..
38. Lueth, T.C. and U. Rembold: "Extensive Manipulation Capabilities and Reliable Behaviour at Autonomous Robot Assembly", IEEE Int. Conference on Robotics and Automation, San Diego, CA,1994
39. Schneider, S., Chen, V. and Pardo-Catellote, G.: "The ControlShell Components - Based Real-Time, Programming System, IEEE Int. Conference on Automation, Nagoya, Japan, 1995.
40 Stewart, D., Khosla, P.: "The Chimera Methodology: Designing Dynamically Reconfigurable and Reusable Real-time Software using Port-Based Objects: Journal of Software Engineering and Knowledge Engineering, June 1996, pp.249-27..
41. Yoerger, D., Newman, J., Slotine: J., "Supervisory Control System for the JASON ROV", IEEE J. Ocean. Eng., vol. OE-11, No. 3, July 1986, pp. 392-399.
42. Saridis, G. and K. Valavanis: "Analytic Design of Intelligent Machines", Automatica, Vol. 24, 1988, pp. 123-133.
43 Po-Robot: "Multi-Purpose Portuguese Flexible Mobile Robot, Working Plan Proposal", Institute of Systems and Robotic, Porto,Portugal,1995;
44 "Vehicle Management System": Po-Robot Task 3 Report, Institute of Systems and Robotic, Porto, Portugal, 1997.
45 Ostroff, J.S.: "Synthesis of Controllers for Real-time Discrete Event Systems", Procs. of the 28th Conference on Decision and Control, Tampa, Florida, December 1989, pp. 138-144
46 Ostroff, J.S. and Wonham W., M.: "A Framework for Real-Time Discrete Event Control", in IEEE Transactions on Automatic Control, Vol 35 N. 4, Abril 1990, pp. 386-397.
47 Sousa. J.: "On the Control of Remus Vehicle", Remus Project Report, Institute of Systems and Robotic, Porto, Portugal, 1997.
48 "Functional System": Po-Robot Sub-Task ST2/T3 Report, Institute of Systems and Robotic, Porto,Portugal,1997.

# 15
# DYNAMIC WALKING: A STEP FORWARD?

R. E. REEVE
*Department of Artificial Intelligence*
*University of Edinburgh*
*Edinburgh, U.K.*

## 1. Introduction

After a long time as a junior partner in robotics, walking has seen an explosion in interest and achievements in the last 15 years as computational power unimaginable until recently has been brought to bear on its problems. At the same time we have seen advances in other related fields: non-linear dynamics has helped us understand and model the robotic systems that we are studying; zoology has provided details of the mechanics of animal locomotion to aid in the design of our robots; and neuroscience has developed a deep understanding of how rhythmic movements like walking occur in the natural world, and this has helped us formulate new ideas for robotic controllers.

In this chapter we shall look at the history of walking research. We shall follow how it has progressed from the study of body parts and the construction of clockwork automata to the understanding of neural rhythm generators and the control of robotic somersaulting. We shall see that it has become a focal point of interdisciplinary research between the biological sciences and robotics, and we shall examine what can be gained from this work.

We shall also look at what commercial uses walking robots are currently being put to and whether they hold out hope for the future. We shall see to which applications modern walking robots can be applied, and we shall examine what prospects there are for research in these areas.

## 2. Early research

Fascination with walking goes back millennia, but research began in the 18th and 19th centuries. Interest arose for varying reasons, not all of them scientific.

### 2.1 STUDIES OF GAIT

Some of the earliest work involved the examination of cadavers to investigate how they were constructed (for instance work by J. C. Lavater in the late 18th century), but two significant studies of gait really started the ball rolling.

The first of these was by E. J. Marey who invented a pneumatic recording device in the 1870's which measured stepping patterns through sensors attached to the feet of subjects which moved the pen on a clockwork recorder [41]. This allowed him to make records of step patterns in different gaits.

The second (and far more famous) was Eadweard Muybridge, a photographer who, initially spurred on by a bet to prove whether a horse lifted all of its legs off the ground simultaneously in a gallop, went on to record a huge collection of high-speed photographs of animal gaits (e.g. [49]).

These people took some of the first steps in researching the field and though they came from very different backgrounds, they were looking for the same information. This has always been a problem for walking researchers — it is an attractive subject to investigate, and it has no particular allegiance to any one field. As a result there is a danger of duplication of effort as new disciplines decide locomotion is a field worthy of study. Zoologists and physiologists were amongst the early researchers; mathematicians found the apparent simplicity of these patterns interesting, and group theoreticians examined their properties; neuroscientists began to study the structures which control locomotion, and roboticists have tried to emulate the ability, as well as embodying their theories of intelligence in legged robots; non-linear dynamicists have examined the whole walking system to determine its dynamics and stability; biomechanicists have looked at how to improve running performance, and are trying to stimulate nerves in paraplegics to allow them to walk again; finally, computer graphics researchers have created the illusion of walking for our entertainment in the cinema or on our computers.

All of these people have different contributions to make, but it has become impossible for anyone to keep track of their different approaches and achievements. However, it is very important to legged robot research that some feel for the overall picture is maintained, and that is the purpose of this chapter.

## 2.2 AUTOMATA

In the mid 18th century, even before these early studies of walking in animals, automata were being constructed to mimic life as closely as possible and were being demonstrated at fairs and exhibitions. One of their greatest creators was J. de Vaucasson, whose aim was to produce the perfect artificial person ([18]). He produced a flautist which could imitate the sound of the instrument and move in a lifelike fashion. It was incredibly intricate and was controlled by hundreds of bellows and levers. He went on to make a mechanical duck which could flap its wings, walk, and even eat, drink and defecate by means of a mechanical stomach. His project was continued in the 1770s by P.-J. Promond and H. Lois who made increasingly humanoid automata one of which (the Draftsman) could make writing movements and follow them with its head, and another could even reportedly play the harpsichord. Of course it is impossible to authenticate these reports and various automata were exposed as frauds, but incredibly complex machines were certainly being made around this time. The makers boasted of mimicking life itself.

In the 1850s Chebyshev invented mechanical linkages which connected joints so that they moved together. This allowed walking to be developed much more easily as linkages could be designed which would make the body move horizontally by moving the feet and legs in a fixed pattern.

One of the difficulties with this early approach however was that the rigid mechanical linkages which made walking possible totally fixed the movement of the legs. This meant that no alterations in the gait could be made, for instance to change gaits and move over uneven terrain. Subsequent decades were spent designing better linkages in an attempt to produce suitable stepping motions to generate stable locomotion. However, to solve this problem some means of control would have to be devised, and there the problem lay for half a century.

## 2.3 PERIPHERAL VS. CENTRAL CONTROL

During this time Neuroscience began to take a serious interest in locomotion, debating how control of locomotion and other rhythmic movements occurred in humans and other animals. Two competing hypotheses came to the fore: Peripheral Control and Central Control.

The former claimed that these movements were achieved through sensory feedback; a reflex chain existed where each phase of the motion cycle provided the sensory cues which triggered the correct timing of the next in a repeating loop. Thus the behaviour would be disrupted by a lack of sensory feedback.

The latter claimed that the Central Nervous System (CNS) does not require sensory feedback to provide the proper timing, but rather that there is a neural pacemaker providing the rhythm which, though it may be modulated by feedback, was essentially independent of it.

Both hypotheses had early supporters, but experimental evidence was scarce until the 1930's and 40's, when much support was found for the importance of sensory feedback.

### 2.3.1 Peripheral Control

Some of the best work supporting Peripheral Control was done by Sir James Gray and his colleagues (e.g. [23], [24]); they showed that there were behaviours in both invertebrates and vertebrates which seemed to consist of chains of reflexes. They also claimed that the same experiments, when repeated on deafferented[1] animals, produced no observable rhythmical motions. These last results have been largely refuted by more careful recent experiments, but the early behaviours they describe, such as reflex walking in the spinal toad[2] when it is held against a moving surface, have been widely repeated.

Other work on a variety of animals showed that particular sensory inputs can disrupt or completely arrest normal motor output, for instance in one experiment a bivalve scallop with its shell bound shut was found to completely stop contracting and relaxing its adductor muscle [15].

Results like these were taken at the time to support the Peripheral Control hypothesis; however, in reality Central Control did not preclude them, except those

[1] sensory (afferent) nerves are cut so no feedback is received by CNS

[2] a toad whose spinal cord has been severed just below the brain

which were later shown to be in error; it only said that sensory feedback was *not necessary* to create the rhythm, not that it did not play a significant part.

This became important when, in the 1960's and 70's, evidence came in which proved conclusively that rhythmical motions could occur without sensory feedback.

### 2.3.2 Central Control

A great body of evidence was gathered by various researchers (for a table see [15], page 494), which showed that creatures from right across the animal kingdom could carry out rhythmical actions when their nervous systems were completely isolated[3], deafferented, or when their muscles were paralysed. In all these cases there was no feedback to allow reflexes to generate the movement. That these results were not gathered before was in some cases largely a result of poorer experimental techniques and equipment which failed to pick up the rhythms, and in others because the experimental procedures caused too much extraneous damage for the nervous system to be able to continue to operate normally.

On the basis of this evidence, there was no doubt that central mechanisms did generate many rhythmical motions, and the idea that all meaningful output had to be driven by specific sensory stimulation died. However, the central control hypothesis fell far short of explaining what was happening in these pacemakers that it proposed, and certainly observations like the reflex walking referred to earlier needed to be put into the structure of central control; indeed some more recent work had also shown results which did not fit in with a strong central theory. For example, Grillner and Wallén say in [25] that a spinal dogfish paralysed by curare will, if the tail is moved at a frequency different from its natural swimming rhythm, show coordinated bursts in its motoneurons at the imposed frequency. Also, as in scallops, the swimming movement can be suppressed by strong stimuli like holding the body tightly.

Experiments like this led to the realisation that a more sophisticated explanation would have to be found to replace the rather simplistic Central Control hypothesis.

## 2.4 EARLY ROBOTS

Meanwhile, back in robotics, the first true walking robots had already been built. It was the arrival of logic circuits and later computers which made this possible by allowing a control mechanism to be built which would make variation in the walking patterns possible.

Nevertheless, one of the first ideas was to use a human as the controller, as this was still early days for computing. General Electric built one such vehicle in the 1960's (the 'versatile walking truck' in [48]) which, under the control of its operator, was capable of up to 5 mph and could climb over large obstacles. This was really ignoring the control problem however, and it was not until the late 60's that truly independenlegged robots began to appear.

One of the first of these was built by Frank and McGhee and was called the Phony Pony ([45]); each joint was controlled by a finite state machine made from digital logic

---

[3] removed from the body

circuits with each of four states triggering the next in a fixed loop. This was actually very restrictive, and made the robot behave in a fixed manner very similar to the automata it replaced. However it opened the door to the gait being computer controlled, and thus changeable in software.

After this, many research projects were started into computer controlled walking, and indeed the first commercial product, Big Muskie, which was a walking dragline used for strip mining, was produced by the Bucyrus-Erie company in 1969.

The late 1970's and 80's saw a succession of simple computer-controlled statically stable robots whose patterns of locomotion were very simple and inspired by insects (e.g. [28], [29]). They remained balanced in static equilibrium all the time, and moved surplus legs to new positions where they could in turn be used for support. They were the first computer controlled walking robots, and they moved very slowly (as they always had to keep 3 legs on the ground for support). Their controllers tended to be reactive (i.e. they would move to a new state based on the sensory feedback and the state they were in), and parallels can easily be drawn to the Reflex Controllers mentioned earlier. This strategy allowed fairly robust movements across terrain, and so long as a state was described in the controller (e.g. leg moving forward hits obstacle), it could be dealt with by the robot (move back, raise leg, and repeat). However, this approach does require the enumeration of possible states, which is far from ideal. These controllers were time-independent, in the sense that the system could be frozen and restarted at any point, or run at a different speed, and there would be no effect on the walking pattern, so it didn't matter if the computer had to sit and calculate for a bit to work out what to do next.

The next step forward was the arrival of dynamically stable (or actively balanced) walking, which arrived as computers became more powerful, and capable of dealing with the complexities involved in staying upright when not always in a stable posture (e.g. [43], [47], and [53]). Controllers for this kind of robot were time-dependent, that is to say that there were points in the motion cycle when stopping or changing speed would be fatal (imagine stopping moving your legs in the middle of a fast step), so the controller had to be able to keep up with what was going on in real time.

## 3. Recent developments

At this point walking was still poorly understood. Physiologists had studied the bodies of animals, neuroscientists had argued about how they walk, and roboticists had created slow, clumsy walkers, but things had really yet to take off. The pace of change was soon to speed up however.

### 3.1 FIRST STEPS WITH DYNAMIC STABILITY

The first attempt at a dynamically stable robot was by Ogo et al. [50], where a biped with huge feet walked in a quasi-dynamically stable fashion. They avoided the problems of time-dependence by only having a small non-statically stable phase where the biped 'fell' from one foot to the other in a controlled fashion.

True dynamic stability was not long in coming however. At the same time as the above Matsuoka in [43] built a robot capable of running in a 2 dimensional world, and not long after Miura and Shimoyama [47] developed the first actively balanced dynamically stable walker, the *stilt biped*, which was supposed to model the behaviour of a person walking on stilts.

Raibert and his team at Carnegie-Mellon University (it was later to move to MIT) then started to produce his walking robots. Initially he designed a 3D one-legged hopping machine ([53]); shortly after this he progressed to bipeds and quadrupeds using the same basic algorithm for each ([54]). They had finite state controllers, but within each state the controller was algorithmic, calculating the desired joint angle and the required joint forces. This was a more centralised control than the statically stable robots and the robots went at impressive speeds (for instance the one-legged robot had a top speed of 4.5 mph).

On a different front, McGeer built a dynamically stable passive walker which he called the 'biped glider' [44]. This could walk stably down gentle slopes without any form of control, and could in theory be pumped to walk on the flat or on other terrain. This has been taken further by Goswani and colleagues at INRIA [22] who looked at passive walking with a very simple gait, and then at the mimicking of passive control with an active mechanism to enlarge the natural basin of attraction of the passive limit cycles and to create new gaits.

### 3.2 CENTRAL PATTERN GENERATORS

Much progress has also been made by neuroscientists studying rhythmical controllers. Following on from the ideas of Central Control, during the 1970's the idea grew that the motoneurones (and hence rhythmical movements) in vertebrates were driven by central networks of interneurones that generated the essential features of the motor pattern, but also that sensory feedback signals played a crucial role in the control system, namely to turn a stereotyped unstable pattern into the coordinated rhythm of the natural movement. The networks were referred to as Central Pattern Generators (CPGs), and evidence showed that every part of the body which makes cyclical movements has its own individual CPG (for a summary, see [15]). Experimental evidence has recently proven this to be true in the lamprey ([63]), where neurophysiologists actually mapped out the neurones making up the CPG, and it is accepted in other animals as well.

In vertebrates the spinal cord contains these neuronal networks. The CPGs, when stimulated, have their own dynamics which set up oscillations in outputs between the neurones. When they are connected to the motoneurones, these generate the characteristic rhythmical behaviour associated with the system. Central Pattern Generator is a general term: for instance, in walking each muscle can be considered to have its own CPG, but opposing muscles CPGs can be considered collectively to form a joint CPG, and similarly for limb CPGs. Feedback from the muscles keeps the CPGs in phase with the limb, so the pattern does not break down. In the same way, limb CPGs maintain the coordination between legs to generate the overall behaviour (i.e. walking) by their cross connections. The system is self-correcting, so that if any part loses synchronisation, the dynamics of the whole system forces it back into the original rhythm. Without the cerebellum, movements are coarser and there are some problems with equilibrium and coordination, but in essence the pattern remains unchanged; thus spinal mammals and birds have been shown to make walking movements very similar to those of intact animals ([26]). On the other hand, without proprioceptive feedback the pattern can break down [15], as it can be very important for adaptation to actual conditions. Even with this feedback the rhythm can break down though if it is knocked too far from equilibrium, as the movement now falls outside the basin of stability of the system.

As a result of work by Sten Grillner and his colleagues in examining and simulating CPGs in the lamprey ([27], [63]), a great deal is now known about the structure and behaviour in this creature; however there is still much to be learnt in other animals — of particular interest to us are legged vertebrates: we know that each muscle has its own CPG and that this is essentially a very simple network, and we know that these are connected together inside each leg to generate a stepping motion, and between legs to generate a stable rhythm. We also know that each of these rhythms (gaits) is stable over a certain range of speeds, so as the legs speed up, the basin of stability for one gait shrinks until a bifurcation occurs and that gait becomes unstable, and walking moves to a new gait with a different attractor [34]. We also know that the cerebellum and other higher centres are highly connected to the CPGs, receiving efference copies of the signals sent from the CPGs to the motoneurones, as well as the afferent feedback from the muscles [26]. This allows fine-tuning of the coordination without which the walking looks a bit rough and is more likely to break down. However, we do not know the details of how the neurones are connected, either inside a CPG or inside or between limbs, and certainly not how the cerebellum works (though theories have been put forward).

What we do know is that a low level dynamical system exists which links the muscles to produce walking and which has a highly regular structure consisting of several very similar neuronal networks (CPGs). These are strongly interconnected to maintain their rhythm and are controlled in turn by higher centres to maintain their stability.

Much of the discussion in this section has been about dynamical systems and it is the contributions of researchers in this field that we shall discuss next.

### 3.3 GROUP THEORY, NON-LINEAR DYNAMICS AND COORDINATION

Researchers from various areas of mathematics have studied walking. Collins and Stewart used group theory to analyse the properties of various coupled non-linear oscillators. They predicted that fixed CPGs should be capable of changing between gaits by varying very few parameters ([11], [12]). Collins then went on to test this with a selection of CPG models and found that it was generally possible to make simple CPGs produce different gaits by varying only a few internal parameters whilst leaving the connectivity unchanged ([13]). This was a simpler solution than many previously proposed which suggested, for instance, that different coordinating neurones might be needed for different gaits ([26]).

There has also been work by non-linear dynamicists like Kelso. He has studied how walking systems behave, what happens to the coordination between legs at gait transitions, and how the system converges to its stable attractor (a particular gait). He particularly stresses that gaits are selected (and are most stable) in animals when they are the most efficient for travelling at the desired speed, though some hysteresis stops switching back and forth during transition ([34]). He suggests that we should build this kind of nonlinearity and multistability into robots to help eliminate problems with redundancy. He also points out that transitions normally occur very smoothly, but in animals with their higher centres removed instability and critical fluctuations occur and gaits switch back and forth near bifurcations.

Gallagher and Beer look directly at evolved locomotion controllers from a dynamical systems perspective, examining controllers evolved to be reflex chains, central controllers (without feedback) and CPGs [19]. They examine the basins of attraction of the limit cycles and fixed points. They discover that the reflex chain controllers on their own have no limit cycles, but rather whatever the current state of the system, it is attracted to a fixed point forward in the walking cycle; this results in the whole (controller-body) system having a stable walking limit cycle, but this will break down if the feedback is interrupted. This behaviour is much like finite state machine walking controllers. The other two types of controllers both have inherent limit cycles which correspond to walking in the whole system, but whereas the central controller has a crude limit cycle which cannot adapt, the CPG is entrained to the frequency of the rhythmic feedback from the legs, so adaptation to different conditions does not have to be learnt, but rather emerges from the dynamics of the controller-body system. In a more heavily mathematical analysis, Cheng and Lin in [10] look at the stability of a biped using a linearized Poincaré map and discuss the robustness of the locomotion.

## 3.4 EVOLVING ROBOT CONTROLLERS

As more people started to built legged robots, they began to look for ways to automate development of the controller for them. The most popular method has been a form of simulated evolution[4].

As many different forms of evolution have been tried as there are researchers, but generally they tend to evolve parameters for some kind of neural controller for the robot: normally the strengths of connections, but sometimes other internal parameters of the neurons as well[5]. In general this has proved a step backwards for the sophistication of walking controllers with most of the controllers evolved being statically stable. However, a popular way of evolving controllers is to simulate the robot concerned and evolve a controller for it on the computer, and it is a very computer intensive process to model a robot dynamically, so it is possible that this is partly to blame. If so, as computer power increases and since fast dynamic simulation software is now available (e.g. [46]), hopefully this problem will go away.

---

[4] for a description of Genetic algorithms, see [21].

[5] for a general review of this, see [36].

Lewis et al. in [38] evolved controllers for a hexapod robot (Rodney) which they had built. The controller was evaluated on the real robot and learnt to walk with insect-like gaits after a staged evolution where it was encouraged a bit at a time towards the final goal of walking.

Beer and Gallagher evolved the parameters of a dynamic neural network to control a (statically stable) insect walking in simulation [7]. They then went on [20] to evolve a statically stable controller for a simulated robot, and then transferred the evolved controller onto a real robot with no problems. Spencer [58] used Genetic Programming[6] to evolve the architecture as well as the parameters of a similar robot.

Karl Sims, on the other hand, (in [56]) evolved dynamic controllers for robots whose morphologies were evolved at the same time. This produced arbitrary shaped robots with controllers which would allow them to perform behaviours like tumbling, sliding, jumping, and swimming. They had the advantage of not having to remain balanced as is usual for walking robots, and it is not clear how they would have performed if that had been what was evolved; but they are the most visually impressive result of this general approach.

## 3.5 VERTEBRATE LOCOMOTION AND DEVELOPMENT

Meanwhile, zoologists have been looking for common ground between different animals when walking, examining which criteria appear to be being optimised, and how locomotion in vertebrates has developed over time. They call this first problem the inverse optimality problem — what was being optimised to produce this?

Alexander has written extensively about locomotion: in [1] he looks at locomotion in reptiles, birds and mammals. He concludes that locomotion with a similar Froude number (a dimensionless measure — speed$^2$/(gravity*hip height)) produces dynamically similar movements in general right across different species (0.1 walk, 1 trot/pace, 2-3 asymmetric). He also looks at the inverse optimality problem, and concludes that for turtles displacement (i.e. roll, etc.) is minimised, but humans minimise work. Tendon elasticity helps in this regard by storing strain energy rather than allowing it to be dissipated as heat. In [2] he proposes the use of springs in robots to replace tendons in their job of minimising energy loss, and also on the feet to soften impact slightly to reduce 'chattering' and hence improve grip. In [3] he looks at how energy is saved in terrestrial locomotion, through tendons, but also through aligning joints to minimise the amount of work that has to be done and the amount of conflict between muscles.

Eilam, in [16], studies how movements change both across species and during ontogeny (development in individuals of one species), and points out that there is the same consistent progression from simple lateral movements of the trunk, to use of limbs, through to vertical movements of the trunk across vertebrates in each.

Work being done on development of locomotion includes [59], where Vaal et al. detail a research agenda for studying human locomotion and gaining insights into its development. They claim that very little work has been done on the ontogeny of

[6] see [37].

locomotion, and they want to identify the crucial subsystems and their interactions, and how these develop to produce an adult locomotory pattern. They discuss CPGs and the importance of feedback, and then move on to the development of walking, starting with precursors like reflex stepping going right through to integrated walking. Unlike most authors, they disapprove of the concept of optimality criteria as they consider them arbitrary, and believe the introduction of functionality constraints should be sufficient — walking must work in a wide variety of situations. They also mention the usefulness of using muscles when modelling locomotion, as they claim that such models have inherently better stability properties than force control for joints.

Van Soest and van Galen look at how animals reduce redundancy problems in multi-joint movements by imposing constraints in [61]. They divide these into physical and self-imposed constraints (they believe the latter are imposed specifically to help solve the problem). And finally, Assaiante and Amblard look at the ontogeny of balance control [4].

### 3.6 FUNCTIONAL NEUROMUSCULAR STIMULATION

One of the most exciting developments recently has been work done to restore the ability to walk to paralysed patients by electrically stimulating their muscles (e.g. [42]). Research done by Yamaguchi recently in [66] shows that generating the appropriate patterns for walking is essential for optimal use of muscles which have been weakened by the paralysis, and that ankles are a particular weak spot which might well benefit from some kind of orthosis.

## 4. Where are we now?

We have now looked at research up to the present day, and in this section we shall look at what commercial products have come out of this, and what ongoing research uses legged robots.

### 4.1 COMMERCIAL ROBOTS

In recent years various projects have been proposed for which walking robots would be 'invaluable'. However, the number which have actually been built and used for anything other than walking research is much more limited.

Walking machines have long been proposed for travel over rough terrain that even a tracked vehicle could not navigate, and for dangerous environments where human would be at risk.

To that end, Dante and Dante II [65] were designed to descend into volcanoes to study conditions in an environment too dangerous for humans and too rough for other robots. They were not unqualified successes. Dante II was the more successful of the two — it was an extremely stable eight legged robot which moved slowly down into Mount Spurr in Alaska. Unfortunately on its way back out the ground collapsed under it, and it was not able to cope and had to be airlifted back out.

NERO was a climbing robot funded by Nuclear Electric in Britain and designed for work on a nuclear reactor pressure vessel where humans have huge safety problems

([40]). It was designed to climb the outside walls of the pressure vessel and inspect its condition, clean it or even install equipment on it, but was too light too do heavy work. It had drawbacks (it travelled at a speed of 0.1 m/min, and had to be placed on the wall of the pressure vessel, it couldn't walk there), but it was successfully used for these purposes. Nuclear Electric together with Electricité de France, CERN and the Italian electricity board then funded work on a next generation of climbing robots which would be more versatile, stronger, and capable of walking to the reactor vessel and performing the floor to wall transition itself (called Robug, see [39]), but there have been problems getting these to work to specification (for instance Robug III is too heavy to support its own weight for long when climbing).

Other commercial robots include a 'walking forest machine' built by PlusTech Oy. in Finland (see [52]). This robot has been build to replace wheeled vehicles for forestry work as it doesn't damage the ground and hence minimises the risk of soil erosion on steep slopes. It also has a high and variable ground clearance and so can move over obstacles which would block most other vehicles. It is driven by a driver in a cabin on top of the robot.

Much more recently some Japanese firms have come into the market. Honda recently revealed their biped robot. This can walk semi-autonomously or be teleoperated and can carry out simple tasks ([31]). This a very impressive robot, and presumably we will see more from them in the future.

In a completely different field, Sony have released some information about one of their experiments — a robot puppy ([57]). This was shown playing with another robot and performing a series of stereotyped actions like rolling over and standing on its hind legs and boxing the air. It seems destined to find a way into the market as a pet, but no further details were available.

One other fun set of robots, which are currently being finished, are the walking dinosaurs being constructed for use in museums round Europe. The project is called Palaiomation and is funded by the EU[7].

## 4.2 RESEARCH ROBOTICS

There is a lot of research currently studying the problems of coping with difficult terrain and avoiding falling and recovering from it. For instance, Boone and Hodgins have recently been looking at how bipeds can recover from slipping or tripping ([8], [9]) despite having little information about what is gong wrong. Yoneda and colleagues [67] have built a robot (Titan VI) capable of travelling at a reasonable speed on flat ground (over 2 mph), but which can move over obstacles, and of course Honda's robot [31] can walk up and down steps and avoid obstacles very effectively.

[7] Brite/Euram Craft CR 1651

### 4.3 MODELLING HUMANS AND OTHER ANIMALS

The other area where legged robots are being used is in modelling work, examining how animals move by modelling them in simulation (or sometimes on robots), and also just modelling their behaviours to create impressive graphics for films.

Grillner and colleagues did some very careful modelling of CPGs in lampreys to see whether what they were finding could explain the behaviour they saw (it could). The simulation work is described by Ekeberg in [17]. Other work includes that by Cruse and others at Bielefeld in modelling stick insects on a real robot ([14]), where they examine their neural network based model of the insect controller and see how its behaviour compares.

Research in Edinburgh is examining CPGs to see how well they can be modelled artificially for controlling swimming movements [32], and simulating a variety of legged robots to see what general principles can be used to build controllers for them [55].

Human modelling has become an attractive target, and several people have investigated it. For example, Hodgins has simulated a 30 degrees of freedom humanoid figure, and got it to run in a fairly biomechanically accurate fashion ([30]). Playter has looked at gymnastics, and has modelled different manoeuvres to see how stable they are, and what is the best way of stabilising them in [51]. He even managed to get a robot to do a layout somersault! Jalics and others at Ohio State University have started investigations into dancing with a simple planar model of a biped, examining how to keep track of the rhythm of the music and move the body in time with it ([33]).

This is moving more towards modelling for the fun of it, and much work has been done in this field (Jurassic Park for instance!). Most of it was animation work, producing a walking-like gait without seeing whether the creature would stay upright if it really moved like that. However, recently some workers in the field decided that the motion would look more believable if they could get it dynamically correct as well. Work like [35] looks at how to generate the right movements to stay stable whilst also maintaining the realistic gait produced by the animation, and this is crucial as the gait produced must look like what we perceive as a normal walking pattern for it to be useful for their purposes.

## 5. The future

In this section we shall look at what applications are being proposed for walking robots, and how practical they are. We shall also discuss what areas of legged research look most promising at the moment, and where useful collaboration should be possible between different disciplines.

### 5.1 USEFUL ROBOTS

There are a lot of different proposals for walking robots to be used in different fields, but they tend to divide into two categories: robots for hazardous environments and robots to operate around humans.

### 5.1.1 Hazardous environments

Proposals for robots in hazardous environments include fire safety robots which would enter fire damaged buildings to investigate the safety of the environment before human crews entered (e.g. Baudoin [6]). It should be possible to build robots to carry out this task, but it is likely that they will be quite slow and very heavy (and thus more likely to destroy the building they are investigating than people would be).

Fire fighting robots would almost certainly be teleoperated, but Baudoin also proposes a more optimistic aim of building robots for humanitarian mine removal. These would have to be fairly autonomous robots so they could travel around on their own detecting mines, and since autonomy is an area which is still being looked at in conventional robots, this is clearly a long term project.

Other proposals include an underwater mine sweeping robots [5] which would be put out to clear a piece of coast before a landing, and various planetary rovers. These would both require a great deal of robustness as well as autonomy, and so will not be seen in the near future (consider that the designers of Dante II, who are planning on landing the first private expedition on the Moon in 1998, have decided to use a wheeled robot for the task).

### 5.1.2 Human environments and interactions

There is much potential for robots which can interact with humans in their own environment, and firms like Honda certainly look set to exploit this. It is worth noting, however, that in the past it has always proved cheaper and easier to have simple single function devices than a robot which can do everything, and it is not obvious that the situation has changed.

There are walking robots which would be single function devices of course, and perhaps proposals like walking wheelchairs (which would be able to climb stairs) will be a possibility [64]. Also robots specifically designed for amusement may be popular (if very expensive) — the success of projects like Palaiomation and Sony's puppies will have to be watched closely.

## 5.2 MODELLING

Modelling roughly divides itself between creating computer graphics and simulating real locomotion. In the former, animating robots, people and animals is becoming more involved with simulation as the computational expense of doing the extra modelling becomes less important and techniques become more sophisticated. In the latter, simulating real animal and human locomotion helps us gain insight into how control of movement happens in the body and provides a source of ideas for robot controllers.

### 5.2.1 Modelling creature dynamics

Until recently most graphics for games and films were produced using kinematic techniques, and simulation has been largely ignored as a waste of time, especially as the results were much cruder than could be achieved by kinematic techniques, but researchers are beginning to realise that it is no longer quite so time consuming, and work has been done to combine the techniques, so gaits which were created in other ways are tuned in a simulator so that they are dynamically plausible, with the expectation that this will improve the appearance of the movement (e.g. [35] and [60]). If this approach works, we can expect a lot of work being done in the field by these researchers.

### 5.2.2 Simulating human and animal locomotion

There has been a lot of collaboration between roboticists and zoologists looking at the energetics and common principles of locomotion, as well as with neuroscientists, and with ethologists studying animal behaviour, and there is every reason to expect that this kind of interdisciplinary research will continue to thrive.

One new element in it may be work on Functional Neuromuscular Stimulation, where researchers like Yamaguchi [66] admit that their current methods for selecting stimulation patterns to create gaits in paraplegic subjects are definitely sub-optimal. It is quite possible that this biomechanics work can be combined with artificial intelligence optimisation techniques, which are already applied to designing controllers for walking robots, to improve the efficacy of this technique.

## 6. Conclusions

This chapter has given an overview of research in the field of legged robotics and walking in general, particularly looking at recent work and applications. Walking research is an extremely popular area at the moment, and much is being done both in simulation and on real robots and animals, a lot of it of a high standard. One point that has become clear about the field is its breadth, spanning the whole spectrum from computer graphics to neurophysiology to nuclear power station inspection, and it is important that some connection be maintained to help the field move forward, and help different disciplines benefit from work carried out elsewhere.

The potential for interdisciplinary work is greater than ever at the moment, and indeed some research can only be done in that manner (e.g. the work at the Royal Institute of Technology in Stockholm on lampreys [63], where some researchers are now looking at walking [62]). This is not confined to the biological sciences either, with insights from mathematics helping to improve the stability and performance of walking, and collaboration is also possible with many other fields.

Hopefully this chapter has raised some readers' awarenesses of other work and has provided enough information to allow them to seek out more detail for themselves.

## 7. References

1. Alexander, R.M.: The gaits of bipedal and quadrupedal animals, *Int. J. Rob. Res.* **3**(2) (1984), 49-59.
2. Alexander, R.M.: Three uses for springs in legged locomotion, *Int. J. Rob. Res.* **9**(2) (1990), 53-61.
3. Alexander, R.McN.: Energy Saving Mechanisms in Walking and Running, *J. Exp. Biol.* **160** (1991), 55-69.

4. Assaiante, C., and Ambland, B.: An ontogenetic model for the sensorimotor organization of balance in humans, *Human Movement Science* **14** (1995), 13-43.
5. Ayers, J.: A Reactive Ambulatory Robot Architecture for Operation in Current and Surge, *Autonomous Vehicles in Mine Countermeasures Symposium*, 1995.
6. Baudoin, Y.: *Colloquium on IT for Climbing and Walking Robots*, October 1996.
7. Beer, R., and Gallagher, J.: Evolving dynamic neural networks for adaptive behavior, *Adaptive Behavior* **1** (1992), 91-122.
8. Boone, G.N., and Hodgins, J.K.: Reflexive Response to Slipping in Biped Running Robots, *IROS'95* **3**, 158-194.
9. Boone, G.N., and Hodgins, J.K.: Slipping and Tripping Reflexes for Bipedal Robots, *Autonomous Robots* **4** (1997), 259-271.
10. Cheng, M.-Y., and Lin, C.-S.: Measurement of robustness for biped locomotion using a linearized Poincaré map, *Robotica* **14** (1996), 253-259.
11. Collins, J.J., and Stewart, I.N.: Coupled non-linear oscillators and the symmetries of animal gaits, *J. Nonlinear Sci.* **3** (1993), 349-392.
12. Collins, J.J., and Stewart, I.: Hexapodal gaits and coupled non-linear oscillator models, *Biol. Cybernetics*, **68** (1993), 287-293.
13. Collins, J.J., and Richmond, S.A.: Hard-wired Central Pattern Generators for quadrupedal locomotion, *Biol. Cybernetics*, **71** (1994), 375-385.
14. Cruse, H., Brunn, D.E., Bartling, Ch., Dean, J., Dreifert, M., Kindermann, T., Schmitz, J.: Walking: a complex behaviour controlled by simple networks, *Adaptive Behavior* **3** (1995), 385-418.
15. Delcomyn, F.: Neural Basis of Rhythmic Behaviour in Animals, Science **210** (October 1980), 492-498.
16. Eilam, D.: Comparative Morphology of Locomotion in Vertebrates, *J. Motor Behavior* **27** (1995), 100-111.
17. Ekeberg, Ö.: Neural Control of Vertebrate Locomotion — A Computer Simulation Study, *Int. Work. on Mechatronical Computer Systems for Perception and Action '93*.
18. Elliott, P. *Personal communication.*
19. Gallagher, J.C., and Beer, R.D.: A Qualitative Dynamic Analysis of Evolved Locomotion Controllers, *From Animals to Animats II*, 71-80.
20. Gallagher, J.C., Beer, R.D., Espenscheid, K.S., and Quinn, R.D.: Applications of Evolved Locomotion Controllers to a Hexapod Robot, *Robotics and Autonomous Systems* **19** (1996), 95-103.
21. Goldberg, D.: *Genetic Algorithms in search, optimisation, and machine learning*, Addison-Wesley, 1989.
22. Goswami, A., Espiau, B., and Keramane, A.: Limit Cycles in a Passive Compass Gait Biped and Passivity-Mimicking Control Laws, *Autonomous Robots* **4** (1997), 273-286.
23. Gray, J., Lissmann, H.W., and Pumphrey, R.J.: J. Exp. Biol. **15** (1938), 408.
24. Gray, J., and Lissmann, H.W: J. Exp. Biol. **17** (1940), 237.
25. Grillner, S., and Wallén, P.: Brain Research **127** (1977), 291.
26. Grillner, S.: Neurobiological bases of rhythmical motor acts in vertebrates, *Science* **228** (April 1985), 143-149.

27. Grillner, S., Wallén, P., and Brodin, L.: Neuronal Network generating locomotor behaviour in the lamprey: Circuitry, transmitters, membrane properties, and simulation, *Ann. Rev. Neuroscience* **14** (1991), 169-199.
28. Gurfinkel, V.S., Gurfinkel, E.V., Schneider, A.Yu., Devjanin, E.A., Lensky, A.V., and Shitilman, L.G.: Walking robot with supervisory control, Mechanism and Machine Theory **16** (1981), 31-36.
29. Hirose, S., and Umetani, Y.: The basic motion regulation system for a quadruped walking vehicle, ASME Conf. on Mechanisms, 1980.
30. Hodgins, J.K.: Three-Dimensional Human Running, *1996 IEEE Int. Conf. on Robotics and Automation*, 3271-3276.
31. Honda Motor Company Ltd., Tokyo:*Honda introduces "Human" Robot*, http://www.honda.co.jp/home/hpr/e_news/robot/index.html
32. Ijspeert, A.J., Hallam, J., and Willshaw, D.: Artificial Lampreys: Comparing Naturally and Artificially Evolved Swimming Controllers, *ECAL'97*, 256-265.
33. Jalics, L., Hemami, H., Clymer, B., and Groff, A.: Rocking, Tapping and Stepping: A Prelude to Dance, *Autonomous Robots* **4** (1997), 227-242.
34. Kelso, J.A.S.: *Dynamic Patterns: the Self-Organization of Brain and Behavior*, MIT Press, 1995.
35. Ko, H., and Badler, N.I.: Animating human locomotion with inverse dynamics, *IEEE Computer Graphics and Applications* **16** (1996), 50-59.
36. Kodjabachian, J., and Meyer, J.-A.: Evolution and development of control architectures in animats, *Robotics and Autonomous Systems* **16** (1995), 161-182.
37. Koza, J.R.: *Genetic Programming: On the Programming of Computers by Means of Natural Selection*, MIT Press, 1992.
38. Lewis, M.A., Fagg, A.H., and Solidum, A.: Genetic Programming Approach to the Construction of a Neural Network for Control of a Walking Robot, *IEEE Int. Conf. on Robotics and Automation*, 1992.
39. Luk, B.L., Cooke, D.S., Collie, A.A., and White, T.S.: Robug III — a semi-intelligent teleoperated walking and climbing robot for disordered and hazardous environments, *European Robotics and Intelligent Systems Conf.*, 1994.
40. Luk, B.L., Collie, A.A., and White, T.: Nero: a teleoperated wall climbing vehicle for assisting inspection of a nuclear reactor pressure vessel, *ASME Int. Comp. Eng. Conf.*, 1993.
41. Marey, E.J.: Animal Mechanisms: a treatise on terrestrial and aerial locomotion, Appleton, New York, 1874.
42. Marsolais, E.B., and Kobetic, R.: Functional walking in paralysed patients by means of electrical stimulation, *Clin. Orthoped. Rel. Res.* **175** (May 1983), 30-36.
43. Matsuoka, K.: A model of repetitive hopping movements in man, Fifth World Congress on Theory of Machines and Mechanisms (1979).
44. McGeer, T.: Passive Dynamic Walking, Int. J. Rob. Res. **9**(2) (1990), 62-82.
45. McGhee, R.B.: Robot Locomotion, Neural Control of Locomotion, Herman et al. (eds.), Plenum, 1980.
46. McMillan, S.: *Computational dynamics for robotic systems on land and under water*, PhD dissertation, Ohio State University, 1994.

47. Miura, H., Shimoyama, I., Mitsuishi, M., and Kimura, H.: Dynamic walk of a biped, Int. J. of Rob. Res. **3**(2) (1984), 60-74.
48. Mosher, R.S.: Testing and evaluation of a versatile walking truck, Off-Road Mobility Research Symposium: Int. Soc. for Terrain Vehicle Systems (1968).
49. Muybridge, E.: Animals in Motion, 1887.
50. Ogo, K., Kanse, A., and Kato, I.: Dynamic walking of a biped walking machine aiming at completion of steady walking, Third Symposium on Theory and Practice of Robots and Manipulators (1980).
51. Playter, R.B.: *Passive Dynamics in the Control of Gymnastic Maneuvers*, PhD Thesis, MIT.
52. Plustech Oy., Finland: *The natural way of walking in uneven terrain*, http://www.plustech.fi/walk.html
53. Raibert, M.H., Brown, H.B., and Chepponis, M.: Experiments in Balance with a 3D One-Legged Hopping Machine, Int. J. Rob. Res. **3**(2) (1984), 75-92.
54. Raibert, M.H.: Legged Robots that balance, MIT Press, 1986.
55. Reeve, R.E., and Hallam, J.: Generating Walking Behaviours in Legged Robots, *2nd Int. Work. on Mechatronical Computer Systems for Perception and Action*, 1997.
56. Sims, K.: Evolving virtual creatures, *SIGGRAPH'94*, 15-22.
57. Sony Corporation:*Video shown at ECAL'97.*
58. Spencer, G.: Automatic Generation of Programs for Crawling and Walking, *Advances in Genetic Programming*, Kinnear, K.E. Jr. (Ed), 1994.
59. Vaal, J., van Soest, A.J., and Hopkins, B.: Modelling the early development of bipedal locomotion: A multidisciplinary approach, *Human Movement Science* **14** (1995), 609-636.
60. van de Panne, M.: Parameterized Gait Synthesis, *IEEE Computer Graphics and Applications* **16**(2) (1996), 40-49.
61. van Soest, A.J., and van Galen, G.P.: Coordination of multi-joint movements: An introduction to emerging views, *Human Movement Science* **14** (1995), 391-400.
62. Wadden, T., Ekeberg, Ö and Lansner, A.: Towards ANN based control of simulated legged locomotion, *Int. Work. on Mechatronical Computer Systems for Perception and Action '93*, 379-382.
63. Wallén, P., Ekeberg, Ö, Lansner, A., Brodin, L., Tråvén, H., and Grillner, S.: A computer-Based Model for Realistic Simulations of Neural Networks.II.The segmental Network Generating Locomotor Rhythmicity in the Lamprey, *J. Neurophysiology* (1992), 1939-1950.
64. Wellman, P., Krovi, V., Kumar, V., and Harwin, W.: Design of a wheelchair with legs for people with motor disabilities, *IEEE Trans. on Rehabilitation Engineering*, **3** (1995), 343-353.
65. Wettergreen, D., Thorpe, C., and Whittaker, R.: Exploring Mount Erebus by Walking Robot, *Robotics and Autonomous Systems* **11** (1993), 171-185.
66. Yamaguchi, G.T., and Zajac, F.E.: Restoring Unassisted Natural Gait to Paraplegics Via Functional Neuromuscular Stimulation: A Computer Simulation Study, *IEEE Trans. Biomed. Eng.* **37** (1990), 886-902.

67. Yoneda, K., Iiyama, H., and Hirose, S.: Intermittent Trot Gait of a Quadruped Walking Machine Dynamic Stability Control of an Omnidirectional Walk, *1996 IEEE Int. Conf. on Robotics and Automation*, 3002-3007.

# 16

# AN OMNIDIRECTIONAL CONTROL ALGORITHM FOR WALKING MACHINES BASED ON A WAVE-CRAB GAIT

MARÍA A. JIMÉNEZ, P. GONZALEZ DE SANTOS
and J. TABERA
*Instituto de Automática Industrial (CSIC)*
*Departamento de Control Automático*
*La Poveda, 28500 Arganda del Rey, Madrid, Spain*

## 1. Introduction

Legged machines are inherently omnidirectional vehicles, i.e., they have the ability to move in any direction by modifying their foot trajectories and/or leg sequence. The bibliography contains diverse omnidirectional control algorithms, depending on the gait [1]-[6].

Some of first work about omnidirectional motion for hexapod-periodic gaits comes from W.J. Lee [2]. This algorithm is based on calculating the time period for each locomotion cycle as a function of how close each leg is to its kinematic limits in time and phase (local phase). To do that, the concept of constraint working volume, which is a subset of the working volume of each leg that ensures static stability, was introduced. The optimal time period determined is such that at least one leg utilizes its constraint working volume fully. If this time period is too short to swing the leg, it should be adjusted to satisfy the minimum return time, which means decreasing body velocity.

According to Messuri [4], the above work is only appropriate for low velocities due to the transition between gaits when the direction of motion changes. This researcher proposes an omnidirectional control algorithm based on a periodic gait which is not limited to the time-sequencing constraints of the wave gaits (duty factor, leg phase and time period). This algorithm distributes the legs in two independent tripods. Thus, this algorithm is only useful for hexapod walking machines.

Hirose and J.K. Lee developed algorithms for quadruped vehicles based on crab walking [1], [6]. Hirose proposed a control algorithm utilizing straight-line crab motion to cross a plane with forbidden zones. This algorithm was based on maintaining the periodic gait for which the deadlock positions are determined. It was proved in simulation, and the authors considered that it could be applied to a quadruped with visual sensors. J.K. Lee proposed a control algorithm based on the wave-crab gait [9] to follow a desired trajectory while keeping static stability. Footholds have to meet three conditions: 1) they must be inside the working area; 2) they must maintain stability while the leg is in the support phase; and 3) the trajectory in the support phase must pass through the center of the working volume. If a selected foothold falls outside the working volume, the leg stroke and the body velocity are decreased. If the stability margin is negative, a method is used to translate the foot trajectory from the center of the working volume. When the crab angle exceeds its critical value, the gait type is changed using a method that makes the change when the machine is stopped. Both algorithms have a shortcoming because they do not consider time-sequencing constraints.

This chapter proposes a technique to change the crab angle for a quadruped machine based on the wave-crab gait with the criterion that all four legs fully utilize their constraint working area. A further criterion is to maintain the module of the velocity constant, as the wave gait is optimally stable when that condition is true and some applications require verifying that condition. It is clear that there is a limit to the change of the crab angle such that the body velocity remains constant. How such a limitation can be removed is outside the scope of this chapter. Therefore, if after increasing the leg stroke, the minimum return time condition is not maintained, it will be necessary to decrease body velocity. However, the present algorithm allows a stability numerical study to be run that is important for designing quadruped machines under the criterion of optimum static stability and constant body velocity. This algorithm finds the transition gait between two gaits with different crab angles. In other words, it establishes the footholds, the return time for each leg in the transfer phase, and the time cycle. During the transition gait, each leg is raised only once; therefore, in this sense it can be considered a periodic gait.

In this approach, a level vehicle is to walk over smooth level terrain. The change of crab angle over irregular terrain can be performed using the adaptive gait control strategy proposed by the authors [11]. Furthermore, among the different types of measurements that can be used to determine the static stability of a walking machine [9], the longitudinal stability margin and lateral stability margin are going to be used in this work. These measurements refer to the body's longitudinal and lateral axes and were used for the first time in [9]. According to this work, these measurements provide a close approximation of the stability margin for quasi-static or static walking, such as in the present case. The longitudinal stability margin is the shorter of the distances from the projection of the center of gravity to the front and rear boundaries measured along the vehicle body's longitudinal axis. The lateral stability margin is the shorter of the distances from the projection of the center of gravity to the front and rear boundaries measured along the vehicle body's lateral axis.

Section 2 approaches the problem of changing crabs with uniform motion and the strategy for performing it. Furthermore, some definitions useful for following developments are established. In Section 3 and Section 4, the leg trajectories in the transition gait are formulated. Also, the return times of each leg are found for finishing the transition gait as a wave gait. Section 5 presents an example of crab angle change to suit the static stability condition. Lastly, the results of an experiment using the RIMHO walking machine to test the validity of the proposed algorithm are given in Section 6.

## 2. Crab Angle Change

### 2.1. APPROACH AND STRATEGY

Let us consider a walking machine moving with the wave gait along a straight line in the $x'$-direction, which forms a crab angle $\alpha_0$ with the longitudinal body axis $x$ (see Fig. 1). Then, the change from $x'$-direction to $x''$-direction ($\alpha_1$ crab angle) maintaining the module of the body velocity or simple velocity, $v$, constant is stated.

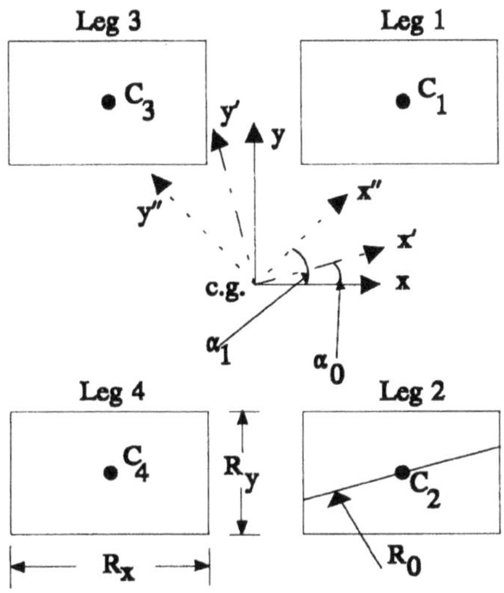


Fig. 1. Body reference system on the $xy$-plane.

The maximum allowable velocity is a function of the leg stroke, $R$, which is related to the crab angle $\alpha$. Then the velocity varies with the direction of motion. From this, it may be inferred that the constraint of maintaining the module of the velocity constant limits the crab angle $\alpha_1$ attainable from an initial angle $\alpha_0$. Therefore, with the condition of uniform motion, this is the worst case for performing the study of changing the crab angle. However, when necessary, the maximum crab angle may be increased by decreasing the body velocity.

Besides keeping the velocity steady, there is another condition consisting in that, at least at the end of the change of the crab angle, the gait must be a wave-crab gait again. The reason for this condition is based on the fact that the wave-crab gait provides the maximum measure of static stability along the longitudinal body axis. Furthermore, it is necessary that the gait be statically stable during the change in direction of motion. This implies, on the one hand, that two legs can never be in the transfer phase at the same time, and on the other hand, that the measurement of static stability must check out to be greater than a given minimum value. Lastly, all four legs have to utilize their working area fully, because static stability increases with the leg stroke. In summary, to change the crab angle, the following conditions have to be met:

1.- Velocity module has to remain steady.
2.- All four legs have to utilize their maximum constraint working volume.
3.- The gait during the change of the crab angle has to check out with two or more lifting events not occurring at the same time, i.e., only one leg can be in the transfer phase at a time.
4.- At the end of the change of the crab angle, the gait has to return to a wave gait.
5.- The motion has to be statically stable all the time, i.e., the gait longitudinal (or lateral) stability margin has to be positive and greater than a given minimum value $SLM_{min}$.

The transition gait between the wave gait with crab angle $\alpha_0$ and the wave gait with crab angle $\alpha_1$ will be found from the basic conditions above. In other words, the sequence of leg motions and the return times for each leg will be determined to verify the conditions specified above.

The time in which the crab angle change starts is given by either the operator or a control system, depending on whether the vehicle is teleoperated or possesses an intelligent control system. When the change of direction of motion starts, the legs in the support phase change their trajectory along the $x'$-direction and move without shifting their foot positions (with respect to an inertial reference system) in the new $x''$-direction. Each leg will remain in the support phase until it reaches its kinematic limits, which are defined by the working area of each leg. When this occurs, the corresponding leg starts its transfer phase to a point inside the working area, situated on the straight line parallel to the $x''$-direction and passing through the center of the working area, $C_i$. Once all four legs are in the support phase over this straight line, the change of the crab angle there will be finished.

On the other hand, a leg that is in the transfer phase when the change of crab angle begins modifies its trajectory so that its foot comes down at a point on the trajectory parallel to the $x''$-direction passing through the point $C_i$. This is always possible in the rectangular working area that it is being considered here. However, in other types of working areas (for example, a convex area), the current foot position has to be considered to determine the trajectory to the new foothold, so the kinematic constraints will be met.

2.2. DEFINITIONS

After approaching the problem of changing the crab angle and before introducing the mathematical formulation to define the transition gait, some definitions needed to solve this problem are given. The trajectory of the legs in the wave gait support phase is a straight line parallel to the direction of the center of gravity, c.g., motion, which passes through the center of its working area $C_i$. When the vehicle moves initially with this gait type along the $x'$-direction, the leg trajectory is called the **initial trajectory** (see Fig. 2). The position of each foot at the start of the change in the crab angle is called the **initial position**. At the end of the change in the crab angle, the vehicle will move with the wave gait along the $x''$-direction, and the leg trajectory is called the **new crab trajectory**.

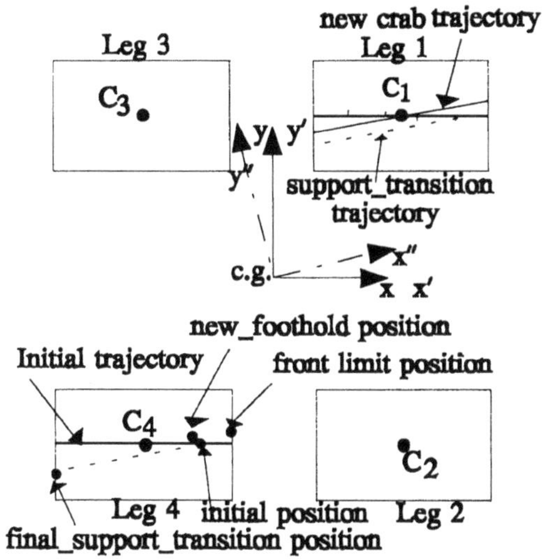


Fig. 2. Definition of trajectories and positions.

The trajectory of a leg in the support phase from the initial position to the intersection with its working area along the $x''$-axis is called the **support-transition trajectory** (see Fig. 2), while the aforementioned intersection point is called the **final-support-transition position**. A point on the new crab trajectory where a foot must be placed to perform the crab angle change is called the **new foothold position**. The trajectory of a leg in the transfer phase between the final-support-transition position and the new-foothold position is called the **transfer-transition trajectory**. A leg in the support phase at the beginning of the change in the crab angle will move along the support-transition and transfer-transition trajectories. The two trajectories are called **initially-support-transition trajectory** (see Fig. 3a). A leg in the transfer phase at the beginning of the change in the crab angle moves along a trajectory called the **initially-**

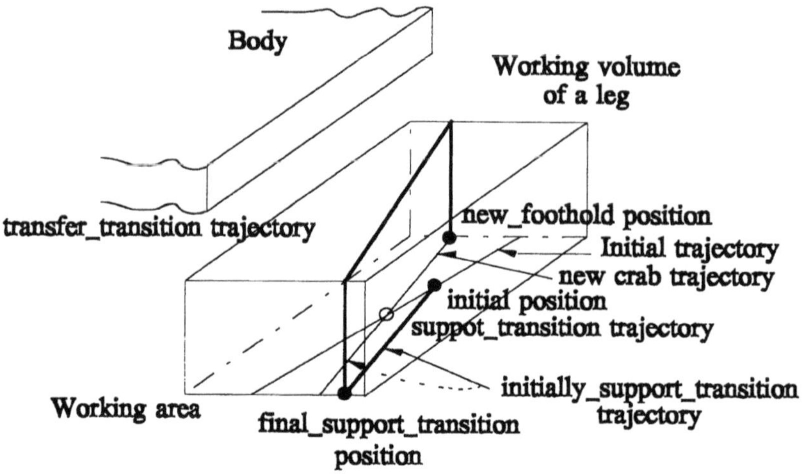


a)

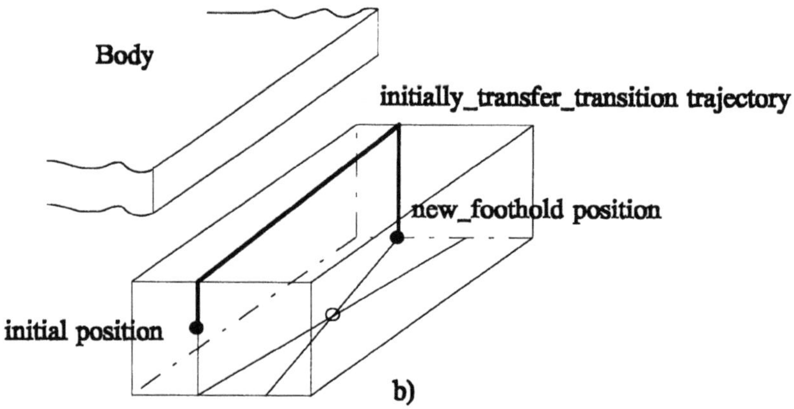


b)

Fig. 3. Transfer-transition trajectory.

**transfer-transition trajectory** between the initial and new-foothold positions (seeFig. 3b).

The sequence of leg motions and the foot trajectories from the start of the change in the crab angle until the four legs move onto the new crab trajectory is called the **transition gait** or simply **transition**. The time between the beginning of the crab angle change and the placement on the new-foothold position of the last leg finishing the transfer-transition trajectory is called the **transition time**.

In the following, the support and transfer transition trajectories and the time of each of them are determined. So, it is possible to establish the return and support times for each leg so that the transition time will satisfy the five conditions stated in Section II. In following Sections, foot positions in the support phase will be given using only the $x$ and $y$ coordinates. Since the body is considered to be level, those coordinates coincide with the vertical projections over a horizontal plane. Note that the $x$ and $y$ coordinates of the foot position do not vary with the terrain slope, since foot trajectories are defined in the xy-plane of the body reference system. However, if the body is not level, the vertical projections over a horizontal plane are different; therefore, in studying static stability, these projections must be calculated using the proper coordinate transformations.

### 3. Support-Transition Trajectory

From the above definitions, the support-transition trajectory is described by a foot between the initial position $[x_{0,i}, y_{0,i}]^T$, and the final position $[x_{f,i}, y_{f,i}]^T$, (see Fig. 2). The trajectory of foot $i$ with respect to the body reference system is given by:

$$\begin{bmatrix} x_{f,i} \\ y_{f,i} \end{bmatrix} = \begin{bmatrix} x_{0,i} - v\ t\ \cos\alpha_1 \\ y_{0,i} - v\ t\ \sin\alpha_1 \end{bmatrix} \tag{1}$$

where $v$ is the body velocity which remains constant; and $t$ is the time passed from the beginning of the change in the crab angle (t=0) that meets $0<t\leq t_{st,i}$, where $t_{st,i}$ is the maximum time that leg $i$ can be in the support-transition phase at the current velocity,

$$t_{st,i} = \frac{\sqrt{(x_{f,i} - x_{0,i})^2 + (y_{0,i} - y_{f,i})^2}}{v} \tag{2}$$

The first necessary condition so that two or more lift-off events in the transition gait do not occur at the same time is for each leg's support transition times to be different. Otherwise, there will be several legs changing from support to transfer at the same time, which would produce a lost of static stability. The order of leg lifting in the transfer-transition trajectory may be obtained by comparing the support times of each leg. According to this leg lifting order, for a case in which the crab angle change occurs with all four legs in the support phase, each leg is denoted by *f*, *s*, *th*, or *fo*, depending on whether the leg is the first, second, third or fourth to lift off the ground. If, the crab angle changes, there are only three legs in the support phase, they are denoted by *f*, *s*, and *th*, while the leg in the transfer phase is denoted by *tl*.

The support transition trajectory of leg *i*, equation (1), and the maximum time of that phase for leg *i*, equation (2), have been expressed as a function of the initial position and the final-support-transition position. Then, to solve the equations completely, both positions are going to be calculated in the Sections below.

### 3.1. INITIAL POSITION

The initial leg trajectory when the c.g. moves with a crab angle $\alpha_0$ is given by [1],[8],[9]:

$$\begin{bmatrix} x_i'(t) \\ y_i'(t) \end{bmatrix} = [R_{z,\alpha_0}]^T \begin{bmatrix} x_{ci} \\ y_{ci} \end{bmatrix} + \begin{bmatrix} \frac{R_0}{2} - \phi_{Li}(t)\frac{R_0}{\beta_0} \\ 0 \end{bmatrix} \quad (3)$$

where $t$ is the normalized time to the time period $T_0$ elapsed from placement of the leg 1; $R_0$ and $\beta_0$ are values of the leg stroke and duty factor for the crab angle $\alpha_0$; $R_{z,\alpha_0}$ is the rotation matrix which furnishes the relationship between the crab reference system and the body reference system; $\phi_{L,i}(t)$ is the local phase providing the normalized time period elapsed from placement of leg *i*; $[x_{ci}, y_{ci}]^T$ provides the coordinates of the center of the working area for each leg with respect to the body reference system.

Let us assume $t_0$ as the time normalized to the time period $T_0$ in which the crab angle is modified. The position of each foot at this time (initial position) in the new crab reference system $x''y''z''$ is given by:

$$\begin{bmatrix} x''_{0,i} \\ y''_{0,i} \end{bmatrix} = [R_{z,\alpha_1-\alpha_0}]^T \begin{bmatrix} x'_{0,i} \\ y'_{0,i} \end{bmatrix} \quad (4)$$

where $R_{z,\alpha_1-\alpha_0}$ is the rotation matrix which furnishes the relationship between the crab reference systems $x''y''z''$ and $x'y'z'$; $[x_{0,i}', y_{0,i}']^T$, which calculated at time $t_0$ using (3), is given by:

$$\begin{bmatrix} x'_{0,i} \\ y'_{0,i} \end{bmatrix} = [R_{z,\alpha_0}]^T \begin{bmatrix} x_{ci} \\ y_{ci} \end{bmatrix} + \begin{bmatrix} \frac{R_0}{2} - \phi_{Li}(t_0)\frac{R_0}{\beta_0} \\ 0 \end{bmatrix} \tag{5}$$

Using (4), (5), and the rotation matrix $R_{z,\alpha_1}$, the initial position in the body reference system is obtained as:

$$\begin{bmatrix} x_{0,i} \\ y_{0,i} \end{bmatrix} = R_{z,\alpha_1} \begin{bmatrix} x''_{0,i} \\ y''_{0,i} \end{bmatrix} = \begin{bmatrix} x_{ci} + (\frac{R_0}{2} - \phi_{Li}(t_0)\frac{R_0}{\beta_0})\cos\alpha_0 \\ y_{ci} + (\frac{R_0}{2} - \phi_{Li}(t_0)\frac{R_0}{\beta_0})\sin\alpha_0 \end{bmatrix} \tag{6}$$

Note that the position calculated thus is only required when working in simulation. When the vehicle is moving, its position is obtained by reading the position sensors when the operator (or a higher level) calls for a change in the crab angle.

### 3.2. FINAL-SUPPORT-TRANSITION POSITION

Final position $[x_{f,i}, y_{f,i}]^T$, is the last position that can be reached by the legs in the support-transition phase to utilize the constrained working area fully. As it was defined before, this position is the intersection point between the straight line passing through the initial position in the $x''$ direction and the boundaries of the working area. These boundaries are denoted by front, rear, right lateral, and left lateral boundary, depending on its location with respect to the center of the working area, $C_i$, (see Fig. 4). The four possible final-support-transition values are the following:

A)

$$x_{f,i} = x_{ci} - j\frac{R_x}{2}$$
$$y_{f,i} = y_{0,i} - (x_{0,i} - x_{f,i})\tan\alpha_1 \tag{7}$$

where $j$ is a factor that can take two different values. When the support leg trajectory intersects with the rear boundary ($\alpha_{p2} \leq \alpha_1 \leq \alpha_{p1}$) then $j$=1. When the support leg trajectory intersects with the front boundary ($\alpha_{p3} < \alpha_1 \leq \alpha_{p4}$ ) then $j$=-1.

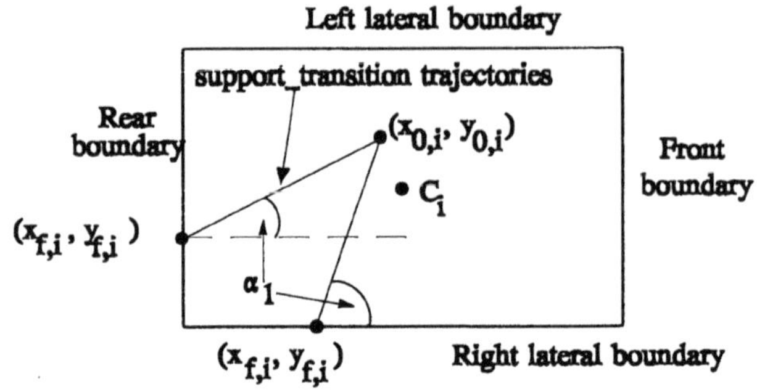


Fig. 4. Intersection between the support-transition trajectory and a working area boundary.

B)

$$y_{f,i} = y_{ci} - k\frac{R_y}{2} \tag{8}$$

$$x_{f,i} = x_{0,i} - (y_{0,i} - y_{f,i})\,cotan\,\alpha_1$$

where $k$ is a factor that can take two values. When the support leg trajectory intersects with the right lateral boundary ($\alpha_{p1} < \alpha_1 \leq \alpha_{p3}$) then $k$=1. When the support leg trajectory intersects with the left lateral boundary ($\alpha_{p4} < \alpha_1 < \alpha_{p2}$ ) then $k$=-1. In all cases $\alpha_{p1}$, $\alpha_{p2}$, $\alpha_{p3}$, and $\alpha_{p4}$ are the angles between the diagonal of the working area of each leg and the longitudinal body axis.

This Section outlines the equations for support-transition time as a function of the initial position, final-support-transition position and velocity, and for making each leg's support-transition times different. In the next Section the constraints derived from the transfer-transition trajectory will be studied.

## 4. Transfer Trajectory in the Transition Gait

Two types of transfer trajectories have been distinguished in the transition gait, depending on whether the leg is in the support phase or the transfer phase at the beginning of the crab angle change. In following Sections, both trajectories will be established.

### 4.1. TRANSFER-TRANSITION TRAJECTORY

The transfer-transition trajectory of a leg initially in the support phase is determined by the final-support-transition position and the new-foothold position. The final-support-transition position was calculated in Section III and is given by (7)-(8) as a function of the leg and the crab angle. The new-foothold position, $[x_{n,i}, y_{n,i}]^T$, will be calculated bellow, where the return time will satisfy certain constraints.

In the transfer trajectory considered here, the foot follows a trajectory made up of three subphases. Considering $h$ to be foot height along the $z$-axis, the lift-off subphase occurs between position $[x_{f,i}, y_{f,i}, z_{f,i}]^T$ and $[x_{f,i}, y_{f,i}, z_{f,i}+h]^T$; the transfer-forward subphase occurs between this position and $[x_{n,i}, y_{n,i}, z_{f,i}+h]^T$; and finally, the placement subphase, between the last position and $[x_{n,i}, y_{n,i}, z_{f,i}]^T$.

### 4.2. INITIALLY-TRANSFER-TRANSITION TRAJECTORY

This trajectory depends on the subphase (lift-off, transfer-forward or placement) in which the leg is at the moment the change in the crab angle begins and will be determined separately for each case. Here, $[x_{s,i}, y_{s,i}, z_{s,i}]^T$ is taken as the position of leg $i$ at the end of the transfer phase.

*Leg in the lift-off subphase:*

When the $z$ coordinate of initial position of leg $i$, $[x_{0,i}, y_{0,i}, z_{0,i}]^T$, meets the condition $z_{s,i} < z_{0,i} \le z_{s,i}+h$, then the leg is in the lift-off subphase. In this case, the initial-transfer-transition trajectory consists of three subphases:

1.- Lift-off subphase between the initial position $[x_{0,i}, y_{0,i}, z_{0,i}]^T$ and the position $[x_{0,i}, y_{0,i}, z_{0,i}+h_1]^T$.
2.- Transfer-forward subphase between the previous position and $[x_{n,i}, y_{n,i}, z_{0,i}+h_1]^T$.
3.- Placement subphase between the last position and $[x_{n,i}, y_{n,i}, z_{0,i}+h_1-h]^T$.

where $h_1$ and $h$ are z-axis displacement in the lift-off and placement subphases, respectively.

*Leg in the transfer-forward subphase:*

Let us consider that the leg in the transfer phase is in the transfer-forward subphase, i.e., it is known that $x_{0,i} \le x_{s,i}$, $y_{0,i} \le y_{s,i}$, $z_{0,i} = z_{s,i}+h$. In this case, the initial-transfer-transition trajectory consists of two subphases:

1.- Transfer-forward subphase between the initial position $[x_{0,i}, y_{0,i}, z_{0,i}]^T$ and position $[x_{n,i}, y_{n,i}, z_{0,i}]^T$.

2.- Placement subphase between the last position and $[x_{n,i}, y_{n,i}, z_{0,i}-h]^T$.

*Leg in the placement subphase:*

Let us suppose that the leg in the transfer phase is in the placement subphase, i.e., that $x_{0,i} = x_{s,i}$, $y_{0,i} = y_{s,i}$, $z_{0,i} > z_{s,i}$ at the beginning of the crab change. In this case, the initially-transfer-transition trajectory will consist of two subphases, as in the last case:

1.- Transfer-forward subphase between the initial position $[x_{0,i}, y_{0,i}, z_{0,i}]^T$ and $[x_{n,i}, y_{n,i}, z_{0,i}]^T$.
2.- Placement subphase between the last position and $[x_{n,i}, y_{n,i}, z_{0,i}-h_2]^T$.

where $h_2$ checks out as $h_2 \leq h$.

There is a minimum return time for executing a step which is function of the distances and maximum velocities of each leg. The minimum return time for each before trajectory is given by the general expression:

$$t_{r\min,i} = 2\frac{d_{z,i}+d_{p,i}}{v_{z\max}} + \frac{d_{t,i}}{v_{xy\max}} \tag{9}$$

where $d_{l,i}$, $d_{t,i}$, $d_{p,i}$ are the displacements in the lift-off, transfer-forward and placement subphases respectively; $v_{zmax}$ and $v_{xymax}$ are the maximum velocities for the $z$-axis and the $xy$-plane.

### 4.3. RETURN TIME VALIDITY RANGE

So far, the minimum return time for both the transfer-transition and the initially-transfer-transition trajectories have been determined. Now, let us find the maximum return time of each leg to avoid having several lift-off events occur simultaneously in the transition gait. In Section III, different support transition times for all legs was established as a condition for avoiding two or more lift-off events. In the same manner, the return time of each leg has to be limited to avoid finding two legs in the transfer phase at the same time. Such an event would produce instability in a machine walking with statically stable gaits.

When the time that a leg remains in the transfer phase is greater than the time another leg remains in the support transition phase, the support leg will begin the transfer phase before the first leg contacts the ground. Then, the maximum return time of a leg $p$, $t_{rmax,p}$, has to be equal to the time that the next leg to begin the transfer phase, $s$, remains in the support transition phase from the time when the leg $p$ lifts off the ground (see Fig. 5). If the times for legs $p$ and $s$ to finish their support transition trajectories are given by $t_{st,p}$ and $t_{st,s}$, then the time that leg $s$ has to remain in support after leg $p$ finishes its support-transition is $t_{st,s}-t_{st,p}$. Therefore, the maximum return time of leg $p$ has to be

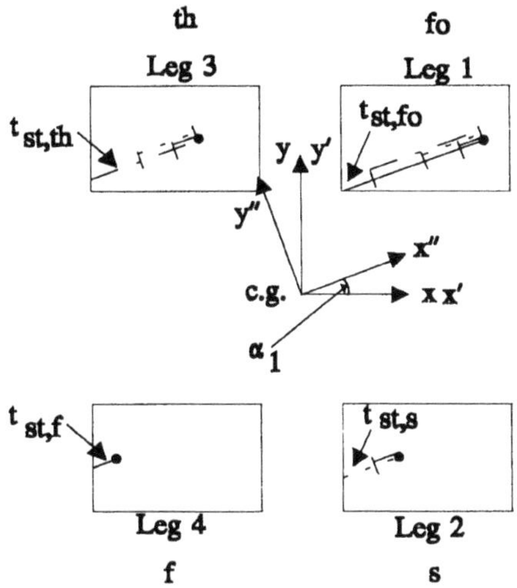


Fig. 5. Support-transition trajectories, duration and leg lifting sequence.

equal to the previous amount.

In general, the maximum return time for each leg is:

$$t_{r\max,k} = t_{st,l} - t_{st,k} \tag{10}$$

where $k$ and $l$ depend on the number of legs in the support phase at the beginning of the crab angle change. So, we consider two cases:

CASE A) $k = f, s, th$ and $l = s, th, fo$, when four legs are in the support phase
CASE B) $k = lt, f, s$ and $l = f, s, th$ (note that $t_{st,lt}=0$), when three legs are in the support phase.

The maximum return time condition for the last leg to begin the transfer phase ($t_{r,fo}$ in case A and $t_{r,th}$ in case B) is implicit in the wave gait condition that will be shown later. Note that such a gait is statically stable, and therefore two lift-off events will never occur at the same time.

Knowing the minimum and maximum return time for each leg, the return time for each leg, $t_{r,k}$, remaining within cases A and B above, is as follows:

CASE A)

$$t_{r\min,k} \le t_{r,k} \le t_{r\max,k}$$
$$t_{r\min,fo} \le t_{r,fo} \tag{11}$$

where $k = f,\ s,\ th.$

CASE B)

$$t_{r\min,k} \le t_{r,k} \le t_{r\max,k}$$
$$t_{r\min,th} \le t_{r,th} \tag{12}$$

where $k = lt, f, s$. The return time $t_{rmin,l}$ is calculated using (9), (14), (17), (19).

In this way, constraints are established for the return time of each leg in order for the transition gait to be statically stable. However, to find return time, it is necessary to know the foot position at the end of the transfer phase. The requirements for the new foothold position will be studied bellow.

### 4.4. APPROACH TO THE NEW FOOTHOLD POSITION

The new foothold position has to ensure that the minimum return time is lower than or equal to the maximum return time for each leg. In a first approach, the front limit position can be considered as the new foothold position which is given by

$$\begin{bmatrix} x''_{fl,i} \\ y''_{fl,i} \end{bmatrix} = [R_{z,\alpha_1}]^T \begin{bmatrix} x_{ci} \\ y_{ci} \end{bmatrix} + \begin{bmatrix} \frac{R_1}{2} \\ 0 \end{bmatrix} \tag{13}$$

with respect to the crab reference system $x''y''z''$ and by

$$\begin{bmatrix} x_{fl,i} \\ y_{fl,i} \end{bmatrix} = R_{z,\alpha_1} \begin{bmatrix} x''_{fl,i} \\ y''_{fl,i} \end{bmatrix} = \begin{bmatrix} x_{ci} + \frac{R_1}{2}\cos\alpha_1 \\ y_{ci} + \frac{R_1}{2}\sin\alpha_1 \end{bmatrix} \tag{14}$$

with respect to the body reference system. $R_1$ is the leg stroke for the crab angle $\alpha_1$.

Considering the last equation to give the new foothold position, the minimum return time for each leg is found. If for a leg $i$ this minimum return time is greater than the maximum, the leg cannot be placed in the front limit position. Then, another position is looked for in a second approach.

As the return time has to be lower than the maximum return time, the condition $t_{rmin,i}=Kt_{rmax,i}$ is imposed for the new-foothold position, where $K$ is a real number such that $0< K \leq 1$. The greater the value of $K$, the nearer the new foothold position to the front limit position. Knowing that the minimum return time is related with foot displacement in the transfer phase and the maximum velocities through (9), (14), (17), (19), the above condition may be used to obtain the distance between the final-support-transition and new-foothold positions as (see Fig. 6):

$$d'_{t,i} = (t_{r\min} - \frac{d_z}{v_{z\max}}) v_{xy\max} \tag{15}$$

On the other hand, the new foothold position, $[x_{nf,i}, y_{nf,i}]^T$, has to satisfy the following system of equations (see Fig. 6):

$$\begin{aligned} d'^2_{t,i} &= (x_{nf,i} - x_{f,i})^2 + (y_{nf,i} - y_{f,i})^2 \\ y_{nf,i} - y_{rl,i} &= (x_{nf,i} - x_{rl,i}) \tan\alpha_1 \end{aligned} \tag{16}$$

where $[x_{rl,i}, y_{rl,i}]^T$ is the right lateral position.

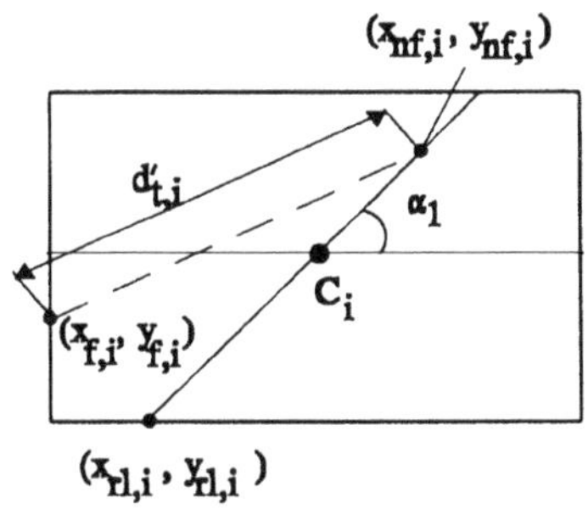


Fig. 6. New foothold position.

Solving this system of equations:

$$x_{nf,i} = \frac{-b \pm \sqrt{b^2 - 4ac}}{2a}$$
$$y_{nf,i} = y_{rl,i} + (x_{nf,i} - x_{rl,i}) \tan\alpha_1 \qquad (17)$$

where a, b and c are coefficients given by:

$$a = -(1 + \tan^2\alpha_1)$$
$$b = 2(x_{f,i} + x_{rl,i}\tan^2\alpha_1 - (y_{rl,i} - y_{f,i})\tan\alpha_1) \qquad (18)$$
$$c = d_{t,i}^{\prime 2} - x_{f,i}^{\;2} - (y_{rl,i} - y_{f,i})^2 - x_{rl,i}^{\;2}\tan^2\alpha_1 + 2x_{rl,i}(y_{rl,i} - y_{f,i})\tan\alpha_1$$

Equation (17) has a real solution whenever $a \neq 0$ and $b^2\text{-}4ac \geq 0$ are true. In this case, there are two possible solutions, and the solution closer to the front limit position must be chosen. Note that otherwise the static stability will be less.

Once the new foothold has been determined, the condition of returning to a wave gait at the end of the transition gait will be formulated.

### 4.5. WAVE-GAIT CONDITION

Once a leg is placed in its new foothold position, it will move along a straight line with a crab angle of $\alpha_1$ that passes through this point and the center, $C_i$, of the working area. This trajectory in body reference system coordinates is given by:

$$\begin{bmatrix} x_{nt,i}(t) \\ y_{nt,i}(t) \end{bmatrix} = \begin{bmatrix} x_{nf,i} - v(t - t_{st,i} - t_{r,i})\cos\alpha_1 \\ y_{nf,i} - v(t - t_{st,i} - t_{r,i})\sin\alpha_1 \end{bmatrix} \qquad (19)$$

where the new foothold position $[x_{nf,i}, y_{nf,i}]^T$ is given by (9), (14), (17), (19) or (9), (14), (17), (19), depending on whether the new foothold position is calculated in a first or second approach; $t$ is such that,

$$t_{st,i} + t_{r,i} \leq t \leq t_T \qquad (20)$$

where $t_{st,i}$ the support transition time of leg $i$ given by (2); $t_{r,i}$ the return time of leg $i$ whose value is being sought; and $t_T$ is the transition time. From the definition of $t_T$ given in Section II, it may be shown that this time is the maximum of the support times, $t_{st,k}$, plus the return time of the same leg, $t_{r,k}$:

$$t_T = t_{st,k} + t_{r,k} \tag{21}$$

where k=th or k=fo depending on the number of legs initially in the support phase.

The following equation

$$\begin{bmatrix} x_i'(t) \\ y_i'(t) \end{bmatrix} = [R_{z,\alpha}]^T \begin{bmatrix} x_{ci} \\ y_{ci} \end{bmatrix} + \begin{bmatrix} \frac{R}{2} - \phi_{Li}(t)\frac{R}{\beta} \\ 0 \end{bmatrix} \tag{22}$$

provides the trajectory, in the crab reference system, of a leg in the support phase which moves according to a periodic gait along a straight line through the center $C_i$ of the working area at a given crab angle. Performing the corresponding transformation to the body reference system, the expression below, equivalent to the trajectory given by (9), (14), (17), (19), is may then be obtained:

$$\begin{bmatrix} x_{nt,i}(t_1) \\ y_{nt,i}(t_1) \end{bmatrix} = \begin{bmatrix} x_{fl,i} - v\phi_{L,i}(t_1)T_1\cos\alpha_1 \\ y_{fl,i} - v\phi_{L,i}(t_1)T_1\sin\alpha_1 \end{bmatrix} \tag{23}$$

where the new foothold $[x_{n,i}, y_{n,i}]^T$ is the front limit position given by (9), (14), (17), (19); $t_1$ is the normalized time since the placement of the reference leg (leg 1); $\phi_{L,i}(t_1)$ is the local phase of leg $i$ at time $t_1$; $T_1$ is the new time period of the locomotion cycle which maintains the velocity $v$ and the duty factor $\beta_0$ when varying the stroke from $R_0$ to $R_1$ (stroke in the new direction of motion defined by $\alpha_1$), i.e., $T_1 = R_1T_0/R_0$. Of course, this time period has to meet the condition $T_1 \geq t_{rmin}/(1-\beta)$. If not, it will be necessary to increase the time period using a velocity control [11] in order to perform the change in crab angle, which involves decreasing the velocity of the legs in the support phase.

When $t = t_T$, all four legs are in the new crab trajectory. At that instant, the time $t_1$ that appears in (23) is the time normalized to $T_1$ in which the placement of leg *fo* (or *th*) occurs when initially four legs are in the support phase (or there are three legs). Since the placement of a leg with respect to leg 1 is defined by the leg phase $\phi_i$, then $t_1 = \phi_{fo}$ (or $t_1 = \phi_{th}$) where $\phi_{fo}$ (or $\phi_{th}$) depends on the crab angle and the leg number. Comparing (9), (14), (17), (19) and (23), under these conditions, the ratio between the return times of legs $i$ and *fo* (or *th*) in order to make the gait a wave gait at the end of the transition is obtained:

$$t_{r,k} - t_{r,i} = \frac{\sqrt{(x_{fl,i} - x_{n,i})^2 + (y_{fl,i} - y_{n,i})^2}}{v} - t_{st,k} + t_{st,i} + \phi_{Li}(\phi_k) T_1 \tag{24}$$

where $k = fo$ and $i = f, s, th$ if the four legs are initially in the support phase; on the other hand, $k = th$ and $i = lt, f, s$ if there are only three legs in the support phase. Note that when leg $i$'s new-foothold position is equal to the front limit position then:

$$t_{r,k} - t_{r,i} = -t_{st,k} + t_{st,i} + \phi_{Li}(\phi_k) T_1 \tag{25}$$

Each leg's return time is now defined by a set of equations, (24) or (11), (12), (25), and by four inequalities, (11), (12), (25) or (11), (12), (25) depending on whether the case fits A or B. Thus the number of unknowns is four, if there is a solution, it is not unique. In the following Section, a set of solutions for the return time of each leg are sought, and then a solution is chosen using a minimization criterion.

### 4.6. RETURN TIME CALCULATION

To calculate return times, let us assume that all four legs are initially in the support phase. Note that the system of equations is:

$$\begin{aligned} t_{r,fo} - t_{r,f} &= t_1 \\ t_{r,fo} - t_{r,s} &= t_2 \\ t_{r,fo} - t_{r,th} &= t_3 \end{aligned} \tag{26}$$

where $t_1$, $t_2$, and $t_3$ are constants in each case whose values are determined by the right side of (24) or (11), (12), (25). As these equations are linear, it is possible to obtain the minimum and maximum of $t_{r,fo}$ using the minimum and maximum of the other unknown return time that appears in the equation. The three first inequalities in (11), (12), (25) provide the minimum and maximum constraints of the return times $t_{r,f}$, $t_{r,s}$, and $t_{r,th}$. By substituting these values in the above set of equations, we obtain three ranges of solutions for $t_{r,fo}$. Each of these ranges guarantees the existence of one solution for $t_{r,f}$, $t_{r,s}$, or $t_{r,th}$. However, the point is to find one interval of solutions of $t_{r,fo}$ that guarantees the existence of solutions for $t_{r,f}$, $t_{r,s}$, and $t_{r,th}$ simultaneously. This means finding the intersection between these closed intervals together with the open interval on the right, which provides the last condition of (11), (12), (25) (the minimum constraint of $t_{r,fo}$).

Let us consider that the intervals of values of $t_{r,fo}$ ($[t_{rmin,fo1}, t_{rmax,fo1}]$, $[t_{rmin,fo2}, t_{rmax,fo2}]$ and $[t_{rmin,fo3}, t_{rmax,fo3}]$) guarantee a solution for $t_{r,f}$, $t_{r,s}$ and $t_{r,th}$, respectively. The interval $[t_{rmin,fo4}, \infty)$ represents the lower boundary condition that $t_{r,fo}$ has to meet. The intersection of the above four intervals, if not empty, provides the following interval:

$$[\max(t_{rmin,fo1}, t_{rmin,fo2}, t_{rmin,fo3}, t_{rmin,fo4}), mim(t_{rmax,fo1}, t_{rmax,fo2}, t_{rmax,fo3})] \tag{27}$$

To minimize each leg's return time, the lower limit of the above interval is chosen as the value of $t_{r,fo}$, i.e.,

$$t_{r,fo} = \max(t_{rmin,fo1}, t_{rmin,fo2}, t_{rmin,fo3}, t_{rmin,\ fo4}) \tag{28}$$

The linear relationship between the return times of the four legs forces to the values for $t_{r,f}$, $t_{r,s}$ and $t_{r,th}$ to be minimums too.

The return time of a leg is related to the velocities of each axis during the transfer phase through the expression:

$$t_r = 2\frac{h}{v_z} + \frac{d_x}{v_{xy}} \tag{29}$$

where $v_z$ is the velocity of the $z$-axis; $v_{xy}$ is the velocity in the $xy$-plane; $h$ is the height of the step, which is equal to the distance that the $z$-axis moves when the ground is flat and the vehicle is level; $d_x$ is the distance that the foot runs in the $xy$-plane. Therefore, the criterion of minimizing each leg's return time drives the leg in the transfer phase to a greater velocity.

When the number of legs at the beginning of the change in the crab angle is three, the same procedure will be followed to calculate return times, considering that the minimum and maximum constraints in this case are given by (11), (12), (25).

Until here, the transition gait has been determined (sequence of leg movements, transition cycle time, return times and foothold positions) in order to satisfy these conditions: 1) the velocity module remains constant, 2) the four legs utilize their constraint working area to the fully, 3) lifting events do not coincide, and 4) at the end of the change in the crab angle, the gait is a wave gait. However, it is also necessary for the transition gait to be statically stable, i.e., its static stability margin must be greater than a minimum value. If this last condition is not met, it will not be possible to change the crab at a time $t_0$ for the angles $\alpha_0$ and $\alpha_1$. In Fig. 7, the algorithm for determining if it is possible to perform the crab angle change and finding the transition gait is shown. The following Section will study the static stability condition.

## 5. Transition Gait Stability

This Section addresses one necessary condition for performing the change in the crab angle: the transition gait must be statically stable. To determine this, it is necessary to

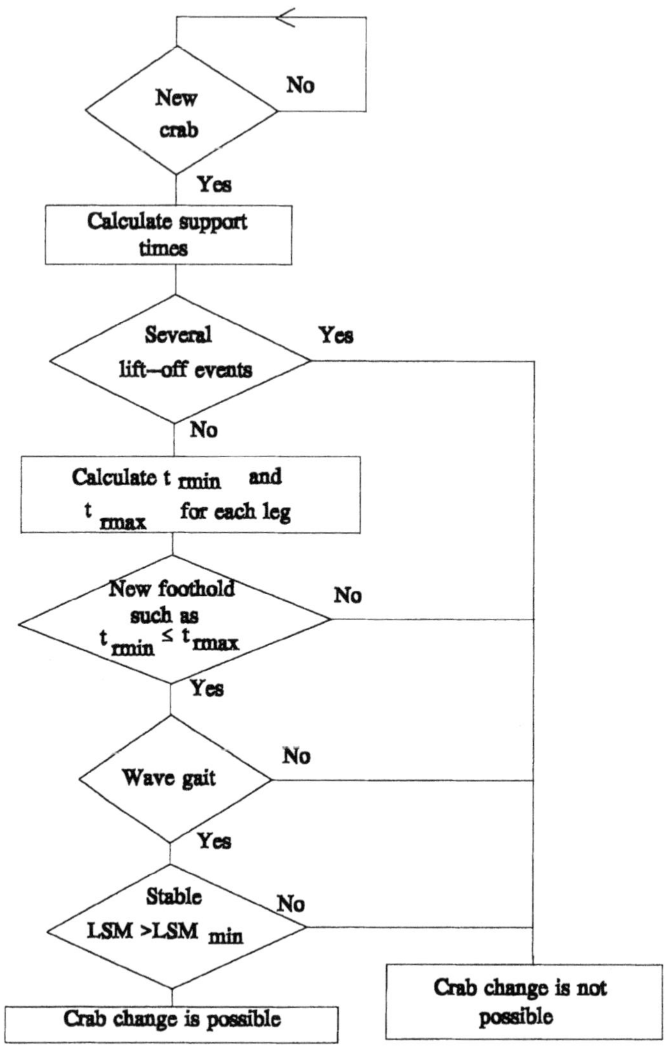


Fig. 7. Flow chart of the omnidirectional control algorithm to determine the new crab angle and the transition gait.

calculate the longitudinal (or lateral) stability margin throughout the transition. Such a stability measure is known analytically as function of $R$, $\beta$, $\alpha$, $P_x$ and $P_y$ for a quadruped that uses the wave gait with constant crab [9]. However, in the current case, it is not possible to secure a similar expression due to the complexity introduced by the trajectory of each leg passing through the center of the working area and the variety of parameters appearing when the crab angle is not constant. These parameters are: the initial crab angle, the final crab angle, the return time of each leg, the foot position at the beginning of the change in the crab angle and the final-support-transition position (both positions are also functions of $R$, $\beta$, $\alpha$, $P_x$ and $P_y$). Therefore, it seems more suitable to perform a numerical study to find how the longitudinal stability margin (LSM) varies with the direction of motion.

The algorithm of crab angle change presented in this chapter makes it possible to perform a numerical study of stability during the transition gait as a function of the time in which the crab change occurs, the initial and final crab angle, the duty factor, the leg working area dimensions and the distance between the centers of these areas. This can prove most useful for designing quadruped walking machines for maximum static stability and mobility.

The example shown in this Section has been performed considering a constraint working area measuring $R_x$=500 mm by $R_y$=200 mm. Furthermore, the distances between centers are $P_x$=700 mm and $P_y$=1100 mm. The change in direction occurs when leg 1 ends its support phase. The duty factor is $\beta$=0.8. In the case presented here, the interval of possible crab angles is going to be found from an initial crab angle keeping a given constant velocity and meeting the rest of the conditions in Section II (see also Fig. 7). Furthermore, the minimum longitudinal (or lateral) stability margin for each transition is presented. Note that the minimum longitudinal (or lateral) stability margin considered is $LSM_{min}$=0 mm. Therefore, if this value is greater, the maximum crab angle within reach from an initial crab angle will decrease.

In the example shown in Fig. 8, the initial crab angle is $\alpha_0$=0° and the direction of motion is modified to increase the crab angle by 1° to a final crab angle $\alpha_1$=57°, which coincides with $\alpha_{c1}$. The type of wave-crab gait is determined by the range of angles; in this case, it is type $+x$. The figure shows the transition gait's minimum longitudinal stability margin associated with crab angle $\alpha_1$. The asterisks represent the crab angles that can be reached from a particular angle and the LSM corresponding to that transition gait. For example, from $\alpha_0$=0° is possible to modify the direction of motion to $\alpha_1$=1°, 2°, 3°,.., 12°, and the LSM for these transition gaits (0° to 1°, 0° to 2°, etc.) is associated with the corresponding $\alpha_1$. Each circle represents the maximum crab angle $\alpha_1$ reachable from an initial angle keeping vehicle velocity and the other aforementioned conditions constant. Furthermore, it represents the new initial crab angle, $\alpha_0$, from which the direction of motion will change. To increase this maximum angle, velocity would have to be decreased, as mentioned in Section II, which will be illustrated with an example. Note that the maximum crab angle reachable with an initial velocity of 9 mm/s and an initial angle of 0° is $\alpha_1$=12°. With 12° as the new initial crab angle, the direction of motion can be modified to $\alpha_1$=18° keeping the same velocity, v=9 mm/s. As can be seen in the figure, the last angle that can be reached at this velocity is 27°. The blanks are the break points, i.e., angles that cannot be reached from the before angle while keeping the same velocity. For example, it is not possible to change the angle from 27° to 28° with v=9 mm/s. The break points are 28°, 34°, 42° and 56°. These points disappear when the velocity is decreased.

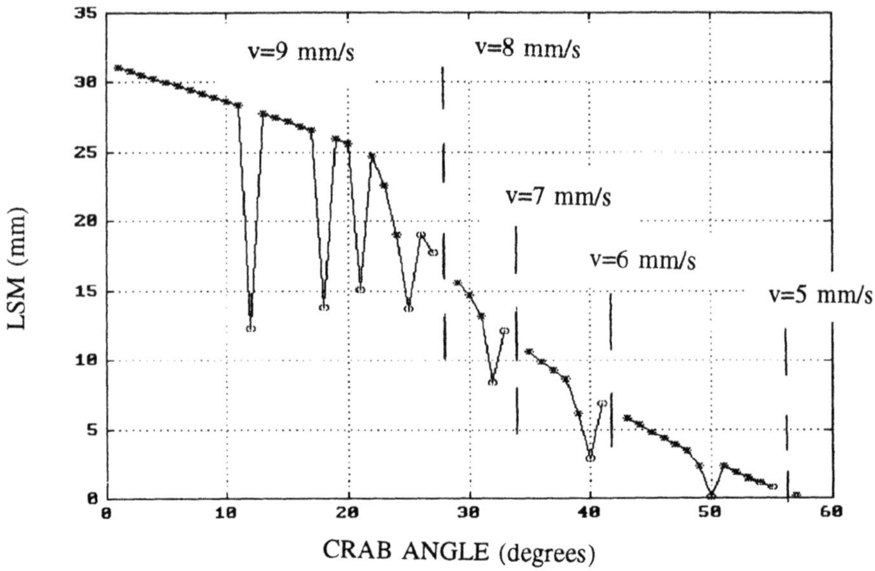


Fig. 8. Longitudinal stability margin as function of the crab angle for the range [0°, 57°] and initial velocity v=9 mm/s.

## 6. Experimental Results

This Section discusses some experiments evaluating the crab angle change algorithm. The results presented here are the trajectory and velocity of the RIMHO walking machine moving in different directions. The 3D location system [12] was used to obtain these data. The wave gait used in these experiments is characterized by a duty factor $\beta$=0.8. The commanded velocity is v=6 mm/s. The instance at which the change of crab is performed coincides with the instant in which the forth leg starts the transfer phase.

The RIMHO (see Fig. 9) is a quadruped walking robot that may be classified as an insect type. Its four legs are based on a three dimensional cartesian pantograph mechanism. The pantograph leg amplifies the foot extension and foot speed, but at the cost of decreasing the torque. This leads to the use of high gearing when using electrical actuators, as in this case. Therefore, it results in a slow walking machine with a maximum cruise speed on level ground without changing direction of about 1 cm/s.

The Fig. 10 shows the theoretical trajectory (straight line) and the current trajectory for the crab angles $\alpha_0$=-5° and $\alpha_1$=10°. The change in the crab angle occurs when the trajectory direction changes. Observe that the current trajectory approaches the theoretical trajectory quite closely, considering that there is no orientation control.

Fig. 9 The RIMHO walking machine.

Fig. 11 shows the velocity as function of the time when the vehicle moves according to the above trajectories. The average velocity is 6.2 mm/s. The velocity peaks are due to position changes when a leg lifts off [13]. Furthermore, it is necessary to consider the change in the initial velocity due to the attitude/position control [13].

## 7. Conclusions

This chapter proposes an omnidirectional control algorithm for periodic gaits based on maintaining the body velocity module constant. This condition stams from two considerations: 1) periodic gaits are intrinsically slow and 2) some applications require this condition. This algorithm provides a transition gait to change the direction of motion while keeping the body orientation unchanged. However, previous work on these gaits was based on controlling the time period by using the distance each leg has to cover in the support phase. In such work, the gait is maintained, but the body velocity is decreased.

First, the problem of changing a crab angle with uniform motion and the strategy for performing it were approached. Then, some definitions useful for the mathematical developments of the transition gait were established. The leg trajectories in the transition gait, initially-support-transition and initially-transfer-transition, were formulated. Also, the return times were found to ensure a wave gait at the end of the transition gait.

An example of crab angle change that shows the LSM as a function of the crab

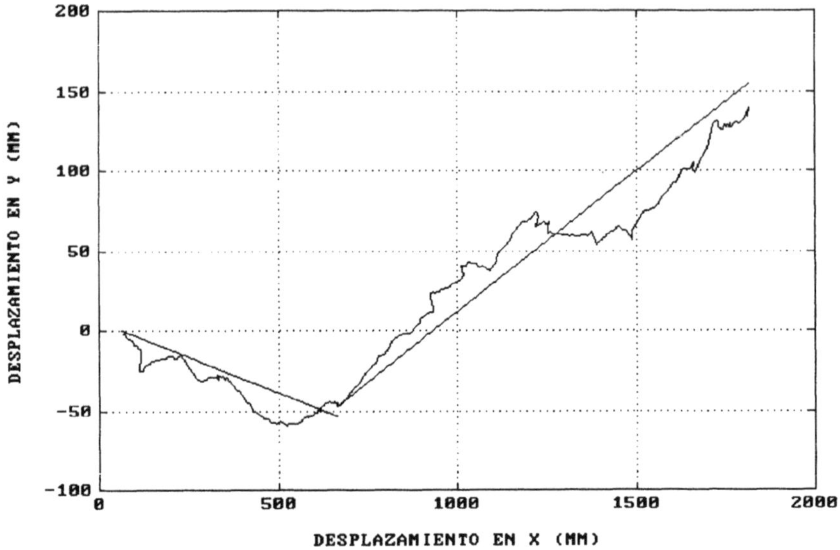


Fig. 10. Change in crab angle from $\alpha_0$=5° to $\alpha_1$=10°.

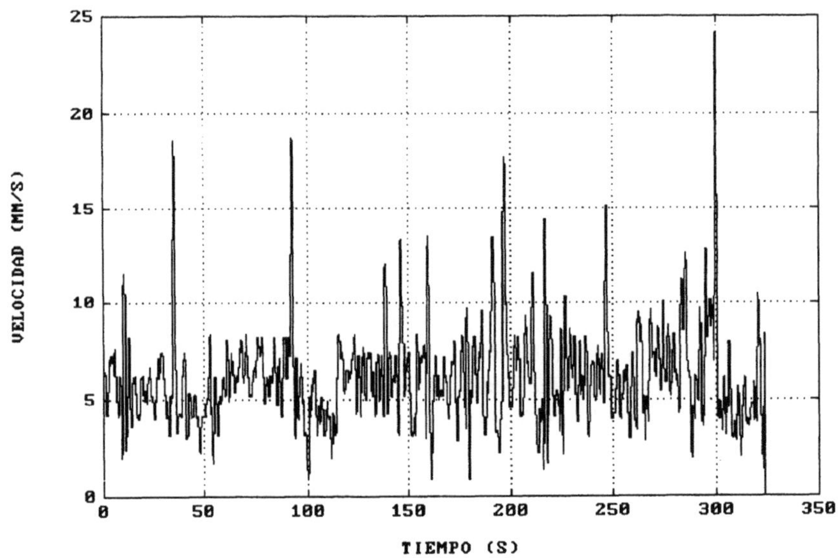


Fig. 11. Velocity as function of the time for the change in crab angle from $\alpha_0$=5° to $\alpha_1$=10°.

angle and body velocity was shown. At the initial velocity of v=9 mm/s and the initial crab angle 0°, the maximum crab angle is 12°. Also, at the same velocity, the last reachable angle is 27°. After that, there is a break angle. Decreasing the initial velocity to 5 mm/s, greater crab angles can be reached. If the initial velocity is 5 mm/s, the break points disappear.

Finally, some experiments were performed to show the effectiveness of the proposed algorithm. They involve body trajectory and body velocity. To grasp the position results, it must be understood that there is no orientation control. Also, velocity results are due to the attitude/position control and changes in body position when a leg lifts off.

**References**


1. S. Hirose, "A Study of Design and Control of a Quadruped Walking Vehicles", The International Journal of Robotics Research, Vol. 3, No. 2, pp. 113-133, 1984.
2. W.J. Lee, A COMPUTER SIMULATION STUDY OF OMNIDIRECTIONAL SUPERVISORY CONTROL FOR ROUGH-TERRAIN LOCOMOTION BY A MULTILEGGED ROBOT VEHICLE, Ph.D. dissertation, Ohio State University, 1984.
3. V.R. Kumar and K.J. Waldrom, "Adaptive Gait Control for a Walking Robot", Journal of Robotics Systems, 6 (1), pp. 49-76, 1989.
4. D.A. Messuri, OPTIMIZATION OF THE LOCOMOTION OF A LEGGED VEHICLE WITH RESPECT TO MANEUVERABILITY, Ph.D. dissertation, Ohio State University, 1985.
5. M. Russell and S.A. Boulevard, "ODEX I: The First Functionoid", Robotics Age, Vol. 5, No. 5, Sep/Oct, 1983.
6. J.K. Lee and S.M. Song, "Path Planning and Gait of Walking Machine in an Obstacle-Strewn Environment", Journal of Robotic Systems, Vol. 8, No. 6, pp. 801-827, 1991.
7. R.B. McGhee and A.A. Frank, "On the Stability Properties of Quadruped Creeping Gaits", Mathematical Biosciencies, Vol.2, No. 1/2, pp. 67-84, 1986.
8. S.M. Song and K.J. Waldron, "An Analytical Approach for Gait Study and Its Applications on Wave Gaits," The International Journal of Robotics Research, Vol. 6, No. 2, pp. 60-71, Summer, 1987.
9. C. Zhang, and S.M. Song, "Stability Analysis of Wave-Crab Gaits of a Quadruped," Journal of Robotics System, 7(2), pp. 243-276, 1990.
10. M.A. Jiménez, P. González de Santos, M.A. Armada, "A Four-legged Walking Test Bed," 1st IFAC International Workshop on Intelligent Autonomous Vehicles, April 18-21, University of Southampton, Hampshire, United Kingdom, 1993.
11. M.A. Jiménez and P. González de Santos, "Terrain Adaptive Gait for Walking

Machines", Int. J. of Robotics Research, Vol. 16, No. 3, pp. 320-339, June 1997.

12. J.M. Martín, R. Ceres, L.A. Pérez, and L. Calderón, "3-D Dynamic Location System by Ultrasonic Techniques," Technical Report, IAI-CSIC, July, 1993.

13. M.A. Jiménez, GENERATION AND IMPLEMENTATION OF WAVE GAITS FOR QUADRUPED ROBOTS, Ph.D. Thesis, University of Cantabria, Department of Electronic, 1994.

# 17
# DYNAMIC MODELLING AND LOCOMOTION CONTROL OF BIPED AUTONOMOUS ROBOTS

**T.E. KRIKOCHORITIS, C.S. TZAFESTAS and S.G. TZAFESTAS**
Intelligent Robotics and Automation Laboratory
Department of Electrical and Computer Engineering
National Technical University of Athens
15773, Zografou Campus, Athens, Greece

## 1. Introduction

The field of legged autonomous robots has received over the years a great attention, and a number of interesting types of biped and multilegged robots are now available [1,2]. Bipeds represent a class of legged robots that attempt to imitate the human style of locomotion. The simplest model which has been used for the study of some basic features of human locomotion is the inverted pendulum [3,4]. More sophisticated models, with many degrees of freedom were introduced and used after 1980 [5-13]. However, we are still far away of designing biped robots that possess the dexterity and flexibility of human walking which is controlled by nearly three hundred and fifty muscle pairs. Some examples of actually walking biped robots are : BIPER-3 robot [7], KENKYAKU-1 [8,9], KENKYAKU-2 [10], BLR-G1 [11], BLR-G2 [12] and WASEDALEG-12 [13]. BIPER-3 is a simple 3-link robot (torso and two simple legs), KENKYAKU-1 is a 5-link robot, KENKYAKU-2 (weight 40kg, height 1.10m) is a 7-link biped, and BLR-G1/BLR-G2 are 9-link bipeds with interesting anthropomorphic features (BLR-G2 has 25Kg weight and 0.97m height). WASEDALEG-12 belongs to the Waseda University biped series and has a weight of 107kg and height 1.80m . This biped is equipped with sophisticated force sensors on its feet and joints, and has realized a maximum stride of 0.30m and a minimum step time of 1.3sec.

Three fundamental characteristics of biped locomotion are the following :

i) The presence of an unpowered degree of freedom (d.o.f.). All joints are equipped with actuators to control the motion of the corresponding links, except for the joint (d.o.f.) formed by the contact of the foot and the ground surface. This d.o.f. cannot be controlled directly by an actuator, but only indirectly via a proper motion of the rest of the system (i.e. by the appropriate actions of the powered d.o.f.s).
ii) A permanent change of the situations when the mechanism is supported on one foot (single-leg support phase) or when both feet are in contact with the ground (double-

leg support phase) or both feet are in the air (when the leg-support-exchange is performed in an instant).

iii) The repeatability problem. For a 2-leg locomotion, the period in which the movements are repeated over and over again represents "one step". This means that the positions and velocities at the beginning and the end of each step are the same, and only the motion that satisfies these conditions is acceptable for bipeds. These conditions are known as the *repeatability conditions*. The repeatability conditions are imposed by the nature of legged motions and introduce additional constraints on the allowed solutions in the motion synthesis process.

The purpose of this chapter is to present a short review of some recent results derived by the authors for the modelling and locomotion control of biped robots. The first attempt was made by considering a 5-link model [14] which was then extended to a 9-link model [15,16]. Here the 9-link biped robot will be considered which includes the 5-link model as a special case. Section 2 deals with the walking pattern and the kinematic and dynamic biped-robot modelling which is a prerequisite for the application of any control method. Section 3 presents the sliding-mode robust control method, and Section 4 derives the adaptive locomotion controller. Section 5 provides a set of representative results obtained by these two types of control, and, finally, Section 6 gives some concluding and critisism remarks.

## 2. Biped Robot Modelling

The biped modelling involves three subproblems : (i) walking pattern modelling, (ii) kinematic modelling, and (iii) dynamic modelling.

### 2.1 WALKING PATTERN MODELLING

Here the basic properties which are required for a biped robot to emulate the 'most significant' gaits, will be presented. In the past thirty years an enormous amount of data has been acquired describing the kinematics, dynamics and energetics of walking. Most theoretical approaches have been limited to calculating from experimental kinematic data the variables of gait which are inaccessible to direct measurement, including the moments at the joints and the energy levels of each of the segments of the body. Very few analytical models of walking starting from first principles have been attempted, perhaps because of the complexity of the human muscle-skeletal system.

Among the analytical models, one hypothesis which has been suggested is that walking is performed in a way that requires the least expenditure of energy. Indeed, experimental findings suggest that human walking is a learned activity which aims to keep energy consumption as low as possible. To achieve this, the body has to take advantage of the external forces (gravity and ground reactions) to minimize the work that the muscles have to provide. This means, for example, that muscles will be turned off when gravity is acting in the direction of the desired motion. It is therefore, plausible to expect that whenever possible, the body or parts of it will behave like the coupled links in a compound pendulum. Simon Mochon [5], in particular, showed that the swing phase of human gait (the portion of the walking cycle in which only one of the legs is in contact with the ground and it starts at toe-off and ends at heel strike) can be described as a ballistic motion and he developed a series of mathematical models of increasing complexity ('ballistic walking models'), which can emulate the motion of the human walking in an optimal way, always based on the important property of the *least expenditure of energy* during walking. In other words, these mathematical models are developed on the basis of the assumption that no muscular moments are applied to any of the joints of the leg during the swing period. Thus the body moves throughout this period under the influence of gravity alone.

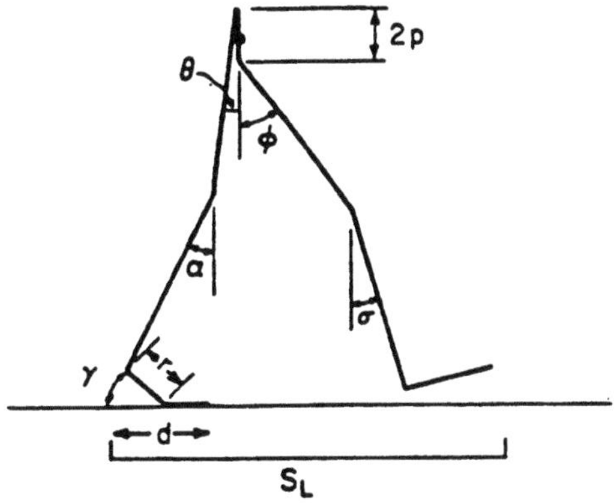


*Fig.1 A ballistic walking model*

In our 9-link biped robot model, it is assumed that at the middle of the supporting leg period, i.e. when the swing leg moves before the supporting leg, a new phase of the gait exists. This is the *kick phase*, where an ankle motion of the supporting leg is achieved, so that a maximum of the vertical force just before the collision appears. Furthermore, it is assumed that the torque, applied to the knee joint is zero, the desired trajectories of the angles $\theta$ and $\gamma$ are specified as a function of $\alpha$, and in addition the reference signals are chosen so as to use the effects of the gravity in a way that increases the angular momentum during the single leg supporting phase (see Fig.1). Therefore, trying to utilize the gravity effect skillfully, the following walking pattern is adopted : 1. The body is always kept upright. 2. The knee of the supporting leg extends straight, and, as a result, the first assumption is satisfied, since the relation between the thigh angle $\theta$ and the shank angle $\alpha$ is $\theta=\alpha$. 3. The ankle and foot joint of the supporting leg is free except for the kick-phase. 4. The foot of the swing leg is kept parallel to the ground. 5. The leg-support-exchange is done in an instant. 6. At the touchdown, the knee joint of the swing leg is kept in bending state. 7. The touchdown of the swing leg is assumed to occur in two stages. Firstly, the toes of the swing leg take a collision with the ground, and then the collision of the heel follows. 8. The same reference signals are supplied at each step repeatedly.

The reference signals shown in Fig.6 (thin line) describe the desired change of the angular position of the robot joints during the first two steps. It is obvious, that the reference signals during the first step are a little different, since the robot starts walking from the upright posture. The signals of the second step are recurrently used in every step.

2.2 KINEMATIC MODELLING

The 9-link biped under consideration is shown in Fig.2. It includes the *trunk* (link 5) and four links in each leg which represent the *thigh* (links 4 and 6), the *shin* (links 3 and 7), the *heel* (links 2 and 8) and the *metatarsal* (links 1 and 9). The links labeled $l_i$ (i=1,...,9) are joined together at ideal pin joints. Hence, it has two hip joints (joints 4 and 5), two knee joints (joints 3 and 6), two ankle joints (joints 2 and 7) and two metatarsal joints (joints 1 and 8), which are assumed to be ideal (without friction) rotational joints (with one d.o.f each of them) driven by independent electric DC motors. At each joint, except the one which contacts the ground, there is an ideal torque $\tau_i$. Since the biped motion is constrained to be on the sagittal plane, for a definite description, we use as generalized variables the set of the angles of each link i with the vertical, which are denoted as $\theta_i$. The direction of the $\theta_i$ is as shown in Fig.2.

There are four parameters for each link : the mass of the link $m_i$, its moment of inertia about the c.o.g. $I_i$, the length of the link $l_i$, and the distance between the c.o.g. and the lower joint $d_i$. Fig.3 shows these parameters for the i-th link. For the heel, the notation is somewhat different (Fig.4). The numerical values of these parameters are listed in Table1, and they are taken from [6].

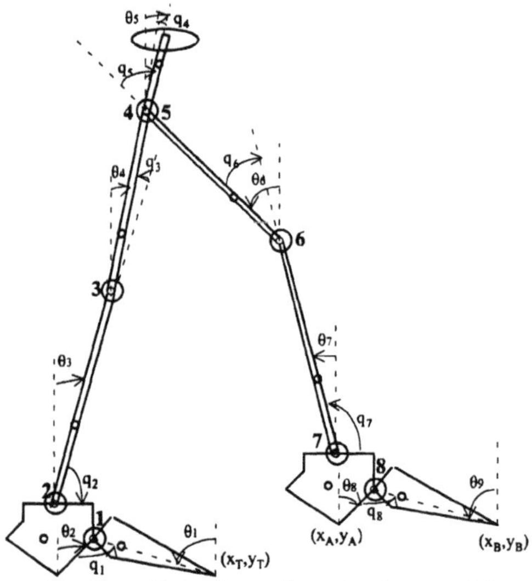
*Fig.2 9-link planar biped robot model*

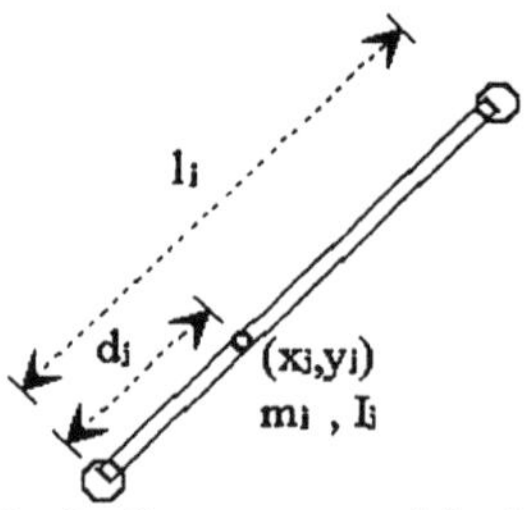
*Fig.3 Parameters of the i-th link*

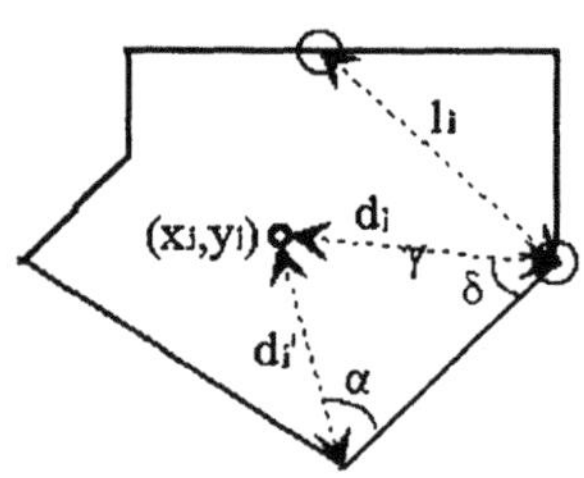
*Fig.4 Parameters of the heel*

*Table 1 Nine-Link Biped Model Parameters*
(Heel angles (rads) : $\alpha$=0.523, $\varepsilon$=2.4436, $\gamma$=1.396, $\delta$=0.872)

| Link | Inertia $I_i$ (kg·m) | Mass $m_i$ (kg) | Radius $d_i$ (m) | Length $L_i$ (m) |
|---|---|---|---|---|
| 1 | 1/6000 | 0.6 | 0.0828 | 0.1077 |
| 2 | 1/6000 | 0.6 | 0.05($d_2$) 0.056($d_2'$) | 0.05 |
| 3 | 0.105 | 4.550 | 0.267 | 0.502 |
| 4 | 0.089 | 7.630 | 0.247 | 0.431 |
| 5 | 2.350 | 49 | 0.280 | 0.827 |
| 6 | 0.089 | 7.630 | 0.247 | 0.431 |
| 7 | 0.105 | 4.550 | 0.267 | 0.502 |
| 8 | 1/6000 | 0.6 | 0.05($d_8$) 0.056($d_8'$) | 0.05 |
| 9 | 1/6000 | 0.6 | 0.0828 | 0.1077 |

The kinematic model which describes the relation between the velocity of the foot of the swing leg and the change of generalized variables, is described by the following equations (the details are given in [16]):

$$V_A=\begin{bmatrix}\dot{x}_A\\ \dot{y}_A\end{bmatrix}=\begin{bmatrix}-l_1\cos\theta_1\\ -l_1\sin\theta_1\end{bmatrix}\dot{\theta}_1+\begin{bmatrix}-l_2\cos(\theta_2+\gamma)\\ l_2\sin(\theta_2+\gamma)\end{bmatrix}\dot{\theta}_2+\begin{bmatrix}l_3\cos\theta_3\\ -l_3\sin\theta_3\end{bmatrix}\dot{\theta}_3+\begin{bmatrix}l_4\cos\theta_4\\ -l_4\sin\theta_4\end{bmatrix}\dot{\theta}_4$$
$$+\begin{bmatrix}l_6\cos\theta_6\\ l_6\sin\theta_6\end{bmatrix}\dot{\theta}_6+\begin{bmatrix}l_7\cos\theta_7\\ l_7\sin\theta_7\end{bmatrix}\dot{\theta}_7+\begin{bmatrix}l_8\cos(\theta_8+\gamma)-d_8\cos(\theta_8+\delta)-d_8'\cos(\theta_8-\alpha)\\ -l_8\sin(\theta_8+\gamma)-d_8\sin(\theta_8+\delta)+d_8'\sin(\theta_8-\alpha)\end{bmatrix}\dot{\theta}_8$$
$$V_B=\begin{bmatrix}\dot{x}_B\\ \dot{y}_B\end{bmatrix}=V_A^{1-7}+\begin{bmatrix}l_8\cos(\theta_8+\gamma)\\ -l_8\sin(\theta_8+\gamma)\end{bmatrix}\dot{\theta}_8+\begin{bmatrix}l_9\cos\theta_9\\ l_9\sin\theta_9\end{bmatrix}\dot{\theta}_9 \tag{1}$$

## 2.3 DYNAMIC MODELLING

**Non-kick action in single-leg-supporting phase**

Here, the dynamic equations are studied when the biped robot has one supporting leg and there is no raising of the heel. Based on the fact that the motion of the robot is constrained to be on the sagittal plane, the only generalized variables that are required for a full description of the system, are the angular shifts $\theta_i$ (i=1...9). A necessary assumption is that we consider the friction of the ground to be large enough for preventing slippage. Applying the Lagrange dynamic equation

$$\frac{d}{dt}\left\{\frac{\partial K}{\partial\dot{\theta}_i}\right\}-\frac{\partial K}{\partial\theta_i}+\frac{\partial U}{\partial\theta_i}=0$$

where K and U are the total kinetic and dynamic energy of the system respectively, the equations of motion take the following closed form, for the case of no-kick phase :

$$\mathbf{D}(\boldsymbol{\theta})\ddot{\boldsymbol{\theta}}+\mathbf{C}(\boldsymbol{\theta},\dot{\boldsymbol{\theta}})\dot{\boldsymbol{\theta}}+\mathbf{G}(\boldsymbol{\theta})=\mathbf{T}_{\boldsymbol{\theta}} \tag{2}$$

where $T_{\theta i}$ is the generalized torque which corresponds to the variable $\theta_i$, $\mathbf{D}(\boldsymbol{\theta})$ is the positive symmetric 9x9 inertia matrix, $\mathbf{C}(\boldsymbol{\theta},\dot{\boldsymbol{\theta}})$ is the 9x9 matrix (with zero diagonal terms) which includes terms from the centrifugal and Coriolis torques, and $\mathbf{G}(\boldsymbol{\theta})$ is the 9-dimensional vector which represents the gravitational torques. The exact form of these matrices can be found in [16].

Introducing in the model (2) the transformations

$$q_1 = \theta_1 + \theta_2 \qquad q_5 = \theta_5 + \theta_6$$

$$q_2 = -90^0 + \varepsilon + \theta_2 - \theta_3 \qquad q_6 = \theta_6 - \theta_7$$

$$q_3 = \theta_3 - \theta_4 \qquad q_7 = -90^0 + \varepsilon + \theta_8 + \theta_7$$

$$q_4 = \theta_4 - \theta_5 \qquad q_8 = \theta_8 + \theta_9$$

$$\tau_i = \sum_{j=1}^{9} T_{\theta j} \frac{\partial \theta_j}{\partial q_i} \qquad i = 1, \ldots, 8$$

where $q_i$ is the joint angular position, and $\tau_i$ is the real driving torque exerted by each independent actuator to each joint of the biped robot (the torque at the toes of the supporting leg is zero because of the existence of one unpowered d.o.f.), we get the following closed form of dynamic equations :

$$\mathbf{D(q)\ddot{q} + C(q, \dot{q})\dot{q} + G(q)} = \boldsymbol{\tau} \qquad (3)$$

which will be actually used for the control of the robot.

**Kick action in single-leg-supporting phase**

As mentioned before, since our biped robot has a 2-link foot we can adopt the biped locomotion with kick-action (only in the single leg support phase) which was firstly employed in Kenkyaku-2 [10]. A very good example of natural and dynamic biped locomotion is human walking. In fact, human walking utilizes the gravity effect very skillfully and does not depend on the ankle torque of the supporting leg so much. For example, natural stop-motion is achieved by reducing its angular momentum using the gravity effect. However, as seen from the shape of human being foot, the ankle torque of the supporting leg can decrease the walking speed but cannot increase it. Since the reduction of the speed causes an energy loss, and according to the first assumption that the biped robot has to keep the properties of a ballistic model (to keep its energy at a constant level), in the continuous walking of this study the ankle joint of the supporting leg is set to be free except for the kick phase, when raising of the heel exists. In this case, an additional variation of the angle $\theta_2$ occurs, while the angle $\theta_1$ keeps on a constant value.

Based again on the Lagrange dynamic model, the equations of motion, during the kick phase, take the following closed form (similar to (2)) :

$$\mathbf{D}(\boldsymbol{\theta})\ddot{\boldsymbol{\theta}} + \mathbf{C}(\boldsymbol{\theta}, \dot{\boldsymbol{\theta}})\dot{\boldsymbol{\theta}} + \mathbf{G}(\boldsymbol{\theta}) = \mathbf{T}_{\boldsymbol{\theta}}$$

where the matrices **D**, **C**, **G** are similar to the corresponding matrices of the previous phase, with some additional terms caused by the raising of the heel (see [16]).

**Leg-support-exchange**

The walking pattern adopted in this study implies that the leg-support-exchange is done in an instant. In this way, the double-leg-supporting phase is omitted. Then, just after the touchdown of the swing leg, the exchange of the supporting leg occurs. Hence, we assume that the biped robot is instantly, just before the collision, on the air. As a result, at the time of the swing leg collision with the ground, the constraint $x_T = y_T =$constant, which exists during the single leg supporting phase, is lost. In this case, given the fact that the motion of the biped robot is constrained on the sagittal plane, unless $\theta_i$

(i=1,...,9), two more variables (the coordinates $x_T$, $y_T$ of the supporting leg toes) are required for an exact description of the position of the biped robot. The elements of the inertial matrix $\mathbf{D_a}$ for this case can again be found in [16].

**Collision of the swing leg with the ground**

For a mobile robot, the collision with the environment is an ordinary affair. The collision of a robotic system with the environment may be handled as an impact problem. There are two aspects related to the impact problem. The first is how the collision changes the generalized states of the system, which are externally observable. The second is how the collision brings impulsive forces to a robotic system. These forces occur internally and are not observable. One of the effects of robot collision with the environment is the abrupt change of the joint angular velocities. Hence, the following mathematical model is derived to establish the quantitative relationship between this abrupt change and the severity of the collision. The latter is represented by either an external impulsive force or by the instantaneous change of the linear velocity of the contact point :

$$\Delta\dot{\mathbf{q}} = \mathbf{D}^{-1}(\mathbf{q})\mathbf{J}^{T}\left(\mathbf{J}\mathbf{D}^{-1}(\mathbf{q})\mathbf{J}^{T}\right)^{-1}\Delta\dot{\mathbf{x}} \tag{4}$$

This quantitative relation between the velocity changes and the impact has been established such that the joint angular velocities immediately after the collision can be computed. These quantities along with the joint positions, which remain unchanged during the short duration of the collision, constitute initial values of the system states for analysis of the collision process and control of the robot after the collision [17].

Here, after the instant that the 9-link biped robot is on the air, the touchdown of the foot of the swing-leg follows, and, as a result, the exchange of the supporting leg occurs. As mentioned in the walking pattern, the collision with the ground occurs in two stages. Firstly, the toes (B) of the swing leg take a collision with the ground and then the collision of the heel (A) follows. According to the above relation we have

$$\Delta\dot{\boldsymbol{\theta}} = \mathbf{D_a}^{-1}\,{}^{B}\mathbf{J_a}^{T}\left({}^{B}\mathbf{J_a}\,\mathbf{D_a}^{-1}\,{}^{B}\mathbf{J_a}^{T}\right)^{-1}\Delta\dot{\mathbf{x}}_{\mathbf{B}} \tag{5}$$

where $\mathbf{D_a}$ is the inertia matrix of the robot model when it is instantly on the air, and ${}^{B}\mathbf{J_a} = \left[\partial\mathbf{x_B}/\partial\boldsymbol{\theta}\right]$ is the Jacobian matrix with elements :

$${}^{B}J_a(1,1) = -l_1\cos\theta_1 \qquad {}^{B}J_a(2,1) = -l_1\sin\theta_1$$

$${}^{B}J_a(1,2) = -l_2\cos(\theta_2+\gamma) \qquad {}^{B}J_a(2,2) = l_2\sin(\theta_2+\gamma)$$

$${}^{B}J_a(1,3) = l_3\cos\theta_3 \qquad {}^{B}J_a(2,3) = -l_3\sin\theta_3$$

$${}^{B}J_a(1,4) = l_4\cos\theta_4,\ {}^{B}J_a(1,5) = 0 \qquad {}^{B}J_a(2,4) = -l_4\sin\theta_4,\ {}^{B}J_a(2,5) = 0$$

$${}^{B}J_a(1,6) = l_6\cos\theta_6 \qquad {}^{B}J_a(2,6) = l_6\sin\theta_6$$

$${}^{B}J_a(1,7) = l_7\cos\theta_7 \qquad {}^{B}J_a(2,7) = l_7\sin\theta_7$$

$${}^{B}J_a(1,8) = l_8\cos(\theta_8+\gamma) \qquad {}^{B}J_a(2,8) = -l_8\sin(\theta_8+\gamma)$$

$${}^{B}J_a(1,9) = l_9\cos\theta_9 \qquad {}^{B}J_a(2,9) = l_9\sin\theta_9$$

$${}^{B}J_a(1,10) = 1,\ {}^{B}J_a(1,11) = 0 \qquad {}^{B}J_a(2,10) = 0,\ {}^{B}J_a(2,11) = 1$$

After the first collision, the velocity of the toes (B) of the swing leg vanishes, hence

$$\underset{\text{after}}{\dot{\boldsymbol{\theta}}_{\mathbf{B}}} = \underset{\text{before}}{\dot{\boldsymbol{\theta}}_{\mathbf{B}}} + \mathbf{D}_{\mathbf{a}}^{-1}\,{}^{\mathbf{B}}\mathbf{J}_{\mathbf{a}}^{\mathrm{T}}\left({}^{\mathbf{B}}\mathbf{J}_{\mathbf{a}}\,\mathbf{D}_{\mathbf{a}}^{-1}\,{}^{\mathbf{B}}\mathbf{J}_{\mathbf{a}}^{\mathrm{T}}\right)^{-1}(-\underset{\text{before}}{\dot{\mathbf{x}}_{\mathbf{B}}}) \tag{6}$$

Then, the collision of the heel (A) of the swing leg occurs. Hence, the general relation takes the form

$$\Delta\dot{\boldsymbol{\theta}} = \mathbf{D}_{\mathbf{a}}^{-1}\,{}^{\mathbf{A}}\mathbf{J}_{\mathbf{a}}^{\mathrm{T}}\left({}^{\mathbf{A}}\mathbf{J}_{\mathbf{a}}\,\mathbf{D}_{\mathbf{a}}^{-1}\,{}^{\mathbf{A}}\mathbf{J}_{\mathbf{a}}^{\mathrm{T}}\right)^{-1}\Delta\dot{\mathbf{x}}_{\mathbf{A}} \tag{7}$$

where ${}^{\mathbf{A}}\mathbf{J}_{\mathbf{a}} = [\partial\mathbf{x}_{\mathrm{A}}/\partial\boldsymbol{\theta}]$ has the elements

$${}^{\mathrm{A}}\mathrm{J}_{\mathrm{a}}(1,\mathrm{i}) = {}^{\mathrm{B}}\mathrm{J}_{\mathrm{a}}(1,\mathrm{i}) \quad \mathrm{i}=1,\dots,7 \qquad {}^{\mathrm{A}}\mathrm{J}_{\mathrm{a}}(2,\mathrm{i}) = {}^{\mathrm{B}}\mathrm{J}_{\mathrm{a}}(2,\mathrm{i}) \quad \mathrm{i}=1,\dots,7$$

$${}^{\mathrm{A}}\mathrm{J}_{\mathrm{a}}(1,8) = l_8\cos(\theta_8+\gamma) - d_8\cos(\theta_8+\delta) - d_8'\cos(\theta_8-\alpha) \qquad {}^{\mathrm{A}}\mathrm{J}_{\mathrm{a}}(2,8) = -l_8\sin(\theta_8+\gamma) + d_8\sin(\theta_8+\delta) + d_8'\sin(\theta_8-\alpha)$$

$${}^{\mathrm{A}}\mathrm{J}_{\mathrm{a}}(1,9) = 0,\ {}^{\mathrm{A}}\mathrm{J}_{\mathrm{a}}(1,10) = 1,\ {}^{\mathrm{A}}\mathrm{J}_{\mathrm{a}}(1,11) = 0 \qquad {}^{\mathrm{A}}\mathrm{J}_{\mathrm{a}}(2,9) = 0,\ {}^{\mathrm{A}}\mathrm{J}_{\mathrm{a}}(2,10) = 0,\ {}^{\mathrm{A}}\mathrm{J}_{\mathrm{a}}(2,11) = 1$$

Similar to the first case, after the second collision, the velocity of the heel (A) goes to zero. Hence

$$\underset{\text{after}}{\dot{\boldsymbol{\theta}}_{\mathbf{A}}} = \underset{\text{before}}{\dot{\boldsymbol{\theta}}_{\mathbf{A}}} + \mathbf{D}_{\mathbf{a}}^{-1}\,{}^{\mathbf{A}}\mathbf{J}_{\mathbf{a}}^{\mathrm{T}}\left({}^{\mathbf{A}}\mathbf{J}_{\mathbf{a}}\,\mathbf{D}_{\mathbf{a}}^{-1}\,{}^{\mathbf{A}}\mathbf{J}_{\mathbf{a}}^{\mathrm{T}}\right)^{-1}(-\underset{\text{before}}{\dot{\mathbf{x}}_{\mathbf{A}}}) \tag{8}$$

where $\underset{\text{before}}{\dot{\boldsymbol{\theta}}_{\mathbf{A}}}$ is equal to $\underset{\text{after}}{\dot{\boldsymbol{\theta}}_{\mathbf{B}}}$ which is computed from the first collision.

From this relation, one can compute the angular velocities of the joints after the collision and the exchange of the supporting leg. As it is well known, each joint of a robot is equipped with an angular velocity sensor. An instantaneous change of the sensor output will indicate the occurrence of a collision. To recover the motion from a collision, a special effort should then be made by the controller. The angular velocities of each joint immediately after the impact are considered as initial states for the new control strategy which either resumes the planned path as soon as possible or completely stops the motion of a robot. The amount of change is directly proportional to the magnitude of the impulsive force. The latter may also cause the deformation of a robotic system with a severe collision. One should then establish a threshold of $\Delta\dot{\boldsymbol{\theta}}$. If $\Delta\dot{\boldsymbol{\theta}}$ is greater than the threshold, an undesired severe collision has happened. On the other hand, as long as $\Delta\dot{\boldsymbol{\theta}}$ is under the threshold, the collision is acceptable.

## 3. Sliding-Mode Biped Locomotion Control

The locomotion activity and gait in particular belongs to the class of highly automated motions. When a man is walking in a steady regime or in an environment imposing small disturbances, the central nervous system is not involved. When large disturbances occur, the system actions are directed only to the preservation of the system overall stability, i.e., towards preventing the system from falling down. This requirement is of primary importance in biped locomotion. Hierarchy is a basic issue on which the control of large scale systems is in general based. This also holds for biped robots. The hierarchical structure of the control system is most often vertical, so that each control level deals with some wider aspects of the overall system behavior than the lower level. A higher control level always refers to the lower ones, and controls those system parameters that vary more slowly. A higher level communicates with a lower level, giving it instructions and receiving from it relevant information required for the decision-making. After obtaining the information from a lower level, each hierarchical level makes decisions taking into account general instructions obtained from a higher level and forwards them to the lower level for execution.

To solve the stability problem, the control synthesis may be performed in two stages. At the first stage, called the *stage of nominal regimes*, the control has to be designed so as to ensure the system's motion along the exact nominal trajectories calculated in advance in the absence of any disturbance. It has to satisfy the conditions of both the desired gait type and overall system equilibrium. The nominal trajectories ensure that the system is in equilibrium under ideal conditions, i.e. when the model of the system is perfect, the trajectories are perfectly realized, and no perturbations are acting upon the system. Since these conditions are never fulfilled, additional control is necessary which will ensure tracking of these nominal trajectories and preserve the system equilibrium when disturbances are present. Hence at the second stage, called the *stage of perturbed regimes*, only the deviation of the actual state vector from its nominal value is considered and additional control is applied to force the system state to its nominal. Therefore, various approximate methods have been proposed. The most frequently technique used in the control of biped robots is the simple decentralized controller, which assumes that each joint is controlled independently from the rest of the system. However, a decentralized controller cannot stabilize the entire system because of the strong dynamic coupling between the joints, and the necessity to indirectly control unpowered d.o.f. and to preserve the system equilibrium in the perturbed regimes. Therefore additional feedback loops have to be applied. In this paper, we consider in detail the possibility of implementing global control by on-line computation of the dynamic forces [2]. This means : (i) that the control computer has to compute dynamic moments around the biped joints at every 1-2 ms in order to achieve sampling rate compatible with the system dynamics (in particular, if the algorithms are programmed in *assembly*, the computation time is of the order of 3-4 msec, but using suitable fast inverse dynamics algorithms or parallelizing the computations, this figure can go down to less than 1-2 msec - another improvement can be obtained if all the trigonometric functions are prestored and called from a ROM memory), (ii) because the tasks of a biped robot involve large range and / or high speed motions, nonlinear control may be necessary to achieve the desired performance. Therefore, the computed torque control law has been proposed, which is a special type of feedback linearization for rigid robots. This nonlinear feedback technique transforms the highly coupled, nonlinear robot dynamics into equivalent, decoupled linear systems (one for each d.o.f.) and then the well known and powerful linear design techniques can be used to complete the control design. These results were based on the assumption that an accurate model of the biped robot is known, a condition that is never fulfilled because of the existence of structured uncertainties which correspond to inaccuracies in the model parameters (e.g. imprecision on the mass properties or the loads or the robot geometry, inaccuracies on the torque constants of the actuators). When the trajectory control problem is addressed without any a priori knowledge of the system's properties, such parameter uncertainty can be gradually reduced on-line by an adaptation or estimation mechanism [2].

In this section, the dynamic model of the 9-link biped robot is assumed to be not exactly known and the estimation error on unknown (and probably time-varying) parameters is assumed to be bounded by some known function. Hence, a sliding mode controller is designed which provides a systematic approach to the problem of maintaining stability and consistent performance in the face of modeling imprecisions and, at the same time, succeeds to quantify the resulting modeling/performance trade offs (in particular, the effect on tracking performance of discarding any particular term in the dynamic model).

The dynamic model of the biped robot with n d.o.f. has the general form (see (3)) :

$$\mathbf{D}(\mathbf{q})\ddot{\mathbf{q}} + \mathbf{C}(\mathbf{q},\dot{\mathbf{q}})\dot{\mathbf{q}} + \mathbf{G}(\mathbf{q}) = \boldsymbol{\tau} \tag{9}$$

By a closer look at the physics of the biped robot model, one can see that the matrix $(\dot{\mathbf{D}} - 2\mathbf{C})$ is skew-symmetric. Furthermore note that since $\mathbf{D}$, and therefore $\dot{\mathbf{D}}$, are symmetric matrices, the skew-symmetry of the matrix $(\dot{\mathbf{D}} - 2\mathbf{C})$ can also be written as $\dot{\mathbf{D}} = \mathbf{C} + \mathbf{C}^T$. Now, define a time-varying surface S(t) in the state-space $\Re^n$ by the vector equation **s(q;t)=0**, with

$$s_i = \left(\frac{d}{dt} + \lambda_i\right)\tilde{q}_i \tag{10}$$

where $\tilde{q}_i = q_i - q_{d_i}$ is the tracking error at the i-th joint, and $\lambda_i$ is a strictly positive constant which symbolize the control bandwidth. Hence, the vector **s** is defined as,

$$\mathbf{s} = \dot{\tilde{\mathbf{q}}} + \boldsymbol{\Lambda}\tilde{\mathbf{q}} \tag{11}$$

where $\boldsymbol{\Lambda}$ is a symmetric positive definite matrix, or more generally a matrix such that $\boldsymbol{\Lambda}$ is Hurwitz.

Furthermore, we may interpret **s** as a 'velocity error' term

$$\mathbf{s} = \dot{\mathbf{q}} - \dot{\mathbf{q}}_r \tag{12}$$

where $\dot{\mathbf{q}}_r = \dot{\mathbf{q}}_d - \boldsymbol{\Lambda}\tilde{\mathbf{q}}$. The vector **s** conveys information about the boundedness and convergence of $\mathbf{q}$ and $\dot{\mathbf{q}}$, since equation (11) of **s** can also be viewed as a stable first-order differential equation in $\tilde{\mathbf{q}}$, with **s** as an input. *Hence, given initial conditions* $\mathbf{q}_d(t=0) = \mathbf{q}(t=0)$, *the problem of tracking* $\mathbf{q} \equiv \mathbf{q}_d$ *is equivalent to that of remaining on the surface S(t) for all t>0.* Indeed, $\mathbf{s} \equiv \mathbf{0}$ represents a set of linear differential equations whose unique solution is $\tilde{\mathbf{q}} \equiv \mathbf{0}$, given the previous initial condition. The simplified, first-order problem of keeping the scalar $s_i$ at zero can now be achieved by choosing the control law such that outside of S(t) the following condition holds :

$$\frac{1}{2}\frac{d}{dt}s_i^2 \leq -\eta_i |s_i| \tag{13}$$

where $\eta_i$ is a strictly positive constant. Essentially, (13) states that the squared 'distance' to the surface, as measured by $\mathbf{s}^2$, decreases along all system trajectories. Thus, it constraints trajectories to point towards the surface S(t).

To address the robust trajectory control problem, let us define the following Lyapunov function :

$$V(t) = (1/2)[\mathbf{s}^T \ \mathbf{D} \ \mathbf{s}] \qquad (14)$$

Differentiating V(t) yields $\dot{V}(t) = \mathbf{s}^T(\mathbf{D}\ddot{\mathbf{q}} - \mathbf{D}\ddot{\mathbf{q}}_r) + (1/2)\mathbf{s}^T\dot{\mathbf{D}}\mathbf{s}$. Then, substituting $\mathbf{D}\ddot{\mathbf{q}}$ from the system dynamics: $\mathbf{D}\ddot{\mathbf{q}} = \boldsymbol{\tau} - \mathbf{C}\dot{\mathbf{q}} - \mathbf{G} = \boldsymbol{\tau} - \mathbf{C}(\mathbf{s} + \dot{\mathbf{q}}_r) - \mathbf{G}$ one obtains: $V(t) = \mathbf{s}^T(\boldsymbol{\tau} - \mathbf{D}\ddot{\mathbf{q}}_r - \mathbf{C}\dot{\mathbf{q}}_r - \mathbf{G})$, where the skew-symmetry of $(\dot{\mathbf{D}} - 2\mathbf{C})$ has been used to eliminate the term $(1/2)\mathbf{s}^T\dot{\mathbf{D}}\mathbf{s}$. Now, define the control input to have the form

$$\boldsymbol{\tau} = \hat{\boldsymbol{\tau}} - \mathbf{K}\,\mathrm{sgn}(\mathbf{s}) \qquad (15)$$

where $\mathbf{K}\mathrm{sgn}(\mathbf{s})$ is defined as the vector with components $K_i\,\mathrm{sgn}(s_i)$, and $\hat{\boldsymbol{\tau}}$ is the control input vector which would make $\dot{V}$ equal 0 if the dynamics were exactly known ($\hat{\mathbf{D}} = \mathbf{D}$, $\hat{\mathbf{C}} = \mathbf{C}$, $\hat{\mathbf{G}} = \mathbf{G}$) : $\hat{\boldsymbol{\tau}} = \hat{\mathbf{D}}\ddot{\mathbf{q}}_r + \hat{\mathbf{C}}\dot{\mathbf{q}}_r + \hat{\mathbf{G}}$. Then, one obtains

$$\dot{V} = \mathbf{s}^T\left[\tilde{\mathbf{D}}(\mathbf{q})\ddot{\mathbf{q}}_r + \tilde{\mathbf{C}}(\mathbf{q},\dot{\mathbf{q}})\dot{\mathbf{q}}_r + \tilde{\mathbf{G}}(\mathbf{q})\right] - \sum_{i=1}^{n} K_i |s_i|$$

Because of the parametric uncertainty, the only available *a priori* knowledge are the bounds of the estimation errors $\tilde{\mathbf{D}}$, $\tilde{\mathbf{C}}$, and $\tilde{\mathbf{G}}$, where $\tilde{\mathbf{D}} = \hat{\mathbf{D}} - \mathbf{D}$, $\tilde{\mathbf{C}} = \hat{\mathbf{C}} - \mathbf{C}$, $\tilde{\mathbf{G}} = \hat{\mathbf{G}} - \mathbf{G}$. From the above, we can see that choosing the components $K_i$ of the vector $\mathbf{K}$ such that

$$K_i \geq \left|[\tilde{\mathbf{D}}(\mathbf{q})\ddot{\mathbf{q}}_r + \tilde{\mathbf{C}}(\mathbf{q},\dot{\mathbf{q}})\dot{\mathbf{q}}_r + \tilde{\mathbf{G}}(\mathbf{q})]_i\right| + \eta_i \qquad (16)$$

where the constants $\eta_i$ are strictly positive, allows one to satisfy the sliding condition

$$\dot{V} \leq -\sum_{i=1}^{n} \eta_i |s_i| \qquad (17)$$

The above sliding condition guarantees that the surface $\mathbf{s}\equiv\mathbf{0}$ is reached in a finite time, and that once on the surface, the trajectories remain on the surface, and therefore tend to $\mathbf{q}_d(t)$ exponentially. However, as one can see from (16), in order to account for the presence of modeling imprecision, the control law has to be discontinuous across S(t). Since the implementation of the associated control switchings is necessarily imperfect, this leads to chattering which is undesirable in practice. To avoid this, $\boldsymbol{\tau}$ is suitably smoothened in a thin boundary layer neighboring the switching surface

$$B(t) = \left\{\mathbf{q},\ |s_i(\mathbf{q},t)| \leq \Phi_i\right\} \quad \boldsymbol{\Phi}>0$$

where $\boldsymbol{\Phi}$ is the boundary layer thickness. In other words, outside B(t) the control law is chosen as before, which guarantees that the boundary layer is attractive and hence invariant (all trajectories starting inside B(t=0) remain inside B(t) for all $t \geq 0$), and then inside B(t) $\boldsymbol{\tau}$ is interpolated, replacing the term $\mathrm{sgn}(s_i)$ by $s_i/\Phi_i$. Thus, instead of simply requiring (17) to be satisfied outside the boundary layer, we now require

$$|\mathbf{s}| \geq \boldsymbol{\Phi} \Rightarrow (1/2)(d/dt)[\mathbf{s}^T\mathbf{D}\mathbf{s}] \leq (\boldsymbol{\Phi} - \boldsymbol{\eta})|\mathbf{s}| \qquad (18)$$

In order to satisfy (18), the quantity $-\Phi_i$ is added to the control discontinuity gain $K_i$. Accordingly, the control law $\boldsymbol{\tau}$ becomes :

$$\boldsymbol{\tau} = \hat{\boldsymbol{\tau}} - \overline{\mathbf{K}}\mathrm{sat}(\mathbf{s}/\boldsymbol{\Phi}) \qquad (19)$$

where $\overline{\mathbf{K}} = \mathbf{K} - \mathbf{\Phi}$.

The effect of control interpolation in a boundary layer can be carried on further, and guide the selection of the design parameters $\lambda_i$ and $\Phi_i$. Hence, the ranges $\Phi_i$ of the interpolation could be chosen so that the maximum value of each $\overline{K}_i/\Phi_i$ is approximately equal to the control bandwidth $\lambda_i$.

For our 9-link biped robot, we have assumed a maximum uncertainty of 45% on the mass properties (mass $m_i$, rotational inertia $I_i$) of the biped robot, and 10%, 20% on the link length $l_i$ and on the radius $d_i$ of the c.o.g. from the lower joint, respectively, i.e.:

$$|m_i - \hat{m}_i| \leq 0.45\hat{m}_i \quad |I_i - \hat{I}_i| \leq 0.45\hat{I}_i \quad |l_i - \hat{l}_i| \leq 0.1\hat{l}_i \quad |d_i - \hat{d}_i| \leq 0.2\hat{d}_i$$

Based on the estimations $\hat{m}_i, \hat{I}_i, \hat{l}_i, \hat{d}_i$ of the parameters, and on the limits em, eI, el, ed of the uncertainty, one can find the bounds of the modeling errors :

$$\begin{aligned}\tilde{D}_{11} \leq & [(1+eI)\hat{I}_1 - \hat{I}_1] + [(1+em)\hat{m}_1(1+ed)^2\hat{d}_1 - \hat{m}_1\hat{d}_1^{\,2}] \\ & +[(1+em)(\hat{m}_3+...+\hat{m}_9)(1+el)^2\hat{l}_1^{\,2} - (\hat{m}_3+...+\hat{m}_9)\hat{l}_1^{\,2}] \\ & = eI\,\hat{I}_1 + \hat{m}_1\hat{d}_1^{\,2}[(1+em)(1+ed)^2-1] + (\hat{m}_3+...+\hat{m}_9)\hat{l}_1^{\,2}[(1+em)(1+el)^2-1]\end{aligned}$$

The other terms can be computed in the same way.

A first key question is to determine how large $\mathbf{\Lambda}$ can be chosen. Although the tuning of this single scalar may in practice be done experimentally, considerable insight on the overall design can be obtained by explicitly analyzing the various factors limiting $\lambda_i$. In mechanical systems, for instance, given clean measurements, $\lambda_i$ is typically limited by three factors : (i) structural resonant modes, (ii) neglected time delays, (iii) sampling rate. In our problem, there is no a priori information about the first two factors, but only about the third one. Hence, through a choice of 0.5 KHz (T=2ms) for the sampling rate, we achieve a maximum desired control bandwidth about 100rad/sec (since $\lambda \geq (1/5)\,\nu_{\text{sampling}}$) [19]. This is a simple limit for $\lambda_i$, since the ideal values are computed through trial and error in simulations. The parameter $\eta_i$ must be selected sufficiently large in order to achieve fast trajectory tracking outside the boundary layer, but at the same time small enough (compared to the average value of $\overline{\mathbf{K}}$) so as to fully exploit the available knowledge on the structure of the parametric uncertainty. Finally, the bound of $\Phi_i$ must be selected small enough to ensure that the tracking errors $\dot{\tilde{q}}_i$ and $\tilde{q}_i$ for each joint take small values, and at the same time sufficiently large to avoid chattering outside the boundary layer.

## 4. Adaptive Biped Locomotion Control

In this section, we apply a robust-adaptive control scheme to the case of the 9-link biped robot which aims at minimizing the sensitivity of the system performance under the presence of large and rapid time-variations in the robot parameters. This control scheme ensures that all signals of the robot system are bounded and that the mean tracking error is of the order of the parameter variations which are not required to be small [18].

The dynamic equations of a biped robot whose parameters may explicitly depend on time, have the following form :

$$\mathbf{D}(\mathbf{q},t)\ddot{\mathbf{q}}+\frac{\partial \mathbf{D}(\mathbf{q},t)}{\partial t}\dot{\mathbf{q}}+\mathbf{C}(\mathbf{q},\dot{\mathbf{q}},t)\dot{\mathbf{q}}+\mathbf{F}(\mathbf{q},t)=\boldsymbol{\tau} \tag{20}$$

where $\mathbf{F}(\mathbf{q},t)=\frac{\partial P(\mathbf{q},t)}{\partial t}$, and $P(\mathbf{q},t)$ is the potential energy of the system.

The above dynamic equations are easily derived by using the Euler-Lagrange equations:

$$\frac{d}{dt}\frac{\partial L}{\partial \dot{\mathbf{q}}}-\frac{\partial L}{\partial \mathbf{q}}=\boldsymbol{\tau}$$

where L is the Lagrangian defined as L=K-P, the difference between the kinetic energy K and the potential energy P of the form :

$$K=\frac{1}{2}\dot{\mathbf{q}}^T\mathbf{D}(\mathbf{q},t)\dot{\mathbf{q}} \qquad P=P(\mathbf{q},t)$$

A key feature of the biped robot (20) is that the inertia matrix $\mathbf{D}(\mathbf{q},t)$ and the potential energy $P(\mathbf{q},t)$ are explicitly time-dependent, which takes into account the effect of changes in environment of the robot system or changes of the robot dynamics with time.

Moreover, the important skew-symmetry property of the matrix $(\dot{\mathbf{D}}-2\mathbf{C})$, because of parameters time-variations, is written in the form :

$$\dot{\mathbf{q}}^T\left[\left(\frac{\partial D_{ij}}{\partial \mathbf{q}}\right)^T\dot{\mathbf{q}}-2\mathbf{C}(\mathbf{q},\dot{\mathbf{q}},t)\right]\dot{\mathbf{q}}=0 \qquad \forall \mathbf{q}\in\Re^n \tag{21}$$

where $D_{ij}$ is the ij-th element of $\mathbf{D}(\mathbf{q},t)$, and

$$\frac{d}{dt}\mathbf{D}(\mathbf{q},t)=\frac{\partial \mathbf{D}(\mathbf{q},t)}{\partial t}+\frac{\partial D_{ij}}{\partial \mathbf{q}}\dot{\mathbf{q}} \tag{22}$$

since the mass and the moment of inertia of each link ( $D_{ii}$ ) should not be explicitly dependent on $\mathbf{q}$.

The control objective is : *For a given reference signal* $\mathbf{q}_d(t)$, *generate the applied torque* $\boldsymbol{\tau}$ *for the biped robot (20) with unknown and time-varying parameters such that all signals in the robot system are bounded and the joint position* $\mathbf{q}$ *tracks* $\mathbf{q}_d$ *as close as possible.*

To achieve such an objective we first use the transformation technique developed in the adaptive control scheme for a biped robot to parameterize the biped robot model (20). Quite similarly, we express the biped model as

$$\mathbf{D}(\mathbf{q},t)\dot{\mathbf{s}}+\mathbf{C}(\mathbf{q},\dot{\mathbf{q}},t)\mathbf{s}=\boldsymbol{\tau}-\mathbf{D}(\mathbf{q},t)\ddot{\mathbf{q}}_r-\mathbf{C}(\mathbf{q},\dot{\mathbf{q}},t)\dot{\mathbf{q}}_r-\mathbf{F}(\mathbf{q},t)-\frac{\partial \mathbf{D}(\mathbf{q},t)}{\partial t}\dot{\mathbf{q}}$$

$$=\boldsymbol{\tau}-\mathbf{D}_K\ddot{\mathbf{q}}_r-\mathbf{C}_K\dot{\mathbf{q}}_r-\mathbf{F}_K-\mathbf{Y}(\mathbf{q},\dot{\mathbf{q}},\dot{\mathbf{q}}_r,\ddot{\mathbf{q}}_r,t)\ \mathbf{a}(t)-\frac{\partial \mathbf{D}(\mathbf{q},t)}{\partial t}\dot{\mathbf{q}} \tag{23}$$

where $(-\mathbf{D}_K\ddot{\mathbf{q}}_r-\mathbf{C}_K\dot{\mathbf{q}}_r-\mathbf{F}_K)$ is the term of the model with known parameters, $\mathbf{Y}(\mathbf{q},\dot{\mathbf{q}},\dot{\mathbf{q}}_r,\ddot{\mathbf{q}}_r,t)$ is a nxr matrix of known functions for some r>0, and $\mathbf{a}(t)\in\Re^r$

contains parameters which may be time-varying. In (23) the regressor $\mathbf{Y}(\mathbf{q},\dot{\mathbf{q}},\dot{\mathbf{q}}_r,\ddot{\mathbf{q}}_r,t)$ is bounded for bounded $\mathbf{q}, \dot{\mathbf{q}}, \dot{\mathbf{q}}_r$ and $\ddot{\mathbf{q}}_r$.

The following assumptions are made about the biped robot model (20) :

(i) $\|\mathbf{a}(t)\| \le \rho_0$ , $\|\dot{\mathbf{a}}(t)\| \le \rho$ for some constants $\rho_0 > 0$, $\rho > 0$. This assumption requires only the boundedness of the robot parameters and their derivatives, not the smallness of the time-variations of the parameters.

(ii) $\left\|\frac{\partial \mathbf{D}(\mathbf{q},t)}{\partial t}\right\| \le \gamma\, f(\mathbf{q})$ for some constant $\gamma>0$ and known bounded $f(\mathbf{q})$ for bounded $\mathbf{q}$.

(iii) $\mathbf{q}_d(t)$, $\dot{\mathbf{q}}_d(t)$, $\ddot{\mathbf{q}}_d(t)$ are bounded.

The new characteristics of this biped robot model are : (i) the presence of the unknown and time-varying vector $\mathbf{a}(t)$ (which is going to be estimated), and (ii) at the same time the existence of the unknown term $\frac{\partial \mathbf{D}(\mathbf{q},t)}{\partial t}\dot{\mathbf{q}}$. If the parameters of the terms $\mathbf{D}(\mathbf{q},t)$ and $P(\mathbf{q},t)$ were known, then $\mathbf{a}(t)$ and $\frac{\partial \mathbf{D}(\mathbf{q},t)}{\partial t}\dot{\mathbf{q}}$ could be calculated so that the control law

$$\boldsymbol{\tau}(t) = \mathbf{Y}(\mathbf{q},\dot{\mathbf{q}},\dot{\mathbf{q}}_r,\ddot{\mathbf{q}}_r,t)\ \mathbf{a}(t) + \frac{\partial \mathbf{D}(\mathbf{q},t)}{\partial t}\dot{\mathbf{q}} - \mathbf{K}_D\ \mathbf{s} \tag{24}$$

could be implemented (like the computed torque methodology), which guarantees global stability and asymptotic tracking. However, for unknown $\mathbf{D}(\mathbf{q},t)$, $P(\mathbf{q},t)$ we have to develop an adaptive control scheme (control law - update law) which is robust with respect to the time-variation of $\mathbf{a}(t)$ and $\frac{\partial \mathbf{D}(\mathbf{q},t)}{\partial t}\dot{\mathbf{q}}$.

A first feedback controller suggested by (24) could include a 'feedforward' term $\mathbf{D}_K\ddot{\mathbf{q}}_r + \mathbf{C}_K\dot{\mathbf{q}}_r + \mathbf{Y}\hat{\mathbf{a}}$ , a simple PD term $\mathbf{K}_D\ \mathbf{s}$ , and a term which causes robustness with respect to the time-variations of $\frac{\partial \mathbf{D}(\mathbf{q},t)}{\partial t}\dot{\mathbf{q}}$.

A first type of the parameter update law could include a term $\dot{\hat{\mathbf{a}}} = -\boldsymbol{\Gamma}^{-1}\mathbf{Y}^T\mathbf{s}$, and at the same time a term $-\boldsymbol{\Gamma}^{-1}\sigma(t)\hat{\mathbf{a}}(t)$ suggested by the σ-modification technique which takes care of the robustness to the time-variations of $\mathbf{a}(t)$.

Taking into account the above considerations, the following structures for the controller and the update law are proposed :

**Controller**

$$\boldsymbol{\tau}(t) = \mathbf{D}_K(\mathbf{q},t)\ddot{\mathbf{q}}_r(t) + \mathbf{C}_K(\mathbf{q},\dot{\mathbf{q}},t)\dot{\mathbf{q}}_r(t) + \mathbf{Y}(\mathbf{q},\dot{\mathbf{q}},\dot{\mathbf{q}}_r,\ddot{\mathbf{q}}_r,t)\hat{\mathbf{a}}(t) + \boldsymbol{\tau}_R(t) - \mathbf{K}_D\mathbf{s}(t) \tag{25}$$

where $\boldsymbol{\tau}_R(t)$ is the term which is going to cause robustness with respect to $\frac{\partial \mathbf{D}(\mathbf{q},t)}{\partial t}\dot{\mathbf{q}}$.

**Update law**

$$\dot{\hat{\mathbf{a}}} = -\boldsymbol{\Gamma}^{-1}\mathbf{Y}^{T}(\mathbf{q},\dot{\mathbf{q}},\dot{\mathbf{q}}_{r},\ddot{\mathbf{q}}_{r},t)\mathbf{s}(t) - \boldsymbol{\Gamma}^{-1}\mathbf{s}(t)\hat{\mathbf{a}}(t) \tag{26}$$

where $\sigma(t)$ is the switching signal :

$$\sigma(t) = \begin{cases} 0 & \text{if } \|\hat{\mathbf{a}}(t)\| < M \\ \sigma_0\left(\dfrac{\|\hat{\mathbf{a}}(t)\|}{M} - 1\right) & \text{if } M \leq \|\hat{\mathbf{a}}(t)\| < 2M \ (\sigma_0 > 0) \\ \sigma_0 & \text{if } \|\hat{\mathbf{a}}(t)\| \geq 2M \end{cases} \tag{27}$$

which uses the a priori information that $\sup\|\mathbf{a}(t)\|$ is upper bounded by M.

Let us now consider, the positive definite function

$$V(\mathbf{s},\tilde{\mathbf{a}}) = \frac{1}{2}\left(\mathbf{s}^{T}\mathbf{D}\mathbf{s} + \tilde{\mathbf{a}}^{T}\boldsymbol{\Gamma}\tilde{\mathbf{a}}\right)$$

where $\mathbf{D}$=$\mathbf{D}(\mathbf{q}(t), t)$ and $\tilde{\mathbf{a}}(t) = \hat{\mathbf{a}}(t) - \mathbf{a}(t)$. Differentiating, and using (22), (23) and the property which is described by (21), we get :

$$\begin{aligned}\dot{V}(\mathbf{s},\tilde{\mathbf{a}},t) &= -\frac{1}{2}\mathbf{s}^{T}\frac{\partial \mathbf{D}(\mathbf{q},t)}{\partial t}\left(\dot{\mathbf{q}}(t) + \dot{\mathbf{q}}_{r}(t)\right) + \mathbf{s}^{T}\boldsymbol{\tau} - \mathbf{s}^{T}\left[\mathbf{D}_{K}\ddot{\mathbf{q}}_{r}(t) + \mathbf{C}_{K}\dot{\mathbf{q}}_{r}(t) + \mathbf{Y}\mathbf{a}(t)\right] \\ &\quad + \tilde{\mathbf{a}}^{T}(t)\boldsymbol{\Gamma}\dot{\hat{\mathbf{a}}}(t) - \tilde{\mathbf{a}}^{T}(t)\boldsymbol{\Gamma}\dot{\mathbf{a}}(t) \\ &= -\frac{1}{2}\mathbf{s}^{T}\frac{\partial \mathbf{D}(\mathbf{q},t)}{\partial t}\left(\dot{\mathbf{q}}(t) + \dot{\mathbf{q}}_{r}(t)\right) + \mathbf{s}^{T}\boldsymbol{\tau}_{R}(t) - \mathbf{s}^{T}\mathbf{K}_{D}\mathbf{s} - \tilde{\mathbf{a}}^{T}(t)\sigma(t)\mathbf{a}(t) - \tilde{\mathbf{a}}^{T}(t)\boldsymbol{\Gamma}\dot{\mathbf{a}}(t) \\ &\leq \frac{1}{2}\|\mathbf{s}^{T}(t)\|\left\|\frac{\partial \mathbf{D}(\mathbf{q},t)}{\partial t}\right\|\left\|\left(\dot{\mathbf{q}}(t) + \dot{\mathbf{q}}_{r}(t)\right)\right\| + \mathbf{s}^{T}\boldsymbol{\tau}_{R}(t) - \mathbf{s}^{T}\mathbf{K}_{D}\mathbf{s} - \tilde{\mathbf{a}}^{T}(t)\sigma(t)\mathbf{a}(t) - \tilde{\mathbf{a}}^{T}(t)\boldsymbol{\Gamma}\dot{\mathbf{a}}(t)\end{aligned}$$

Then choosing :

$$\boldsymbol{\tau}_{R}(t) = -\left(k_0\left\|(\dot{\mathbf{q}}(t) + \dot{\mathbf{q}}_{r}(t))\right\| f(\mathbf{q})\right)^2 \mathbf{s}(t)$$

the above inequality becomes :

$$\dot{V} \leq -\left(m(t)\|\mathbf{s}(t)\| - \frac{\gamma}{4k_0}\right)^2 + \frac{\gamma^2}{16k_0^{\,2}} - \mathbf{s}^{T}\mathbf{K}_{D}\mathbf{s} - \tilde{\mathbf{a}}^{T}(t)\sigma(t)\mathbf{a}(t) - \tilde{\mathbf{a}}^{T}(t)\boldsymbol{\Gamma}\dot{\mathbf{a}}(t) \tag{28}$$

where $m(t)= k_0\|\dot{\mathbf{q}}(t) + \dot{\mathbf{q}}_{r}(t)\| f(\mathbf{q})$ and $\boldsymbol{\tau}_{R}(t) = -m^2(t)\mathbf{s}(t)$.

Let us now work on the fourth and fifth terms of the inequality (28). To this end, in analogy to the previous methodology, where we want to make some negative definite functions so that $\dot{V}\leq 0$ in a region $V\geq V_0$, and because of the switching signal $\sigma(t)$ which takes care of the robustness to time-variation of $\mathbf{a}(t)$, we make the following calculations, omitting any elaboration of the three first terms of the inequality (28) :

$$\begin{aligned}\dot{V} &\leq \ldots - \tilde{\mathbf{a}}\sigma(t)\hat{\mathbf{a}} - \hat{\mathbf{a}}^{T}\boldsymbol{\Gamma}\dot{\mathbf{a}} + \mathbf{a}^{T}\boldsymbol{\Gamma}\dot{\mathbf{a}} \\ &\leq \ldots - \tilde{\mathbf{a}}\sigma(t)\hat{\mathbf{a}} + \tilde{\mathbf{a}}\sigma_0\hat{\mathbf{a}} - \tilde{\mathbf{a}}\sigma_0\hat{\mathbf{a}} + \|\hat{\mathbf{a}}(t)\|\|\boldsymbol{\Gamma}\|\rho + \rho\rho_0\|\boldsymbol{\Gamma}\| \\ &\leq \ldots - \left(\sigma(t) - \sigma_0\right)\tilde{\mathbf{a}}^{T}\hat{\mathbf{a}} - \tilde{\mathbf{a}}\sigma_0\hat{\mathbf{a}} + \mathbf{a}^{T}\sigma_0\hat{\mathbf{a}} + \|\hat{\mathbf{a}}(t)\|\|\boldsymbol{\Gamma}\|\rho + \rho\rho_0\|\boldsymbol{\Gamma}\|\end{aligned}$$

$$\leq \ldots -(\sigma(t)-\sigma_0)\tilde{\mathbf{a}}^T\hat{\mathbf{a}} - \sigma_0\|\hat{\mathbf{a}}(t)\|^2 + \|\hat{\mathbf{a}}(t)\|(\sigma_0\rho_0 + \|\mathbf{\Gamma}\|\rho) + \rho\rho_0\|\mathbf{\Gamma}\|$$

$$\leq \ldots -(\sigma(t)-\sigma_0)\tilde{\mathbf{a}}^T\hat{\mathbf{a}} - \sigma_0\|\hat{\mathbf{a}}(t)\|^2 + 2\|\hat{\mathbf{a}}(t)\|\sigma_0\left(\frac{\rho_0}{2} + \frac{\|\mathbf{\Gamma}\|\rho}{2\sigma_0}\right) + \rho\rho_0\|\mathbf{\Gamma}\|$$

$$\leq \ldots -(\sigma(t)-\sigma_0)\tilde{\mathbf{a}}^T\hat{\mathbf{a}} - \sigma_0\left[\|\hat{\mathbf{a}}(t)\| - \left(\frac{\rho_0}{2} + \frac{\|\mathbf{\Gamma}\|\rho}{2\sigma_0}\right)\right]^2 + \left(\frac{\rho_0}{2} + \frac{\|\mathbf{\Gamma}\|\rho}{2\sigma_0}\right)^2\sigma_0 + \rho\rho_0\|\mathbf{\Gamma}\|$$

Hence, the inequality (28) takes the form :

$$\dot{V} \leq -\left(m(t)\|\mathbf{s}(t)\| - \frac{\gamma}{4k_0}\right)^2 + \frac{\gamma^2}{16k_0^2} - \mathbf{s}^T\mathbf{K_D}\mathbf{s}$$

$$-(\sigma(t)-\sigma_0)\tilde{\mathbf{a}}^T\hat{\mathbf{a}} - \sigma_0\left[\|\hat{\mathbf{a}}(t)\| - \left(\frac{\rho_0}{2} + \frac{\|\mathbf{\Gamma}\|\rho}{2\sigma_0}\right)\right]^2 + \left(\frac{\rho_0}{2} + \frac{\|\mathbf{\Gamma}\|\rho}{2\sigma_0}\right)^2\sigma_0 + \rho\rho_0\|\mathbf{\Gamma}\| \qquad (29)$$

In the inequality (29) we observe that since $\gamma$, $\rho$ and $\rho_0$ are positive constants, then the first, third and fifth terms are negative definite functions and as a result have a negative contribution to $\dot{V}$, while the other terms are positive definite. Moreover, since $\sigma(t)$ defined in (27) satisfies the inequality :

$$\left|(\sigma(t)-\sigma_0)\tilde{\mathbf{a}}^T(t)\hat{\mathbf{a}}(t)\right| \leq 12\sigma_0 M^2 \qquad (30)$$

it follows that $\dot{V}\leq 0$ in a region $V\geq V_0$, for $\hat{\mathbf{a}}(t)$, $\mathbf{s}(t)$ outside a certain bounded set so that the first, third and fifth terms predominate (contributing negatively) in this inequality. The bound $V_0$ is dependent on the other positive definite terms. Therefore, this robust-adaptive control scheme causes the boundedness of $V(\mathbf{s},\tilde{\mathbf{a}},t)$ and as a consequence the signals $\mathbf{s}(t)$ and $\hat{\mathbf{a}}(t)$ are bounded, which, in view of (11), (12) and (25) implies that $\mathbf{q}(t)$, $\dot{\mathbf{q}}(t)$ and $\tau(t)$ are also bounded.

Now the tracking error $\tilde{\mathbf{q}}(t)$ will be computed. Note that $\tilde{\mathbf{a}}(t)\sigma(t)\hat{\mathbf{a}}(t) \geq 0$ because of the switching signal form. Hence, this term contributes negatively to the upper bound of $\dot{V}$, like the first term. Therefore, one could omit these two terms from the inequality (29) in order to compute the bound of the tracking error for the worst case. Using (29), one obtains :

$$\dot{V}\leq -\mathbf{s}^T(t)\mathbf{K_D}\mathbf{s}(t) + \frac{\gamma^2}{16k_0^2} + k_1\rho$$

for some constant $k_1 = \|\tilde{\mathbf{a}}^T(t)\|\|\mathbf{\Gamma}\| > 0$. Since V(t) is bounded, the above inequality yields:

$$\int_{t1}^{t2}\|\mathbf{s}(t)\|^2 dt \leq \alpha_1\left(\frac{\gamma^2}{16k_0^2} + \rho\right)(t2 - t1) + \beta_1 \qquad (31)$$

for some constants $\alpha_1 > 0$, $\beta_1 > 0$ and any $t2 > t1 \geq 0$.

It is known that $\mathbf{s}(t) = \dot{\tilde{\mathbf{q}}}(t) + \mathbf{\Lambda}\tilde{\mathbf{q}}(t)$, where $\mathbf{\Lambda}$ is a positive definite symmetric matrix (Hurwitz). Therefore, denoting by **H** the linear operator or the impulse-response matrix from $\mathbf{s}(t)$ to $\tilde{\mathbf{q}}(t)$ as the case may be, i.e.

$$\tilde{\mathbf{q}}(t) = \int_0^t \mathbf{H}(t-\tau)\mathbf{s}(\tau)d\tau = \int_0^t e^{-\Lambda(t-\tau)}\,\mathbf{s}(\tau)d\tau\,,$$

it follows that

$$\begin{aligned}\int_{t1}^{t2}\left\|\tilde{\mathbf{q}}(t)\right\|^2 dt &= \int_{t1}^{t2}\left\|\int_0^t \mathbf{H}(t-\tau)\mathbf{s}(\tau)d\tau\right\|^2 \\ &= \int_{t1}^{t2}\left\|\int_0^{t1}\mathbf{H}(t-\tau)\mathbf{s}(\tau)d\tau + \int_{t1}^{t}\mathbf{H}(t-\tau)\mathbf{s}(\tau)d\tau\right\|^2 dt \\ &\le 2\int_{t1}^{t2}\left\|\int_0^{t1}\mathbf{H}(t-\tau)\mathbf{s}(\tau)d\tau\right\|^2 dt + 2\int_{t1}^{t2}\left\|\int_{t1}^{t}\mathbf{H}(t-\tau)\mathbf{s}(\tau)d\tau\right\|^2 dt \\ &\le 2\left\|\mathbf{H}\right\|_2^2\int_{t1}^{t2}\left\|\mathbf{s}(t)\right\|^2 dt + 2\int_{t1}^{t2}\left\|\mathbf{H}(t-t1)\tilde{\mathbf{q}}(t1)\right\|^2 dt \end{aligned} \qquad (32)$$

where $\left\|\mathbf{H}\right\|_2$ is the induced operator norm. Since the operator **H** is exponentially stable and $\tilde{\mathbf{q}}(t)$ is bounded, both $\left\|\mathbf{H}\right\|_2$ and $\int_{t1}^{t2}\left\|\mathbf{H}(t-t1)\tilde{\mathbf{q}}(t1)\right\|^2 dt$ are finite for any $t2>t1\ge 0$.

Hence, from (31) and (32) one gets

$$\int_{t1}^{t2}\left\|\tilde{\mathbf{q}}(t)\right\|^2 dt \le \alpha_0\left(\frac{\gamma^2}{k_0^{\,2}} + \rho\right)(t2-t1) + \beta_0 \qquad (33)$$

for some constants $\alpha_0 > 0$, $\beta_0 > 0$, and $t2>t1\ge 0$.

To implement the controller (25), we need the knowledge of $f(\mathbf{q})$ to generate the bounding signal $m(t)$. A more sophisticated choice of $f(\mathbf{q})$ admits a wider class of $\frac{\partial \mathbf{D}(\mathbf{q},t)}{\partial t}$, but it may make the implementation $m(t)$ more complicated. Note also that the above design does not need the knowledge of the bounds $\gamma$ and $\rho$. Furthermore, for a chosen $f(\mathbf{q})$, different choices of $k_0$ in generating $m(t)$ may have different effects on the tracking performance, while increasing $k_0$ may reduce the effect of $\gamma$ in the mean error.

Finally, let us note that for the signal boundedness and the mean tracking error (33), the parameter variations characterized by $\gamma$ and $\rho$ are not required to be small.

For the 9-link biped robot, we have assumed the presence of time variations in the trunk parameters of the robot (mass $m_5$, rotational inertia $I_5$, distance $d_5$ from the hip joint to the mass center of the trunk) (see Fig.5). This is a very reasonable assumption.

Therefore, here, the vector $\mathbf{a}(t) \in \Re^4$ which contains all the unknown and rapidly time-varying trunk parameters, takes the form

$$\mathbf{a}(t) = \left[I_5,\ m_5 d_5,\ m_5 d_5^{\ 2},\ m_5\right]$$

and as a result, for both walking phases, the 8x4 matrix $\mathbf{Y}(\mathbf{q}, \dot{\mathbf{q}}, \dot{\mathbf{q}}_r, \ddot{\mathbf{q}}_r, t)$ of the linear parameterization of the system dynamics and of the known functions takes a special form (see [16]).

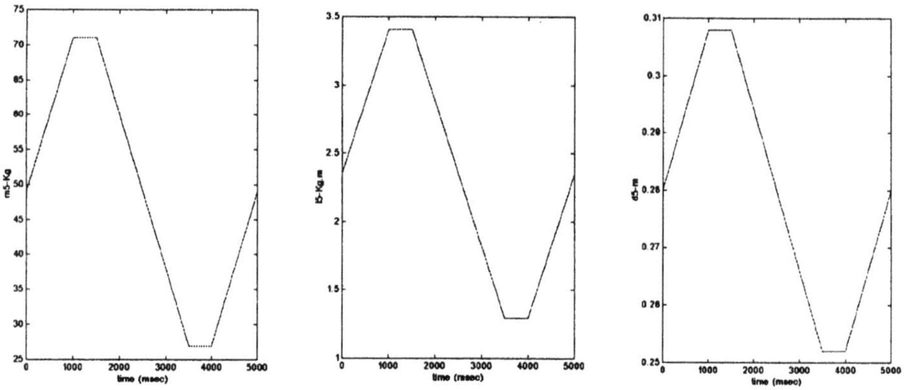


*Fig.5 The time-variations of the parameters* $m_5$, $d_5$, *and* $I_5$

Clearly the only thing one has to know about the time-varying parameters are the upper bounds of the variations so that he has an estimation of M. Here, we have considered the time-variations for the parameters $m_5$, $d_5$, and $I_5$ shown in Fig.5, which are not really known. As it is shown, the variation is such that the parameters $m_5$, $d_5$, and $I_5$ take values in the region (26.95kg, 71.05kg), (0.252m, 71.05m), and (1.2925kgm, 3.4075kgm), respectively, i.e., the mass parameters $m_5$, $I_5$ vary 45% around the constant value that they would have if no time-variation existed, and the parameter $d_5$ varies 10% around the corresponding value. Thus

$$|a_1(t)| = |I_5| \le 3.4075 = \varphi_1$$

$$|a_2(t)| = |m_5 d_5| \le 71.05 \cdot 0.308 = 21.88 = \varphi_2$$

$$|a_3(t)| = \left|m_5 d_5^{\ 2}\right| \le 6.74 = \varphi_3$$

$$|a_4(t)| = |m_5| \le 71.05 = \varphi_4$$

$$M = \sup\|a(t)\| = \sup\sqrt{\varphi_1^{\ 2} + \varphi_2^{\ 2} + \varphi_3^{\ 2} + \varphi_4^{\ 2}} = 74.72 \approx 75$$

Furthermore, there exist constants $\gamma > 0$, $\rho > 0$ such that

$$\|\dot{\mathbf{a}}(t)\| \le \rho \qquad \left\|\frac{\partial \mathbf{D}(\mathbf{q}, t)}{\partial t}\right\| \le \gamma\, f(q)$$

As mentioned before, these constants affect only the mean tracking error and have no influence on the design of control and update law. Hence

$$|\dot{a}_1(t)| = |\dot{I}_5| = 1.057dt = \rho_1$$

$$|\dot{a}_2(t)| = |\dot{m}_5 d_5 + m_5 \dot{d}_5| \leq 8.78dt = \rho_2$$

$$|\dot{a}_3(t)| = |\dot{m}_5 d_5^2 + m_5 2 d_5 \dot{d}_5| \leq 3.32dt = \rho_3$$

$$|\dot{a}_4(t)| = |\dot{m}_5| = 22.05dt = \rho_4$$

$$\rho = \sqrt{\rho_1^2 + \rho_2^2 + \rho_3^2 + \rho_4^2} \approx 24dt$$

where dt is the sampling rate. Here, f(**q**) is a known function which contains trigonometrical terms of angular position $q_i$, and as a result it is bounded for bounded **q**. In particular, an upper bound is f(**q**)=1.

For the exact application of the control, one has to compute the positive definite matrices $\mathbf{\Lambda}$, $\mathbf{K}_D$, $\mathbf{\Gamma}$. If the model dynamics was known and no time-varying parameters existed, then the feedback controller structure, omitting the robustness term, would be similar to that of the computed torque, and the last term of the PD regulator would be sufficient to lead the trajectories on the surface **s≡0**, which, given the filter-like definition of **s**, in turn shows that the tracking errors tend to **0** with time which is dependent on the Hurwitz matrix $\mathbf{\Lambda}$. Therefore, we choose $\mathbf{\Lambda}$, $\mathbf{K}_D$ to take values similar to those we should have chosen in the computed torque control case, in order to achieve the ideal convergence. It is remarked that this approach does not necessarily estimate the unknown and time-varying parameters exactly, but simply generates values that allow the desired task (bounded tracking error) to be achieved. 'Sufficient richness' conditions on the desired trajectory indicate how demanding the desired trajectory should be for tracking convergence to necessarily require parameter convergence. For example, in case of a constant desired trajectory, it would be more difficult to achieve parameter convergence. Hence, using our a priori knowledge of the desired trajectories of the robot joints, we choose different values for the terms of $\mathbf{\Gamma}$ for the cases that these signals don't include enough information. Also, the tracking error does not merely tend asymptotically to zero, but for all practical purposes, converges within finite time constants determined for a given trajectory by the values of the gain matrices $\mathbf{\Lambda}$, $\mathbf{K}_D$, $\mathbf{\Gamma}$, themselves limited by the presence of high-frequency unmodeled dynamics and measurement noise.

## 5. Simulation Experiments

The 9-link biped robot, initially at upright posture, is commanded a desired trajectory similar to that being synthesized by the reference signals adopted in the walking pattern. The corresponding angular positions and position errors, during the first two steps (in a 3.5-sec interval), are plotted in Fig.6 and 7, respectively. These diagrams show clearly the very good tracking of the desired reference signals, achieved despite the presence of the uncertainty. Something that is also obvious from the fact that the average tracking error for the first and second step is 0.017 rads and 0.062 rads (applying sliding-mode control), 0.037 rads and 0.09 rads (applying the robust-adaptive control technique), respectively. Moreover, Fig.8 depicts the time profiles of the driving torques of the biped in the same interval, where one can observe: (i) the increased values of these torques required in order to be able to face the existing parametric uncertainty, and (ii) the expected chattering outside the boundary layer (at the beginning of each phase) of the sliding-mode control. Finally, the walking mode of the 9-link, human-sized biped on the horizontal surface in a 5-sec interval has the form of Fig.9, where one can easily see that this walking mode satisfies all the necessary properties that have been adopted in Section 2 in order to utilize the gravity effect skillfully.

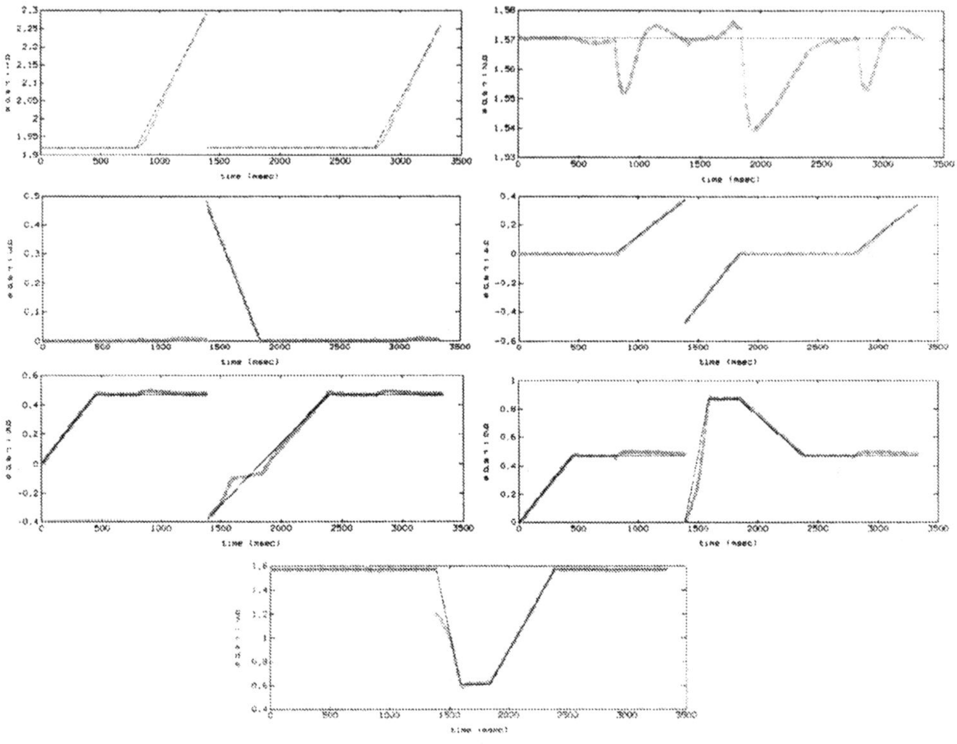

*a) applying sliding-mode control*

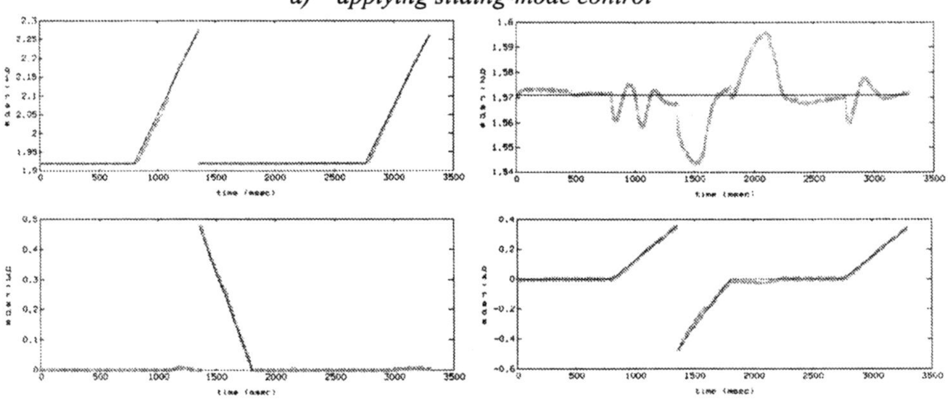

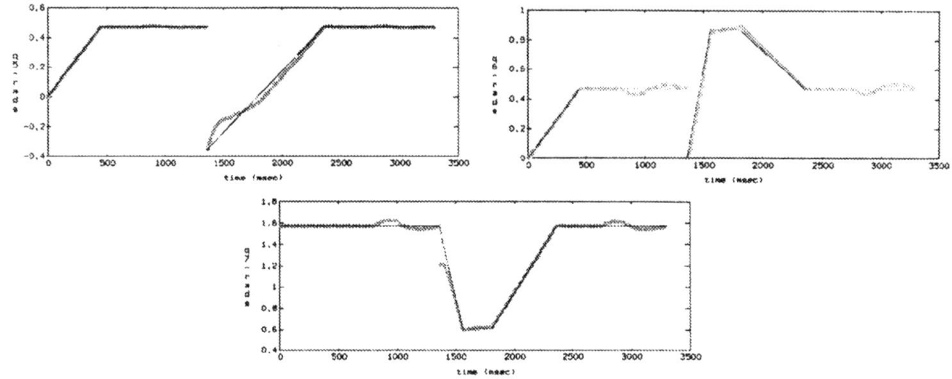

*(b) applying the robust-adaptive control technique*

*Fig.6 Angle displacements and reference signals of the 9-link, human-sized biped joints*

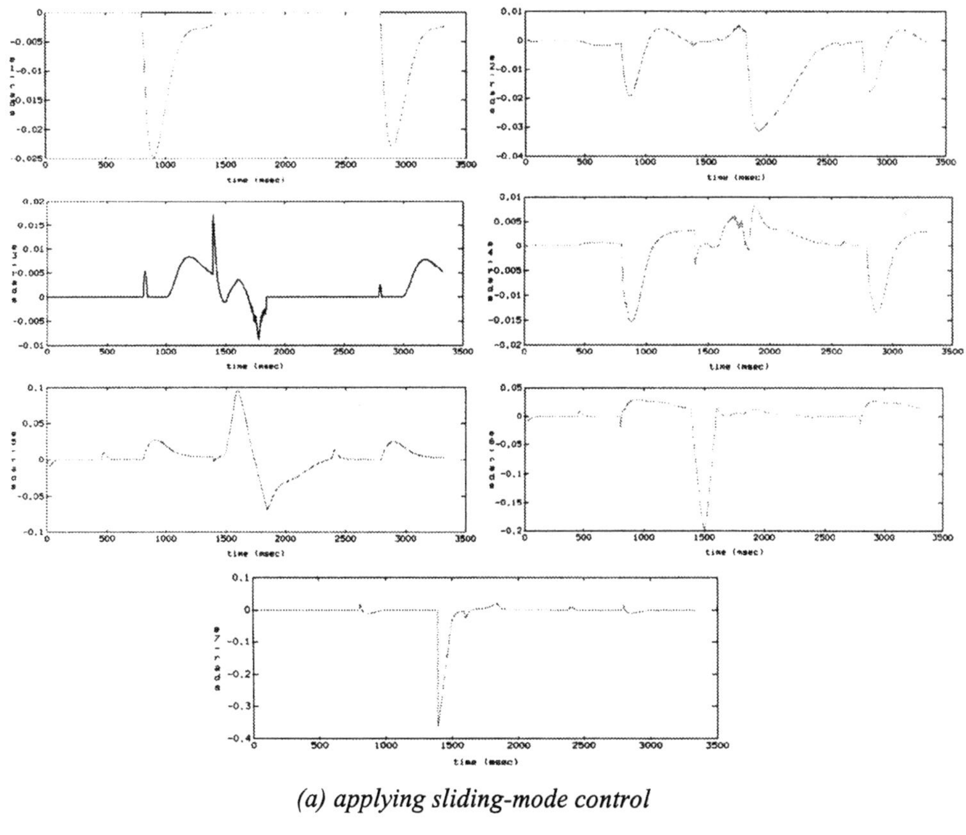

*(a) applying sliding-mode control*

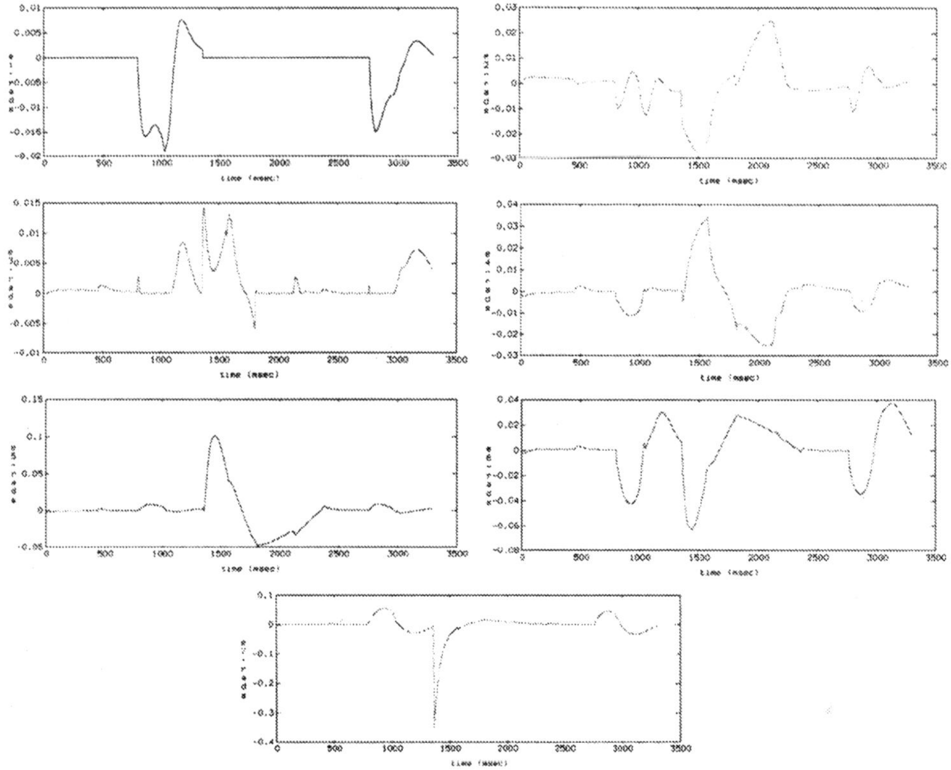

*(b) applying the robust-adaptive control technique*
*Fig.7 Angle tracking errors of the 9-link, human-sized biped joints*

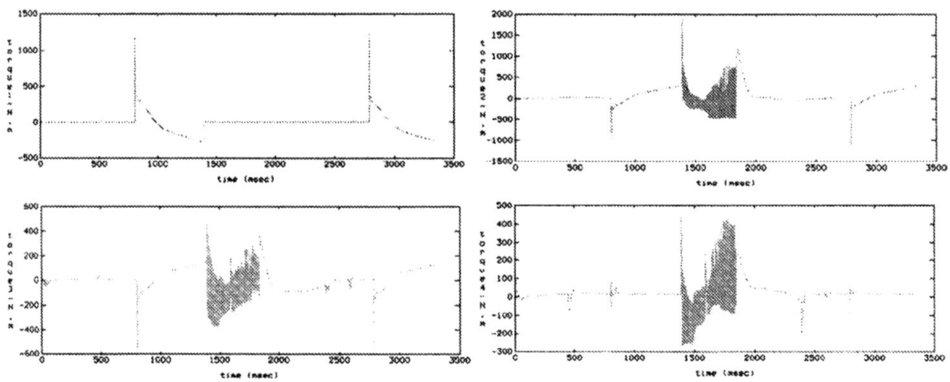

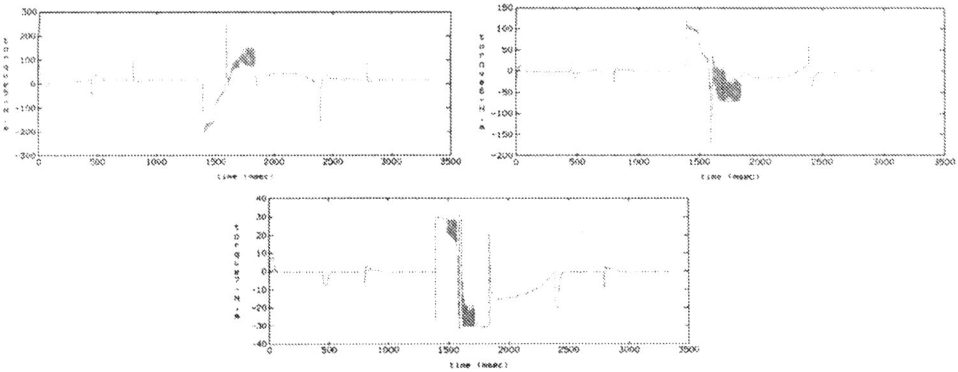

*(a) applying sliding-mode control*

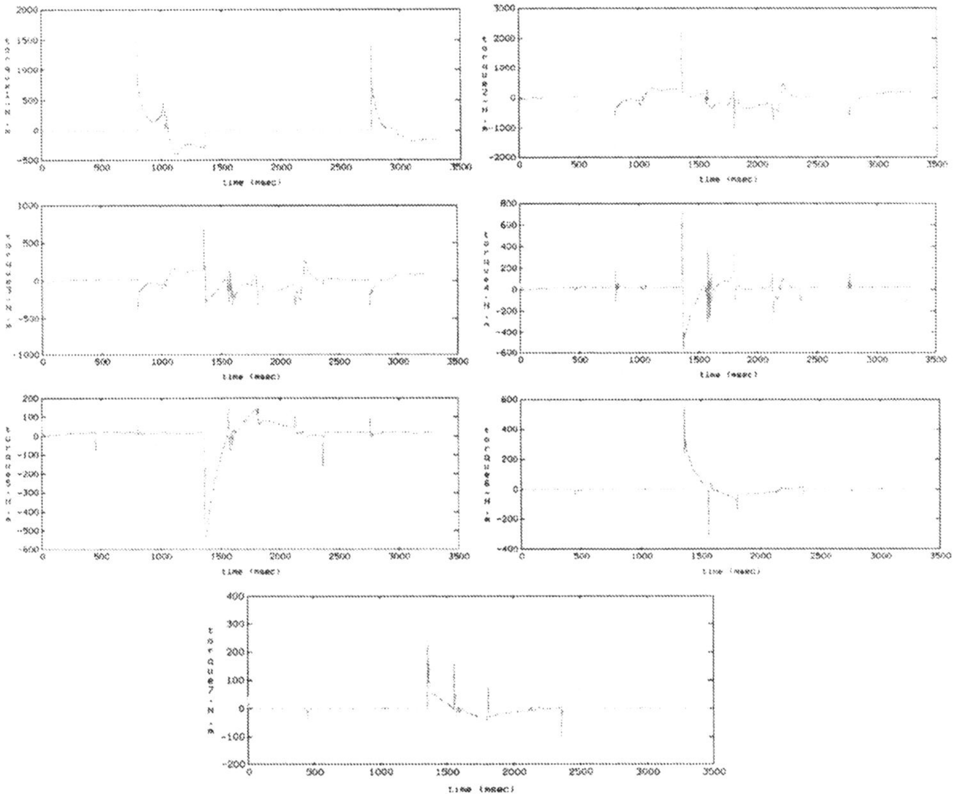

*(b) applying the robust-adaptive control technique*
*Fig.8 Driving torques of the 9-link, human-sized biped joints*

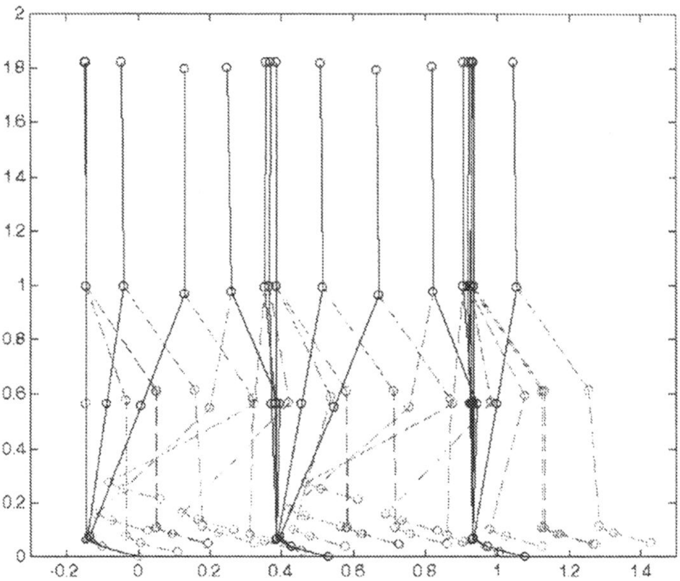


*Fig.9 Locomotion mode of 9-link biped (stick diagram)*

## 6. Conclusions

The study and control of the locomotion of a biped robot has always been a challenging problem. The mechanical complexity of biped robots, the existence of an unpowered degree of freedom and the peculiarity of their motion due to the repeatability of movements and a permanent change of situations during one step, synthesize a complex problem. At the same time, undesired collisions with its environment, transport of unknown loads, or simplification of some dynamic characteristics, cause uncertainty (structured or not) which because an accurate model of the biped robot is not available, makes more difficult the control design problem.

In this chapter a biped robot model and its control were studied. This robot model was assumed to have nine-links or, equivalently, eight degrees of freedom that correspond to two hip, two knee, two ankle, and two metatarsal joints. The walking pattern of the biped considered is based on the least expenditure of energy, and the motion was constrained on the sagittal plane. The phases of the walking mode are : (i) single-leg-support with no kick action, (ii) single-leg-support with kick action, (iii) leg-support-exchange, and (iv) collision of the swing leg with the ground (at first the toes touch the ground and then the touching of the heel follows).

The kinematics and the dynamics of the biped during the above walking phases were shortly presented. Then, two modern control schemes of the biped locomotion were provided. The first is the sliding-mode control scheme which can face parametric and model uncertainty of many kinds [19, 20]. The second is a robust-adaptive scheme which aims at minimizing the sensitivity of the system performance under the presence of large and rapid time variations in the robot parameters. This control scheme, which is based on the σ-modification parameter-updating law, ensures that all signals of the biped robot system are bounded, and the mean tracking error is of the order of the parameter variations which are not required to be small [21]. A set of simulation experiments were presented for the case where the variations occur in the trunk parameters of the biped. These experiments have verified the expected capabilities of the proposed gait control techniques (to face model variations and uncertainties) which are good candidates for practical application.

Some points that need further investigation are : (i) to explore the effectiveness of robust-adaptive control to handle the situation where one or more robotic arms are attached on the body of the biped, (ii) to explore the benefits obtained by using parallel computation on the control implementation, and (iii) to explore the performance of alternative robust and adaptive control schemes.

**References**


1. M.Raibert, *Legged Robot that Balance*, MIT Press, Cambridge, MA, 1986.
2. M.Vukobratovic, B.Borovac, D.Surla and D.Stokic, *Scientific Fundamentals of Robotics 7-Biped Locomotion*, Springer, Berlin, 1990.
3. H.Hemami, C.Wil, and G.L.Goliday, The Inverted Pendulum and Biped Stability, *Math. Biosci.*, Vol.34, 1977, pp.95-110.
4. H.Hemami and A.Katbab, Constrained Inverted Pendulum Model for Evaluating Upright Postural Stability, *ASME J. Dyn. Syst. Meas. Contr.*, Vol.104, 1982, pp.343-349
5. S.Mochon, A Mathematical Model of Human Walking, *Lectures on Mathematics in Life Sciences*, vol.14, Amer. Math. Soc., New York, 1981.
6. H.Hemami, Y.F.Zheng and M.Hines, Initiation of Walk and Tiptoe of Planar Nine-Link Biped, *Mathem. Biosci.*, Vol.61, 1982, pp.163-189.
7. H.Miura and I.Shinoyama, Dynamic Walk of a Biped, *Intl. Journal of Robotics Research*, Vol.3, No.2, 1984, pp.60-74.
8. J.Furusho and M.Masubuchi, Control of a Dynamical Biped Locomotion System for Steady Walking, *ASME J.Dyn. Syst., Meas. Contr.*, Vol.106, No.2, 1984, pp.111-118.
9. J. Furusho and M.Masubuchi, A Theoretically Motivated Reduced-Order Model for the Control of Dynamic Biped Locomotion, *ASME J. Dyn. Syst., Meas. Contr.*, Vol.109, 1987, pp.155-163.
10. M.Yamada, J.Furusho and A.Sano, Dynamic Control of Walking Robot with Kick-Action, *Proc. 1985 Intl. Conf. on Advanced Robotics (ICAR'85)*, Tokyo, 1985, pp. 405-412.
11. A.Sano and J.Furusho, 3D Steady Walking Robot with Kick Action, *Proc. 1988 USA-Japan Symp. on Flexible Automation*, Vol.II, Mineapolis, July 1988.

12. T.Mita, T.Yamguchi, T.Kashiwase and T.Kawase, Realisation of a High Speed Biped Using Modern Control Theory, *Intl. J. Control*, Vol.9, No.2, 1984, pp.107-119.
13. A.Takanishi, Y.Egusa, M.Tochizawa, M.Takeya, and I.Kato, Realization of Dynamic Biped Walking Stabilized with Trunk Motion, *Proc. Robotics and Manufacturing Systems '88*, September, 1988.
14. S.G.Tzafestas, M.Raibert and C.S.Tzafestas, Robust Sliding-Mode Control Applied to a 5-Link Biped Robot", *J. Inteligent. and Robotic Systems.*, Vol.15, No.1, pp.67-133, 1996.
15. S.G.Tzafestas, T.E.Krikochoritis and C.S.Tzafestas, Robust Sliding-Mode Control of Nine-Link Biped Robot Walking, *J. Intelligent and Robotic Systems,* Vol.20, No.2-4, 1997, pp.375-402.
16. S.G.Tzafestas, T.E.Krikochoritis and C.S.Tzafestas, Robust-Adaptive Gait Control of a 9-Link Biped Robot, *Syst. Anal. Modelling Simul.,* Vol.31, No.4, pp.247-304, 1998 .
17. Y.F.Zheng and H.Hemami, Impact Effects of Biped Contact with the Environment, *IEEE Trans. Systems, Man and Cybernetics*, Vol.14, No.3, 1984, pp.437-443.
18. Gang Tao, On Robust Adaptive Control of Robot Manipulators, *Automatica*, Vol.28, No.4, 1992, pp.803-807.
19. J.J.Slotine and W.Li, *Applied Nonlinear Control*, Prentice Hall, 1992.
20. J.J.Slotine, The Robust Control of Robot Manipulators, *Intl. J. Robotics Res.,* Vol.4, No.2, 1985.

# 18

# A NOVEL ROBOTIC SNAKE

K. J. KYRIAKOPOULOS, K. SARRIGEORGIDES
and G. MIGADIS
*Control Systems Laboratory*
*Department of Mechanical Engineering*
*National Technical University of Athens*
*Athens 10682, Greece*

## 1. Introduction

In a number of industrial environments, such as nuclear, chemical, power plants etc., the need for an efficient means for inspection and minor repair which can easily approach any part of the plant in remote or inaccessible sites is eminent. Similar needs arise in tasks such as for fire-fighting reconnaissance, maneuvering through the rubble to look for survivors after an earthquake, investigation or minor repair of leaks or other problems in municipal sewer systems or utility tunnels etc. A robotic mechanism could both help in minimizing risks for human personnel and increasing the "in between repairs" time by contacting inspection, surveillance and minor repairs operating in higher temperatures, contaminated environments or narrow spaces of an industrial installation. Additionally, such a robot could be used to efficiently move inside medium and large size pipes during inspection, clean up or minor repair tasks.

The motion and size requirements needed for those kind of tasks are met by snakes, in nature, the shape and motion mechanism of which inspired the design of a number of researchers. In [1] the design and motion planning of a mechanical snake is presented. In [2] the kinematics of hyperredudant robot locomotion over uneven solid terrain are considered (see also [3]).

By following those snake characteristics as generic design guidelines but at the same time by making major shortcuts and additions we designed a novel robotic snake [4]. In this chapter, an analysis of the basic mechanism design along with most of the component specifications is presented. Furthermore, we present its planar kinematic analysis, as a first step towards controlling it and we deal with the simplest case of steering this "snake" robot on the plane. Thus, our robot is like a car with two trailers. The nonholonomic motion planning for several car-like mobile robots has been extensively investigated. In [5], the example of a car pulling

$n$ trailers with 2 inputs is considered. In these developments, the control system is converted into a chained form for which the solution to the steering problem is straightforward. In [6] the kinematic model is locally converted into, a nilpotent, chained form. In [7] the machinery of exterior differential forms is developed in order to convert the 2 input $n$ trailer system into the Goursat normal form which is dual to the chained form. These results were extended to systems with three inputs by [8] and to systems with multiple inputs [9].

The examples used though in those developments consider axle-to-axle hitching between the trailers. Our robot can be considered as a 2 trailer system with 2 kingpin hitches, where the axles are connected by a kingpin between the bodies. This kind of system cannot be converted into chained form [10] due to the fact that sush a system is not flat [11]. We overcome this problem by adding 2 more controls in the 2 trailers.

In section 2 we present the design requirements that make such a design unique. A literature survey of rpior designs is presented in section 3. In section 4, the generic design of our snake is presented . We present some of the fundamental details of it in section 5 while, in section 6, we state the control problem associated with our model and derive the modeling equations of the motion on a plane. In Section 7, we convert our model to the extended Goursat normal form. A method of steering the system is presented in section 8, while simulation results are presented in section 9. Finally, in section 10 we discuss our current state and present our near future steps. In the appendix we provide the reader with the neccesary background on exterior differential forms.

## 2. Design Specifications

The design procedure was preceded by a number of specifications mandated primarily by the tasks nature. Those are divided into the three groups:

(i) Specifications of motion capabilities

- Reaching areas difficult to approach, in a complex environment.
- Entering and moving into pipes of medium and large size.
- Moving among obstacles in two and three dimensions (Flexibility).
- Moving Forward with adequate speed, in order to cover fast long distances when necessary.
- Ascending or descending stairs.
- Carrying cameras, instruments or tools, properly attached to it, in order to carry out the necessary tasks.

(ii) Specifications of structural characteristics

- Small width and height.

- Robust structure & precise motion mechanism.
- Easily assembled.
- Easy access of all components of the robot for repairs.

(iii) Other specifications

- Use of standard (available in the market) motion devices and machine elements.
- Space for placing proper feedback devices to all the degrees of freedom.

## 3. Other Robots with Similar Characteristics

The idea to design a snake-like articulated robot is not novel. In the past there have been researchers who built robots based more or less on the shape and the movements of the natural snakes. All of those designs focused at the comparative advantages with respect to other mechanisms like the quadruped or the simple wheeled mobile ones. Hiroses group pioneered the area and made thorough studies of the biomechanics snakes motion [12]. They designed four different articulated mobile robots since 1974 :

The Active Cord Mechanism Model 3 (ACM III) [12, 3], is 2 meters long and has 21 links, joined with revolute joints driven by DC motors. It moves forward only in two dimensions by performing undulations with its body, similar to those that real snakes perform.

The next model, ACM IV, has 8 cylindrical links, each one supported by two free wheels similar to the previous one. They are connected with each other by oblique swivel joints, which allow for 3-D motion. This model also propels itself forward by undulating its body.

Both of those served only as "simulation" platforms to test models for snake-like forward motion.

Hirose designed two more articulated body mobile robots, KRI and KRII [12, 13]. Both of them have six cylindrical links with vertical direction. Each one can both rotate and slide vertically relatively to the next. The dimensions of these robots, 1391x206x393 mm (Length x Height x Width) for KR I and 3300x1080x460 mm for KR II, do not allow motion in narrow spaces. On the other hand they are capable of carrying several pieces of equipment inside the free space of the links. They move forward using motor driven crawler tracks and wheels respectively. Both robots perform well in various cases, like passing over obstacles, ascending stairs, stepping over floor gaps etc., but they always needed enough space to move in it.

A different approach to the subject of designing a robotic snake was made by Yansong Shan and Yoram Koren at the University of Michigan, Ann Arbour [1]. They designed two robots, MS-1 and MS-2, which have 7 links and a camera placed at the front. The small height and width of the links combined with the rotational degrees of freedom around the vertical axes of the joints make them look like snakes crawling on the floor. Both of them can crawl on a smooth surface by bending their bodies with a certain sequence, following the irregularities of the terrain. Only MS-2 can lift its first link to pass over an obstacle.

The crawling movement these robots adopt needs some free space around their body in order to bend the links and the velocity of forward progression is low, compared with wheeled robots.

We designed a robotic snake according to the specifications set at the preceding section. Those are more general and include the capability of in pipe motion. Such a design is undoubtedly more complicated.

## 4. Description of the NTUA Robotic Snake

Our robotic snake has a shape similar to that of real snakes, as shown in fig. 1. Its cylindrical body has a diameter of 100mm and total length 1.65m. It consists of 7 links which are connected with 6 revolute joints. Inside the first link at the front, which is 100mm long, there is free space where several instruments or devices can be fitted in order to carry out the jobs previously mentioned. The maximum load of instruments it can carry is 500 gr. Inside the rest of links all the necessary motion devices are placed in order to execute the motions that this robot is expected to perform. There are three sets of wheels along its body to enable forward motion. The total height of the body from the ground is 11.5 mm and the minimum internal diameter of a pipe in which the robotic snake can easily move is 15 cm. The main characteristic of the body is that it consists of three modules of two links each, placed repeatedly along it, as it is shown in fig. 2. Each one of the modules is 541 mm long and weights 5,500 gr. It has six degrees of freedom, which are necessary in order to achieve efficient operation. Two of these degrees of freedom are the relative rotations $\theta_1, \theta_2$ between two successive links round a horizontal and a vertical axis respectively, as shown in figure 2, No.1,2. These are similar to the two main motions made by the vertebrae of the snakes which, in conjunction to the big number (100 - 300) of them, result to the proverbial flexibility of them. The maximum value of $\theta_1$ and $\theta_2$ to either sides is 90 degrees. The combination of those elementary motions enables the active change of its shape while moving either on a plain surface or also in three dimensions. So it can avoid obstacles, pass through narrow passages and holes and reach certain places of interest to carry out the assigned work. In those cases where the robotic snake has to pass over an obstacle or reach a point that is higher than the ground level, it can lift the front module of its body and the link that carries the equipment by rotating joint no 1. The maximum height of an obstacle that it can overpass is 25 cm. This capability is also useful whenever it has to go up- or downstairs in areas where human personnel usually moves. The front link lifting enables the access of areas placed up to 70 cm higher from the rolling surface. In order to realize the forward motion, two wheels are placed on each module, in the middle of one of its two links (fig. 1). Each of those can turn around a vertical axis wheel (fig. 2, No.6), just like the way a front wheel of a car turns. Such a mechanism can directly

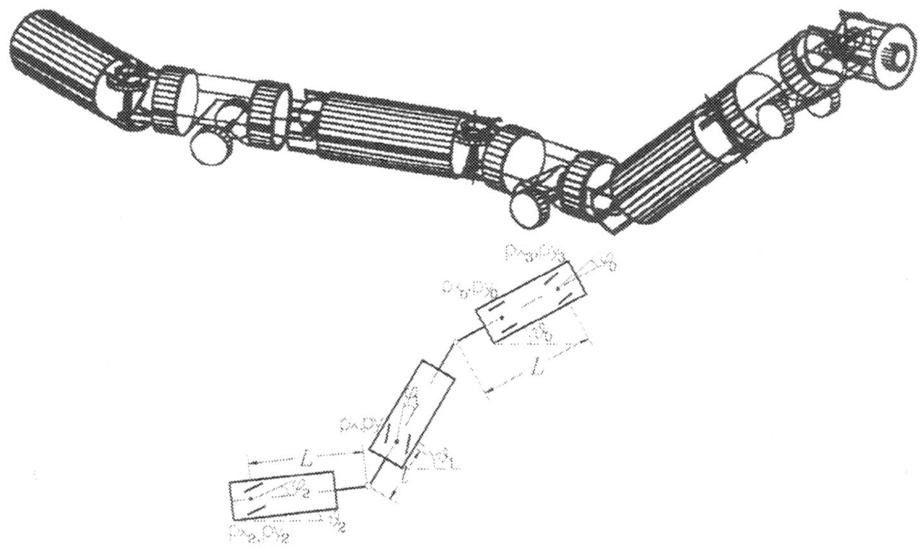

Figure 1: a) the "snake" robot b) model of the "snake" robot

and precisely control the turning of each wheel with minimum torque requirements from its actuator. The turning of a wheel is independent from that of the other one at the same module. This ability helps the robotic snake to change the motion direction while rolling on a surface, making possible the arbitrary determination of its course by the motion controller. So, combining properly the angles of rotation between the links and those of the turning wheels, a center of rotation of the whole mobile robot can be defined, when moving on a bend. The rolling of the wheels of a module (fig. 2, No.5) is also independent for each one. This allows the differentiation of their angular velocity between the inner and the outer wheel on a bend, so as the wheels not to slip and loose grip with the ground. If the wheels slipped this would cause disturbance of the forces exerted from the ground on the body of the robotic snake yielding to (nonintegrable) errors.

Using those wheels, the robotic snake can move easily with adequate velocity compared to the propulsion techniques adopted by real snakes [14, 12]. These are quite complicated to be realized, since the only way with which snakes can move is by bending their flexible body in certain ways. The use of wheels is extremely advantageous in cases when long distances have to be covered by the robot in order to approach a certain point of interest. Such a case is when moving inside long industrial pipes.

At the sides of every link where the rolling wheels are placed, two more rotational degrees of freedom (fig. 2, No.3,4) are realized. They rotate the whole body of the snake around its central axis. There are no limits to the angles of rotation $\theta_3$, $\theta_4$. These degrees of freedom enable the easier compensation of the

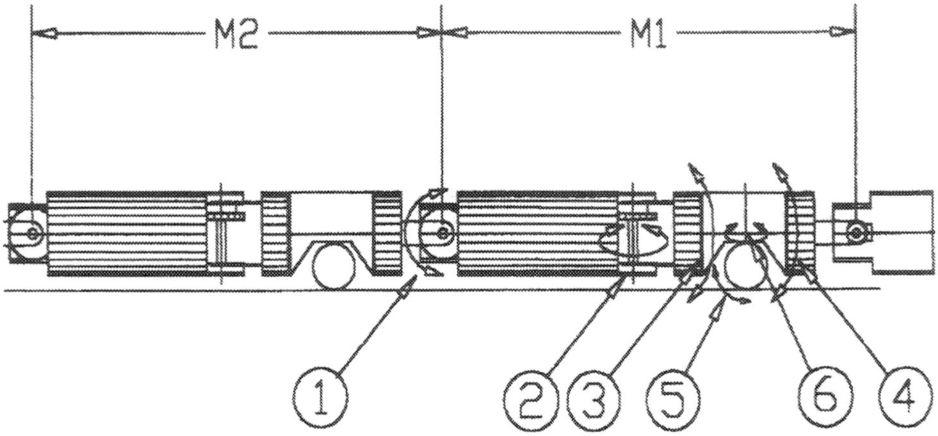


Figure 2: Modules and Degrees of Freedom of the Robotic Snake

irregularities of the rolling surface by the wheels. To achieve this, the degrees of freedom should rotate simultaneously at the same angle, so that the wheels take the desired position, while the rest of the robotic snake's body does not alter its previous position.

It is also possible to rotate a bigger part of the body relatively to the rest by activating a single degree of freedom of that kind. A motion like this can rotate either only the first link, offering the equipment the capability of having a certain orientation while approaching the inspection points, or the whole front module while it is lifted off the ground, which allows it to move in a hemispherical area.

The proper combination of all those degrees of freedom enables the robotic snake to move efficiently inside various areas and reach certain points in them with the proper direction using its snake-like flexibility.

## 5. Details of the Robotic Snake Design

In figure 3 the internal layout and the principal dimensions of the two links which compose the module of this robotic snake's body are shown in order to demonstrate how every degree of freedom is realized. The first of those implements $\theta_1$ and $\theta_2$. The implementing mechanisms are placed at the ends of the link. A 22 W brushed DC motor (2) supplies the necessary power to each one of them. The output speed of the motor is reduced in three stages : first, by a worm gear-crown mechanism (1) and then, by two stages of spur gears placed after it. The last and biggest gear (3) is connected with the neighboring link transmitting the required torque through a properly manufactured steel laminate. The maximum required torque at the joints is 147 kp.cm, based on which all the relevant machine parts have been designed. Two of those laminates from each link form the rotational joint round a steel tube. The average speed of rotation of the last gear is 7.5 RPM. It is low enough so as not to generate significant inertial forces and overload the

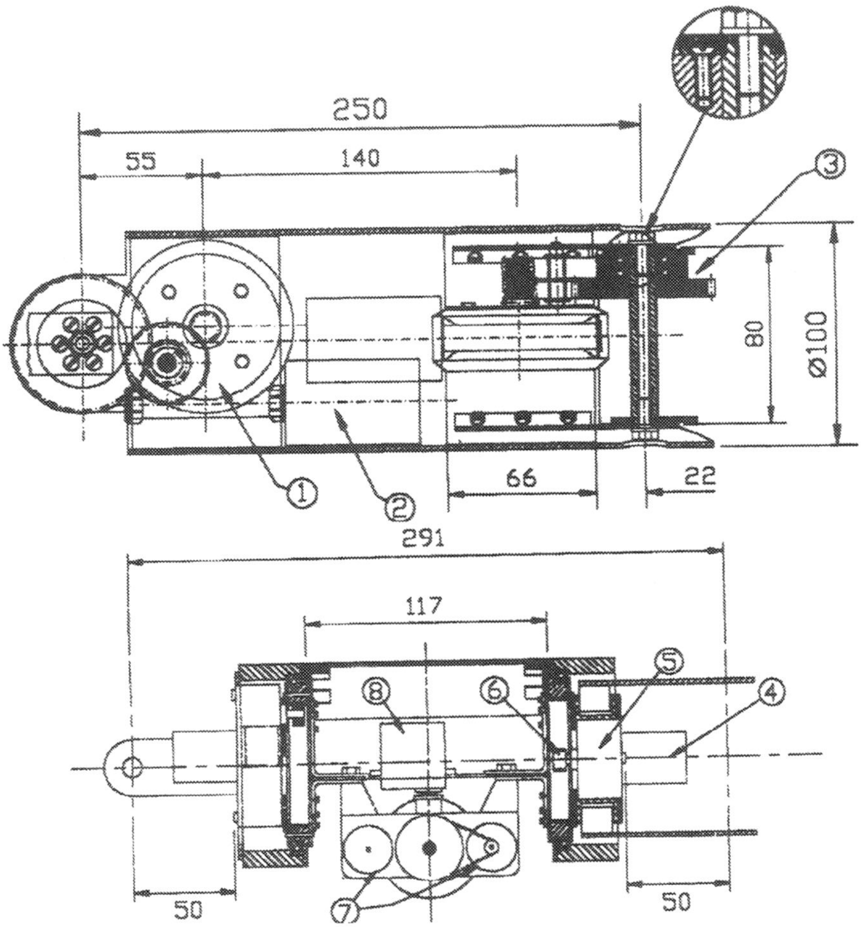


Figure 3: Internal Structure of Modules

elements of the links. Both mechanisms of this link are placed properly inside the 2.1 mm thick tubular cover, made of hard PVC (polyvynilchloride), in a way that enough space is left free. Through this space the power and signal transferring cables are placed. In the same free space the feedback devices for every degree of freedom will be placed. The second link of the module implements each one of $\theta_3$ and $\theta_4$ using small 6V brushed DC motors (4) through geared speed reducers (5), each with reduction ratio n=102 rotating a pinion gear (6). They are interlaced with internal spur gears which are firmly connected with the middle part of the link. So the rotation of the pinion gears cause relative rotation between the every side of the link and the middle part of it. The three parts of this link are connected with fixed section large diameter ball bearings, which are mainly used in robotic applications. They are placed inside 12 mm thick aluminum rings, which also form the external cover of the side parts of the link. These bearings are capable of transferring the forces exerted on the robot's body along the link with the minimum angular displacement due to bending moment, and offer smooth and frictionless rotation between the parts of this link.

At the center of the second link the wheels that move forward the robotic snake are supported by a specially designed frame made of hard plastic. At the sides of it are placed two brushed DC motors (7) used for the independent rolling of the wheels. Such motors are mainly used for the propulsion of small car and boat models, with very good performance. Each wheel is coupled with its motor through a toothed belt. On this frame the wheel turning mechanisms are also placed. Those include two small sized servomotors with a built-in feed back system (8), used to turn and keep each wheel at the optimum angle every time the robotic snake is moving on a bend. The frame itself is bolted on the steel laminates that form the chassis of this middle part of the link. At the sides of this chassis two 20 mm holes are located at the center of its transverse laminates through which all necessary cables are placed subject to minimum strain due to the relative rotation there.

All the necessary technical specifications of the parts used and the manufacturing processes needed for the construction of this robotic snake, as well as instructions for its assembly are presented in [15].

## 6. Modeling and Motion Control Problem Statement

We want to derive the kinematic model of our system (see fig. 1). The states of the system is $x = [p_x\ p_y\ \theta_0\ \theta_1\ \theta_2\ \phi_0\ \phi_1\ \phi_2]^T$ where : $\theta_i$ is the orientation of the $i$-th module of the robot with respect to the horizontal axis of the inertial frame, $\phi_i$ is the steering angle of the $i$-th module, $(p_x, p_y)$ are the cartesian coordinates of the rear axle of the second module.

The distance between the axle and joint of each module is $L$ and the distance between the joint of one module and the axle of the next module is $l$. This gives the holonomic constraints:

$$p_{x_0} = p_x + L\cos\theta_1 + l\cos\theta_0 \quad (1)$$

$$p_{y_0} = p_y + L\sin\theta_1 + l\sin\theta_0 \quad (2)$$

$$p_{x_3} = p_x + L\cos\theta_1 + L\cos\theta_0 \quad (3)$$

$$p_{y_3} = p_y + L\sin\theta_1 + L\sin\theta_0 \quad (4)$$

$$p_{x_2} = p_x - l\cos\theta_1 - L\cos\theta_2 \quad (5)$$

$$p_{y_2} = p_y - l\sin\theta_1 - L\sin\theta_2 \quad (6)$$

In order to simplify our kinematic model, each pair of wheels is modeled as a single wheel centered at the midpoint of the axle. In fact the two wheels have different angles and their normals intersect at a single point [16]. The nonslipping requirement of the wheels gives the four linear velocity constraints :

$$\dot{p}_x\sin(\phi_0+\theta_0) - \dot{p}_y\cos(\phi_0+\theta_0) - L\dot{\theta}_1\cos(\phi_0+\theta_0-\theta_1) - L\dot{\theta}_0\cos\phi_0 = 0 \quad (7)$$

$$\dot{p}_x \sin\theta_0 - \dot{p}_y \cos\theta_0 - L\cos(\theta_1 - \theta_0)\dot{\theta}_1 - l\dot{\theta}_0 = 0 \quad (8)$$

$$\dot{p}_x \sin(\phi_1 + \theta_1) - \dot{p}_y \cos(\phi_1 + \theta_1) = 0 \quad (9)$$

$$\dot{p}_x \sin(\phi_2 + \theta_2) - \dot{p}_y \cos(\phi_2 + \theta_2) + l\dot{\theta}_1 \cos(\phi_2 + \theta_2 - \theta_1) + L\dot{\theta}_2 \cos\phi_2 = 0 \quad (10)$$

Due to rolling our mechanical system satisfies the nonholonomic constraints of the type

$$\omega_i(x) \cdot \dot{x} = 0 \qquad i = 1, 2, \ldots k \quad (11)$$

where $x \in \Re^n$ is the state of the system and $\omega_i(x) \in \Re^n$ are row vectors. We assume that vectors $\omega_i$, $i = 1, 2, \ldots k$ are linearly indepedent and smooth. The corresponding codistribution $\Omega(x) = \text{span}\{\omega_1(x), \ldots, \omega_k(x)\}$ has dimension $k$. Therefore, we can find an $(n-k)$-dimensional distribution $\Delta(x) = \text{span}\{g_1(x), \ldots, g_{n-k}(x)\}$ with $g_i(x) \in \Re^n$, $i = 1, \ldots, n-k$ such that $\Delta = \Omega^\perp$ i.e. $\omega_i(x) g_j(x) = 0 \; \forall \omega_i \in \Omega, \; g_j \in \Delta$. Then the system with the above velocity constraints can be represented as a system with inputs $u_i$

$$\dot{x} = g_1(x)u_1(t) + \ldots + g_{n-k}(x)u_{n-k}(t) \quad (12)$$

The constraints can be written in the form of (11) if

$$\begin{aligned}
\omega_1(x) &= [\sin(\phi_0 + \theta_0) \quad -\cos(\phi_0 + \theta_0) \quad -L\cos\phi_0 \quad -L\cos(\phi_0 + \theta_0 - \theta_1) \quad 0 \quad 0 \quad 0 \quad 0] \\
\omega_2(x) &= [\sin\theta_0 \quad -\cos\theta_0 \quad -l \quad -L\cos(\theta_1 - \theta_0) \quad 0 \quad 0 \quad 0 \quad 0] \\
\omega_3(x) &= [\sin(\phi_1 + \theta_1) \quad -\cos(\phi_1 + \theta_1) \quad 0 \quad 0 \quad 0 \quad 0 \quad 0 \quad 0] \\
\omega_4(x) &= [\sin(\phi_2 + \theta_2) \quad -\cos(\phi_2 + \theta_2) \quad 0 \quad l\cos(\phi_2 + \theta_2 - \theta_1) \quad L\cos\phi_2 \quad 0 \quad 0 \quad 0]
\end{aligned}$$

Since the corresponding codistribution $\Omega(x)$ is 4-dimensional and the state space is 8-dimensional, we can find a 4-dimensional distribution

$$\Delta(x) = \Omega^\perp(x) = \text{span}\{g_1(x)\; g_2(x)\; g_3(x)\; g_4(x)\}$$

where :

$$g_1(x) = \begin{bmatrix} \cos\theta_0 + A \cdot L\sin\theta_1 + l\frac{\sin\theta_0 \tan\phi_0}{L-l} \\ \sin\theta_0 - A \cdot L\cos\theta_1 - l\frac{\cos\theta_0 \tan\phi_0}{L-l} \\ \frac{\tan\phi_0}{L-l} \\ A \\ B \\ 0 \\ 0 \\ 0 \end{bmatrix}$$

$$\begin{aligned}
g_2(x) &= [0\,0\,0\,0\,0\,1\,0\,0]^T \\
g_3(x) &= [0\,0\,0\,0\,0\,0\,1\,0]^T \\
g_4(x) &= [0\,0\,0\,0\,0\,0\,0\,1]^T
\end{aligned}$$

where

$$\begin{aligned} A &= \tfrac{1}{L}\sin(\theta_0-\phi_1-\theta_1)\sec\phi_1 - \tfrac{l}{L(L-l)}\cos\quad(\phi_1+\theta_1-\theta_0)\sec\phi_1\tan\phi_0 \\ B &= \tfrac{1}{L}\sin(\theta_0-\phi_2-\theta_2)\sec\phi_2 - \tfrac{l}{L(L-l)}\cos\quad(\phi_2+\theta_2-\theta_0)\sec\phi_2\tan\phi_0- \\ &\qquad -A\tfrac{L+l}{l}\sec\phi_2\cos(\phi_2+\theta_2-\theta_1) \end{aligned}$$

One can easily write now the model in the form (12) with inputs : $u_1$ the forward driving velocity of the rear axle of the first module and $u_2$, $u_3$, $u_4$ the steering velocities $\dot\phi_0$, $\dot\phi_1$, $\dot\phi_2$, respectively. Obviously, our system has 4 inputs and 8 states.

The motion planning problem can be stated as: find a control law $u = (u_1(t), \ldots \ldots, u_{n-k}(t))$ to steer the system from the initial state $x(0) = x_0$ to $x(T) = x_f$ on the time interval $[0, T]$. We want to convert the system into multi-chain single-generator form since, as stated in the introduction, there exist methods for steering such systems.

We plan to convert the above system into the multi-chained form. To do so we need tools from exterior algebra and Pfaffian systems. Those tools are presented in Appendix A.

In the next section we apply the proofs of the theorems 2 and 3 in order to convert the control system of the mobile robot into the extended Goursat normal form.

## 7. Converting the "snake" robot system into Extended Goursat Normal Form

Our mechanical system is depicted in 1. In order to derive the Pfaffian system associated with the above system we write the nonholonomic constraints, taking into consideration the holonomic ones, as the one forms:

$$\begin{aligned} a^0 &= \sin\theta_0 dp_x - \cos\theta_0 dp_y - l d\theta_0 - L\cos(\theta_0-\theta_1)d\theta_1 \\ a^1 &= \sin(\phi_1+\theta_1)dp_x - \cos(\phi_1+\theta_1)dp_y \\ a^2 &= \sin(\phi_2+\theta_2)dp_x - \cos(\phi_2+\theta_2)dp_y + L\cos\phi_2 d\theta_2 + l\cos(\theta_2+\phi_2-\theta_1)d\theta_1 \\ a^3 &= \sin(\phi_0+\theta_0)dp_x - \cos(\phi_0+\theta_0)dp_y - L\cos\phi_0 d\theta_0 - L\cos(\theta_0+\phi_0-\theta_1)d\theta_1 \end{aligned}$$

The derived flag of the Pfaffian system is:

$$I = I^{(0)} = \{a^0, a^1, a^2, a^3\} \quad I^{(1)} = \{a^0\} \quad I^{(2)} = \{0\}$$

Calculations of the exterior products give:

$$\begin{aligned} da^1 \wedge \alpha &= h_1(\theta_0,\theta_1,\theta_2,\phi_0,\phi_1,\phi_2)d\phi_1 \wedge ds \\ da^2 \wedge \alpha &= h_2(\theta_0,\theta_1,\theta_2,\phi_0,\phi_1,\phi_2)d\phi_2 \wedge ds \\ da^3 \wedge \alpha &= h_3(\theta_0,\theta_1,\theta_2,\phi_0,\phi_1,\phi_2)d\phi_0 \wedge ds \end{aligned}$$

where $\alpha = a^0 \wedge a^1 \wedge a^2 \wedge a^3$, $ds = d\theta_0 \wedge d\theta_1 \wedge d\theta_2 \wedge dp_x \wedge dp_y$. Thus the Pfaffian system $\{I^{(0)}, d\theta_0\}$ is integrable. We also have $da^0 \wedge d\theta_0 = 0$ and therefore $\{I^{(1)}, d\theta_0\}$ is integrable, too. According to theorem 3 our system can be converted to extended Goursat normal form. We now follow the proof of theorem 3 in order to find a basis $I$ which satisfies the Goursat congruences (31).

The basis of the last nontrivial system $I^{(1)}$ is $a^0$. The definition of the derived flag $I^{(2)}$ and the fact that $\{I^{(1)}, d\theta_0\}$ is integrable imply that

$$da^0 \not\equiv 0 \bmod I^{(1)} \qquad da^0 \equiv 0 \bmod \left\{I^{(1)}, d\theta_0\right\} \tag{13}$$

Combining those equations we obtain the congruence $da^0 \equiv d\theta_0 \wedge \beta^0 \bmod I^{(1)}$ for some $\beta^0 \not\equiv 0 \bmod I^{(1)}$. From this and the definition of the derived flag $da^0 \equiv 0 \bmod I^{(0)}$ we obtain $\beta^0 \in I^{(0)}$. Applying this procedure to our system we obtain $\beta^0 = \bar{a}^1 = a^3 \csc\phi_0$ such that $da^0 \equiv d\theta_0 \wedge \bar{a}^1 \bmod a^0$. We choose $\bar{a}^3 = a^1$ such that $\{a^0, \bar{a}^1, a^2, \bar{a}^3\}$ is a basis of $I$ which is not only adapted to the derived flag, but also satisfies the Goursat congruences:

$$\begin{array}{ll} da^0 \equiv d\theta_0 \wedge \bar{a}^1 \bmod a^0 & da^2 \not\equiv \bmod I^{(0)} \\ d\bar{a}^1 \not\equiv \bmod I^{(0)} & d\bar{a}^3 \not\equiv \bmod I^{(0)} \end{array} \tag{14}$$

We see that the form $a^0$ has rank 1 since $da^0 \wedge a^0 \neq 0$ and $da^0 \wedge da^0 \wedge a^0 = 0$. Therefore, according to theorem 1 there are $f_1$, $f_2$ such that $da^0 \wedge a^0 \wedge df_1 = 0$, $a^0 \wedge df_1 \neq 0$, $a^0 \wedge df_1 \wedge df_2 = 0$, $df_1 \wedge df_2 \neq 0$. Integration of these equations gives: $f_1 = \theta_0$,

$$f_2 = z_3^1 = p_x \sin\theta_0 - p_y \cos\theta_0 + L\sin(\theta_0 - \theta_1) \tag{15}$$

So the form $a^0$ can be written as : $a^0 = df_2 - g \cdot df_1 = dz_3^1 - z_2^1 dz^0$, with

$$\begin{array}{l} z^0 = d\theta_0 \\ z_2^1 = l + p_x\cos\theta_0 + p_y \sin\theta_0 + L\cos(\theta_0 - \theta_1) \end{array} \tag{16}$$

From the Goursat congruences (14) we can use $a^0$ and $\bar{a}^1$ in order to put $\bar{a}^1$ in the form $dz_2^1 - z_1^1 dz^0$. From the equation $\tilde{a}^1 = b_0 a^0 + b_1 \bar{a}^1$ we obtain $b_0 = -\cot\phi_0$ and $b_1 = 1$. So we have $\tilde{a}^1 = dz_2^1 - z_1^1 dz^0$ and

$$z_1^1 = (L - l)\cot\phi_0 + p_y\cos\theta_0 - p_x\sin\theta_0 - L\sin(\theta_0 - \theta_1) \tag{17}$$

We can modify $a^2, \bar{a}^3$ using forms from $I$ in order to put them in the form $dz_2^2 - z_1^2 dz^0$ and $dz_2^3 - z_1^3 dz^0$, respectively. We choose

$$z_2^2 = p_y, \qquad z_2^3 = \theta_2 \tag{18}$$

We set $\bar{a}^2 = k_0 a^0 + k_1 \bar{a}^1 + k_2 a^2 + k_3 \bar{a}^3 = dz_2^2 - z_1^2 dz^0$ and easily obtain $k_0$, $k_1$, $k_2$, $k_3$. Thus

$$z_1^2 = \csc\phi_0 \sec\phi_1 \sin(\phi_1 + \theta_1)\left[L\cos\phi_0\cos(\theta_0 - \theta_1) - \; -l\cos(\phi_0 + \theta_0 - \theta_1)\right] \tag{19}$$

In the same way we set $\tilde{a}^3 = c_0 a^0 + c_1 \bar{a}^1 + c_2 a^2 + c_3 \bar{a}^3 = dz_2^3 - z_1^3 dz^0$ and choose $z_2^3 = \theta_2$. Algebraic calculations provide $c_0$, $c_1$, $c_2$, $c_3$ and we obtain

$$\begin{array}{rl} z_1^3 = & \frac{1}{L}\cot\phi_0 \sec\phi_1 \sec\phi_2 \cdot \{\, l\cos(\phi_2 - \theta_1 + \theta_2)\sin(\phi_1 - \theta_0 + \theta_1) + \\ & L\cos(\theta_0 - \theta_1)\sin(\phi_1 - \phi_2 + \theta_1 - \theta_2)\} + \\ & \frac{l}{L^2}\{\, l\cos(\phi_2 + \theta_0 - \theta_1 + \theta_2)\sin(\phi_0 - \phi_1 - \theta_1) - \\ & L\cos(\phi_0 + \theta_0 - \theta_1)\sin(\phi_1 - \phi_2 + \theta_1 - \theta_2)\} \cdot \csc\phi_0 \sec\phi_1 \sec\phi_2 \end{array} \tag{20}$$

So we converted our control system in the four input, three-chained system:

$$\begin{array}{ll} \dot{z}^0 = v_1 & \dot{z}_1^2 = v_3 \\ \dot{z}_1^1 = v_2 & \dot{z}_2^2 = z_1^2 v_1 \\ \dot{z}_2^1 = z_1^1 v_1 & \dot{z}_1^3 = v_4 \\ \dot{z}_3^1 = z_2^1 v_1 & \dot{z}_2^3 = z_1^3 v_1 \end{array} \tag{21}$$

## 8. Steering the"snake" robot

There are a number of algorithms that can be used to steer systems in chained form. In [5, 8] the method of sinusoidal inputs is used to bring the system to the desired state. In [9] an "all at once" sinusoidal method for multiple-input systems was proposed. Another steering method uses polynomial inputs and is presented fot the multiple-input case in [9]. In [17] the method of piecewise constant inputs is used. The method of steering nonholonomic systems using piecewise constant inputs was first introduced in [18] as multirate digital control. All of these algorithms steer the system from the initial to the desired state. The resulting trajectories may look "nicer" for some of those methods compared to others. Through simulations we found that the trajectories generated using the method of multirate controls are reasonable.

The theoretical framework that we adopted is based on [19] since it is powerful for systems admitting an exact discretization, because in those cases, an exact point-to-point trajectory can be generated.

We consider our system: $\dot{x} = \sum_{i=1}^{4} g_i(x)u_i$ driven by[1] $v_i(t) = v_i^D(k)$ $t \in [k\delta, (k+1)\delta]$, $k \geq 0$, $i = 1, \ldots 4$. If the operator $X_i = v_1^D L_{g_1} + v_{i1}^D L_{g_2} + v_{i2}^D L_{g_3} + v_{i3}^D L_{g_4}$ is defined and the choice $v_{33}^D = v_{23}^D$, $v_{32}^D = v_{22}^D$ is made, one can proove [19, 17] that

$$z(k+1) = e^{\bar{\delta}X_1} \circ e^{\bar{\delta}X_2} \circ e^{\bar{\delta}X_3}(I_d)|_{z(k)} \tag{22}$$

where $e^X \equiv \sum_{k>0} \frac{1}{k!} X^k$ is the exponential Lie operator, $\bar{\delta} = \delta/3$. $I_d$ is the identity function and "$\circ$" is the composition operator. We choose the controls:

$$\begin{array}{rcl} v_1(t) & = & v_1^D \quad t \in [k\delta,\ (k+1)\delta] \\ v_2(t) & = & \begin{cases} v_{11}^D & t \in \left[k\delta,\ (k+\frac{1}{3})\delta\right] \\ v_{21}^D & t \in \left[(k+\frac{1}{3})\delta\ (k+\frac{2}{3})\delta\right] \\ v_{31}^D & t \in \left[(k+\frac{2}{3})\delta,\ (k+1)\delta\right] \end{cases} \\ v_3(t) & = & \begin{cases} v_{12}^D & t \in \left[k\delta,\ (k+\frac{1}{3})\delta\right] \\ v_{22}^D & t \in \left[(k+\frac{1}{3})\delta,\ (k+1)\delta\right] \end{cases} \\ v_4(t) & = & \begin{cases} v_{13}^D & t \in \left[k\delta,\ (k+\frac{1}{3})\delta\right] \\ v_{23}^D & t \in \left[(k+\frac{1}{3})\delta,\ (k+1)\delta\right] \end{cases} \end{array} \tag{23}$$

Using (22) with $\bar{\delta} = \delta/3$, the discretization of the system in chained form coordi-

[1] symbol $v_i^D(k)$ is used to demonstrate the "discretized" nature of the input

nates is exact, given by:

$$
\begin{array}{lcl}
z^0(k+1) & = & z^0(k) + 3\bar{\delta} v_1^D \\
z_1^1(k+1) & = & z_1^1(k) + \bar{\delta}(v_{11}^D + v_{21}^D + v_{31}^D) \\
z_1^2(k+1) & = & z_1^2(k) + \bar{\delta}(v_{12}^D + 2v_{22}^D) \\
z_1^3(k+1) & = & z_1^3(k) + \bar{\delta}(v_{13}^D + 2v_{23}^D) \\
z_2^1(k+1) & = & z_2^1(k) + 3\bar{\delta} v_1^D z_1^1(k) + \\
 & & \frac{\bar{\delta}^2}{2!}(5v_{11}^D + 3v_{21}^D + v_{31}^D)v_1^D \\
z_2^2(k+1) & = & z_2^2(k) + 3\bar{\delta} v_1^D z_1^2(k) + \frac{\bar{\delta}^2}{2!}(5v_{12}^D + 4v_{22}^D)v_1^D \\
z_2^3(k+1) & = & z_2^3(k) + 3\bar{\delta} v_1^D z_1^3(k) + \frac{\bar{\delta}^2}{2!}(5v_{13}^D + 4v_{23}^D)v_1^D \\
z_3^1(k+1) & = & z_3^1(k) + 3\bar{\delta} v_1^D z_2^1(k) + \frac{9}{2!}\bar{\delta}^2 (v_1^D)^2 z_1^1(k) \\
 & & + \frac{\bar{\delta}^3}{3!}(19v_{11}^D + 7v_{21}^D + v_{31}^D)(v_1^D)^2
\end{array}
\tag{24}
$$

The motion planning problem is recasted now to finding the controls $v_1^D, v_{11}^D, v_{211}^D, v_{31}^D, v_{12}^D, v_{22}^D, v_{13}^D, v_{23}^D$ to lead the system $z(t) = [z^0\ z_1^1\ z_2^1\ z_3^1\ z_1^2\ z_2^2\ z_1^3\ z_2^3]$ from an initial state $z(k) = z_0$ to the desired state $z(k+1) = z_f$. This can be easily done by simple algebra.

## 9. Simulation results

The simulation of the "snake" robot is performed using the system in the chained form. Then the actual state $x$ and input $u$ trajectories are obtained from $z$ and the $v_i$s via the inverse of transformations (15,16,17,18,20).

We considered a "snake" with $L = 1$, $l = 0.5$". The case study we chose is a transition from an initial state $x(0) = x_0 = [0\ 0\ 0\ 0\ 0\ 0.2\ 0\ 0]^T$ to the desired $x_f = [6\ 4\ \pi/2\ 1.5\ \pi/2\ 0.07\ 0\ 0]^T$. Notice that we chose $\phi_0(0) = 0.2 \neq 0$ to avoid singularity of certain coordinate transformations. The corresponing chained form coordinates are $z(0) = z_0 = [0\ 2.46658\ 1.5\ 0\ 0\ 0\ 0.25\ 0]^T$ and $z_f = [\pi/2\ 1.06045\ 5.4974\ 6.07\ 7.13\ 4\ -0.5085\ \pi/2]^T$

Substituting these to the (24) and solving we obtain:

$$
\begin{array}{lll}
v_{11}^D = 1.74597 & v_{12}^D = 0.339106 & v_{13}^D = 2.00568 \\
v_{21}^D = -2.55373 & v_{22}^D = 3.39584 & v_{23}^D = -1.3821 \\
v_{31}^D = -0.59837 & &
\end{array}
$$

States $x, y, \theta_0$ of the system can be shown in fig. 4, 5.

## 10. Discussion - Issues of further research

In this chapter, a new idea in the field of multi-articulated snake-like robotic devices is presented. The final product of this work succeeds in meeting all the specifications set. It is a robot that can be easily assembled. The relatively small dimensions of the cross section of its body and the great flexibility its various degrees of freedom provide comparative advantages compared to other similar mechanisms. Its capability of passing over obstacles and moving efficiently inside

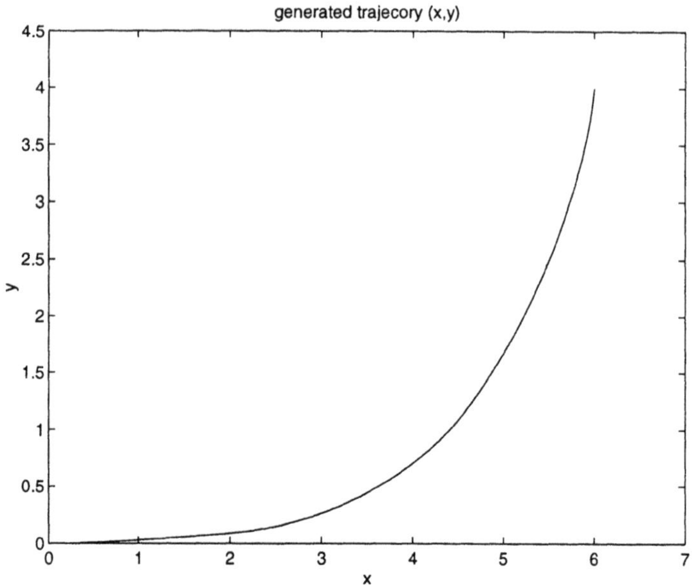


Figure 4: a) x-y trajectory of the middle link

small spaces, including pipes, make it ideal for carrying out important inspection tasks inside industrial installations.

The kinematic modeling and steering control were presented as a first step towards developing proper control algorithms that can guide it effectively to perform various useful tasks. Extensive work is currently being carried out on this field, trying to controlling it in different working environments and tasks.

We plan to address a number of issues following the problem that we tackled. Stabilization around a point, trajectory tracking and motion planning in a maze-like environment are our immediate step while steering the robot on $SO(3)$ is our ultimate task.

The control problem will also provide the appropriate feedback to make structural improvements and modifications. Our current interests include the installation of the appropriate control hardware.

## APPENDIX : Pfaffian systems and Goursat normal forms

We present some issues on exterior algebra and Pfaffian systems and the theorems providing the conditions to convert a Pfaffian system into the extended Goursat normal form ([20]).

A nonholonomic system can be defined by a codistribution $I = \{a^1, \ldots a^k\}$ on the cotangent space to the configuration manifold, specifying the directions that the system is not allowed to move. This codistribution generates a Pfaffian system, and can be analyzed using exterior differential systems. This formulation is dual

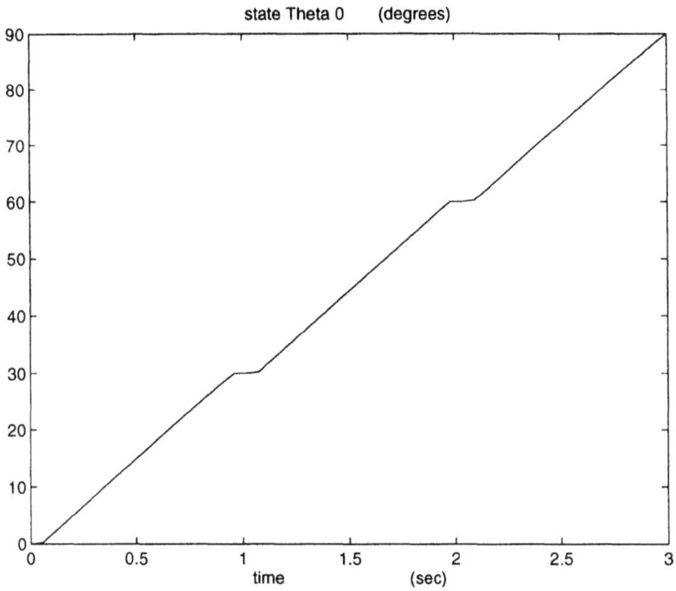


Figure 5: Orientation $\theta_0$ of the front link w.r.t time

of this of sec. 4 in the sence that codistribution $I$ annihilates distribution $\Delta$, i.e. $I = \Delta^{\perp}$ or, $a^i(g_j) = 0 \; \forall i, j$. A basis for $I$ is written in coordinates by taking each $a^i = 0$ as the $i$-th nonholonomic constraint.

We will need the notion of *congruence modulo* of a Pfaffian system. We say that

$$\eta \equiv \xi \bmod I \quad \text{if} \quad \eta \equiv \xi + \sum_{i=1}^{k} \vartheta^i \wedge a^i \tag{25}$$

for some forms $\vartheta^i \in \Omega(M) \equiv \oplus \Omega^P(M)$ (the module of forms on the n-dimensional differentiable manifold M). The *derived flag* is defined to be the nested chain of codistributions given by $I^{(0)} = I$ and

$$I^{(k+1)} = \left\{ \omega \in I^{(k)} : \; d\omega \equiv 0 \bmod I^{(k)} \right\} \tag{26}$$

The construction is assumed to terminate at some $N$, when $I^{(N)} = I^{(N+1)}$.

One way to find integral curves for Pfaffian systems is to transform the system into a normal form. If such a normal form can be found, then there is a 1-1 correspodence between the integral curves of the normal form and the original system. A pertinent theorem is:

**Theorem 1** *Pfaff's Theorem. Suppose $a$ is a one-form on $\Re^n$ which satisfies $(da)^{r+1} \wedge a = 0$, $(da)^r \wedge a \neq 0$. Then there exist local coordinates $z$ such that*

$$a = dz_1 + z_2 dz_3 + \ldots + z_{2r} dz_{2r+1} \tag{27}$$

In the case $r = 1$, the proof [20] shows that there exist two functions $f_1$, $f_2$ which satisfy the p.d.e

$$da \wedge a \wedge df_1 = 0 \quad a \wedge df_1 \neq 0 \tag{28}$$
$$a \wedge df_1 \wedge df_2 = 0 \quad df_1 \wedge df_2 \neq 0 \tag{29}$$

If suitable $f_1$, $f_2$, are found, $a$ can be scaled so that

$$a = df_2 + g \cdot df_1 \tag{30}$$

In this chapter we consider the case of multi-steering [8, 21, 9], with a Pfaffian system of codimension four. We have the following definition:

**Definition 2** *Extended Goursat Normal Form. A Pfaffian system $I$ on $\Re^{n+m+1}$ of codimension $m+1$ is in extended Goursat normal form if it is generated by $n$ constraints of the form:*

$$I = span\left\{dz_i^j - z_{i+1}^j dz^0 \ : \ j = 1, \ldots m; i = 1, \ldots s_j\right\}$$

We note that all solution trajectories of $I$ are determined by the $m+1$ functions $z^0(t), z_1^1(t), \ldots z_1^m(t)$ and their derivatives with respect to time $t$. Sufficient and necessary conditions for converting a two-input nonholonomic system into Goursat normal form are given in [22]. For the multi-input case there are conditions due to [23] for converting a system into extended Goursat normal form. The proof of the next theorem can be also found in [21].

**Theorem 3** *Extended Goursat Normal Form. [23] Let $I$ be a Pfaffian system of codimension $m+1$. If there exists a set of generators $\left\{a_i^j \ : \ j = 1, \ldots m; i = 1, \ldots s_j\right\}$ for $I$ and an integrable one-form $\pi$ such that*

$$\begin{array}{l} da_i^j \equiv -a_{i+1}^j \wedge \pi \bmod I^{s_j - i} \ i = 1, \ldots s_j - 1 \\ da_{s_j}^j \not\equiv 0 \bmod I \end{array} \tag{31}$$

*for all $j$, then there exists a set of coordinates $z$ such that $I$ is in Goursat normal form ,*

$$I = \left\{dz_i^j - z_{i+1}^j dz^0 \ : \ j = 1, \ldots m; i = 1, \ldots s_j\right\} \tag{32}$$

The conditions (31) are called *Goursat congruenses.*

An algorithm for converting systems to extended Goursat normal form is given in [24]. There is also another set of conditions for converting a system into Goursat normal form, which are easier to check, since they do not require finding a basis which satisfies the Goursat congruences but only one which is adapted to the derived flag. A basis of one-forms $a^j$ for $I$ is said to be adapted to the derived flag if a basis for each derived system $I^{(i)}$ can be taken to be some subset of the $a^j$'s. One special case of this theorem is proved in [25]. The proof can also be found in [21].

**Theorem 4** *Extended Goursat Normal Form. A Pfaffian system I of codimension $m+1$ on $\Re^{n+m+1}$ can be converted to Goursat Normal Form if and only if there exists a one-form $\pi$ such that $\{I^{(k)}, \pi\}$ is integrable for $k = 0, \ldots N-1$.*

Consider all the one-forms in $I$ which are integrable modulo the entire codistribution: $\{a \,:\, da = 0 \bmod I\}$. If this set is the entire codistribution I, then I is integrable. The extended Goursat normal form is the dual of the multi-input, single-generator chained form presented in [8]. Sufficient conditions for converting the system to the above form are given in [8].

# References


[1] Y.Koren and Y.Shan. Design and motion planning of a mechanical snake. *IEEE Transactions on Systems, Man and Cybernetics*, 23(4), July / August 1993.

[2] G.S. Chirikjian and J.W.Burdick. Kinematically optimal hyper-redudant manipulator configurations. *IEEE Transactions on Robotics and Automation*, 11(6), December 1995.

[3] S. Hirose and A. Morishima. Design and control of mobile robot with an articulated body. 9(2):99–114, April 1990.

[4] G. Migadis and K. J. Kyriakopoulos. Design and forward kinematic analysis of a robotic snake. In *1997 IEEE International Conference on Robotics and Automation*, April 1997.

[5] Murray R.M. and S. Sastry. Nonholonomic motion planning : Steering using sinusoids. *IEEE Transactions on Robotics & Automation*, May 1993.

[6] O.J. Sordalen. Conversion of the kinematics of a car with n trailers into a chained form. In *1993 IEEE International Conference on Robotics and Automation*, pages 382–387, April 1993.

[7] R. Murray D. Tilbury and S. Sastry. Trajectory generation for the n-trailer problem using goursat normal forms. 40(5), May 1995.

[8] L. Bushnell, D. Tilbury, and S.S. Sastry. Steering three input chained form nonholonomic systems using sinusoids:the firetruck example. In *European Control Conference*, pages 1432–1437, 1993.

[9] L.Bushnell D. Tilbury, O. Sordalen and S. Sastry. A multi-steering trailer system :conversion into chained form using dymamic feedback. In *IFAC Symposium on Robot Control*, September 1994.

[10] L. Bushnell. An obstacle avoidance algorithm for a car pulling trailers with off-axle hitching. In *34th IEEE Conference on Decision and Control*, December 1995.

[11] M. Fliess. Flatness,motion planning,and trailer systems. In *32th IEEE Conference on Decision and Control*, pages 2700–2705, December 1993.

[12] S. Hirose. *Biologically Inspired Robots.* Oxford University Press, 1993.

[13] E.F. Fukusima S. Hirose and S. Tsukagoshi. Basic steering control methods for the articulated mobile robot. pages 5–14, February 1995.

[14] C. Gans. How snakes move. 222:82–96, 1970.

[15] G. Migadis and G. Karavokyros. Design and forward kinematic analysis of a robotic snake. Technical Report Dipl. Thesis (in greek), 1996. Control Sys. Lab., Mech. Eng. Dept., NTUA.

[16] J. Alexander and J. Maddocks. On the kinematics of wheeled mobile robots. *International Journal of Robotics Research*, pages 15–27, October 1989.

[17] D. Tilbury and A. Chelouah. Steering a three-input nonholonomic system using multirate controls. In *European Control Conference*, pages 1428–1431, 1993.

[18] S. Monaco and D. Normand-Cyrot. An introduction to motion planning under multirate digital control. In *1992 IEEE Conference on Decision and Control*, pages 1780–1785, 1992.

[19] A. Chelouah and M. Petitot. Finitely discretizable nonlinear systems :concepts and definitions. In *34th IEEE Conference on Decision and Control*, December 1995.

[20] R.L. Bryant, S.S. Chern, R.B. Gardner, H.L. Goldschmidt, and P.A. Griffiths. *Exterior differential systems.* Springer, 1991.

[21] D. Tilbury and S. Sastry. On coursat normal forms ,prolongations and control systems. In *IEEE Conference on Decision and Control*, December 1994.

[22] R.M. Murray. Nilpotent bases for a class of nonholonomic distributions with applications to trajectory generation for nonholonomic systems. Technical Report CIT/CDS 92-002, California Institute of Technology, 1992.

[23] R.M. Murray. Applications and extensions of goursat normal forms to control of nonlinear systems. In *IEEE Conference on Decision and Control*, pages 3425–3430, 1993.

[24] L. Bushnell, D. Tilbury, and S.S. Sastry. Extended goursat normal forms with applications to nonholonomic motion planning. In *IEEE Conference on Decision and Control*, pages 3447–3452, 1993.

[25] S.L. Sluis. *Absolute Equivalence and its Applications to Control Theory.* PhD thesis, University of Waterloo, 1992.

# PART 3

# APPLICATIONS

# 19

# AUTONOMOUS AGENTS IN CELLULAR MANUFACTURING

E. S. TZAFESTAS
*Intelligent Robotics and Automation Laboratory*
*National Technical University of Athens*
*Zographou 15773, Athens, GREECE*
*brensham@softlab.ece.ntua.gr*

## 1. Introduction

The problems of the design and the design automation of modern cellular manufacturing systems cannot be tackled independently of the underlying system model and architecture. Indeed, often enough, choice of the appropriate model, makes the solution to the design problem transparent. In search of such principles and models that would simplify design, we have been investigating the option of autonomous agents and we have been trying to understand to what degree are those agents suitable for cellular manufacturing tasks and what additional techniques are required. Up to now, we have limited our investigation to the process control/production management domain, which among all industrial tasks involves maximal use of non-formal, qualitative knowledge and where various artificial intelligence techniques have been used in the past.

The use of autonomous agents generally inverses the knowledge representation problem by altogether dismissing the traditional symbolic representation view and relying instead on "know how" algorithms that themselves "represent" naturally the solution to the given problem. This is possible whenever a holistic interactionist view of the problem/problem-solver system is a better alternative to a piece-by-piece modeling and design. We present a case study on tool-management and regulation where this is precisely the case -a typical specimen of the family of tasks that modern manufacturing is interested in. The possibility of simplified holistic modeling becomes obvious once the agent and the manufacturing process in question are regarded as coupled dynamical systems with conflicting steady states, that constantly perturb each other -a motivationally autonomous agent behaves then as an intelligent controller ensured to function within certain operational bounds.

In section 2, we are reviewing and discussing those aspects of current cellular manufacturing technology that contribute to the applicability of autonomous agents, while in section 3 we are summarizing the autonomous agents philosophy. Next, we are describing the example process -tool management- and give a comparative analysis of various motivational schemes for the tool manager robot. Some experiments and thoughts on agent specialization are also presented, as well as a deeper discussion of operational principles and requirements behind the feasibility of motivated industrial robotics. The chapter concludes with a brief overview of recent autonomous agents research directions with potential applicability to cellular manufacturing problems.

## 2. Cellular manufacturing

During the last fifteen years, research trends in manufacturing systems shifted, from the study of tasks that necessitated high precision and accuracy (such as assembly, welding and other mechanical operations), toward the study of system integration, production flexibility, process adaptivity to unforeseen events, uncertainty handling and fault-tolerance [1-4]. The concept of cellular manufacturing has emerged of social and organizational considerations [5] and constitutes a unifying paradigm that captures the need for product-oriented rather than process-oriented layout of the production plant [3].

Cellular manufacturing relies on allocating the complete manufacturing of a family of parts to a cell, i.e. a group of machines including robots. Such individual cells have to respond to local operational demands and are subject to the constraints of flexibility and adaptivity. *Flexibility* involves the possibility to reinitialize/reconfigure the system on a new set of operational demands, including deadlines to be met, cost limits, order and timing of input etc., while *adaptivity* corresponds to the possibility of recovering from failures of various types. Flexibility therefore concerns mainly the connection of a cell to the other components of the plant, while adaptivity refers to its local self-sustainance.

Those constraints have led to a reconsideration of the two fundamental assumptions of conventional manufacturing systems: The first is the *almost-perfect-knowledge assumption*, according to which all or almost all of the parameters that affect the operation and the performance of the system are known in advance - for example, scheduling presupposes that the number, types, durations etc. of jobs are known as well as the number, capacities and possibilities of the processors to which those jobs will be allocated. The second is the *centralized-decision assumption*, according to which -and since there is almost perfect knowledge about the task- the designer of the system can derive a method of controlling it to meet the required operational goals, or else make decisions in advance about the operation of the system.

The first assumption has been relaxed over the last decade since no complete information about the overall process is available in advance and the optimization principle has gradually lost ground in favor of more qualitative and heuristic approaches that may manipulate less formal knowledge about the process; those techniques are mainly borrowed from the artificial intelligence domain and their integration into manufacturing domains [6,7] has paralleled the emergence of the intelligent control field that combined conventional control with artificial intelligence techniques [8]. On the other hand, doubts on the centralized decision have only recently begun to find their way into the scientific arena and some manufacturing systems that involve multiple tasks, robots and resources have started being modeled in terms of concurrently acting cooperative or competitive components, usually called agents (see for example, [9]). In such manufacturing systems, decisions have to be made separately at almost every point during operation of the system, since the exact overall state cannot be thoroughly known in advance and long-term planning/scheduling soon becomes obsolete. Local rules or heuristics at various points become more important and complicated analytical pre-processing loses value. Introducing autonomous agents at those decision/redecision points -i.e. points where some local intelligent, or for that purpose autonomous, "processing" is necessary- looks beneficial. The advantageous consequences of such decomposition of the overall process into a number of interacting agents are *ease of design* (through complexity multiplication during synthesis) and *reconfigurability/scalability*. This multi-agent view of manufacturing systems is in line with Baldwin's view of autonomous manufacturing systems, where goals and intelligence are distributed across all entities (robots, tools, resources etc.) participating in a manufacturing plant [10].

## 3. Autonomous agents

During mid-eighties and in an attempt to provide an efficient alternative to classical planning methods (a good collection of papers appears in [11]), the paradigm of reactive systems has emerged. Whereas planning relies on the existence and update of a perfect global world model, reactive systems have adopted the other extreme view of total absence of any internal model. Instead, every action is in fact a direct reaction to a perceived stimulus (stimulus-response or S-R scheme). However simple in itself this might appear, its power grows exponentially when several reactive components are connected and used within the same system. Interactions between such reflexes may result in what appears to an observer concrete activity or even purposeful reasoning. For the case of a single robot, if S stands for its perceptions and R for the commands finally fed to the actuators, it appears as though something exists in between (S-x-R). This "something" is nothing else but a *Reactive Algorithm* (RA). Reactive because as explained it does not maintain an explicit model of the world nor of the global task (although it *does* have some knowledge about its conditions of operation and about its operation itself) and algorithm because it appears as such and because it was designed so (S-RA-R scheme).

This reactive approach has been found to work best in partially unknown environments, where sensor readings are noisy or unreliable and operation has to be extremely real-time and responsive to unexpected events [12,13]. Reactive agents are equipped with a set of competences that they use in order to "survive" in the environment they are put into. Completion of the designers' task occurs as a side-effect of the agents' struggle for survival. In this context, reactive algorithms precisely *minimize processing* of sensor data and accelerate it by exploiting parallelism on fixed topology networks. Another feature of reactive algorithms is that sensor readings or other "stimuli" may arrive *at any point* inside it, what may serve both to monitor command output and to ease design. Other advantages of the use of reactive algorithms include: low implementation cost, design simplicity, enhanced robustness (due to minimal processing and crude data representations), finally potential adaptivity, since a designer may decide to adapt one of the parameters the algorithm uses -there are not too many of them and they are usually visible since the RAs are rather primitive.

Other desirable properties of the reactive, behavior-based paradigm are [12,13]: no central world representation bottlenecks, minimization of time lags between perception and actuator level command output and fault-tolerance. In one word, the power of the behavior-based paradigm lies in its simplicity and it is left to the interaction with the world to give rise to potentially complex phenomena. On top of these and by exploiting implicit knowledge about the world and the task, we can significantly simplify our life as designers and by using such reactive autonomous agents we can build low-cost, robust, fault-tolerant systems that demonstrate conceptual as well as pragmatic ease of design.

Although the defining property of autonomous agents is to be independent of human intervention and control and thus responsive or reactive to the dynamic environment in which they are situated, it is now widely accepted that other architectural properties are equally necessary, such as adaptivity and learning, modularity and reconfigurability (see for example [14]). Adaptivity is the ability of the agent to modify its behavior locally to fit the environmental conditions so as to better fulfil its goals and satisfy its needs. Learning is longer-term adaptivity and may be permanent or reversible according to the task at hand. Finally, modularity and reconfigurability concern the ability of the designer to scale or trim down the agent's behavior and configure it to specific requirements. Note that autonomous agents research is driven in its majority by biological metaphors, analogies and principles [15,16].

Actual use of autonomous agents to industrial contexts has already been sketched (see, for instance, [9,17-19]). Such autonomous agents, that need to comply with flexibility and adaptivity constraints, have come closer to the recently appeared service robots/agents [20].

## 4. Case study : A regulation problem

### 4.1. PLANT SETUP

We assume [21,22] an industrial plant that consists of four service units (for instance, assembly cells) that issue demands for tool use. The tool-manager robot goes from unit to unit responding to those demands. Each service unit increments demands periodically up to an upper limit steady state (which corresponds to a unit demanding maximal tool use). Beyond this point, the unit's demand does not increase anymore (the production of this unit is temporarily stopped). Obviously, this arrival process alone tends to saturate the system (all units tend to maximum demand). Our goal is to find suitable reactive robot behavioral specifications that will delay this saturation as long as possible, so as to maximize service units' production (assuming that the robot can't be quick enough to service all units at no time). Demands represent service times. The robot stands initially in its rest place in the center; it repeatedly goes to a unit, services it and returns to the center. Each move from the center to a unit or vice versa needs $T_M$ time units, while each service lasts at most $T_S$. All the results reported are averages over 10 runs, each lasting 1000 time cycles, with $T_M$=10 and $T_S$=3. All simulations have been conducted without and with noise; addition of noise means that with a certain probability -which in the present experiments has been set to 5%- the demand is incremented by 1. This corresponds to failures or other unexpected events in the service unit, that actually delay production and augment tool demand. Most of the experiments used a period of 3 to 5, in which case the default saturation of the system is not too quick and the operational consequences of different motivational control schemes are therefore more visible.

Based on previous experience with reactive robots, we have created one service task for each unit. The tool-manager robot's actuators systems repeatedly evaluate those tasks' motivations and "choose" for execution the one that expresses maximal preference. For this to work, all maximal demands have been normalized to a value $d_{max}$, which for the present experiments has been set to 40. Each task's motivation is a function of the corresponding unit's demand, such that when demand = 0, motivation = 0. This way, the robot's "nirvana" is when all its motivations are 0; every time its motivations are non-zero the robot tries to act in a way that will bring them back to 0. The presence of the process, that tends to saturate those demands, perturbs the motivational state of the robot and causes it to act/react. The robot control programs are written using a cellular control architecture [22,23], while the actual high-level pseudo-code run by each service task is:

- If the robot is at the corresponding service unit, service it for a computed amount of time.
- If the robot is at center, go to service unit.
- If the robot is elsewhere, go to center.

The tasks' motivations are being evaluated after every such step (move or service). It should be stressed that the navigation and perception systems of the robot have been written once and for all in the beginning and never changed since. Sole intervention in the motivational level of each task has allowed drastic increases in the performance to be observed, as will be shown below.

Note that this is *not* a usual resource allocation problem (see, for example, several papers in [24]): we assume that the individual service units possess a local inventory of tools that they use at will and there is an additional set of tools that need to be shared across units and whose use is sporadic. The tool-manager robot regulates the units' demands for those more "rare" tools, which it is more cost-effective to share rather than possess local instances of them. For this to work, we also assume that individual service units make optimal local use of those tools, for instance by regrouping in time operations that depend on these tools. This flexibility of operation is generally achieved through use of distributed representations of the industrial process/product, such as the assembly Petri nets used in [25] and whose properties have been explored in more detail in [26]. Such a degree of local intelligence in the service units enables us to afford a tool-manager robot that will decide alone when to retreat from a given unit, in which case it will withdraw all tools previously allocated to it, to make them available to other units. Therefore, the effective operation of the units-manager system depends on the existence of such a cooperative coupling. We can certainly proceed to more complex allocation cases, however, by only defining different motivations for different instances of tools that will be therefore treated and allocated independently. In the following section, we investigate the general principles of use of a motivationally autonomous reactive robot in the simplest case described above.

## 4.2. MOTIVATION AND OPERATIONALITY

Reactivity has been opposed to the traditional planning approach in an attempt to account for real-time behavior that would not suffer from brittleness of traditional knowledge representation techniques. We believe that time has come for reactivity to be dissociated from the stimulus-reflex concept (originally promoted as the foundation of real-time responsiveness) and to be regarded as purely phenomenal, while in reality it will preserve its real-time character by relying on well-chosen internal "representations" that will ensure operationality. We propose here the animal motivational control theory (a good survey may be found in [27]) as a valid domain of inspiration when it comes to combining reactivity and autonomy with operationality. Related work to this investigation of motivational control schemes includes [28-30].

***Persistence.*** The first set of experiments were conducted using a naive motivation for each task.

`motivation=demand`

This led to the robot never servicing a unit, because at each evaluation of the tasks' motivations, and due to asynchronicity, another one was selected. The solution to this problem is to introduce some persistence: once a task is selected, it should persist a little more beyond the point it has absolute (naive) priority. This mechanism is necessary because the switching costs from task to task are very high relative to the satisfaction gained from each task (here, the $T_M$ cannot be neglected, otherwise the satisfaction state would be immediately attainable).

The first persistence mechanism involved an additive factor to be applied only to the selected task.

`motivation=demand+factor`

Experiments with different values for the factor showed that the robot was often stuck to the first unit it serviced, never going to the other ones. Analyzing the above expression for the motivation, we see that this can only happen whenever

$$\text{factor}-T_S+(T_S \text{ div period})>0$$

(the second and the third term of the left hand are the quantities of demand serviced and renewed respectively, while a task is active). For period>$T_S$, the above is true whenever factor>$T_S$. Using a factor lower than that only yields a minor improvement, since the robot still abandons a selected task too quickly. A few of the results are summarized in Table 1.

Next, a multiplicative factor was introduced, again to be applied only to the selected task.

`motivation=demand*factor`

Obviously, that factor should be greater than 1. Experiments with different values for the factor showed that the performance of the system improved substantially - compared to the additive persistence factor- and stabilized with a descending tendency after `factor=period`. In the worst case (when all service units are saturated with their maximum demand value $d_{max}$), it makes sense to have the robot satiate units in turn. In that case, a selected task should persist for as long as its demand is non-zero, hence factor>$d_{max}$. Comparative results for `factor=period` and factor=1+$d_{max}$ are given in table 2.

| | *Period=3* | *Period =4* | *Period =5* |
|---|---|---|---|
| *factor=2* | 110(57)<br>36.74 | 109(57.75)<br>35.336 | 44(23.3)<br>35.019 |
| *factor=3* | 110(57)<br>36.74 | 65.8(34.95)<br>35.808 | 1(0.5)<br>35.844 |

Table 1: *Comparative results for the additive persistence case (without noise): number of moves (average service time per unit), average demand per unit. Note that the units are generally saturated, so that even in the case where the robot is stuck (lower right case) the average performance is not much different from the other cases.*

| | *Period=3* | *Period =4* | *Period =5* |
|---|---|---|---|
| *factor= period* | 37(156.75)<br>30.893 | 35.9(156.25)<br>26.498 | 37(155.93)<br>22.127 |
| *factor=* $1+d_{max}$ | 29(177)<br>26.975 | 31(171.5)<br>23.205 | 34(164)<br>19.342 |

Table 2: *Comparative results for the multiplicative persistence case (without noise): number of moves(average service time per unit), average demand per unit. Compare with table 1 to see the substantial improvements from the additive case.*

Persistence is always imperative, even if the system is not heavily loaded, because it greatly reduces switching costs and hence enhances performance. Results in the case of period=25 or 30 are evidence to this (see table 3).

| | *Period=25* | *Period =30* |
|---|---|---|
| *Without persistence* | 90(17.825)<br>11.301 | 88.8(19.25)<br>7.415 |
| *With persistence* | 82(37.5)<br>1.793 | 84(31.25)<br>1.507 |

Table 3: *Comparative results for high periods with or without persistence (without noise): number of moves(average service time per unit), average demand per unit. Note that in such high periods the units saturate more slowly than in lower ones.*

To write the corresponding cellular control program, however, we needed a continuous function for the persistence factor, so we adopted the exponential time-convergent

$$\text{factor}=1+d_{max}(1-e^{-t/T_p})$$

which rises from 1 to $d_{max}+1$ (t is the time that has elapsed since selection of the task). Since this rise should happen almost immediately after a task is selected, the $T_p$ constant should be such that $1+d_{max}(1-e^{-1/T_p})>d_{max}$, which gives $T_p<1/\ln(d_{max})$. For the present experiments we used a safe value of $T_p$ approximately equal to $1/2\ln(d_{max})$, $T_p = 0.13$ (since $d_{max} = 40$).

***Autonomy.*** Autonomy means freedom from control. Up to this point, we have introduced persistence mechanisms to make the robot operational in its role. However, the robot is not autonomous yet, since it may be manipulated by vicious/capricious greedy units that will emit maximal demand all the time (this may also be the case of defective units where repeated failures keep the demands saturated). Indeed, simulations have shown that the robot tended to get stuck in such malicious or defective units without ever noticing it was not productive. What appears compulsory is some kind of local "boredom" or decay factor that would make the persistence factor drop gradually back to 1 after a period of elapsed time $T_R$ (for which, $T_R > 2T_M + d_{max}(1+1/period)$ should hold). In terms of motivational systems theory, we need an additional variable representing an internal incentive: this variable is initialized to $T_R$ whenever a task is selected and drops gradually with time.

$$factor = var(t) * (1+d_{max}),$$
$$var(0) = T_R.$$

It may be said that this is a hedonistic variable, i.e. one that expresses expected future pleasure; this expected pleasure drops with time, as though the motivated robot was becoming satiated. For permanent failures, an additional decaying factor further toward 0 was introduced, but this is dependent on the types of failures we want to be able to detect and recover from.

## 4.3. A STUDY ON SPECIALIZATION

The same process has been simulated and studied for multiple robots, in search of self-catalytic models of specialization, that would make each of the robots specialize in a fraction of the service units. Self-catalysis means that the motivation of a task depends on a term that is reinforced on execution of the task. Again, a number of experiments have been conducted with additive, multiplicative, multiplicative with an upper bound and self-regulated factors. The second case was far superior to the first, but motivational factors could rise to a very high numeric value; definition of an upper bound solved this problem but resulted in phenomena similar to those observed in the naive persistence case. Self-regulation instead of self-catalysis of the specialization factor has yielded the best results for multiple robots, but still only marginally better than the equivalent case of multiple robots with only persistence. This hardly detectable improvement of the performance with specialization compared with sole persistence can be attributed to the asynchronicity of the system's operation which induces, for example, slightly better regulation of a single-robot 2-unit plant than a 4-unit one that is managed by two robots.

We then proceeded to examine the case of interdependent service units where the output of one unit is fed as input demand to one or more other units, so that a flow network is built. This study was prompted by the observation [21] that specialization as studied above is probably not beneficial because service units are independent and no genuine "load balancing" is necessary in the system. The network structures studied were multi-level structures with full or 1-1 connectivity between levels. To our surprise, and while the self-regulated specialization mechanism has been found to work correctly (for example, in the case of 3 robots and 9 units, each robot quickly specializes to three of the units), there has been no quantitative improvement to the performance of the overall system.

The lesson that may be drawn from the above is that specialization is not a good idea in this case and the reason appears to be that switching costs between tasks are the same. In all cases persistence alone looks sufficient. We are currently studying the case of diverse switching costs between tasks in both cases of independent units or units in a network.

## 4.4. LESSONS

The above formulation and study of the adopted process control problem allows a number of interesting observations to be made:

If the control system of the reactive robot is decomposed into a number of activities, or else tasks, that define a desirable steady state, the action selection, or else arbitration, problem may be tackled directly at the actuator systems' level and rely on individual tasks' expressed motivations. The details of separate tasks, such as local adaptation mechanisms, can therefore be addressed independently of the motivational system.

Furthermore, discrete solutions, such as mere presence/absence of motivation are suboptimal. What is necessary is continuous functions at all levels of the reactive control system. Time may be used as an additional, active, indirect control parameter, such as in the case of the time-convergent multiplicative persistence; often enough, the only real measure of success/failure a robot can have about itself is the amount of time spent on a certain activity. As designers of a robotic system, we ought to exploit this. Continuity in the process itself also induces noise to the inputs of the robot's control system. Precisely, all the experiments have been conducted with and without noise and it has been found that the relative performance of the system is a little better when there is noise present than when there is not. Actually, noise is a source of variation and it is known that *"the real cause of stability in a distributed system is sufficient diversity"* [31]. Furthermore, noise in the sense of asynchronicity between process events and robot perception has been found necessary to ensure operationality of the robot's motivational system: for example, we simulated the case of demand increments of 2 with a period of 2 and we found that the robot was always stuck to the first serviced unit, because actually the operation of the system and the robot were synchronous, so that the robot could not *see* (perceive) any worthwhile changes in other units and consequently would not service them.

This study also showed that there should be structural/operational coupling between the robot and the process (the robot's world). This coupling has the sense of using certain process parameters (such as $d_{max}$ and $T_M$) while designing the robot's motivational control system. The identification of the exact *laws* of this coupling can result from a worst-case analysis of the overall system. Operationality in the above sense means that the behavior/performance of the robot with a given architectural substrate will be "statistically optimal" for the right parameter setting among all available alternatives.

Finally, and since there is process-robot coupling, the robot may be viewed externally as following a purposeful, roughly preset, highly robust and adaptive plan. However, the robot does not manipulate any plans; it merely reacts to the situation at hand. The plan, if any, is in the designer's head (also in the observer's eye): the role of the robot's creator is then to identify the proper operational parameters and devise laws of robot/world coupling that will ensure task completion in such an apparently purposeful way. We have shown in the above simple case that this is feasible: the robot itself doesn't need to plan, because its designer has done it; the idea is that reactivity is the optimal solution in the noisy system case, but it should be based on enough operational knowledge about the overall system. In this sense, we are not very far away of the on-line versus off-line debate within the planning community: the designer has taken care of the off-line component which concerns the structure of the motivational control system, while interaction with the world will provide the actual data that will be used by the on-line decisional process. One trick of that mechanism is the presence of evaluation points in the motivational system: as is described before, the motivations of the robots are re-evaluated only at certain points and not continuously. On the one hand, this is to account for local noise that might make the robot oscillate between activities, while on the other hand it concerns an implicit system assumption: the robot can afford checking its motivations with considerable lags because the individual activities are adaptive and robust by themselves.

## 5. Conclusions and perspectives

We have briefly exposed observations and experiments that lend support to the proposition that autonomous agents are a conceptually and operationally valid candidate for modeling and design of cellular manufacturing systems.

We studied the case of a production plant based on the modeling of the agent(s) and the controlled process as two coupled dynamical systems with conflicting steady states that constantly perturb each other. The fortunate consequences of such a modeling is that the motivational state space of the robot may be designed in advance given the process characteristics and this design should be based on a worst-case analysis.

Besides modeling and design of cellular manufacturing systems, we also feel there is much room for the use of autonomous agents methodologies at a "meta" level, to support the solution of some problems in an abstract way by a multi-agent system operating in parallel with the real system. One such recent research direction that is directly applicable to the cellular manufacturing domain is the social insect inspired "ant search and optimization" methodology, where a colony of abstract ant like agents solves a search or optimization problem [32]. This methodology has been successful in classical computer science problems, such as the traveling salesman problem [33], as well as in engineering problems, such as load balancing in a telecommunications network [34]. We believe that methodologies like this one are directly applicable to real manufacturing problems, such as adaptive scheduling or flow regulation.

## References

[1] Jones, P.F.: *CAD/CAM: Features, applications and management*, Macmillan Press Ltd., London, 1992.

[2] Ranky, P.G.: *Computer integrated manufacturing - An introduction with case studies*, Prentice-Hall, Englewood Cliffs, NJ, 1986.

[3] Rolstadås, A., Ed.: *Computer-aided production management*, Springer Verlag, Berlin/Heidelberg, 1988.

[4] Wright, P.K. and Bourne, D.A., Eds.: *Manufacturing intelligence*, Addison Wesley, Reading, MA, 1988.

[5] Bjørke, O. Computer-aided part manufacturing, *Computers in Industry*, (1979):3-9.

[6] Almgren, R.: *On knowledge-based planning and programming systems for Flexible Automatic Assembly*, University of Linköping, Studies in Science and Technology, Thesis no. 176, LiU-Tek-Lic-1989:16.

[7] Kusiak, A.: Manufacturing systems : A knowledge- and optimization-based approach, *Journal of Intelligent and Robotic Systems*, 3(1990):27-50.

[8] Meystel, A.: Intelligent control : A sketch of the theory, *Journal of Intelligent and Robotic Systems*, 2:97-107, 1989.

[9] Fischer, K.: Modelling autonomous systems in a flexible manufacturing system, *Proceedings of the IJCAI '93 Workshop on Dynamically Interacting Robots*, Chambéry, France, August 1993, pp. 109-116.

[10] Baldwin, K.E.: Autonomous manufacturing systems, *Proceedings of the 1989 IEEE International Symposium on Intelligent Control*, Albany, NY, September 1989, pp. 214-220.

[11] Allen, J.F., Hendler, J. and Tate, A., Eds.: *Readings in Planning*, Morgan Kaufmann, San Mateo, California, 1990.

[12] Brooks, R.A.: Intelligence without Representation, *Artificial Intelligence, Special Issue on the Foundations of Artificial Intelligence*, 47(1-3):139-159, 1991.

[13] Brooks, R. A.: Intelligence without Reason, *M.I.T. AI Memo No. 1293*, April 1991, also *Computers and Thought Workshop, IJCAI-91*.

[14] Maes, P.: Modeling Adaptive Autonomous Agents, *Artificial Life*, Vol 1(1&2):135-162, 1994.

[15] Maes, P., Ed.: *Designing autonomous agents: Theory and practice from biology to engineering and back*, MIT/Bradford Press, Cambridge, MA, 1991.

[16] Steels, L., Ed. (1993). *The biology and technology of autonomous agents*, Springer Verlag, 1993.

[17] Arkin, R.C. and Murphy, R.R.: Autonomous navigation in a manufacturing environment, *IEEE Transactions on Robotics and Automation*, Vol. RA-6(4):445-454, August 1990.

[18] Doty, K.L. and Van Aken, R.E.: Swarm robot materials handling paradigm for a manufacturing workcell, *Proceedings of the 1993 IEEE International Conference on Robotics and Automation*, Atlanta, GA, May 1993, pp. 778-782.

[19] Vaario, J. and Ueda, K.: Self-Organization in Manufacturing Systems, *Proceedings 1996 Japan-USA Symposium on Flexible Automation*, July 1996, Boston, MA, pp. 1481-84.

[20] Schraft, R.D., Degenhart, E. and Hägele, M.: Service robots: The appropriate level of automation and the role of operators in the task execution, *Proceedings of the 1993 IEEE Systems, Man and Cybernetics Conference*, pp. 163-169.

[21] Tzafestas, E.S.: Reactive robots in the service of production management, *Journal of Intelligent and Robotic Systems*, 21(2):179-191, 1998.

[22] Tzafestas, E.S.: *Vers une systémique des agents autonomes : Des cellules, des motivations et des perturbations,* Ph.D. Thesis, Université Pierre et Marie Curie, Paris, December 1995.

[23] Tzafestas, E.S.: A cellular control architecture for autonomous robots, *Proceedings of the 1994 International Workshop on Intelligent Robotic Systems*, Grenoble, July 1994, pp. 70-79.

[24] Desrochers, A.A. and Silva, M., Eds.: *IEEE Transactions on Robotics and Automation, Special Issue on Computer Integrated Manufacturing*, 10(2), April 1994.

[25] Tzafestas, S.G. and Tzafestas, E.S.: The blackboard architecture in knowledge-based robotic systems, pp. 285-317, *Expert Systems and Robotics*, *NATO ASI Series Vol. F71*, T. Jordanides and B. Torby (Eds.), Springer Verlag, Berlin/Heidelberg, 1991.

[26] Suzuki, T., Kanehara, T., Inaba, A. and Okuma, S.: On algebraic and graph structural properties of assembly petri nets, *Proceedings of the 1993 IEEE International Conference on Robotics and Automation*, Atlanta, GA, May 1993, pp. 507-514.

[27] Toates, T.: *Motivational systems*, Cambridge University Press, Cambridge, UK, 1986.

[28] Balkenius, C.: Motivation and attention in an autonomous agent, *Presentation Workshop on Architectures Underlying Motivation and Emotion - WAUME 93*, University of Birmingham, 1993.

[29] McFarland, D. and Bösser, T.: *Intelligent behavior in animals and robots*, MIT Press, Cambridge, MA, 1994.

[30] Halperin, J.P.R.: Cognition and emotion in animals and machines.

[31] Hogg, T. and Huberman, B.A.: Better than the best: The power of cooperation, in L. Nadel and D. Stein (Eds.), *SFI 1992 Lecture Notes in Complex Systems*, pp. 163-184, Addison-Wesley, 1993.

[32] Dorigo, M., Maniezzo, V. and Colorni, A. The Ant system: Optimization by a colony of cooperating agents, *IEEE Transactions on System, Man and Cybernetics*, Part B, 26(1996):1-13.

[33] Dorigo, M. and Gambardella, L.M.: Ant colony system: a cooperative learning approach to the traveling salesman problem, *IEEE Transactions on Evolutionary Computation*, 1(1997):53-66.

[34] Schoonderwoerd, R., Holland, O., Bruten, J. and Rothkrantz, L. Ant-based load balancing in telecommunications networks, *Adaptive Behavior*, 5(1997):169-207.

# 20

# MAN-MACHINE COOPERATION FOR THE CONTROL OF AN INTELLIGENT POWERED WHEELCHAIR

G. BOURHIS and Y. AGOSTINI
*Laboratoire d'Automatique des Systèmes Coopératifs,*
*Université de Metz, Ile du Saulcy,*
*BP 80794*
*57012 METZ Cedex 01, France*

## 1. Introduction

The usefulness of electric wheelchairs no longer needs to be demonstrated. Of course they bring mobility autonomy to people with severe motor disability, but also help the cognitive development of disabled children [1]. However their use may be difficult or impossible for some persons due to excessively weak residual physical capacities, a serious spasticity or cognitive impairments. For instance, some people with tetraplegia only have access to an on-off sensor. This makes controlling a wheelchair particularly difficult, especially for delicate manoeuvres: the only possibility consists in selecting the choices "forward-backward-left-right" on screen or on a LED scanning device.

For usual users of electric wheelchairs the addition of some automatic functionalities, although not indispensable, can bring extra driving comfort, especially during delicate manoeuvres such as narrow door passages. This contribution is similar to the installation of driving assistance on a car.

However, a simple improvement of the classical powered wheelchair can often palliate difficulties of control. A lot of progress has been made to this purpose in wheelchair ergonomy as well as automatisms. Many works on the conception of specific interfaces are also described: adaptation of a control joystick [2], development of head movement sensors [3], isolated-word speech recognition [4], scanning devices [5], ...

If these improvements are not sufficient, we can consider the addition of intelligent primitives resulting from research led in mobile robotics. They are of two types depending on whether or not they necessitate a global knowledge of the environment the robot moves in:

*Local primitives*

- Obstacle detection: it is generally a simple collision prevention or, possibly, a fall prevention in non-protected staircases. This last functionality implies installing sensors positioned towards the ground. These sensors are therefore dedicated to this function.

- Obstacle avoidance: the obstacles are detected and the robot goes round them according to a strategy based only on sensor information,
- Wall following: this primitive may be particularly interesting in long corridors for persons having difficulties holding a straight path due to an important spasticity for example,
- Person following: without any effort from its pilot, the robotized wheelchair follows a valid person,
- Passing through a door: this primitive is a fine manoeuvre assistance in narrow passages,
- Docking: a feedback control at the end of the motion may help to achieve some maneuvers like typically positioning the wheelchair at a table or workstation.

*Global primitives*

- Automatic path following: these may be recorded, generally through learning, or planned in a global model of the environment,
- Localization aid: this concerns persons suffering from cognitive trouble and having thereby difficulties in planning their way. Realizations of this type are not yet installed on wheelchairs but are on the other hand studied for persons moving about without mechanical assistance [6].

The various functionalities described above constitute an almost exhaustive list of what is being done or could be done at present concerning mobility assistance. However, real needs in this area remain very badly defined: few studies to this purpose have been undertaken, essentially through lack of reference tools. Indeed there is currently no polyvalent long-term home or rehabilitation centre system. We only get replies to questionnaires a priori on what it would be good to achieve. In [7], for example, we can note that aspirations concerning assistance to displacement are very varied depending on the type of disability of the potential users or even on their age.

In this chapter, after a background on smart wheelchairs, we are going to define the specificities of this application of mobile robotics. From this study we deduce the interest of conceiving a system in which man and machine cooperate to achieve the mobility task. An example of a system allowing this cooperation is described to illustrate this purpose.

## 2. Background

Most functionalities implemented on robotized powered wheelchairs described in the literature are local servo-control primitives. The simplest servo-control consists in following a line on the ground. In [8] the pilot of the wheelchair positions a magnetic line sensor placed under the footrests and engages an automatic line following. Stopping is performed at predetermined places located by infra-red sensors. If an obstacle is met on the path the wheelchair stops and waits for the way to be cleared.

The " Navchair" system developed at Michigan University is a semi-autonomous electric wheelchair equipped with a belt of ultrasonic sensors [9]. For its navigation it uses two original algorithms in order to deal with the ultrasonic data. The EERUF algorithm allows fast sensor reading while being free of the crosstalk phenomena. From a local model of the environment the VFH algorithm looks for the best forward direction according to the obstacles and to the direction pointed by the joystick. Wheelchair control is therefore performed both by the human operator and the machine.

Let's note that a rather similar realization has recently been described in [10]. The authors emphasize the low cost of their system and on its standard M3S interface, which allows various rehabilitation devices to be connected together [11].

The "Smart Wheelchair" of the Edimburgh CALL Center integrates various functionalities concerning mobility (contact detection, automatic obstacle avoidance, line following) and communication (varied man-machine interface sensors, speech synthesizer, scanning device) [12]. This wheelchair is therefore widely configurable. More than a mobility assistance, it is a tool to develop the cognitive abilities of young children with motor disabilities (environment learning, social behavior). For this purpose the robot's architecture is conceived on a reactivist model.

In the framework of the OMNI European project [13], a German team implements local primitives (obstacle avoidance, docking) and global tasks such as recorded path following on an omnidirectional wheelchair. These paths are defined by infra-red beacons or by lines on the floor. It uses ultrasonic sensors whose reliability is insured by the coupling to infrared sensors. An environment control and a communication assistance may also be connected to the system through a standard M3S bus.

Another European project, the SENARIO project, has begun recently [14]. Two control modes of a robotized wheelchair are proposed. A semi-autonomous mode consists in moving forward with prevention of collision, and an autonomous mode allows the wheelchair to follow recorded paths. The accent is put here on risk prevention. Various sensors are used to this end: ultrasonic sensors, inclinometers, odometry.

In [15] a navigation assistance by path learning is described. The electrical wheelchair is equipped with a PC microcomputer, a dead-reckoning sensor and a vision system composed of two cameras. The wheelchair user's usual home paths are recorded by having them executed by a person who controls the vehicle well. Mobile location is ensured by recognition of beacons made up of marks stuck on the walls associated with odometric data.

Let's note finally in [16] an original project which concerns a "move to that point" primitive implementation. The wheelchair pilot chooses a destination on an image coming from a camera, which allows target tracking during the motion. The local obstacle avoidance is carried out thanks to ultrasonic sensors.

We can observe that the application of mobile robotics to the assistance to people with disabilities has recently lost its marginal character. Two points are nevertheless to be underlined: first of all we notice that there still is no return of long term evaluation of systems described in the literature. The above described "Smart Wheelchair" is an exception but its objective is essentially educative. Furthermore the man-machine relation is in general only briefly evoked in the description of the different systems when it is not totally eluded. This aspect nevertheless appears as essential for the success of projects of this type.

## 3. Specificities of this application

### 3.1. GENERAL FRAMEWORK

It is usual to divide the field of robotics in two sub-fields, industrial robotics and service robotics. Rehabilitation robotics and, more specifically, intelligent wheelchairs, are closely connected to this last category. Two important characteristics follow from that: on the one hand the robot shares its workspaces with the humans, on the other hand the user is not, a priori, competent in the technological field. This last point, associated to the fact that the user is a person with disabilities, brings to light the importance of the man-machine interface.

Moreover by many aspects this application is also close to the teleoperation framework with an important difference: the operator, located on the mobile base, has a direct visual return of the robot's environment. The man-machine cooperation problematics are however very similar.

Finally the robot's moving environment is generally indoor, therefore partly structured by walls and doors that constitute natural beacons. Nevertheless in many real situations numerous modifications of the modelled environment may appear: a closed door is an impassable obstacle, objects, persons or other wheelchairs may hamper or forbid passage.

### 3.2. SPECIFIC CONSTRAINTS

#### 3.2.1. *Versatile and modular system*

The great diversity of users and situations is another important characteristic of the application. The environment may first of all be the user's home or a public structure such as a rehabilitation center or a hospital. Potential path clutter as well as distances to cover are therefore very variable. Wheelchair pilots also have very varied cognitive and physical possibilities.

The man-machine interface of a robotized powered wheelchair therefore has to be modular and configurable. This goes for its software as well as its control sensors. The latter, often described in the literature [17,18], may be classified from a functional point of view into three categories:

- Proportional sensors: they are generally hand controlled sensors like a joystick or, less frequently, a chin controlled sensor. However other anatomical sites like, for example, the nape of the neck [19] or the foot [20], may also be used.

- A single on-off sensor: it is a simple switch controlled by head movements for example. In order to command a complex technical aid or the environment it is necessary to associate it with a scanning of choices on a screen or a board of LEDs.
- Two (or more) on-off sensors: we can then combine an action order and a safety stop order. This may be achieved for example by a breath sensor where exhaling actuates a movement and inhaling, easier to control, stops it. Note that speech recognition systems functioning with isolated words also come under this category.

Data coming from these different types of sensors may be filtered in particular for tremor disabled people. For instance some authors use neuron networks to distinguish involuntary gestures from voluntary movements [21]. More simply, we may also act on the sensor's mechanics [22].

### 3.2.2. *Cost-performance compromise*

It is not very usual to take into account the implementation cost price in scientific publications. In an application of high technology technical aid for people with disabilities, it is however essential in the initial technological choice due to the tightness of the potential market [23]. In particular it is preferable whenever possible to use equipment non-specific to the application. Unless we bring in a supplementary mechanical functionality, like an omnidirectional characteristic, it is therefore necessary to use a commercialized powered wheelchair as a mobile base. Besides the practical problem of the sensors implementation, this imposes some constraints concerning the robot autonomous navigation: the base has a non-holonomic characteristic on the one hand, and a weak reliability of the odometry on the other hand due to the unequally inflated tires and the non-uniform and variable distribution of the person's weight on the machine.

The choice of sensors also strongly depends on the cost criterion. In this area we generally content ourselves with ultrasonic sensors possibly associated to infrared sensors. The use of cameras is acceptable for the achievement of simple functions which don't need important data processing, a beacon recognition for example.

### 3.2.3. *Psychological acceptability*

The use of a robotized technical aid also poses psychological acceptability problems. Following an evaluation of the Masters robotized arm, in [24] it is emphasized that the robot should respect the user's environment. First of all, this implies taking care of the final realization design. In particular, the environment sensor implementation should be very discreet.

Partly depending on the design, the wheelchair comfort is a parameter which may imply technical problems: a lack of comfort may make the mobile base control difficult, for example if the pilot has trouble in keeping his head straight. For the same reason it is necessary to avoid sudden speed or direction variations during automatic or manual motions. The robot technical conception also has to take into account psychological factors. First of all, the man-machine interface has to be very user-friendly since in principle it addresses non-technician persons. It is also necessary to set up a dialogue necessitating a minimal number of orders for persons with reduced physical capacities.

On the other hand the automatic motions of the wheelchair have to seem natural to the user. The autonomous course has to correspond to the one the pilot would have used in a direct manual control. Paths have therefore to be smoothed and direction changes to be predictable, either because they are natural, or because the robot previously slows down, or again through information returned to the pilot via the man-machine interface.

Still, for this acceptability constraint, it is necessary for the user to constantly remain the system's master. Thus in the automatic movements of the wheelchair he has to be able to intervene if a problem appears, in the worst case by stopping the machine. Furthermore orders have to be sufficiently intuitive to inspire the confidence of the operator.

#### 3.2.4. *Safety and robustness*

Robustness and security aspects are particularly sensitive in such an application where the robot operator is situated on the machine and moreover has, on hypothesis, reduced physical capacities. It is therefore essential to exclude all blocking situations: even if it means degrading the functioning of the robot, for example to a manual mode, the pilot has to be able to resume control at any time. This may be the case if the mobile base is blocked due to a safety area that we would have installed around it or for an important loss of localization in a planned movement.

In any situation a wheelchair stop may be decided by a single action on the command sensor. Taking into account the great diversity of the users this safety command may of course be adaptable to each particular case. However, as is emphasized by several authors [25, 26], total safety and reliability is impossible. So we have to know the maximum acceptable cost/risk ratio according to the expected benefit.

### 3.3. DISCUSSION

Various constraints specific to the application of mobility assistance for people with disabilities have been presented, especially robustness, the necessary cost-performance compromise and psychological acceptability. So that they would not be contradictory it is necessary to attain a cooperation between the man and the robot in the achievement of the mobility task. The person brings his perception and interpretation of the close environment abilities and, unless he has got cognitive troubles, his global comprehension of the situation. The machine brings its autonomous mobility ability of course and, possibly, its knowledge concerning a global model of the environment it moves in.

The man-machine cooperation may take various forms. First, sharing degrees of freedom has often been described in the literature in the framework of manipulator arms teleoperation [27], but also for the control of semi-autonomous wheelchairs [28]: the pilot chooses the forward direction, the machine deals with local obstacle avoidance. Like in numerous studies about complex automatized systems, it's also possible to envision a decision aid from the man to the machine or, conversely, from the machine to the man. We thus use the pilot's and robot's knowledge and competence.

However, before achieving task sharing and trading between the human operator and the robot, we have to know what tasks or what information have to be asked from the man or from the machine but also who distributes these tasks, the man, the machine or both these two entities [29]. The security of the system can for example be ensured by the supervision of the person who visually controls the wheelchair motions, or by the machine thanks to a supervisor which detects control errors or absurd behavior. According to the previously described principles, it seems desirable that in the last resort the man remains the system's master, that is to say that he is always able to impose his choice or, at worst, to take the robot's manual control again. Likewise, task trading may be proposed by the machine, the man taking the final decision.

## 4. Application

### 4.1. THE VAHM ROBOTIZED WHEELCHAIR

#### 4.1.1. *Architecture*

The VAHM (french acronym for autonomous vehicle for people with disabilities) robotized wheelchair conception is based on the above evoked principles [30]. The aim is to achieve an intelligent assistance to electric wheelchair piloting using techniques coming from mobile robotics. Three operating modes are defined for this purpose. In the manual mode we find the classic control of a powered wheelchair with possibly an adaptation of the man-machine interface or an anti-collision security system. In the semi-automatic mode we may call upon local primitives like "wall following" or "obstacle avoidance". Finally, in the automatic mode, the wheelchair performs planned paths with local avoidance of non-modelled obstacles. This system is currently implemented on a Robuter$^{TM}$ mobile base connected to a microcomputer PC486 (Figure 1).

The system architecture is constituted of three levels which depend on the abstraction level of the processed data (Figure 2). The lowest one, the physical level, is composed of the actuators, two DC motors driving the back wheels, and of the sensors, a belt of 14 ultrasonic sensors associated to a dead-reckoning system. So, it processes the robot's numerical data. The intermediary level, the "local level", treats the data according to local knowledge, which depends on the sensors' range for the robot, and on the visual field for the user. The last level, the "global level", adds symbolic information to the local ones in order to set up a communication between the man and the machine.

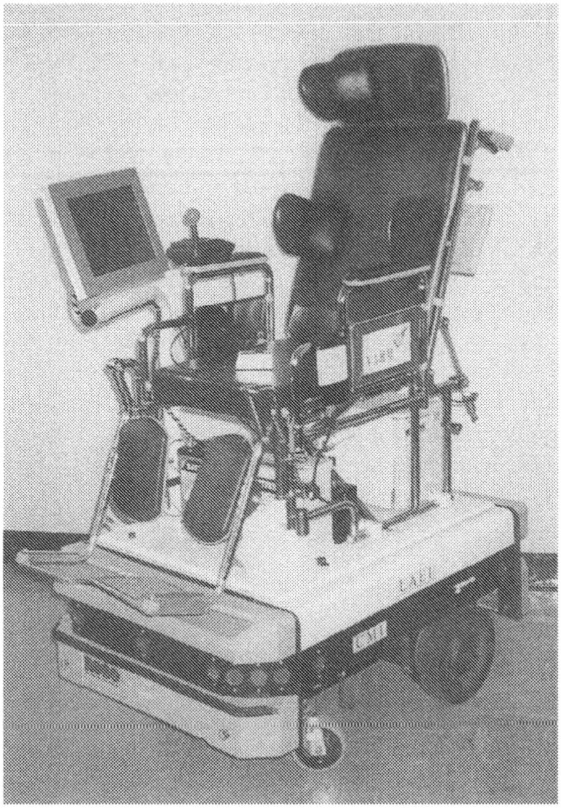

*Figure 1.* The VAHM wheelchair

The tasks situated on the local and global levels are defined by analogy with the main human tasks. The communication task deals with the man-machine information exchange. The perception task processes data research and identification. The navigation task manages the motor activities and, finally, the supervision task deals with the cognitive activities and, more particularly, with the dialogue between the user and the robot. To obtain the wished behaviour, each task has a set of tools and an execution control module.

4.1.2. *Local level*

The second architecture level is composed of the navigation, perception and communication tasks. The navigation task manages the automatic or semi-automatic motions (Figure 3). Automatic movements are obtained by a trajectory planning calculated in a topological model of the local environment [31]. The robot's initial position is defined thanks to the data provided by the dead reckoning system and to possible additional information coming from the perception task. The goal to reach is either a particular point transmitted by the communication task, or a memorized point noticed by a real or virtual beacon transmitted by the architecture global level.

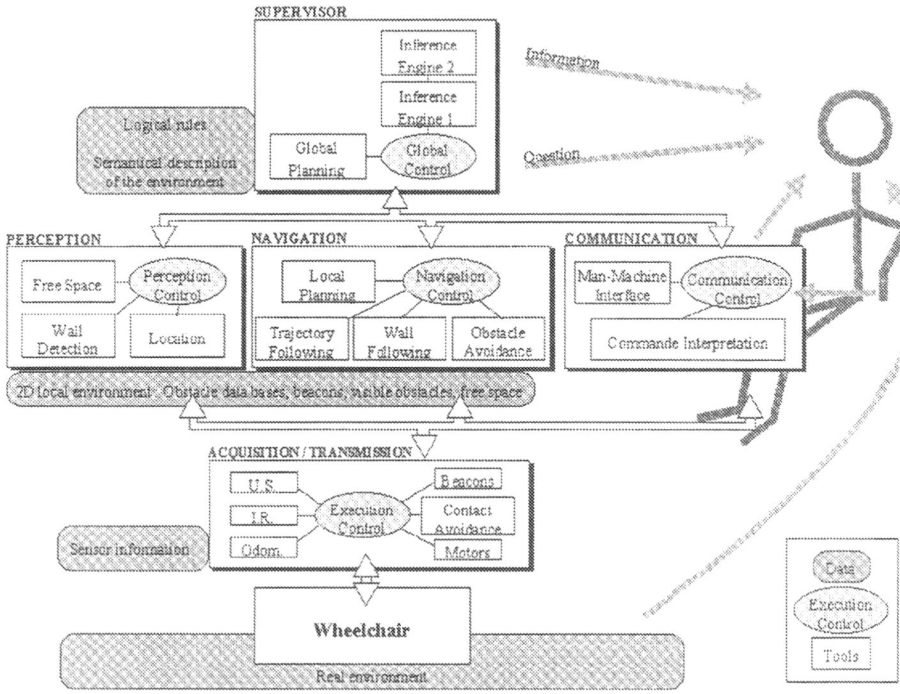


*Figure 2.* The VAHM robot architecture

The semi-automatic motions don't need an environment model. At present it is possible to call upon a "wall following", an "obstacle avoidance" or a "move towards the free space". For the "wall following" primitive, the choice of the wall to follow is transmitted either by the perception task, or by the communication task. The "obstacle avoidance" allows to move along a direction initially determined, avoiding obstacles. The initial orientation and the start or stop orders come from the communication task.

The semi-automatic and automatic motions are achieved locally avoiding the obstacles detected by the ultrasonic sensors.

The perception task brings additional information. It comprises a localization primitive that undertakes a matching between the sensor data and a geometrical model of the environment if this model is available [32]. A second module, whose development is in progress, is entrusted with the detection of the environment's characteristic elements, the walls for example. Finally, a module has to detect the free space.

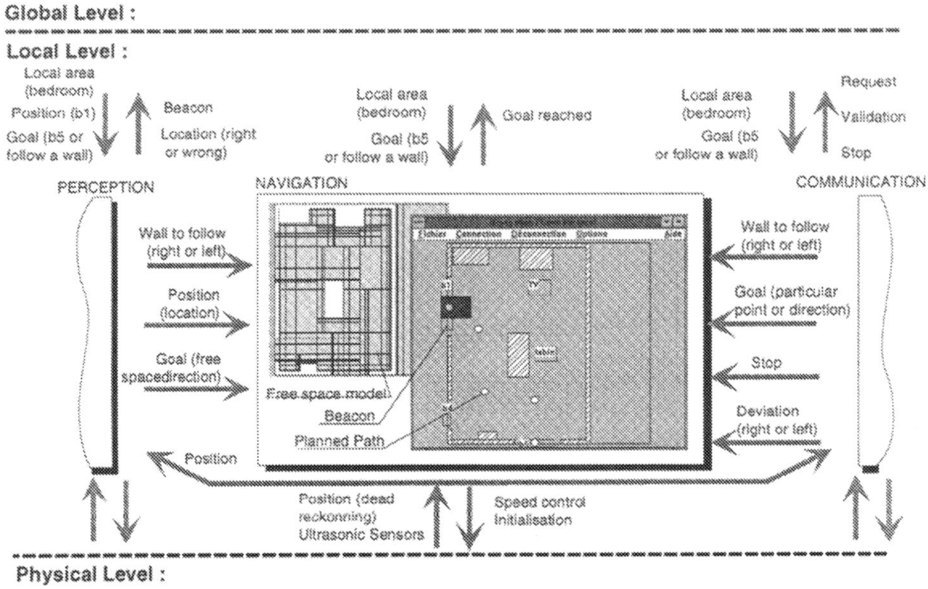


*Figure 3.* Architecture details on the navigation task

The communication task part interprets information or simple orders coming from the wheelchair pilot. For that purpose a conversation is set up through the medium of a screen and of a proportional or an on-off sensor. Moreover, this task chronologically records the man-machine data transfer flow in a "spy file". A second "spy file" is also created to record numerical data such as the ultrasonic data or the speed applied to the wheels. These data allow to evaluate the man-machine system performance and to locate the possible problems such as, for example, too uncomfortable a robot trajectory or an incoherent icon selection on the screen.

### 4.1.3. *Global level*

The highest level of the machine uses global information on the environment to plan an action of mobility, possibly to help the person's decisions and to actuate the local level.

It uses a semantic representation of the environment, for example "the kitchen is limited by b1 and b2 and contains a sink and a fridge" where b1 and b2 are perceivable beacons which show passing points. A unit of knowledge management allows to bring information on the state of the system and to make proposals to the human operator. It is composed of two inference engines. The first manages environment description data. The second inference engine manages general production rules such as "If we are in the living-room and if we are going toward b4 then we are going to the bathroom".

## 4.2. MAN-MACHINE COOPERATION

### 4.2.1. *Behaviour model*

During the performing of mobility action the man and the machine may cooperate at all levels of the previously described architecture. Two cases are to be considered. If we are in the automatic or semi-automatic modes, the machine executes an autonomous movement. Then the man plays the role of supervisor and can help the machine's decision thanks to information provided by the robot and to his own knowledge. Conversely, in the manual mode, the man is the operator and the machine the supervisor.

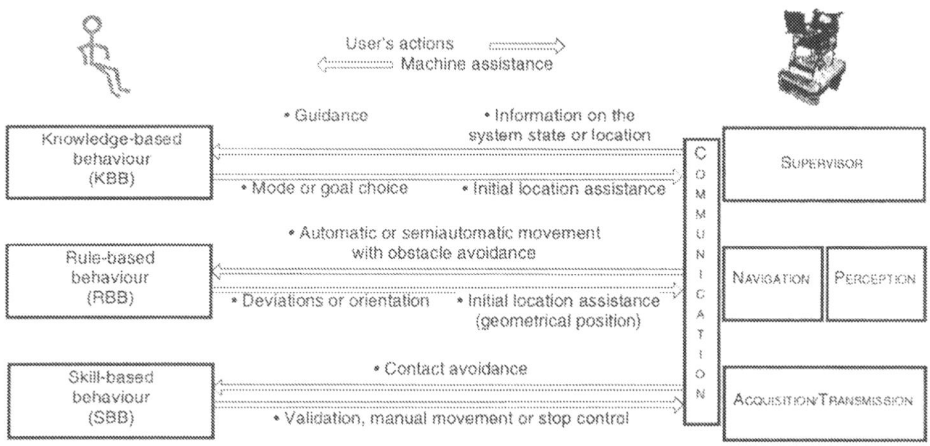


*Figure 4.* Behaviour model

It may be interesting to base the description of the different cooperation modes between the man and the machine on the automated process supervisor model elaborated by Rasmussen [33] (Figure 4). Faced with a problem, three types of behavior are listed. The "Skill based behaviour" is characterized by reflex reactions. The second level corresponds to a "Rules based behaviour". The operator is in a known situation and applies a standard reaction procedure. In the last level of "Knowledge based behaviour" on the other hand, the situation is unknown for the supervision operator. It has therefore to use the totality of its knowledge to establish a strategy of reaction.

### 4.2.2. *Man supervisor*

*Skill based behaviour*: During the performing of an autonomous movement the man as supervisor can react at the three levels of Rasmussen's model. Unless the mobile base is equiped with a prohibitive sensor assortment, the environment perception may be entirely reliable only with the user's assistance. In the circumstances, some obstacles, due to their particular configuration (table with central foot, staircase, ...), are difficult to perceive by ultrasonic sensors. If, approaching these obstacles, the robot does not slow, the pilot can deduce that it has not seen them and then stops the movement. We are then on the reflex level.

*Rules based behaviour*: In the "Rules based behaviour" the human operator helps the machine during the performing of local primitives. In the case of a "wall following" for example, when the robot arrives at a corridor intersection, the person may indicate with a simple order if it has to turn right or left, or stop. Other situations require a human intervention at this same behaviour level: in the performing of a globally planned motion some non-modelled obstacle configurations lead to blocking situations (Figure 5). Thanks to his better environment perception the human supervisor is able to intervene indicating the obstacle avoidance direction. A navigation screen which allows this only requires the display of three choices, right, left and stop.

One last situation brings to light the advantage we have to use the pilot environment interpretation capacity. Whatever operating mode is selected, a safety area is established around the wheelchair thanks to the ultrasonic sensors. Sometimes however this collision avoidance has to be inhibited by the user considering the environment configuration: for example a slightly open door may be pushed by the mobile to free the way. This situation like the ones previously evoked, is simple to evaluate for the pilot whereas it is inextricable for the machine unless prohibitive software and hardware equipment is used.

*Knowledge based behaviour:* The third level described by Rasmussen rests on general knowledge of the supervision operator. In the present case the latter generally has a richer knowledge of the global environment than the machine. Thus, in an automatic movement, the robot indicates the path that it is going to perform by its initial orientation and by displayed or uttered information. If the person is aware of the impossibility of this performing it can stop the progress. Another path is then immediately planned and proposed to the user. At this same level an initial location aid may also be envisaged. When the system starts up, if the size of the modelled environment is great, the location algorithms, which match the model and the sensors data, may fail. Then the machine may call upon the user's help: if he roughly selects the robot's location, for example by pointing the room in which he is situated, the location procedure proves to be really simplified.

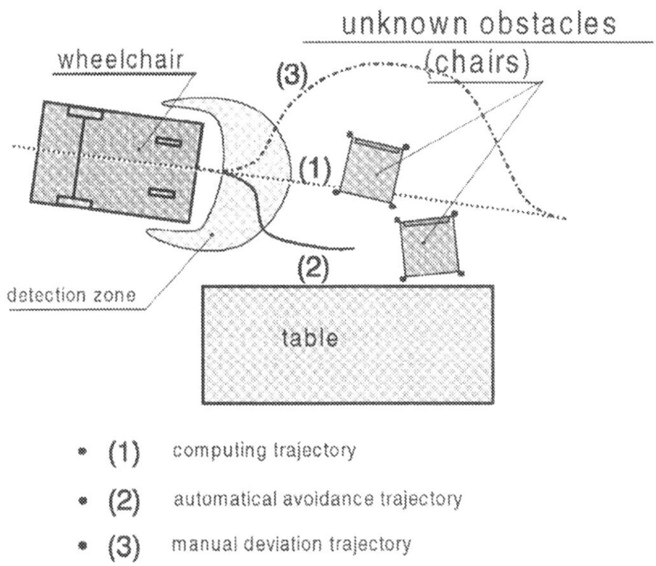


*Figure 5.* Obstacle avoidance

4.2.3. *Robot supervisor*

During the execution of the manual mode which is a direct classical wheelchair control, the machine assists the pilot by decision assistance proposals or by reflex interventions. By analogy with the human operator's behaviour, three types of intervention may also be distinguished for the machine.

*Skill based behaviour*: First of all the robot intervenes on the reflex level when it forbids contact with obstacles or, by a degree of freedom sharing in order to help obstacle avoidance during manual control, which it is not yet implemented on the VAHM robot. If the man-machine interface sensor is a joystick or any other proportional sensor, it will permanently indicate which direction to follow, whereas the environment sensors deal with obstacle avoidance. On the other hand if the wheelchair's pilot only controls an on-off sensor, the command sharing can not be permanent. The trajectory becomes therefore much less stable.

*Rules based behaviour*: At the "Rules based behavior" level the machine may propose operating mode modifications according to the environment, to its performances and to criteria which make up the physical and mental operator workload. These criteria have been defined from a system evaluation [34].

Two main parameters contribute to physical workload. First, the "temporal efficiency criteria" measures the global execution time of an action, including the time required for the orders to be selected on the screen. The effort carried out by the pilot is thus naturally balanced by the measurement duration. The second parameter, the "navigation comfort" criterion, is quantified by the on-line measurement of angular and linear accelerations of the base.

The mental workload is a more difficult notion to estimate objectively [35]. Three criteria seem to strongly participate to this workload. The first criterion, the "validation incoherence", locates erroneous interface uses which imply a certain irritation or a discouragement of the user. The second, the "motion incoherence" criteria, measures the pilot's ability when he establishes and manages a displacement strategy. The last criteria, the "trust" criteria estimates the pilot's stress near an obstacle or, more precisely, his trust in the machine.

*Knowledge based behaviour*: A human decision assistance on the "Knowledge based behavior" supposes that the machine has a better environment knowledge than the man. It is the case if the person, suffering from cognitive trouble, has difficulties in planning a complex action. The robot may then give him indications if he takes a wrong path or, more simply may guide him along the whole path. For that purpose the mobile must at least move in a globally modelled environment. This model is made up of beacons and areas. The beacons are particular elements of the environment such as doors or corridor intersections which allow to roughly characterize a move. The areas are the parts of the environment delimited by the beacons.

## 4.3. EXAMPLE OF USE

### 4.3.1. *Method*

We are going to illustrate our purpose with an example of use. The robot's environment is made up of an apartment reconstructed in our laboratory, a long corridor and an appended room (Figure 6). The whole environment is modelled by beacons which allow to plan a move globally. On the other hand, only the appartment is locally modelled with a topological model allowing trajectory calculations. The pilot is supposed to only be able to use an on-off sensor. So the different choices are effected by a scanning of the screen followed by a validation of the selected icon.

Initially the interface allows to select the choices forward, backward, right or left (Figure 7). This is a classical manual mode where the user has the whole control of the wheelchair's movements. Two graphical windows are managed by the machine. The "Information" window gives simple data on the global system state. During a dialogue, the "Questions" window communicates the robot's questions and suggestions. The environment's important elements (doors, corridor ends, specific positions) are materialized by active infra-red beacons.

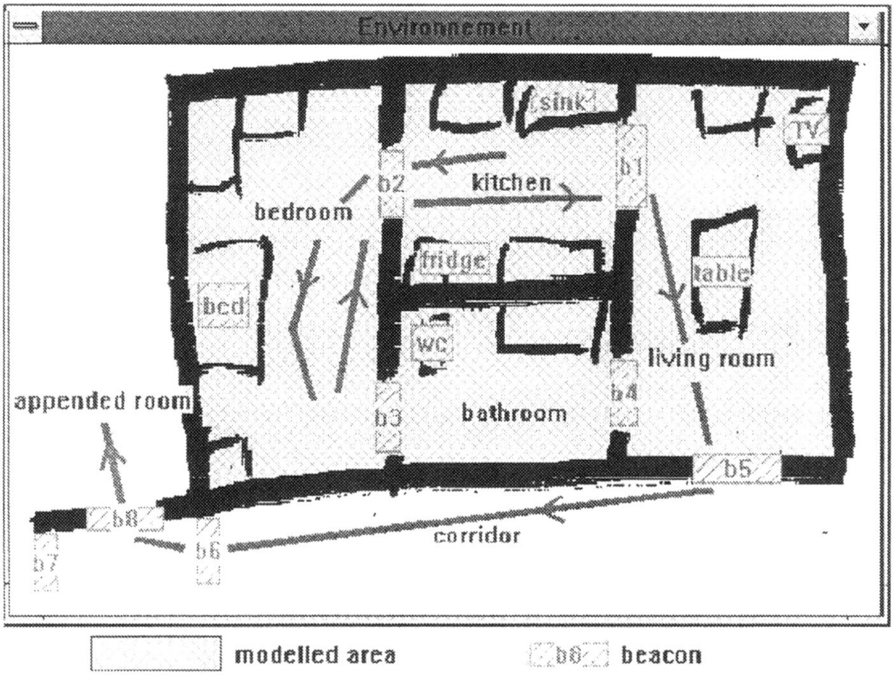


*Figure 6.* Test environment

4.3.2. *Results*

The trial scenario begins in the kitchen.When the pilot enters the bedroom the perception task detects the beacon b2. This information is transmitted to the robot-supervisor. The inference engine of the expert system deduces that the mobile is situated in the bedroom. In the "Information" window, the supervisor proposes the available goals ("activities") from this room. The user selects the "X" icon and then begins a conversation with the supervisor which proposes the possible goals in the "Questions" window. The "corridor" destination being chosen, the automatic mode is engaged. The supervisor plans a global way, b3, b4 then b5, which is, a priori, the shortest one. The expert system deduces that b3 is the washroom door and communicates to the pilot: "We are going to: washroom door". This destination is transmitted to the navigation task which locally calculates a trajectory in the modelled free space, and realizes this trajectory following avoiding unexpected obstacles. The door b3 being closed, the user realizes that this way is not possible. So he stops the motion and the machine proposes another direction, the "bedroom door", that he validates.

So most of the information exchanges between the man and the machine are done in a natural language. Each choice the user has made is moved to the top of the facts base. This symbolic manipulation allows the system to adapt itself to the user's preferences and then to create a short term memory. In particular, it may thus make the more probable choices quicker to select for a person with a severe disability who only uses a switch as a man-machine interface.

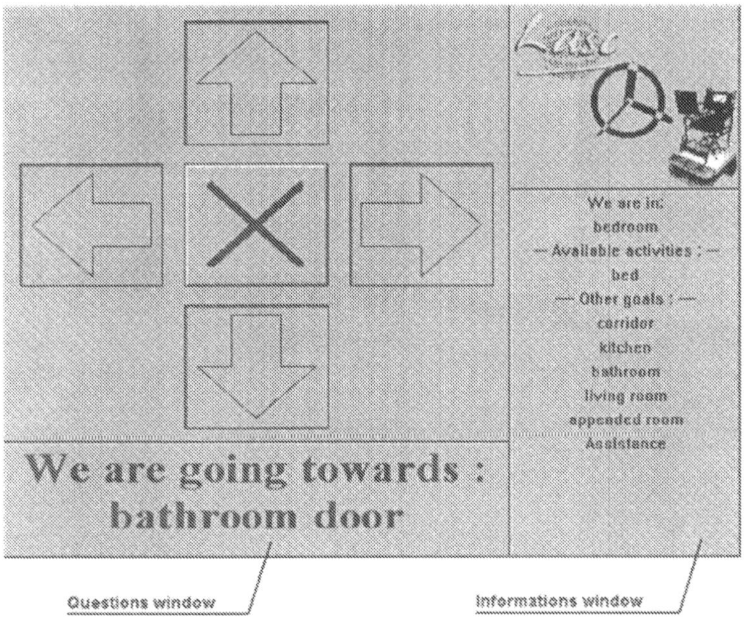


*Figure 7.* Man-Machine interface

When the robot arrives in the corridor, the automatic mode is no longer valid, this area being a non-modelled one. The user chooses to go forward with the "obstacle avoidance" primitive. As the selected initial direction is not well centered on the corridor axis, the wheelchair's trajectory is not very stable. This fact is identified thanks to the "spy files" data which allow to quantify the navigation comfort in real time. The following rule is then applied:

IF the "navigation comfort" criteria is under the threshold,

AND IF the working mode is the "obstacle avoidance" primitive,

AND IF we are in a corridor,

THEN propose the "wall following" primitive.

Indeed, in a non-cluttered corridor, this primitive generally gives a less disrupted motion because of a better definition of the direction to follow.

In the appended room, which is also non-modelled, the pilot moves in the manual mode. The communication task monitors the commands coming from the man-machine interface. When an incoherent motion is detected, two opposite consecutive actions for example, it informs the supervisor which applies the rule: "If incoherent motion then propose assistance". In a non-modelled area, the navigation assistance essentially calls upon primitives like "wall following", "free space searching" or "obstacle avoidance". The mode selection depends on the information given by the perception task (wall, door or free space detection), by the navigation task (the realized movement is uncomfortable). If there is no such information, the suggestions always depend on the user's preferences. In the present case, as the pilot often uses the obstacle avoidance, this primitive is proposed at the top of the list.

### 4.4. DISCUSSION

Some of the above described functionalities have been tested on potential users in a rehabilitation center : automatic mode, navigation primitives, manual mode [36]. These trials have clearly shown this chapter's basic principle: unless we build a very complex system from a software and hardware point of view, which would not be so realistic in such an application, an intelligent mobility aid machine will only be reliable if we optimize the man-machine cooperation. In fact the environment in which such a system may move is only weakly structured: numerous non-modelled obstacles appear on the mobile's trajectory, other persons on wheelchairs use the same space, ...

On the other hand the "cooperative architecture" such as it has been described has only been tested on non-disabled persons [37]. Its validation will be effective after trials in real situations. A transfer on an electrical wheelchair (Power-Push™) is in progress for that purpose. The system's modular conception allows indeed to test it at its different development stages. As has been emphasized in [38], it is one of the keys to success in the rehabilitation robotics field: we may thus better involve the various actors of the project, users and clinicians, at each development stage.

## 5. Conclusions

This chapter mainly focuses on the application to navigation assistance for people with motor disabilities of methods coming from mobile robotics. The specificity of such a study imposes conflicting constraints, in particular small cost and great working robustness. An acceptable compromise between these two requirements consists in optimizing the collaboration between the person with weak physical but with generally great cognitive possibilities, and the machine with complementary capacities.

Defining the man's part thus appears as an essential factor in designing an intelligent powered wheelchair. The user has to be taken into account throughout the study and must not appear as an element grafted at the final stage. So, to this end, it seems interesting to develop objective and, when possible, on-line evaluation tools for the man-machine system.

Let's finally note that, apart from the technical interest of the man-machine collaboration optimization, it is psychologically important to involve the user in the system's control. In such an application field of mobile robotics, the robot's complete autonomy doesn't seem to be the ideal end to reach.

## References


1. Nisbet, P. D., and Craig, I.: Mobility and mobility training for severely disabled children: results of the "smart" wheelchair project, in: *RESNA 17th Annual Conference*, Nashville, 1994, 341-343.
2. Anttczak, J., Al-Temen, I. Balfour, L., and Naumann, S.: The development of an heavy duty joystick, in: *RESNA 14th Annual Conference*, Kansas City, 1991, 154-156.
3. Jaffe, D., Harris, H.L., and Leung, S.K.: Ultrasonic head controlled wheelchair interface: a case study in development and technology transfer, in: *RESNA 13th Annual Conference*, Washington DC, 1990, 23-24.
4. Amori, R.D. : VOCOMOTION-An intelligent voice-control system for powered wheelchairs, in: *RESNA 15th Annual Conference*, 1992, 421-423.
5. Shire, B.: Microcomputer-based scanning interface for powered wheelchair, in: *RESNA 10th Annual Conference*, San Jose, CA, 1987, 541-543.
6. Jaffe, D.L., Sabelman, E.E., and Curtis, G.E.: Responsive environment project Transparent navigation assistant, in: *RESNA 15th Annual Conferencex*, 1992, 176-178.
7. Nelson, P.J., Verburg, G., Gibney, D., and Korba, L.: The smart wheelchair. A discussion of the promises and pitfalls, in: *RESNA 13th Annual Conference*, Washington DC , 1990, 307-308.
8. Wakaumi, H., Nakamura, K., and Matsumura, T.: Development of an automated wheelchair guided by a magnetic ferrite marker lane, *Journal of Rehabilitation Research and Development*, **29**(1) (1992), 27-34.
9. Bell, D.A., Borenstein, J., Levine, S.P., Koren, Y., and Jaros, L.: An assistive navigation system for wheelchairs based upon mobile robot obstacle avoidance, in: *IEEE International Conference on Robotics and Automation*, San Diego, 1994, 2018-2022.
10. Alanen, J., Karasti, O., Mattila, P., Santï, V., and Virtanen, A.: A drive assistant for a wheelchair, in: *CESA'96, IMACS Multiconference*, Lille, France, 1996, 376-381.
11. Linnman, S.: M3S : The local network for electric wheelchairs and rehabilitation equipment, *IEEE Transactions on Rehabilitation Engineering*, **4**(3) (1996), 188-192.
12. Craig, I., and Nisbet, P.: The smart wheelchair: an augmentative mobility toolkit, in: *2nd ECART Conference*, Stockholm Sweden, 1993, 13.1
13. Hoyer, H., and Hoelper, R.: Intelligent omnidirectional wheelchair with a flexible configurable functionality, in: *RESNA 17th Annual Conference*, Nashville, 1994, 353-355.
14. Katevas, N., Tzafestas, L. et al.: SENARIO-The autonomous mobile robotics technology for the locomotion handicap: operational and technical issues, in: *2nd TIDE Congress*, Paris, 1995, 371-374.
15. Yoder, J.D., Baumgartner, E., and Skaar, S.B.: Reference path description for an autonomous powered wheelchair, in: *IEEE International Conference on Robotics and Automation, San Diego*, California, 1994, 2012-2017.

16.Trahanias, P.E., Lourakis, M.I.A., Argyros, A.A., and Orphanoudakis, S.C.:Navigational support for robotic wheelchair platforms: an approach that combines vision and range sensors, in: *IEEE International Conference on Robotics and Automation*, Albuquerque, USA, 1997, 1265-1270.

17. Smith, R.O., and Vanderheiden, G.C.: Teaching about simple switches: not so simple, in: *RESNA 15th Annual Conference*, 1992, 549-550.

18.Roy, D.M., Panayi, M., Harwin, W.S., and Fawcus, R.: Advanced input methods for people with cerebral palsy: a vision of the future, in: *RESNA 15th Annual Conference*, 1992, 99-101.

19.Ferrario, M., and Lodola, M.: A multifunctional nape joystick, in: *RESNA 15th Annual Conference*, 1992, 524-526.

20.Audet, J., Lozac'h, Y., Montiglio, M., Mauger, D., and Giasson, Y.: The sliding disk control interface ; A successful technology transfer, in: *RESNA 15th Annual Conference*, 1992, 418-420.

21.Cooper, R.A.: Intelligent control of power wheelchairs, in: *IEEE Engineering in Medicine and Biology magazine*, 1995, 423-43.

22.Hendriks, J.L., Rosen, M.J., Berube, N.L.J., and Aisen, M.L.: A second-generation joystick for people disabled by tremor, in: *RESNA 14th Annual Conference*, 1991, 248-250.

23.Commission of the European Communities, Technology Initiative for Disabled and Elderly People, 1993-1994 Workplan.

24.Didier, J.P., Gras, P., Casillas, J.M., Busnel, M., and Leclaire, G.: Robotics applied to rehabilitation. What does the physician expect ? in: *3rd French-Israeli Symposium on Robotics*, 1995, Hertzelia, Israël, 42-44.

25.Harwin, W., and Rahman, T.: Safe software in rehabilitation mechatronic and robotics design, in: *RESNA 15th Annual Conference*, 1992, 100-102.

26.Van Der Loos, H.F.M., Lees, D.S., Leifer L.J.: Safety considerations for rehabilitative and human-service robot systems, in: *RESNA 15th Annual Conference*, 1992, 322-324.

27.Yokokohji, Y., Ogawa, A., Hasunuma, H., and Yoshikawa, T.: Operation modes for cooperating with autonomous functions in intelligent teleoperation systems, in: *IEEE International Conference on Robotics and Automation*, 1993, 510-515.

28.Levine, S.P., Borenstein, J., and Koren, Y.: The navchair control system for automatic assistive wheelchair navigation, in: *RESNA 13th Annual Conference*, 1990, Washington, 193-194.

29.Shoval, S., Koren, Y., and Borenstein, J.: Optimal task allocation in task-agent-control state space, in: *IEEE International Conference on Systems, Man and Cybernetics*, Le Touquet, France, 1993, 27-32.

30.Bourhis, G., and Pino, P.: Mobile robotics and mobility assistance for people with motor impairments: rational justification for the VAHM project, *IEEE Transactions on Rehabilitation Engineering*, **4**(1) (1996), 7-12.

31.Pruski, A.: Multivalue coding: application to autonomous robots, *Robotica*, **10**(1992), 125-133.

32.Bourhis, G., Horn, O., and Agostini, Y.: Localization and high level planning for a powered wheelchair, in: *IEEE International Conference on Systems, Man and Cybernetics*, San Antonio, USA, 1994, 2629-2634.

33.Rasmussen, J.: Skills, rules and knowledge; signals, signs and symbols, and other distinctions in human performance models, *IEEE Transactions on Systems, Man and Cybernetics*, **13**(3) (1983), 257-266.

34.Pino, P.: Robot mobile pour handicapés moteurs: évaluation du système homme-machine, Thesis report (in french), University of Metz, France, 1995.

35.Kantowitz, B.H. : Mental workload, Human Factors Psychology, *Advances in Psychology*, P.A. Hancock Eds., North Holland, 1988, 139-185.

36.Pino, P., Agostini, Y., Bourhis, G. et al.: The VAHM project: first results, in: *RESNA 18$^{th}$ Annual Conference*, Vancouver, Canada, 1995, 508-510.
37.Agostini, Y.: Architecture coopérative pour la commande d'un fauteuil roulant robotisé, Thesis report (in french), University of Metz, France, 1996.
38.Van Der Loos, H.F.M.: VA/Stanford rehabilitation robotics research and development program: lessons learned in the application of robotics technology to the field of rehabilitation, *IEEE Transactions on Rehabilitation Engineering*, **3**(1) (1995), 46-55.

# 21

# THE DESIGN, DEVELOPMENT AND OPERATION OF AN AUTONOMOUS ROBOTIC EXCAVATOR


DAVID A. BRADLEY[1] and DEREK W. SEWARD[2]

[1]*SEECS, University of Wales, Bangor*
*Gwynedd, LL57 IUT, UK*

[2]*Engineering Department, Lancaster University*
*Lancaster LA1 4YR, UK*


## 1. Introduction

Foundations, general earthworks and earth removal are all major activities within the mining and construction industries and are tasks involving what would seem to be a series of repetitive operations on the part of the machine operator. Additionally, much of the plant and machinery involved in these activities has a kinematic configuration very similar to that of many industrial robots, as suggested by figure 1 which compares a generalised anthropomorphic robot arm with the two most common forms of excavator geometry.

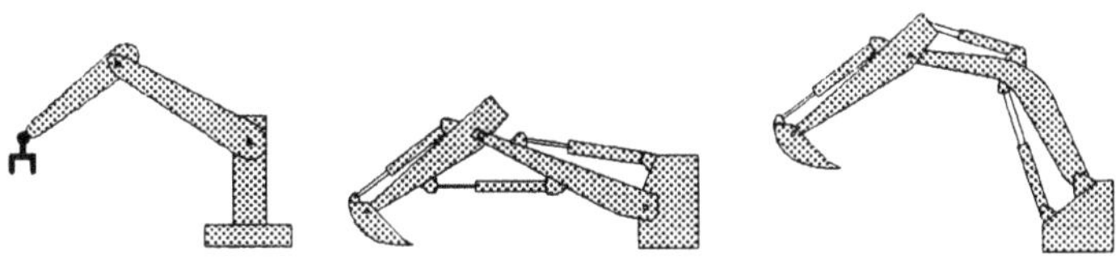

(a) Generalised anthropomorphic robot arm

(b) Type 1 excavator or back-hoe arm

(c) Type 2 excavator or back-hoe arm

Figure 1: *Comparison between a robot arm and excavator arms*

The automation of the excavation process through the adoption of robotics technologies would therefore appear to be an option which would offer the opportunity to improve machine utilisation and throughput. In addition, the automation of the process of earth removal may also provide further benefits such as a reduced dependence on operator skill and a lower operator work load, contributing to a more consistent, and hence higher quality, result. Other possible changes include reductions in the labour force, the removal of operators from hazardous environments and support for remote operation together with improvements in machine maintenance through the introduction of on-board condition monitoring.

In recent years several researchers have contributed work concerned with the automation of the excavation process and a good general review is provided by Singh [1].

In terms of automated and robotic excavators work includes studies into the task planning associated with the achievement of effective operation [2,3], trajectory control and rule-base systems [4,5]. More recently, studies have been carried out into the use of fuzzy logic and artificial neural networks to control the operation of automated and robotic excavators [6,7]. Other areas of activity include the use of a metal detector for locating buried cables and the use of ground penetrating radar to detect underground obstacles [8]. In Japan the majority of the major plant and equipment manufacturers have evaluated some form of automation including teleoperation [9] while in Australia work has been carried out into simulation and operator training [10].

Away from conventional applications in constructions, the use of automated and robotic excavation has been considered in relation to areas such as the handling and disposal of hazardous and toxic wastes [11,12] and space [13,14]. Military applications include a teleoperated system for the repair and maintenance of runways [15] and mine clearance [16].

Based on interviews and the literature, initial considerations of the excavation process suggested that the operation of an autonomous and robotic excavator would depend on an improved understanding of machine behaviour in a number of areas, in particular:

- The motion of the excavator bucket while in the ground, including the mechanics of soil removal.
- The formal identification of the strategies and tactics associated with the removal of soil by an expert operator.
- The inter-relationship between the geometry and kinematics of the machine and the operator controls.

It was also at first believed that there would be a need to provide a wide range of sensor feedback to control the operation of the machine and that, in addition to joint position, there would be a specific need to provide force feedback in order to control the motion of the excavator bucket in different types and conditions of ground. However, experiments using a one-fifth scale model of an excavator arm suggested that control strategies based on those used by conventional industrial robots where the primary concern is with achieving positional accuracy would not be appropriate and that instead an approach to control which imitated the actions of an expert operator might be more effective.

The paper therefore considers these and other related topics in the context of the development of the Lancaster University Computerised Intelligent Excavator (LUCIE) [17,18] and illustrates how the various problems encountered were overcome.

## 2. System Development

Initially, a set of objective criteria were drawn up against which the performance of any automated and robotic excavator would need to be judged. These were:

- It should be capable of working in any type of soil.
- It should excavate with an accuracy of 50 mm.

- It should be autonomously capable of dealing with different types of underground obstacle as appropriate.
- It should achieve a speed of operation comparable with that of an average operator in all conditions.
- It should be capable of integrating its operation with other site systems.

The sequence of tasks associated with the operation of the excavator in digging a trench was also identified [17,18] and the primary goal of the project established as the need to:

'*Fill the bucket as fast and efficiently as possible in all types and conditions of ground*'.

It was originally believed that the starting point for achieving this goal would be an understanding of the excavation process in a form from which the fundamental structures and rules governing the strategic and tactical decisions required for successful operation could be derived. This implied a need for a detailed consideration of the geotechnics and soil mechanics of the excavation process as well as the operating procedures used by skilled operators in achieving their levels of performance. From this starting point it was intended that a rule-based structure would be derived which incorporated geotechnic and soil mechanics data together with operator responses which could be used as the core of an expert system to oversee operation. This structure would also provide the focus for a real-time artificial-intelligence (AI) controller capable of adapting its operation to a wide range of different conditions and environments.

Two immediate problems were encountered with this approach. Firstly, the soil mechanics of the excavation process is not very well understood and where work has been done, it has tended to assume a uniformity of material that would be unrealistic in practice. Secondly, certain of the usual means of knowledge elicitation - interviews, logs, task descriptions and so forth - were not applicable to an environment which was very dependent on physical, manipulative skills for success but within which individuals were not used to articulating the reasons for their actions and decisions. This meant that reliance had to be placed on observation as the primary means of gathering knowledge about the digging process [19-21].

## 2.1 GEOTECHNICS AND SOIL MECHANICS

The geotechnics of the excavation and removal of soil is complicated by the complex interaction of many parameters including:

- The physical properties of the material to be excavated which range from wet clay to hard gravel.
- Changes in the properties of material being excavated along the intended path of the excavation.
- The presence of obstructions such as boulders and tree roots along the intended path of the excavation.

As a result, the controller is required to continually adjust in real-time the trajectory of the bucket through the ground as well as its orientation and it was concluded that a knowledge of the geotechnics of soil removal is almost certainly of less importance than the ability of the controller to detect and respond to changes in the nature and conditions of operation. This means that conventional robot control strategies are not directly applicable to excavation as they are generally concerned with the achievement of positional accuracy and not with the exertion of a force. This suggests that what is needed is a control strategy which continuously monitors and responds to the force that the excavator is exerting on the ground, implying some form of force feedback.

## 2.2 OPERATOR STRATEGIES

Observation of a range of operators suggested that the better operators would typically control up to three degrees-of-freedom simultaneously with expert operators occasionally operating four degrees-of-freedom simultaneously. It was also clear that while vision was the main source of feedback to the operator, they also made use of other performance indicators such as the engine note and the motion of the vehicle in reaching decisions about the sequence of actions to be undertaken.

The major conclusion of the studies of operators was that there were two basic strategies that they used for soil removal; namely 'penetrate and rotate' and 'penetrate and drag', and that these were modified by a wide range of major and minor tactical adjustments based on the conditions encountered, the geometry of the bucket being used and the working environment, as for instance the ability to observe the motion of the bucket directly. These basic strategies may be summarised as follows:

### 2.2.1 *Penetrate and Rotate*

This strategy is used primarily in soft ground such as sand which can easily be penetrated by the bucket. Using this strategy, the operator would first force the bucket into the ground in a near vertical position. Once suitable penetration to a suitable depth had been achieved the bucket would then be rotated to fill the bucket. The filled bucket would then be removed from the line of the trench, emptied and the procedure repeated.

### 2.2.2 *Penetrate and Drag*

The penetrate and drag strategy is used in harder ground which the bucket cannot easily penetrate. Using this strategy, the operator would first penetrate the ground as far as practical and would then adjust the angle of attack of the bucket to 'plane off' material as the bucket was dragged back towards the body of the excavator. This strategy is much more complex in its demands on the operator, requiring the simultaneous control of more degrees of freedom than the penetrate and rotate strategy.

Observation suggested that the major tactical variations associated with the penetrate and drag strategy involved altering the 'angle of attack' of the bucket and the direction in which the bucket was dragged and that these were in turn influenced by bucket geometry.

## 2.3 DEVELOPMENT STRATEGY

As a result of the review of the geotechnics and soil mechanics and the observation of operators, it was apparent that the key problem requiring solution if autonomous excavation is to be achieved is the control of the motion of the bucket through the ground and that an analytical approach based on conventional robot control strategies was unlikely to provide the solution to this problem.

What was initially decided upon therefore was to implement an experimental approach based on the use of a one-fifth scale model of a backhoe arm which could be used both for the investigation of digging techniques and strategies and as a tool for software development [22].

The reasons for the use of a physical model for system development can be summarised as follows:

- There was not sufficient data available to support the construction of a mathematical model of soil conditions and any such model would be limited to conditions unlikely to be encountered in practice. In contrast, the model would operate in a controlled environment enabling performance in a variety of soil types to be readily investigated and evaluated.
- The cost of implementing a model environment was much less than that associated with the purchase of a full size machine and it would also be simpler and cheaper to make changes to the model environment.
- The model would provide a controlled environment for both hardware and software development, with much reduced safety problems.
- The model would provide a means of getting operator feedback on different system implementations.
- The use of a model of a real backhoe would provide a more realistic development environment than any implementation based on the use of a conventional industrial robot of similar geometry, particularly as the latter is designed to be structurally stiffer than a backhoe arm and operates with greater precision. The use of a conventional robot controller was also considered to be restrictive in the development of strategies for movement through ground.
- The use of a computer dynamic simulation model, as an alternative, was discounted because of the difficulties in adequately modelling realistic non-homogenous soil material.

The results achieved throughout the life of the LUCIE programme from 1987 to date have supported the validity of these early considerations, the one-fifth scale model having provided an effective environment for system development at all stages of the programme.

## 2.4 SOFTWARE DEVELOPMENT

From the outset, a clear decision was made to partition the software into two distinct modules. An "Activities Manager" module which contains the AI knowledge-based rules for excavation the output of which consists of a demand to move the excavator bucket at a specified speed in a specific direction. This demand is then converted into the appropriate signals for the electro-hydraulic valves that drive the rams by the lower level controller software.

Trials involving the implementation of 'teach and repeat' cycles and a 'point-to-point' approach to the control of bucket motion provided confirmation of the view that neither of these was suited to the achievement of autonomous and robotic excavation and that an approach more closely modelled the way in which human operators controlled their machines was what was needed. This led to the development of the concepts of velocity vector control and software force feedback as the means of controlling the motion of the bucket through the ground.

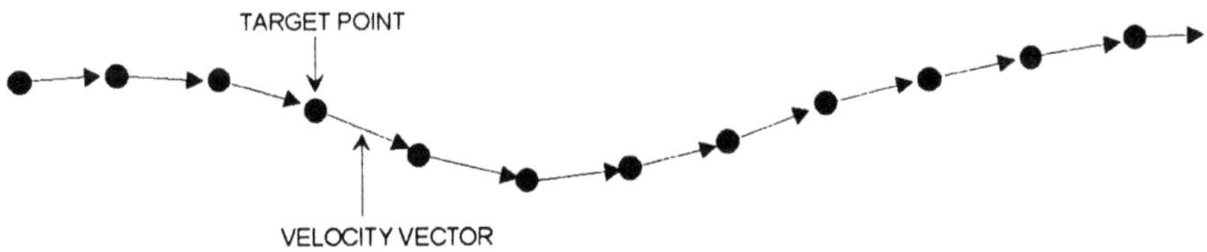


Figure 2: *Generation of a motion profile using a target point and velocity vector*

## 2.5 VELOCITY VECTOR CONTROL

Consider the situation of figure 2 in which a motion profile is generated by moving a reference '*target point*' through space in a controlled manner by generating for each time interval a velocity vector which moves the target point the required distance in the desired direction. The tip of the bucket is then controlled to attempt to track this target point with zero positional error in x and y by generating, for each time interval, a velocity vector which defines the motion of the tip of the bucket in that time interval.

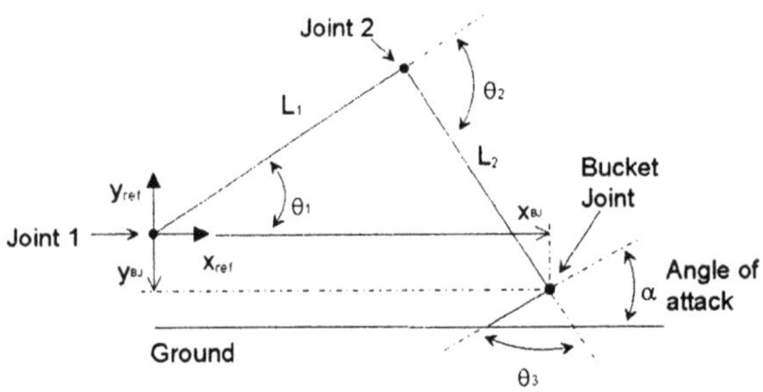


Figure 3: *Excavator kinematics*

Soil removal is also affected by the 'angle of attack' of the bucket as shown in figure 3 and 'relief angle' of the bucket with respect to the ground. The optimum value for the angle of attack depends on the ground conditions and bucket geometry. The parameters defining bucket geometry are contained in a software module while the knowledge base contains rules which use this information to actively adjust or 'tune' the angle of attack to attempt to maximise the rate of soil removal in a manner akin to the way in which a human operator would do so. Changing the bucket for one of a different geometry requires the replacement of the existing module with that for the new bucket geometry.

## 2.6 EXCAVATOR KINEMATICS

The (0, 0) reference point for the co-ordinate system used in performing the kinematic analysis is the point of attachment of the excavator arm to the vehicle with positive motion in the x-direction then being away from the vehicle and positive motion in the y-direction being vertically upwards. Using this co-ordinate system and referring to figure 3, for a constant angle of attack of the bucket, the velocity vector of the bucket joint is identical to that of the bucket tip. This simplifies the control model by removing one degree-of-freedom and separates bucket joint rotation from the inverse kinematic equations.

During excavation, the motion of the excavator arm is constrained to the line of the trench in which case the position of the bucket joint ($x_{BJ}$, $y_{BJ}$) relative to the point of attachment of the arm to the vehicle is given by:

$$x_{BJ} = L_1 \cos\theta_1 + L_2 \cos(\theta_1 - \theta_2) \qquad 1$$

and

$$y_{BJ} = L_1 \sin\theta_1 - L_2 \sin(\theta_1 - \theta_2) \qquad 2$$

From figure 4, the bucket angle ($\theta_3$) may then be obtained since:

$$\phi = \alpha - \Psi \qquad 3$$

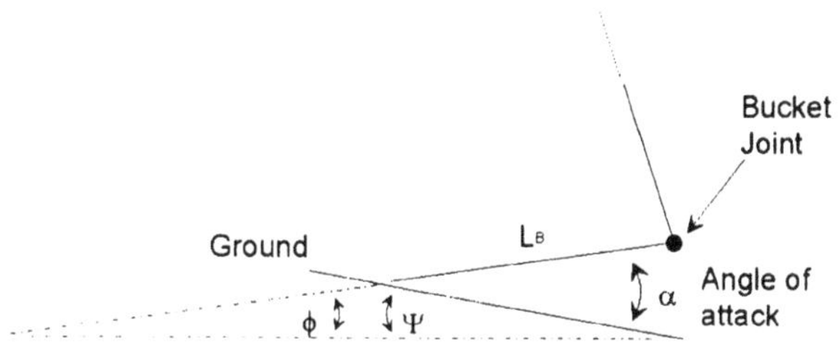


Figure 4: *Bucket position and orientation*

where is the angle of attack of the bucket relative to ground; $\psi$ is the drag angle of the bucket relative to a horizontal reference; and $\phi$ is the effective angle of the bucket relative to the horizontal reference. In which case:

$$\theta_3 = \pi + \theta_1 - (\phi + \theta_2) \qquad 4$$

The co-ordinates of the tip of the bucket are then found as:

$$x_{TIP} = x_{BJ} - L_B \cos\phi \qquad 5$$

and

$$y_{TIP} = y_{BJ} - L_B \sin\phi \qquad 6$$

The inverse kinematic equations for $\theta_1$ and $\theta_2$ are now:

$$\theta_2 = \cos^{-1}\left(\frac{x_{BJ}^2 + y_{BJ}^2 - \left(L_1^2 + L_2^2\right)}{2.L_1.L_2}\right) \qquad 7$$

where $\theta_2$ is in the range 0° to 180° only, and,

$$\theta_1 = \tan^{-1}\left(\frac{y_{BJ}\left(L_1 + L_2\cos\theta_2\right) - x_{BJ}.L_2\sin\theta_2}{x_{BJ}\left(L_1 + L_2\cos\theta_2\right) - y_{BJ}.L_2\sin\theta_2}\right) \qquad 8$$

Refer now to figure 5 in which $P_1$ is the current position of the target point and $B_1$ the current position of the bucket tip. The position, $P_2$, of the target point after the reference time interval $\Delta t$ can then be obtained using the velocity vector of the target point. The required velocity vector of the bucket tip is then that required to bring the bucket tip in line with the target point at the end of the reference time interval with the angle of attack of the bucket correspondingly adjusted. The angular velocities of each joint; $\omega_1$, $\omega_2$ and $\omega_3$, can be obtained using the generalised form of equation 9:

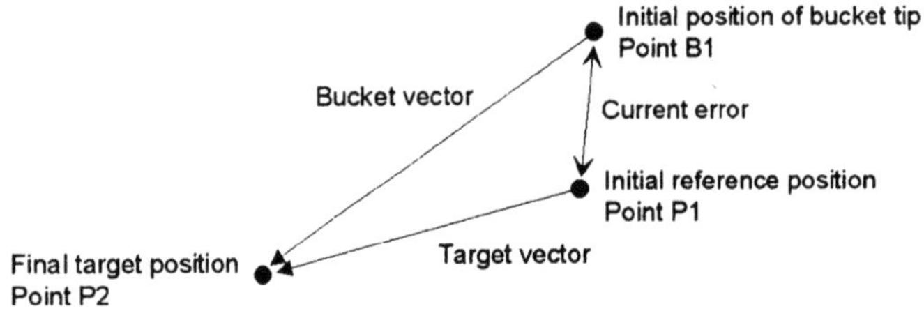


Figure 5: *Generation of bucket velocity vector*

$$\omega_N = \frac{\Delta\theta_N}{\Delta t} = \frac{\theta_{N:2} - \theta_{N:1}}{\Delta t} \qquad 9$$

where N = 1, 2 or 3; $\theta_{N;1}$ is the angle of Joint 'N' at the start of the time interval $\Delta t$ at position $B_1$; and $\theta_{N;2}$ is the required angle of Joint 'N' at the end of the system time interval to achieve the position of point $P_2$.

These rates are then passed to the valves and the valve settings calculated taking into account the actuator geometry as suggested by figure 6, in which case:

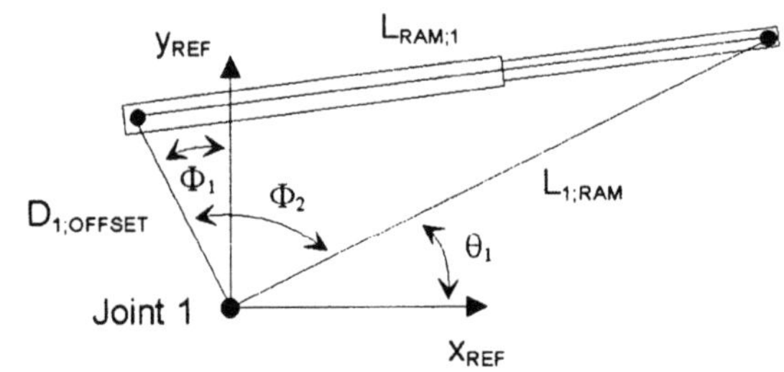


Figure 6: *Actuator geometry - Joint 1*

$$\Phi_2 = \cos^{-1}\left(\frac{L_{1;RAM}^2 + D_{1;OFFSET}^2 - L_{RAM;1}^2}{2.L_{1;RAM}.D_{1;OFFSET}}\right) \qquad 10$$

where $\Phi_2$ is in the range -90° to 90° only.

and

$$\theta_1 = \frac{\pi}{2} - (\Phi_2 - \Phi_1) \qquad 11$$

Using these and other similar equations for the other joints, the required motion can be expressed in terms of the changes in ram length $\Delta L_{RAM;1}$, $\Delta L_{RAM;2}$ and $\Delta L_{RAM;3}$ using the generalised form of equation 12 as:

$$V_{RAM:N} = \frac{\Delta L_{RAM:N}}{\Delta t} \qquad 12$$

where N = 1, 2 or 3; and $V_{RAM;N}$ is the required velocity of Ram 'N' and can be translated directly into the appropriate valve settings by means of a look-up table which relates the valve opening to flow rates and hence to ram velocities.

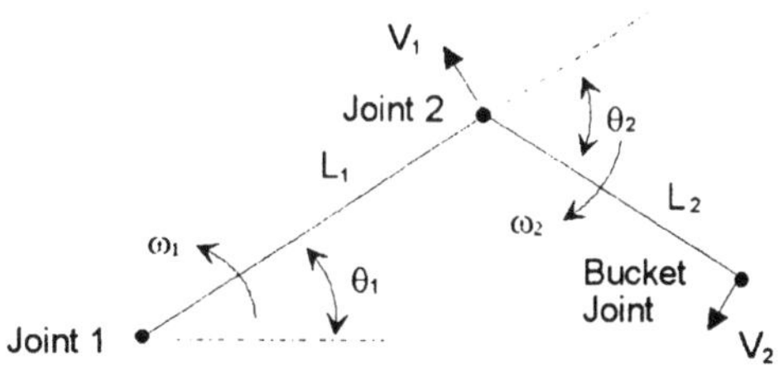


Figure 7: *Joint velocities*

While this approach does not yield a true velocity vector for the bucket tip, it has been found that provided the reference time interval Δt is sufficiently small, of the order of 25 ms, then the effect is that of a true velocity vector as far as machine performance is concerned.

Alternatively, the required velocity vector can be calculated directly. Referring to figure 7, it is seen that the velocity of the bucket joint is made up of two component velocities, $V_1$ due to the rotation of joint 1 and $V_2$ due to the rotation of joint 2 such that:

$$V_1 = L_1.\dot{\theta}_1 = L_1.\omega_1 \qquad 13$$

and

$$V_2 = L_2.\dot{\theta}_2 = L_2.\omega_2 \qquad 14$$

Resolving $V_1$ and $V_2$ into their components in the x and y directions gives:

$$V_{1;x} = V_1.\sin\theta_1 \qquad 15$$

$$V_{1;y} = V_1.\cos\theta_1 \qquad 16$$

$$V_{2;x} = V_2 . \sin(\theta_2 - \theta_1) \qquad 17$$

and

$$V_{2;y} = - V_2 . \cos(\theta_1 - \theta_2) \qquad 18$$

From knowledge of the desired velocity vector, the desired x and y components of bucket joint velocity, $V_{x;REQ}$ and $V_{y;REQ}$, can be obtained as:

$$V_{x;REQ} = V_{1;x} + V_{2;x} = L_1 . \omega_1 . \sin\theta_1 + L_2 . \omega_2 . \sin(\theta_2 - \theta_2) \qquad 19$$

and

$$V_{y;REQ} = V_{1;y} - V_{2;y} = L_1 . \omega_1 . \cos\theta_1 - L_2 . \omega_2 . \cos(\theta_2 - \theta_1) \qquad 20$$

Solving equations 19 and 20 for $\omega_1$ and $\omega_2$ gives:

$$\omega_1 = \frac{V_{x;REQ} \cdot \cos(\theta_2 - \theta_1) + V_{y;REQ} \cdot \sin(\theta_2 - \theta_1)}{L_1 . \sin\theta_2} \qquad 21$$

and

$$\omega_2 = \frac{V_{x;REQ} \cdot \cos\theta_1 - V_{y;REQ} \cdot \sin\theta_1}{L_2 . \sin\theta_2} \qquad 22$$

Knowing $\omega_1$ and $\omega_2$, the ram velocities can be calculated as before.

## 2.7 SOFTWARE FORCE FEEDBACK

As was indicated previously, it was at first believed that force feedback would be required to provide an indicator of machine behaviour while moving the bucket through the ground and consideration was given to the introduction of strain gauges onto the excavator structure and of other possible force measuring systems such as pressure sensors in the hydraulic system. Because of the vulnerability of such systems a means of monitoring machine performance using the information available from the joints themselves as the basis of a force measuring system was evolved.

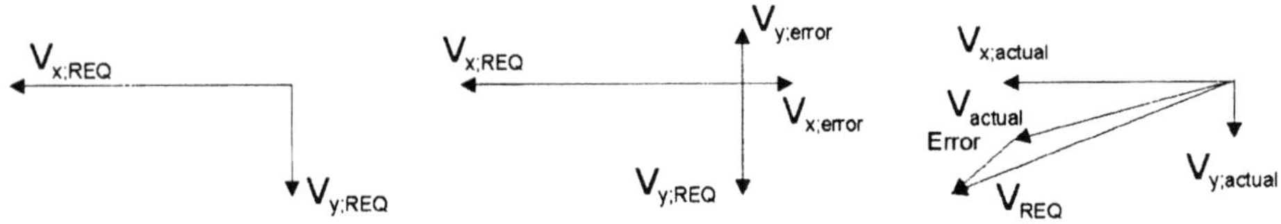


(a) Required conditions (b) Velocity errors (c) Resulting error

Figure 8: *Generation of velocity error*

Consider the situation in figure 8(a). Here, the motion of the bucket tip is expressed in terms of its component velocity vectors, $V_{x:REQ}$ and $V_{y:REQ}$, in the x and y directions respectively. In the absence of any external system of forces, the motion would be as determined by the valve settings derived from equation 12 and the bucket tip would arrive at the desired location at the end of the time interval Δt. If there is resistance to motion, then this appears as position error at the end of the period Δt which can be expressed as changes in the values of the $V_{x:REQ}$ and $V_{y:REQ}$ vectors by the amounts $V_{x:ERROR}$ and $V_{y:ERROR}$ in the x and y directions respectively as shown in figure 8(b). This leads to an effective positional error as shown in figure 8(c).

This position error vector is therefore a measure of the force that the ground is exerting on the bucket and hence of the force required to move the bucket through the ground. By monitoring the magnitude, direction and rate of change of the position error vector, software force feedback is achieved which, in practice, has proved sufficiently sensitive to detect the contact of the bucket with the ground and forms the primary input into the Artificial Intelligence (AI) system. Based as it is entirely on the information provided by the joint encoders, software force feedback eliminates the need for additional force sensors, increasing the robustness of the system and reducing its vulnerability to damage under site conditions.

## 3. Artificial Intelligence Structures

The role of the real-time artificial intelligence is to control the strategy and tactics to be used to remove soil and to generate the appropriate commands to operate the joints of the excavator arm in accordance with those strategic and tactical decisions. In the case of the automated and robotic excavator its operation is based on the concept of a task centred - goal oriented structure in which the basic task is divided into a series of goals. These individual goals are achieve through the combination by the AI system of a series of user defined activities [23].

This structure was implemented in real-time using a production rule based AI format. Using this approach the AI system first identified the current goal and then determined the activities associated with achieving that goal. Each activity is then tested by the AI system at the end of a time cycle and allowed to proceed if appropriate. If it is not possible to proceed with the current activity, then this activity is placed onto an activity stack and a new activity selected. Once an activity is completed, the activity stack is interrogated to see if it contains any uncompleted activities. If this is the case, then the activity at the top of the stack becomes the new current activity and operation proceeds as before. Information is passed to and from the AI system by means of a *blackboard* which carries the current sensor data as input and the required joint angles as output together with other system data as appropriate.

The result is a heuristic approach to automated and robotic excavation incorporating a trial and error feature within the real-time artificial intelligence. It uses a rule base of around 80 rules, the majority of which were derived as a result of observation.

### 3.1 OPERATION

In operation the first, and only predictable, action of the LUCIE system is to extend the arm to its fullest extent and then to lower it until the tip of the bucket contacts the ground. Once the ground has been detected, the excavator attempts to penetrate the ground to a point at which the bucket can be filled by rotation. If this is not possible, then the angle of attack of the bucket is adjusted and a dragging strategy adopted. Initially, this will use an upwardly inclined drag angle which will gradually be reduced as the trench bottom is approached to provide the flat bottom.

During this procedure, the system remembers the previous point of entry into the ground and the swept profile. This is then compared with the desired profile and the point of entry and the drag angle adjusted accordingly for the next pass. At the end of the process, a rectangular trench with square ends is produced.

### 3.2 OBSTACLES

The ability to deal with obstacles such as boulders along the line of the trench is a necessary and important feature for any automated and robotic excavator. In the system as developed the approach adopted is essentially one of repeated trials until it is possible to insert the bucket under the boulder and remove it from the trench. Referring to figure 9, the sequence of operations would be as follows:

1. Initially, the bucket encounters the boulder as in figure 9(a) when its presence is detected by an increase in the error vector of figure 8. When this occurs, the angle of attack and drag angle are adjusted, the bucket raised and the bucket advanced. This procedure is then repeated until the bucket passes over the boulder at which point, the bucket may either be removed from the trench if a minimum fill level is assessed as having been achieved or it will be lowered to continue digging.
2. In the next digging pass, the boulder will again be encountered, though at a lower level as in figure 9(b), when the sequence described in (1) above will be repeated.
3. Eventually, the bucket will reach a point such as that in figure 9(c) at which it is able to penetrate the ground underneath the boulder when provided the obstacle is capable of being lifted by the bucket, it will be removed from the trench.

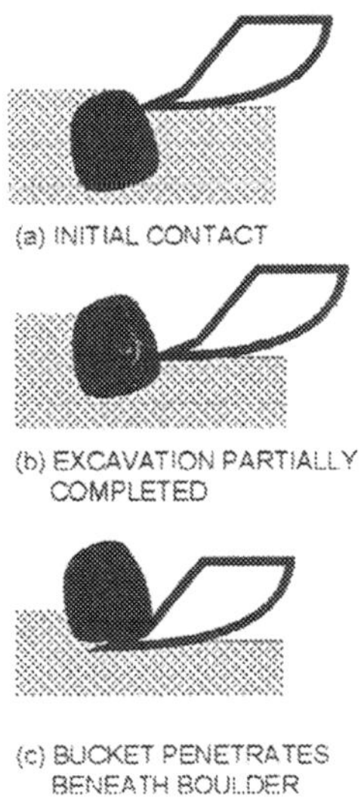


Figure 9: *Obstacle removal*

This *trial and error* approach has been found to work well in most cases .However, with the system set-up as described, the specified depth of the trench sets a limit on the motion of bucket in the vertical plane. In order to remove an obstacle it is necessary for the bucket to penetrate beneath the obstacle before it can be lifted out. Thus if the obstacle extends below the specified depth limit for the trench then it will not be possible for the bucket to pass beneath it. In practice however, it is not possible to provide a precise figure for the degree to which an obstacle must penetrate below the bottom of the trench before it cannot be removed as this is affected by factors such as the condition of the soil and the shape of the object. Thus, in some situations it has been observed that the action of the bucket on the upper part of the obstacles can loosen or tip the obstacle, freeing it to be lifted on a later pass. As the digging tests were carried out in real, unprepared ground there was no control over the nature and shape of the objects and obstacles encountered.

Similarly, the sides of the trench are defined by the vertical plane of motion of the arm when in position over the trench and the width of the bucket. Hence, obstacles that are wider than the trench cannot in general be handled unless the action of the bucket on earlier passes frees them to fall within the boundaries of the trench. Linear obstacles such as tree roots therefore present a particular problem.

## 4. The Lancaster University Computerised Intelligent Excavator - LUCIE

The development of the functions and systems developed using the fifth-scale model arm took the form of an implementation on a JCB801 360° tracked excavator. This transfer of the focus of the work to a full scale system capable of digging trenches of over a metre in depth was considered essential if the concepts for autonomous and robotic excavation developed using the model, and in particular the concepts of velocity control and software force feedback, were to be proven and demonstrated as viable for application on a construction site and elsewhere.

The resulting system, referred to as the Lancaster University Computerised Intelligent Excavator or LUCIE, was first fully demonstrated in 1992 [24] and has since proven to be capable of digging rectangular trenches in a variety of different types of ground and under a range and variety of conditions. The test and experimental strategy used throughout the development process was as follows:

- Initial software development and evaluation using the one-fifth scale model.
- Testing on-board the JCB801 in a laboratory environment to assess motion in the x and y direction as of combinations of motions. This was achieved by video recording the motion against a reference background and by attaching a marker to the bucket joint and recording the motion. Deviations from target linear motion of the order of 15 mm in the x direction and of 25 mm in the y-direction have been achieved, and are thus within the original target specification of 50 mm.
- Following the laboratory tests, diggings test were carried out in unknown, and hence unprepared, ground with conditions ranging from hard to very soft and from 'clean' soil to areas containing significant amounts of small to medium sized stones. In each case, the final trench was evaluated in terms of its depth accuracy, squareness and overall quality. In practice, an accuracy of the order of 20mm in depth were achieved in all types of soil with true, square ends. The quality of the final trench in terms of

finish was in all cases comparable with that achieved by expert operators, though with less errors in depth than with a manually dug trench.

- Cycle times were also a major factor in determining performance. Observation of machine operators on this size of excavator suggested a typical cycle time for an average operator of the order of 15 to 25 seconds depending on the type of ground The cycle times achieved by LUCIE were comparable with those achieved by average operators in the same type of ground across a wide range of ground conditions, particularly in terms of the motion of the bucket through the ground, and were generally superior to those achieved by novice operators..

## 4.1 JOINT SENSORS

The model arm used continuous analogue potentiometers as its joint sensors. However, for an arm such as that of the JCB801 with an effective radius of the order of 2 metres, an error of 0.05% in the measurement of joint angle results in a positional error of 6.283 mm of arc at the end of the arm. This implies a resolution of at least 0.18° or 11 bits for the joint angle sensor. In practice, a resolution of at least 0.05° or 14 bits is needed to allow for noise and other effects. Both continuous analogue sensors and optical encoders have been used satisfactorily.

In a real system, instead of the bucket moving through the ground, the excavator body itself may move. As the reference point for the basic excavator is the point of connection of the arm to the vehicle body, this would be interpreted as a movement of the bucket and under certain conditions it is possible for the body of the excavator to be lifted of the ground, ultimately causing it to overturn. In addition, the vehicle may be positioned on a slope in which case the reference co-ordinates of the arm will need to be adjusted to take account of this. Using only the information from the joint sensors this condition would be interpreted as the continuous motion of the bucket through the ground. A tilt sensor is therefore included to measure the angle of the vehicle and has the dual function of providing a safety feature to detect overturning and of monitoring the vehicle angle to take account of working on a slope [18].

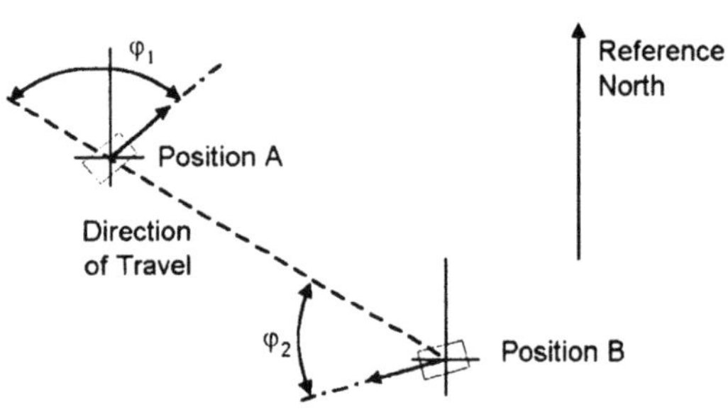


Figure 10: *Motion reference*

## 4.2 POSITIONING AND MOBILITY

The proper location of excavators and other items of plant is an essential requirement if they are to carry out their designated tasks. The introduction of differential Global Positioning Systems (GPS) with centimetre resolution provides the means to accurately determine position [25]. Consider the situation of figure 10 in which it is required to move the excavator from position A to position B before continuing with an excavation. A basic strategy for this movement requires the vehicle to rotated through an angle $\varphi_1$ to align it with the required direction of travel followed by a final rotation of $\varphi_2$ at the target point to achieve the required alignment. During linear motion, errors in heading resulting from any slipping of the wheels or tracks can be compensated for by control of either steering or track speeds. However, despite a total latency for the GPS signal currently of the order of 0.4 seconds there remains a danger of overshoot and subsequent hunting around the target point. For this reason, the concept of an *error circle* around the target point as illustrated by figure 11 has been evolved, the radius ($R_{ERR}$) of which is given by:

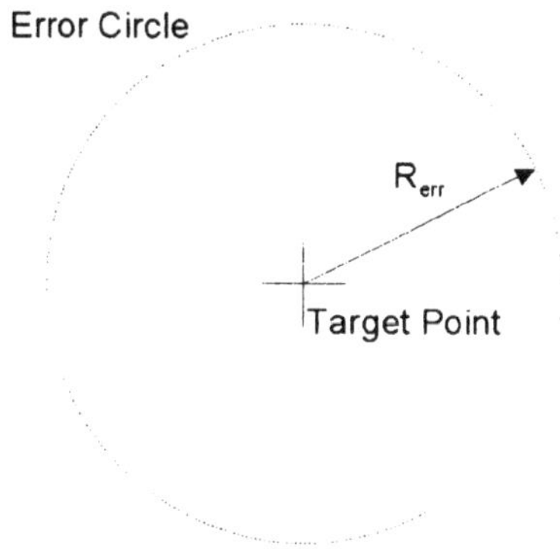


Figure 11: *Error circle*

$$R_{ERR} = V_M . t_{up} \qquad 23$$

where $V_M$ is the vehicle speed and $t_{up}$ is the update rate of the GPS.

If the vehicle position at any update is determined to lie within the error circle for its current speed, the speed will be adjusted so that it will not have overshot the target point at the next update.

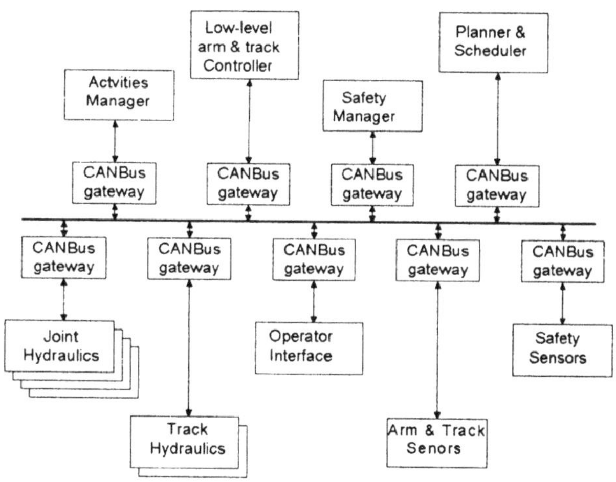


Figure 12: *Modular architecture based on the use of CANBus*

## 4.3 NETWORKING

By replacing the conventional wiring configuration with a bus architecture a more modular form of system architecture can be achieved making it much easier to change the system configuration, for instance to accommodate different machine kinematics. The networking approach adopted for its robustness and the availability of interface devices is that of the Controller Area Network (CANBus).. Figure 12 shows the resulting modular structure of the LUCIE architecture resulting from the use of CANBus. The activities manager, the low-level controller and the safety manager each reside on independent 486 processors in compact PC104 form. Only these main processors currently communicate via the CANBus, with sensors and actuators hard-wired to PC104 interface cards, but it is the long-term intention to use the bus for all communications.

### 4.4 SAFETY

The safety of any large item of plant is a critical and important feature which must be taken into consideration from the earliest stages of the design and development process taking into account all current and future standards and guidelines [26]. In the case of LUCIE, safety is being approached both from a consideration of the system architecture in the form of the *safety manager* module of figure 12 which acts to oversee and sanction actions in line with established risk criteria and also through the analysis and modelling of the safety zones around the vehicle and the performance of different types of sensors and sensor configurations [27].

## 5. **Design Support**

### 5.1 PROTOTYPING

In the design of items of plant and equipment such as the LUCIE system there is an increasing need to provide the designer with the means of rapidly evaluating the operation of the plant within its intended operating environment and of comparing alternative design configurations [28,29]. However, for systems on the scale of a typical item of construction plant, the cost of developing a prototype can be high and the cost in a competitive market of introducing significant structural and kinematic changes into that prototype prohibitive. Plant designers may well therefore tend to be over conservative and hence fail to take effective advantage of the opportunities offered by the application and use of computer technology in the design process [30-32].

There is therefore a need for a low cost approach to the design and prototyping of large and complex plant which can be used from the very earliest stages of the design and development process and which can provide a means of visualising the operation of the plant in a range of possible environments as well as providing support in areas such as software development and user interface design. Any such system must therefore support the reproduction within the model of the major functional requirements of the full scale system. These include:

- The ability to rapidly reconfigure the model, for instance to take account of changes in system geometry and kinematic structure.
- The ability to provide the full range of controlled motions taking into account any operator or system defined limits on the operating envelope.

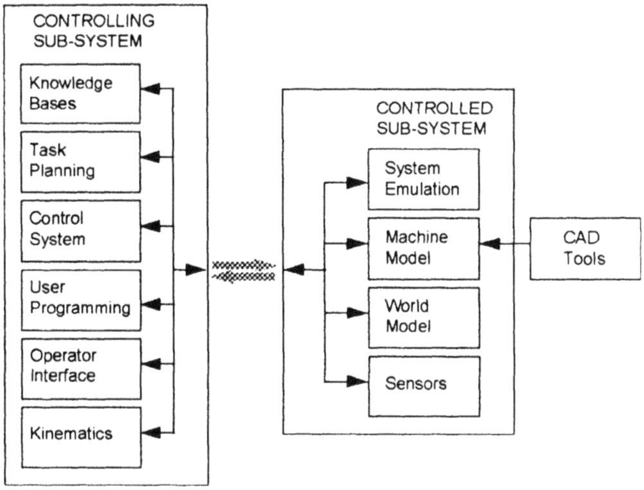


Figure 13: *Distribution of software functions between the controlling and controlled sub-systems*

- Support for the definition by the user of the allowable operating envelope of the machine.
- The investigation and evaluation of different operator strategies including a wide range of input devices.
- The simulation of a wide range of operator display configurations.
- The provision of feedback to the user on both the machine and task status.
- Support for the development of operating software.
- The visualisation of the operation of the system from a range of viewpoints, including that of the operator.

The prototyping facility is therefore based around the following major sub-systems:

- A controlling sub-system dealing with the operator interface, displays, machine kinematic models and control algorithms.
- A controlled sub-system capable of emulating the plant and its operating environment at an appropriate level of resolution and providing performance feedback including visualisation.

The prototyping system that eventually evolved took the form of a pair of interconnected PCs representing the controlling and controlled sub-systems respectively and figure 13 suggests the distribution of software functions between these sub-systems.

Consideration can also be given to the design of the operator interface an this stage with the simulated interface being reproduced on the controlling sub-system display. The controlling sub-system can also be used to investigate and evaluate designs for the operator interface and figure 14 shows an example of such an interface.

The environment has been tried out using the LUCIE system as the target application and has demonstrated its ability to support developments in both hardware and software [33].

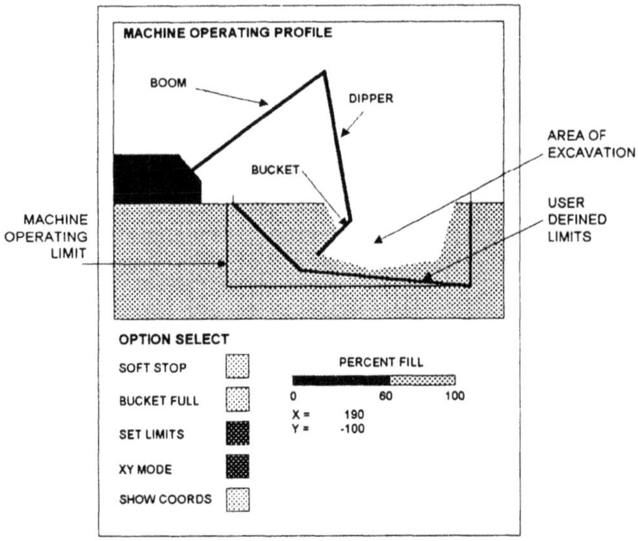


Figure 14: *Operator interface screen*

## 5.2 SENSOR SYSTEMS

In order to ensure the safety of site personnel, of the machine itself and of the other machines and systems with which it must co-exist and collaborate it is necessary to identify potential hazards within the operating envelope of the machine requiring the designer to consider factors such as:

- Identification and elimination of blind spots.
- Identification of areas of sensor overlap.
- The effect of processing time on performance.
- The effect of the failure of an individual sensor on overall performance.
- The positioning of the individual sensors.

The failure of the sensor system to respond to the presence of an object within the working volume of the machine will typical be as a result of one or other of the following:

- A failure to detect the object.
- A system failure before detection.
- A system failure subsequent to detection.

The performance performance of the sensor system can therefore be expressed in terms of the Composite Failure Rate ($CF_{RATE}$) of equation 24:

$$CF_{RATE} = \frac{\text{Number of Failures}}{\text{Total Number of Obstacles}} \qquad 24$$

and the Detection Failure Rate ($DF_{RATE}$) of equation 25:

$$DF_{RATE} = \frac{\text{Number of Detection Failures}}{\text{Total Number of Scans}} \qquad 25$$

Consider a scanning sensor mounted on top of the vehicle. The effect of varying both the scanning period and the processing time is evaluated using a simulator for different combinations of processing time and scan rate in relation to the detection of random dynamic objects entering the working volume of the machine. From the results of the simulations it was found that for objects less than 0.3 m in size the ration between the scanning interval ($t_{SCAN}$) and the processing time ($t_{PROC}$) was significant in determining performance. In particular, when $t_{SCAN}/t_{PROC}$ is below 100, the value of $DF_{RATE}$ was found to increase while no detection failures were found to occur in the simulation when the ratio $t_{SCAN}/t_{PROC}$ was set to be greater than or equal to 100 [34,35].

## 6. Conclusions

The LUCIE system has demonstrated a number of novel concepts in its approach to automated and robotic excavation. In particular, the capability of an 'operator like' goal oriented control strategy with a minimum of pre-planning and analysis, the proving of features such as the use of velocity vector control and software force feedback to control the motion of the bucket through ground and their integration within a modular, rule-based AI structure. These concepts have been successfully demonstrated on a variety of sites in a range of different soil types and conditions at performance and quality levels comparable with, and in some cases better than, those achieved by average operators. On a more generic level, the structure of a real-time artificial intelligence based control architecture using the concept of an activities manager has been defined which is considered to be applicable to a number of different types of construction plant performing a wide variety of tasks and a formal program shell constructed.

Additionally, though the LUCIE programme has to date concentrated the achievement of automated and robotic excavation and the development of a generic real-time AI based controller strategy capable of application to a range of types of construction plant, the techniques developed are also capable of being applied to improve the performance of operator controlled systems, for instance through the use of multi-degree-of-freedom joysticks.

However, though both the theoretical and practical basis for robotic and automated excavation can be considered to have been demonstrated, its practical development to a point where it becomes viable for site use on a large scale will depend on an ability to integrate operation with other site systems and to resolve problems in areas such as operational safety. Other current work is therefore focusing on those topics associated with a wider implementation of the activity based approach to other construction processes and including areas such as the operational safety of all types of automated and robotic construction plant, the integration of all types of plant with site operations, means of task definition and analysis for construction plant, vehicle positioning and orientation systems, rapid prototyping and design systems and operator interface design and performance.

At the time of writing, the LUCIE system is undergoing a major reconfiguration based on the deployment of a modular architecture and the introduction of an on-board local area network (CANBus) for system communications. Though the primary target for the reconfigured system remains that of automated excavation with extensions to the system as originally implemented to cope with different types of obstacles, it is also intended to explore operation in other task regimes such as palletising in order to test the flexibility and adaptability of the modular architecture.

## 7. Acknowledgements

The authors would like to thank all the MEng & MSc students, research students, research associates and research assistants who have contributed to the development of the LUCIE system over the years and especially Peter Green, Jim Mann, Mark Goodwin, Lili Zhang, Gallen Melling, Howard McKee, Frank Margrave, Jason Scott and Steve Quail. In addition, they would like to thank the Science and Engineering Research Council and the Engineering and Physical Sciences Research Council for their support as well as the assistance given by a number of companies, in particular that provided by JCB, Vickers Hydraulics and Danfoss.

## 8. References

1 Singh, S, 1996, *A Survey of Automation in Excavation,* submitted to Journal of Mining and material Processing Institute of Japan

2 Vähä, P.K., Skibniewski, M.J., & Koivo, A.J., 1991, *Kinematics and trajectory planning for robotic excavation*, Proc. Construction Congress 91, pp 787 - 793

3 Apte, S.M. & Oppenheim, I.J., 1994, *Planning intricate robot motions to remove natural materials*, ASCE Conf. On Robots for Challenging Environments, pp 106 - 114

4 Huang, X.D. & Bernold, L.E., 1994, *Control model for robot backhoe excavation and obstacle handling*, ASCE Conf. On Robots for Challenging Environments, pp 131 - 139

5 Koivo, A.J., Thoma, M., Kocaoglan, E., & Andrade-Cetto, J., 1996, *Modelling and controlof excavator dynamics during digging operation*, Journal of Aerospace Engineering, Vol 9, No 1, 10 pages

6 Shi, X., Lever, P. & Wang, F. 1996, *Experimental robotic excavation with fuzzy logic and neural networks,* ,IEEE International Conference on Robotics and Automation, Minneapolis, pp 957 - 962

7 Lever, P., Wang, F., Shi, X. & Chen, D., 1994, *A fuzzy behaviour approach for controlling mining excavator bucket/rock interactions*, IEEE Industrial Applications Society 29th Annual Meeting, Denver, pp 2126 - 2133

8 Herman, H. & Stenz, A., 1994, *Active surface/subsurface perception for autonomous excavation*, Conf. on Sensor Fusion and Aerospace Applications, pp 154 - 167

9 Minamoto, M., Nakayama, K., Aokage, H. & Sako, S., 1994, *Development of a tele-earthwork system*, 11th International Symposium on Automation and Robotics in Construction, Brighton, pp 269 - 275

10 Wakefield, R.R. & O'Brien, J.B., 1994, *A virtual reality type simulation system for construction automation development*, 11th International Symposium on Automation and Robotics in Construction, Brighton, pp 239 - 247
11 Wohlford, W.P., Bode, B.D. & Griswold, F.D., 1990, *New capability for remotely controlled excavation*, Proc. 1990 Winter Meeting of the American Nuclear Society, pp 628 - 629
12 Burks, B.L., Killough, S.M. & Thompson, D.H., 1992, *Remote excavation using the telerobotic small emplacement excavator*, Proc. 1992 Winter Meeting of the American Nuclear Society, pp 559 - 560
13 Bernold, L.E, 1991, *Experimental studies on the Mechanics of Lunar Excavation,* ASCE Journal of Aerospace Engineering, Vol 4, No 1, 9 pages
14 Lever, P., Wang, F. & Chen, D., 1994, *Control model for robot backhoe excavation and obstacle handling*, ASCE Conf. On Robots for Challenging Environments, pp 97 - 105
15 Nease, A.D. & Alexander, E.F., 1993, *Air Force construcxtion automation and robotics*, 10th International Symposium on Automation and Robotics in Construction, Houston, pp 341 - 346
16 Crane, C., Cunningham, S., Armstrong, D., Rankin, A., Novick, D., Wit, J. & Duffy, J., 1996, *Development of an autonomous navigation and teleremote mapping system*, Mechatronics'96 + M$^2$VIP'96, Guimarães, Vol 1, pp 5 - 10
17 Bradley, D.A. & Seward, D.W., 1992, *The Lancaster University Computerised Intelligent Excavator Programme*, IMechE Conf. on Mechatronics - Designing Intelligent Machines, Dundee, September, pp 163 - 168
18 Bradley, D.A., Seward, D.W., Mann, J.E. & Goodwin, M.R., 1993, *Artificial intelligence in the control and operation of construction plant - the autonomous robot excavator*, Automation in Construction, 2, pp 217 - 228
19 Eisenstadt, M. & Brayshaw, M., 1990, *A knowledge engineering toolkit*, Byte, October, pp 268 - 282
20 Ginsberg, M., 1993, *Essentials of Artificial Intelligence*, Morgan Kaufman
21 Green, P., Seward, D.W. & Bradley, D.A., 1990, *Knowledge acquisition for a robot excavator*, 7th International Symposium on Automation and Robotics in Construction, Bristol, pp 351 - 357
22 Seward, D.W., Bradley, D.A., Bracewell, R.H., 1988, *Research models for automatic excavation*, 5th International Symposium on Automation and Robotics in Construction, Tokyo, pp 703 - 709
23 Bradley, D.A., Seward, D.W., Heikkilä, T. & Vähä, P., 1994, *Control architectures and operational strategies for intelligent, high-powered manipulators*, Proc. 11th International Symposium on Automation and Robotics in Construction, Brighton, pp 367 - 372
24 Seward, D.W., Bradley, D.A., Mann, J.E. & Goodwin, M.R., 1992, *Controlling an intelligent excavator for digging in difficult ground*, 9th International Symposium on Automation and Robotics in Construction, Tokyo, pp 743 - 750

25 Seward, D.W., Ward, J., Findlay, J., & Kinniburgh, H., 1996, *The automation of piling rig positioning using satellite GPS,* 13th International Symposium on Automation and Robotics in Construction, Tokyo, pp 223 - 230

26 31EN 292-1:1991, Safety of Machinery - Basic Concepts, general principles for design: Part 1: Basic terminology, methodology
EN 292-2:1991, Safety of Machinery - Basic Concepts, general principles for design: Part 2: Technical principles and specifications

27 Seward, D.W., Bradley, D.A. & Margrave, F., 1994, *Hazard analysis techniques for large mobile robots,* Proc. 11th International Symposium on Automation and Robotics in Construction, Brighton, pp 35 -42

28 Webster, D.A. & Challinor, S.F., 1993, *Long-reach manipulators for decommissioning*, Nuclear Energy, Vol 32, No 3, pp 173 - 176

29 Taylor, A.J., 1993, *Design issues for underwater manipulator systems*, Mechatronics, Vol 3, No 4, pp 419 - 432

30 Thayer, S., Gourley, C., Butler, P., Costello, H., Trivedi, M., Chu Xin Chen & Marapane, S., 1992, *Three-dimensional sensing, graphics and interactive control in a human-machine system for decontamination and decommissioning applications*, The International Society for Optical Engineering, SPIE, Vol 1828, pp 74 - 85

31 Bejczy, A.K., Kim, W.S. & Schenker, P.S., 1994, *The role of computer graphics in advanced tele-operation*, ASCE Conference on Robots for Challenging Environments, Albuquerque, pp 1 - 9

32 Hoffman, R. & Simmons, R., 1994, *Simulation of autonomous robot excavation*, ASCE Conference on Robots for Challenging Environments, Albuquerque, pp 115 - 122

33 Bradley, D.A., Margrave, F., Widden, M.B., Melling, G. & McKee, H., 1997, *The development of a low caost approach to the prototyping of construction plant*, Proc. 14th ISARC, Pittsburgh, pp 299 - 304

34 Zhang, L., Bradley, D. A., & Seward, D.W., 1993, *Safety Critical Systems for Robot Excavation*, Proc. 10th ISARC, Automation and Robotics in Construction X, Elsevier Science Publishers, pp 109-116

35 Zhang, L., Bradley, D.A. & Seward, D.W., 1997, *The simulation and modelling of safety systems for construction robots*, Proc. 14th ISARC, Pittsburgh, pp 224 - 229

# 22

# AUTONOMOUS ROBOTIC EXPLORATION AND GAZE PLANNING USING RANGE SENSING

PAMELA A. M. RENTON
*Department of Mechanical Engineering*
*McMaster University. Hamilton, Ontario, Canada L8S 4L7*

HODA A. ELMARAGHY
*Intelligent Manufacturing Systems (IMS) Centre*
*University of Windsor. Windsor, Ontario, Canada N9B 3P4*

MICHAEL GREENSPAN
*National Research Council of Canada*
*Ottawa, Ontario, Canada K1A 0R6*

HASSEN ZGHAL
*Intelligent Manufacturing Systems (IMS) Centre*
*University of Windsor. Windsor, Ontario, Canada N9B 3P4*

## 1. Introduction

In recent years, a number of robotic applications have appeared in which little is known about the scene structure or the geometry of the robot workspace. It may sometimes be difficult or impossible for humans, due to adverse conditions, to observe and model environments such as nuclear sites with high levels of radioactivity. The examination of these environments and determining the objects present require autonomous sensing and exploration, which in turn involve significant remote and collision-free robotic activity.

The work presented here was initially motivated by the serious problem of remediation of hazardous nuclear waste sites. These radioactive materials were simply stored, often buried in underground tanks, since there was no way of safely destroying them. Unfortunately, these storage tanks have proven to be less durable than the radioactive materials they contain and have started to leak and spread their contaminants uncontrollably. These tanks and the contaminated surrounding soil must be inspected, identified, gathered, and repackaged in more durable containers [1]. Since no record was kept of the contents of the waste tanks and their levels of radioactivity, the tanks' contents must first be determined. It is imperative to do this without collisions and damage to the tanks or equipment. A system for remotely exploring the tanks' contents by teleoperating machinery from a safe distance was needed.

Telerobotics has long been considered the main alternative to operating in hazardous environments, and efficient and safe teleoperation has been an active area of research. When the operators work at a remote site, they often lose the direct visual cues they would normally rely on to perform their tasks. A number of techniques for regaining or substituting these cues with alternate sources of information have been considered (*e.g.* [2,3,4,5]). Most of these techniques have been found inadequate as they do not consider teleoperation simultaneously with collision avoidance. Quantitative models of the environment, which use the rich sensory information about the scene, can be obtained and used to view the scene and select the best viewing position for each required operation.

The approach to autonomous exploration and gaze planning was implemented in a system called Plan-N-Scan which was developed as part of the IRIS/ISDE-3[1] project involving "Integrated Sensing, Planning and Action in Robotics Applications." The collision avoidance scheme used in Plan-N-Scan was developed by Greenspan and Burtnyk [6], and was then integrated into the PERK[2] development system at NRC[3] which uses a PUMA 560 robot as well as a laser scanner developed at NRC [7]. A research team from SPAR[4], AECL[5] and NRC is involved in a PRECARN[6] project called VERO[7] to explore the potential use of the laser range imaging technology. The project aims at developing systems and techniques to clean-up contaminated soil and barrels and to overcome the difficulties of teleoperating heavy hydraulic machinery [8]. SPAR is also developing a robotic arm equipped with range sensors to work within the hazardous waste tanks. The developed Plan-N-Scan system has the potential for future field implementation and testing in this and many other applications due to its generic nature.

Plan-N-Scan involves autonomous scanning and mapping of selected regions of interest in the initially unknown robot workspace, using a wrist-mounted laser range camera [9]. The technique involves the incremental scanning of a sequence of targets that are selected based on the information obtained in the previous scans. The result is a voxel map (3-D occupancy grid) suitable for real-time collision detection, which enables safe teleoperation and path planning for the robotic arm. Two scanning modes have been designed for either scanning towards a given target point or scanning a selected volume. The scanning methods are independent of the robot configuration. The Plan-N-Scan technique provides the ability to safely scan an environment and determine the occupancy of objects in its workspace. It also generates a world representation that allows collision-

[1] IRIS - Institute for Robotics and Intelligent Systems, a network of the federal Centres of Excellence Program in Canada

[2] PERK - Program for the Exploitation of Research Knowledge sponsored by PRECARN

[3] NRC - National Research Council of Canada

[4] SPAR - Spar Aerospace Ltd.

[5] AECL - Atomic Energy of Canada Limited

[6] PRECARN Associates Inc. - PRE-Competitive Advanced Research Network, a not-for-profit corporation managing the IRIS network

[7] VERO -Virtual Environment for Remote Operation

free path planning and teleoperation, enabling repositioning of the manipulator to acquire additional information as needed.

The remainder of this chapter is organized as follows. Section 2 presents a review of the literature on autonomous workspace exploration and gaze planning as well as related issues such as robot vision and sensing, model representations, search algorithms and range sensing technology. Section 3 presents the requirements of the environment model and provides details of the selected representations. The process of target scanning, which is a core element required to map a given volume, is described in Section 4. The details of the target selection procedure used to scan and map a given workspace volume is described in Section 5. The system implementation and some test results are presented in Section 6, and the conclusions are discussed in Section 7.

## 2. Literature Overview

Plan-N-Scan is an inter-disciplinary project, involving a number of research topics related to autonomous workspace exploration. These include: robot vision and sensing, gaze planning for determining where to scan next, based on the previous scans; and the planning of a collision-free path to reposition the sensor to scan selected locations.

### 2.1. ROBOT VISION AND SENSING

Sensor-based motion planning in unknown, dynamic and/or unstructured environment requires the acquisition of scene knowledge of the robot workspace using non-contact robot sensing. 3-D vision or range sensing is considered the best candidate for providing detailed and accurate 3-D information about the scene and has become a vital tool in various robotic applications. It measures the distance from a reference point to objects in the scene; however, raw range data only provides 2 1/2-dimensional information. Hence, knowledge of the camera pose (position and orientation) is required to obtain true 3-D scene information. The common techniques for range sensing include stereo vision, triangulation, structured lighting, radar, laser time-of-flight and ultrasonic. Some of these techniques are direct methods, while others are image-based and require analysis of one or more brightness images to extract range (or depth) information. Laser and other ranging devices have significantly improved in accuracy and data collection speed, and became universally accepted in various applications ranging from 3-D object recognition [10] to reverse engineering [11]. The reader may refer to a review of range sensing by Jarvis [12].

In this work, a triangulation-based laser range camera, designed and built at NRC [7,13], was used to acquire range information about the scene. The NRC's camera uses an auto-synchronized approach and can achieve a large field of view with small triangulation angles without sacrificing precision. This is due to the arrangement of the mirrors causing the instantaneous field of view of the camera to shift in synch with the projected beam of light. The sensing frustum for this camera is a rectangular pyramid with an angular field of view of 30 by 40 degrees. The effective range of its field of view is 0.15 to 0.50 meters from the camera head, with an accuracy of 0.1% of the range reading value. It takes approximately 20 seconds to acquire a single range image.

## 2.2. GAZE PLANNING AND AUTONOMOUS EXPLORATION

Mobile robots and autonomous navigation systems often operate in unknown environments which need to be automatically explored and mapped. Gaze planning is a major issue in autonomous exploration and involves the automatic selection of viewing positions based on the sensory data obtained from the scene. The goal of gaze planning is to incrementally sense the environment guided by previous sensory data, to plan subsequent sensor actions and robot motion. Interest in gaze planning has increased lately due to recent developments in range sensing, robotics and computer control.

One important aspect of gaze planning is the automatic and efficient selection of viewing positions. Currently, the sequence of scans required is usually selected manually. However, some methods for automatically selecting the viewing positions to ensure as complete surface coverage as possible have been proposed. Whaite and Ferrie [4] developed an object mapping system using a least-squares approach to match the range data to super-ellipsoid models. Areas of high uncertainty, where the range data and super-ellipsoid surface match poorly, were targeted for further scanning. Maver and Bajcsy [14] presented another approach to object mapping using occlusions to guide the scanning actions. Mapping starts by taking a range scan from a random viewing direction, and is followed by a set of planned scans in the "scanning plane" which are selected to be perpendicular to the initial direction. Occlusions were used to guide the acquisition of 3-D data of an unknown scene. The occluded regions visible from different directions within the "scanning plane" are analyzed, and a set of scanning directions is selected to cover all the occlusions. Similarly, Banta *et al.* [15] used occlusions to guide scanning as well, but they did not restrict the subsequent scans to be at 90 degrees from the initial scan, enabling the scanning system to efficiently deal with more difficult cases. Kruse *et al.* [16] presented a framework for guiding robot perception for the exploration of an unknown 3-D environment using range sensors. Their sensing action planner uses rating functions (applied to a number of scanning and robot motion constraints) in configuration space to select goal configurations then plan robot paths to them. Most of the object scanning and motion planning systems have some limitations. In fact, their scanning methods consider the scanner to be outside the active workspace and therefore never in danger of collision. Many of these systems require the scanning of certain regions more than once. Furthermore, most methods described in the literature for selecting the viewing positions are somewhat limited since they tend to completely ignore the problem of collision-free sensor placement and do not consider the kinematic feasibility of the viewing position and the ability to plan a path to it.

Mobile robot research generally deals with collision detection but only requires sensor placement and path planning in a 2-D environment, which are far simpler than in the case of a robot manipulator operating in three-dimensional (3-D) space. Due to the simple 2-D map construction (2-D versus 3-D for robotic arms) and simple path planning, (point robot moving in a 2-D world), much of the work on mobile robots is not applicable to mapping the 3-D workspace of a robotic arm. Although methodical approaches to efficiently map the region of interest are proposed, the common representations used with mobile robots, such as visibility graphs [17], become complicated and may break down in the 3-D case. Other methods use 2-D grids and quadtree representations of the environment [18,19]. The implementation of such approaches can be simplified by dividing the problem into target selection and target scanning components [5]. Most existing target selection methods are usually based on either a *breadth-first* or *depth-first* search [17]. In other words, targets are either chosen as the closest unknown points to the original starting position of the robot, resulting in a *breadth-first* approach to workspace scanning, or as the closest unknown space to the current robot position, resulting in *depth-first* exploration. For robotic arms, *breadth-first* target selection techniques are more suitable, as they encourage scanning the space closest to the origin first, then progressing further into the workspace, thus reducing the probability of the arm encountering unknown space.

The Plan-N-Scan technique involves a gaze planning strategy that goes a step further by additionally considering collision-free sensor re-positioning in a full 3-D environment. In addition, it considers scanning a region only once, as it is mainly concerned with determining voxel occupancy and does not have to be as accurate as object scanning for modeling purposes.

### 2.3. COLLISION-FREE MOTION PLANNING

Robot motion planning involves two major components: collision detection to determine acceptable robot configurations; and selection of "the best" amongst the acceptable configurations to achieve the desired goal position. Motion planning is an active area of research, and extensive literature has appeared in this area. A comprehensive survey of motion planning methods was presented by Hwang and Ahuja [20].

Motion planners can be classified as *model-based* and *sensor-based*. *Model-based* motion planners assume that the scene is static and that the scene structure and objects as well as the robot models are known in advance. Collision detection is achieved by determining the configurations which avoid the physical intersection between the robot and obstacle models. On the other hand, *sensor-based* motion planners assume the robot workspace model and scene structure to be completely or partially unknown and often dynamic, hence motions cannot be pre-planned. Classical motion planners, used to find collision-free paths in known and well-structured environments, may not always be useful as they do not take into account the dynamic and unstructured nature of the workspace. Knowledge about the scene structure and robot workspace must be acquired to detect and avoid any potential collision between the robot and obstacles in the scene [2,21,22]. *Sensor-based* motion planning is often handled by heuristics that act in some pre-set manner when obstacles are encountered.

When the workspace is partially known, a system that incorporates the principles of both *sensor-based* and *model-based* motion planning may be desirable. Such system would use the prior knowledge about the scene to get as close to the goal location as possible. Only when additional knowledge is required will the system gather more information and incorporate it into the world model. Therefore, the system can take advantage of pre-existing knowledge while still being able to acquire and use more information when needed, thus allowing motions to be planned anywhere within the mapped regions. A number of issues are dealt with in collision-free motion planning. These include object and scene sensing and representation and search techniques.

### 2.3.1. Object and Scene Representations

Once information about shapes and configurations of objects is acquired, it can be represented in different ways. A number of model representations have been developed and used in robotics and robot motion planning. The common representations used for geometric modeling of solid objects include: polyhedra [23], generalized-cylinder or sweep representations [24], surface boundary representations (B-rep) [25], constructive solid geometry (CSG) representation [25] and spatial occupancy representations [26]. The spatial occupancy representations, which define the object using non-overlapping sub-regions of the 3-D space occupied by an object, appear to be the most suitable approach for scene mapping. Two common examples of such a scheme are the voxel and the octree representations. The voxel representation is a simple method which uses small volumetric elements (fixed-sized cubes) called voxels (or 3-D pixels) to represent the discretized 3-D space. An octree (or quadtree for the 2-D case) is a more compact hierarchical representation of spatial occupancy [27].

Voxel representations may be criticized for the extensive memory requirement, but they have the advantage of quickly and directly storing the acquired range data, hence allowing real-time processing particularly for collision avoidance. Other representations, such as polyhedra, would require extensive processing time to extract the polyhedral models from the range data [23].

The workspace representation selected in this technique consists of the voxel representation for the static elements such as workspace obstacles, combined with the spherical representation for the dynamic elements, such as the robot manipulator links [28]. This voxel-sphere combination results in an efficient collision detection method which will be described below.

### 2.3.2. Collision Detection

Collision detection is generally achieved by testing the intersection of the robot-workspace model pairs and simply preventing the robot (or its model) from getting into occupied space. More sophisticated methods, such as the distance transform methodology, have been developed to increase efficiency [5]. Other methods divide the workspace into an octree structure, so only the objects in the same octant are compared [3]. Other variations on the intersection method use sets of spheres or points for the moving robot, thus simplifying the intersection calculations [29].

The collision detection method adopted in Plan-N-Scan is a real-time technique developed by Greenspan and Burtnyk [6]. This method efficiently tests the robot spheres for collision with the workspace voxels without exhaustively checking all the voxel-sphere pairs. Therefore, it is obstacle-count independent [28], hence it is very efficient.

### 2.3.3. Search Techniques

Robot motion planning is concerned with finding a connected sequence of feasible configurations between the start and the goal configurations. This can be achieved by searching through a graph or tree of possible paths. Several search techniques for path planning have been developed. These mainly include *hill-climbing*, *random search*, and enumerative methods. An example of a *hill-climbing* technique employing the potential field approach is given in [30]. *Random-search* methods such as simulated annealing have also been used for motion planning [31]. Gupta and Guo [32] presented the *sequential* search method combined with the configuration space (C-space) representations [33].

Enumerative search methods are easily and effectively used for grids. They include the *breadth-first, depth-first, best-first* and *A** search techniques [34]. The *breadth-first* search is thorough but time consuming, since it checks all the neighboring positions, then all the neighbor's neighbors and so on. The *depth-first* method selects the best position among the neighbors and checks only its neighbors. The *A** algorithm is a *best-first* algorithm with backtracking, which combines these two approaches by following several promising paths towards the goal at a time, switching between them as different branches become more promising [35].

## 3. Robot and Workspace Representations

The representation of the robot and objects present in the scene should allow real-time collision detection and enable teleoperation. Collision detection must be carried out throughout the mapping stage for the camera to be able to move safely and take scans of the unknown regions in the workspace. As the new information is gradually acquired by scanning the scene, this data must be quickly incorporated into the workspace representation. Therefore, this representation should be flexible enough to handle prior model information as well as sensory data gathered from different sources, and it should clearly distinguish unscanned from scanned regions.

### 3.1. REPRESENTATIONS AND COLLISION DETECTION

The spherical representation of the Puma 560 robot consists of five sets of spheres, each completely encompassing one link of the robot arm. A set of spheres for each link is generated, given the link's surface geometry and the parameters controlling the spacing between the spheres and the volume that they can project past the link surfaces [28]. These parameters can be adjusted to minimize the number of spheres required to represent each link. A total of 509 spheres were used to model the Puma arm, as illustrated in Fig. 1.

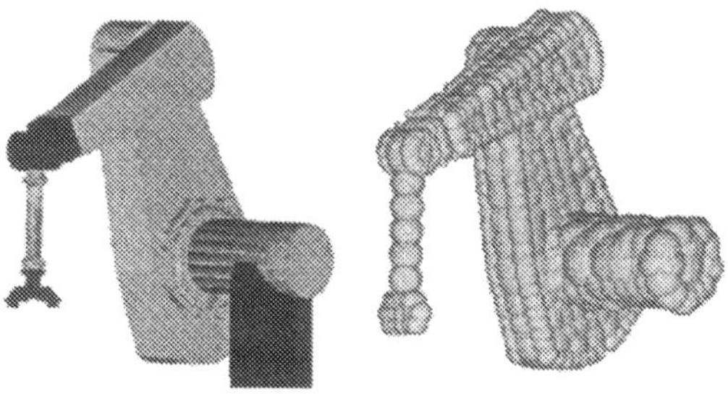

*Fig. 1.* Solid model and spherical representation of the Puma 560 robot.

The workspace representation consists of a voxel map, which is simply a 3-D grid of small cubes called voxels. The voxel size is fixed and depends on the absolute positioning accuracy of the robot. A set of routines has been developed to automatically generate and maintain voxel maps using range images and polyhedras [28]. Incrementally acquired sensor data can be quickly incorporated into the voxel maps, because the data can be added directly to the map without extracting object models. The map can readily be used for collision-free path planning during the scanning process. The reach of the Puma 560 manipulator used in this work is approximately 1 *m*. The resulting workspace can be contained by a 2 *m* cubed box. The robot's positioning accuracy is approximately 2 *cm*, requiring the voxels to be 2 *cm* cubed. This results in a voxel map containing one million voxels. The information necessary for a single voxel can be stored in one byte of data, so the voxel map will only require one Megabyte of memory.

Collision detection for any given robot configuration is performed as follows. The forward kinematics places the sphere sets in the correct configuration. The spheres in each set are then tested for collision by comparing the value of the voxel at the center of each sphere to the sphere's radius as illustrated in Fig. 2. If the voxel's distance value is greater than the sphere's radius, then the voxel is farther from the nearest obstacle than the radial distance, and the sphere is collision-free. If all the spheres are collision-free, then the planned robot move is safe. The difference between the voxel distance value and the sphere radius gives a measure of how far the sphere is from collision. The clearance between the robot and any workspace obstacles should exceed one voxel (equal to the robot positioning accuracy) to be considered collision-free. This efficient method reduces the collision detection computations to a level suitable for real-time applications.

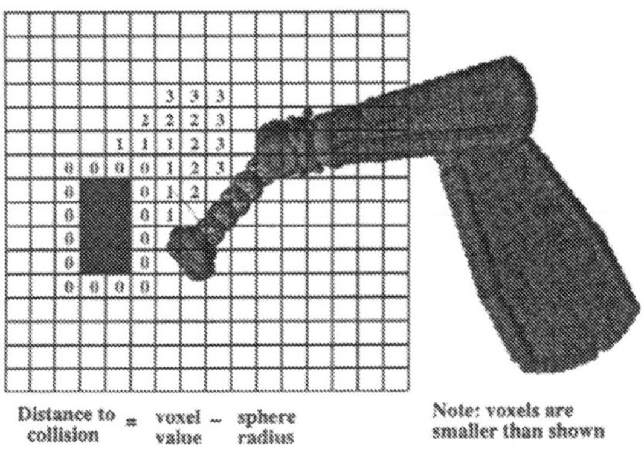


*Fig. 2.* Collision Detection.

## 3.2. DESCRIPTION OF THE VOXEL MAPS

Unlike the VERO prototype developed at NRC, which uses a single voxel map for collision detection, the Plan-N-Scan system developed in the IRIS project uses two types of voxel maps. The Workspace Mapping Voxel Map (WMVM) records the additional required data. The Collision Detection Voxel Map (CDVM) is compatible with the existing collision detection scheme and voxel map maintenance routines.

The WMVM contains information on the status of each voxel, *i.e. unknown*, *obstacle*, *free* or *unscannable*. *Unknown* voxels represent space that has not yet been scanned. *Obstacle* and *free* voxels indicate space that has been scanned and found to contain obstacles or free space, respectively. *Unscannable* voxels represent space occurring behind objects where the camera can not be positioned to view and which the system has tried but failed to scan. The strategy is to use the knowledge of unknown space to indicate where the system still needs to scan and where new information can be gathered.

The CDVM covers the exact same space as the WMVM but records slightly different information and is operated on by different algorithms. The CDVM used for collision detection recognizes voxels with the status of either *obstacle* or *free*. Voxels are treated as *obstacle* if there is no information about their occupancy status (*i.e.* they have not been scanned) or if they are known to contain an obstacle. Known *obstacle* voxels can occur if object models are added into the voxel map, or if range data indicates obstacle presence. *Free* voxels are processed and given a distance value indicating how many voxels they are from the nearest *obstacle* voxel, resulting in a weighted workspace map.

The two voxel maps encode complementary information about the same space, and are therefore not independent. The possible values of the CDVM voxels depend on the values for the corresponding WMVM voxels. *Free* voxels in WMVM correspond to weighted *free* voxels (with distance values) in CDVM, while *obstacle*, *unknown*, and *unscannable* voxels in WMVM correspond to *obstacle* voxels in CDVM. Initially, all the voxels in the WMVM are set to *unknown*, with the exception of the voxels inside a cylinder surrounding the base of the robot manipulator which are set to *free* to allow some arm maneuverability to take initial scans. The CDVM voxels are set correspondingly to *obstacle* and *free* (with a distance value), as shown in Fig. 3. When new data is acquired, it is first encoded into the WMVM, re-assigning the appropriate voxel values from *unknown* to either *free* or *obstacle*. The data is then added to the CDVM, updating the distance values of the *free* voxels [Fig. 3].

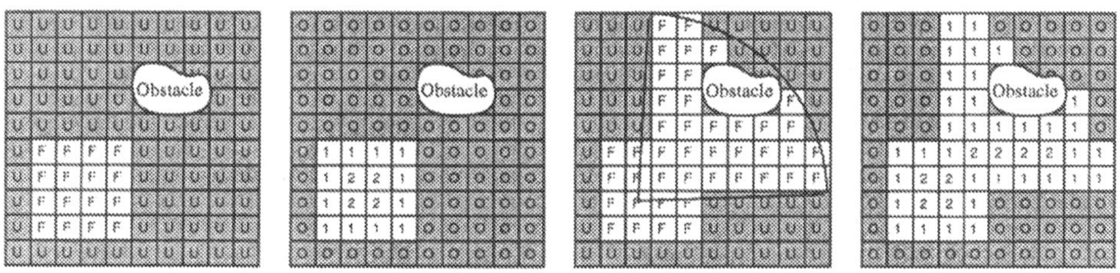


(a) Unitial WMVM with Free and Unknown voxels

(b) Initial CDVM with corresponding Obstacle and Wighted voxels

(c) Scan in acquired, changing some Unknown voxels to Free or Obstacle

(d) The affected voxels are updated

*Fig. 3.* Voxel map correspondence for a 2-D grid.

## 3.3. SCAN ACQUISITION AND VOXEL MAP UPDATING

Each time a scan is taken, the captured range data is incorporated into the voxel maps. The acquired range images generate a dense range map of 256 by 256 points. At the maximum range of 0.5 meters, the beams of laser light are only about 1.5 *mm* apart, resulting in many laser beams passing through a single voxel. This information can be incorporated into the voxel maps faster if a sparser data set is extracted from each image. This is done by reducing the dense 256 by 256 range image to a smaller and sparser array of 32 by 32 patches before entering it into the voxel maps, each patch representing an 8 by 8 set of the original range points. As a range image is acquired, the laser beam passes through free space until it hits an object, for which it returns a range value. The path traversed by the laser beam for each recorded range point is examined, checking each voxel in the WMVM that is along that path and setting *unknown* voxels to either *free* or *obstacle* as appropriate. This method is similar to the model-updating technique suggested by Roth-Tabak and Jain [36].

Whenever a change is made in the WMVM, the CDVM is updated to reflect it, using the lists of voxels that changed status to *obstacle* or *free* in the WMVM. Two procedures are used to efficiently change the status of these lists of voxels from *obstacle* to *free* and vice versa, and to adjust the distance values of the corresponding *free* voxels, without processing the entire voxel map [37]. These routines only visit those voxels in the local neighborhoods of the voxels being changed, called the voxel influence zones (*Voiz*). The *Voiz*, which is equivalent to a Voronoi Region, is limited to those of *free* voxels whose distance values are affected by the voxels being changed.

## 4. Target Scanning

The Plan-N-Scan system is based on the automatic selection and collision-free scanning of targets (3-D points in space) as required to determine the scene spatial occupancy. This system enables successful target scanning even in difficult cases where the target cannot be viewed immediately. This is due to the beneficial use of incrementally acquired scans to allow the system to iteratively progress towards the goal target.

A flow chart of the target scanning system is illustrated in Fig. 4. The target scanning system first selects the viewing position for a target, and either plans collision-free paths to it or determines if additional scan targets are required. As it is generally possible to view a target from many different positions, target scanning is a less constrained objective than reaching an exact final position. The viewing positions should be selected such that the scan rate efficiency is increased when large volumes requiring multiple scans are explored. Target scanning must therefore find a non-occluded viewing position for the specified target, then search for a collision-free path to that viewing position. The WMVM is used to generate a viewing position for the goal and intermediate targets, while the CDVM is accessed to derive a collision-free path to the viewing position.

This system, which is based on the STRIPS method [34], is capable of selecting a sequence of intermediate targets in order to scan the original target. A viewing position is generated for the target, and an *A* Search* is invoked to find a collision-free path to that position. If the viewing position cannot be reached due to potential collisions with unknown space, then intermediate targets may be pushed onto the *scan stack*, which is data structure of target points. The *A* Search* progresses towards the initial goal voxel by successively scanning intermediate targets.

### 4.1. VIEWING POSITION SELECTION

For a given target, a viewing position is calculated as a direction and an offset. If the target voxel resides in a cluttered region of space, then there may only be a small range of unobstructed viewing positions, some of which may be unreachable by the manipulator. Therefore, to find the possible viewing directions, a local filter is applied within the WMVM which examines the target voxel's immediate 26 neighbors. The corresponding reachable (*i.e.* kinematically possible) viewing positions are then calculated by projecting a specified offset from the target along the negative viewing direction.

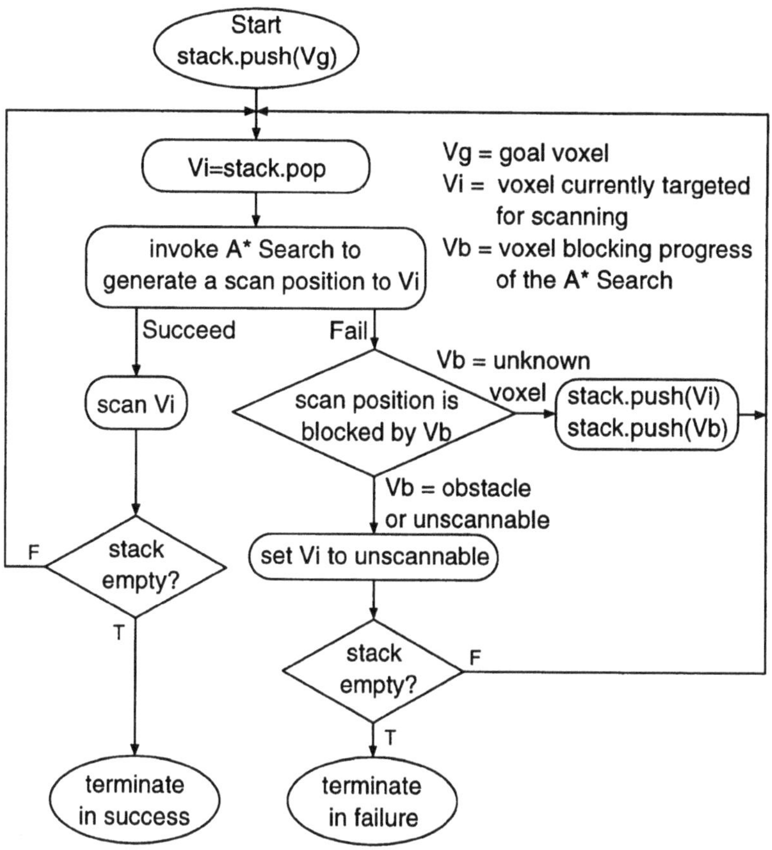


*Fig. 4.* Flow chart of the target scanning system.

The selected reachable viewing position should satisfy two criteria: (1) the line of sight from the camera head to the target should be unobstructed, and the target must be within the camera's field of view [Fig. 5]; when in the viewing position; and (2) the manipulator should be collision-free. If none of the positions meets these criteria, then, depending on the desired scanning thoroughness, the *A* Search* could be used to find an acceptable position, starting with the best potential viewing position as a reasonable first guess, otherwise the target could be deemed *unscannable.* On the other hand, if a number of viewing positions qualify, the one farthest from joint limits is selected. This pre-processing step is not only important for decoupling the viewing position selection from the path search, but also for detecting cases when the target is obviously *unscannable,* thus avoiding fruitless search. The *scan stack,* is initialized to contain the goal voxel. A viewing position, found to be collision-free and with a clear line of sight, will be selected and pushed onto the *scan stack.* A heuristic is included which attempts to increase the amount of new information acquired with each range image. The offset viewing position with the maximum number of *unknown* voxels in sight, when scanned from the chosen direction, is selected and pushed onto the *scan stack.*

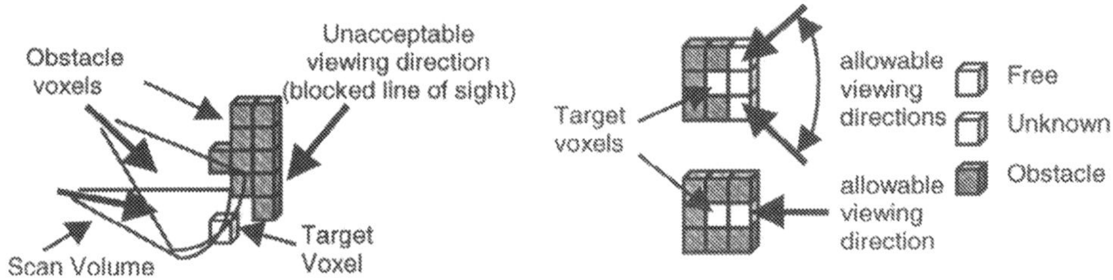


*Fig. 5.* Acceptable scanning positions, and potential viewing directions.

## 4.2. INTERMEDIATE TARGET SELECTION

If the robot cannot reach the viewing position without passing through unknown space, then the *unknown* voxel that is blocking progress towards the viewing position is selected and pushed onto the *scan stack,* becoming an intermediate target. If in the next iteration this voxel is found to be *free*, then the search can resume, progressing beyond this hindrance. If the *unknown* voxel is sensed and found to be an *obstacle*, then an alternative route to the goal target must be found. It is worthwhile to note that this is the main backtracking mechanism, which allows the beneficial use of incrementally acquired images to scan a selected target, even if it is not immediately visible. If either *obstacle* or *unscannable* voxels blocked all attempted paths to a viewing position, then the target voxel is set to *unscannable* and is removed from the *scan stack.* Setting the target to *unscannable* ensures that it will not be subsequently selected as an intermediate target. If, after removing this *unscannable* target, the *scan stack* is empty, then the goal target is *unscannable* and the scanning process terminates in failure. Otherwise, if more targets remain in the *scan stack,* then this failed attempt is only one of a number of alternative solution paths that will be tried in turn.

## 4.3. COLLISION-FREE PATH FINDING USING A* SEARCH

An *A* Search* is used to generate a collision-free path between two specified manipulator configurations. This is accomplished by exploring all possible collision-free configurations and selecting the most suitable one, given the knowledge of the environment as represented by the spherical manipulator and the weighted CDVM. All possible combinations of a discrete translation along, or a discrete rotation around, the positive and negative x, y and z axes are evaluated. The *A* Search* is optimal in terms of a cost function that doggedly pursues the path of least-cost towards the goal, and since it is a global search method, it will not get stuck in local minima. It is also flexible enough to find paths around many different workspace arrangements, and is general enough to be used with many different robots. However, the *A* Search* can be costly in both time and memory if many local minima exist in the cost function space. Under certain circumstances, it may degrade into an expensive *breadth-first* search.

The *A* Search* does not have to find the selected viewing position exactly in order to be successful, given a feasible goal position, but it is rather sufficient to reach a position whose view of the target is unobstructed. This is important when the scanning position does not meet the collision-free or clear-line-of-sight criterion. If such a position is found, the robot is moved along the generated path, a scan is acquired, the voxel map is updated, and the current target is removed from the *scan stack.* If the *scan stack* becomes empty, then the original goal target has been scanned; otherwise, the top *scan stack* target that is still *unknown* becomes the current target and the *A* Search* iterates. *The A* Search* is implemented to search, in a reasonable number of iterations, for the best possible position, rather than searching to find the exact selected viewing position. This is accomplished by evaluating each entry (as it is added to the *solution stack*) using an acceptability cost function.

The *A* Search* produces three distinct results which generate three different responses from the target scanning system: case 1) finding an *acceptable* viewing position; case 2) finding *unknown* voxels blocking the robot from reaching an acceptable viewing position; or case 3) finding *obstacle* voxels obstructing the robot. Since the *A* Search* only looks through collision-free positions, to test if they are case-1 solutions only requires in-view and clear-line-of-sight calculations. If they are not acceptable viewing positions (*i.e.* not case-1), but their distances to collision are small, then they are considered blocked. In order to determine why the robot cannot achieve an acceptable viewing position, the interfering voxel has to be identified as either *unknown* or *obstacle*. This is done by checking neighboring voxels (in the CDVM) and proceeding towards an obstacle until the nearest *obstacle* voxel is reached. The corresponding voxel in the WMVM is checked to see if it is an *obstacle* or *unknown* voxel, which will then distinguish whether the position is case-2 or case-3.

## 5. Workspace Scanning

From the initial set-up of the system, with most of the workspace being unknown, it is desired to thoroughly scan and map a user-specified section of the workspace volume without colliding with any objects that might be present, hence resulting in a voxel map that can be directly used for teleoperation. Workspace scanning involves the target selection pattern and the target scanning system, as shown in Fig. 6. The automatic target selection must be performed such that when the selected targets are scanned, the whole specified volume will be scanned.

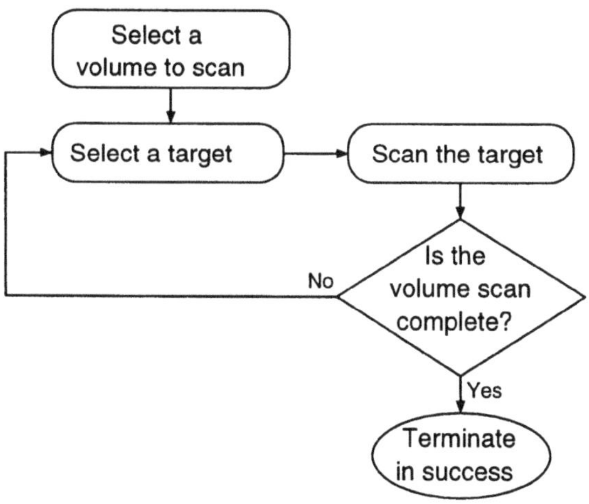


*Fig. 6.* Flow chart of the workspace scanning system.

The volumes specified for scanning can be large, up to the entire reachable workspace, hence requiring multiple scans. In order to test the system within a reasonable amount of time, regions in the Puma's workspace were selected which could be mapped in about twenty to thirty scans. Workspace scanning incrementally maps sections of the volume until it completely covers the region of interest. A volume is covered when the occupancy of all scannable voxels within it is determined. All scannable *unknown* voxels are possible targets for scanning, as information on their occupancy is required. The volume is considered scanned only when no more scannable *unknown* voxels remain in the region of interest, causing the scanning process to terminate.

As the collision detection method is whole-arm sensitive, it is intuitively better to approach workspace mapping for the arm using a *breadth-first* search. This means that the system will always target the *unknown* voxel which is both closest to the origin of the workspace and within the region of interest specified by the user. This will ensure that the space is scanned from the origin outwards, mapping as much free space as possible before the arm navigates through it. In order to select a target voxel to scan, all the voxels in the volume of interest are checked, starting with those closest to the origin of the workspace. The first *unknown* voxel with at least one *free* neighbor voxel is chosen. It is then sent to the target scanning routine which scans it if possible. A new target voxel is chosen after the previous one's occupancy is determined. Rather than processing all the voxels to select the best goal, it is faster to pre-calculate the voxels equidistant from the origin and store them in a set of lists. These target voxels are ordered by increasing Euclidean distance from the origin. Each voxel is therefore considered as a potential target only once rather than $N$ times, where $N$ is the number of scans.

Volume scanning is very efficient initially because the targets are selected by the target selection pattern instead of the backtracking method, and paths can be easily planned. As the robotic arm penetrates into the unexplored workspace, some backtrack points occur at the boundary of the region of interest, as sections of the arm encounter unknown space. This would not be a problem in the "Waste Tank" application since the entire workspace would be scanned. The final scans become less efficient as the system fills in the few remaining unknown sections in the workspace. These sections occur most commonly behind objects and are therefore difficult to scan, accounting for the reduced scanning efficiency. A method of determining when there is enough of a concentration remaining to warrant a scan would improve the system performance in the final stages of volume scanning.

## 6. Plan-N-Scan System Implementation

This section presents an overview of the Plan-N-Scan system implementation, followed by simulation, experimental tests and discussion of the obtained results. Although the target scanning module is used by the volume scanning module, the two sub-systems are tested independently in order to clearly demonstrate the operation of the whole system.

The Plan-N-Scan system is implemented using three main hardware subsystems: a Silicon Graphics (SGI) Indigo workstation which runs the system software; wrist-mounted laser range scanner system; and a Puma 560 robot and its controller. The SGI workstation provides the graphical user interface for Plan-N-Scan and runs the scanning algorithms described above. It must communicate with both the Puma and the laser range camera in order to relay the needed system information. The system software is written in C++, using a public domain GUI design tool called FORMS Designer and other standard graphics libraries. The Puma 560 uses a VAL controller, whose "alter port" accepts small changes of the Cartesian position of the Puma's end-effector, allowing it to follow a desired path as well as facilitating direct teleoperation. In order to ensure that the real-time requirement is met, a personal computer (PC) 486 DX2-66, running the QNX operating system, is used. This PC has an Ethernet connection to the SGI, and uses sockets to create streams for data flow. The SGI sends to the PC paths for the Puma end-effector to follow, while the PC returns the Cartesian position of the Puma's end-effector. The laser range camera's controller is interfaced with the SGI through a Macintosh Mac-II. When scans are required, the SGI sends the Mac a request for scan range data. The Mac initiates data capture, and sends back the range image in the form of Cartesian range points with intensity readings.

The Plan-N-Scan system has been implemented and demonstrated to successfully scan targets and volumes. The set-up of the development and testing system at NRC is shown in Fig. 7. The system initially starts with a cylinder (radius of 0.68 *m*) of free space around the robot's base to give it maneuvering room for taking initial scans. The rest of the workspace is initially unknown. These assumptions are made whenever scanning is initiated, for both target and volume scanning modes.

*Fig. 7.* Development system set-up used in the experiments.

## 6.1. EXPERIMENTATION

### 6.1.1. Target Scanning Experiment

A pole was placed at 0.9 *m* from the center-line of the Puma to simulate the pillars expected to be found in most of the waste tanks. The pole was within the region that was initially assumed to be unknown space. The target was selected to be behind the pole, at 1.0 *m* from the center-line of the robot's base.

The target scanning system was executed on a MIPS R4000 Risk Chip (about 4 MFLOP) in an SGI Indigo machine. It took five (5) scans and lasted over 31 minutes to successfully view the target. Fig. 8 shows the CPU time taken to plan each scan and the number of unknown voxels captured.

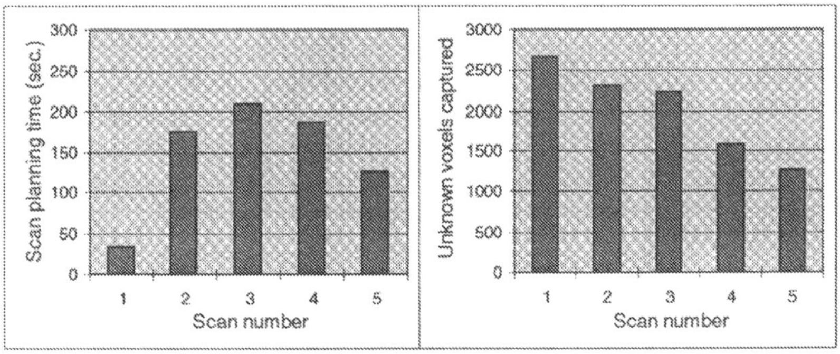


*Fig. 8.* Target scanning results; CPU times and mapped voxel data.

The five viewing locations are shown in Fig. 9, with crosses to indicate the scan targets. Note that the backtrack points (scans 2 to 4) are computationally costly. Scan 5 was also costly (relative to scan 1) because it was required to find a difficult path. The general trend was a drop in the number of unknown voxels captured with each successive scan due to overlapping scan fields. Note also that the planning times do not include the time taken to move the robot, acquire scans and update the voxel map with the scan data.

The sequence of planning actions taken by the target scanning system is shown in Fig. 10. It indicates the results achieved by the *A* Search,* which controls the actions taken by the target scanning system. With the first application of the *A* Search,* a viewing position was found, and a scan was taken. Note that for the clear-line-of-sight calculations, unknown space is optimistically assumed to be *free.* In this case, the assumption was incorrect as the target was in fact occluded by the pole, causing it to remain *unknown* after the first scan, hence remaining on the *scan stack.* The next application of the *A* Search* tried to move around the pole to scan the target, but failed due to obstruction by an *unknown* voxel. Following the target scanning system rules, the obstructing voxel was pushed onto the *scan stack,* becoming a scan target itself. Three such targets were selected and scanned before the original goal target was successfully viewed. This demonstrates the backtrack method at work, which takes advantage of the ability to incrementally acquire scan information in order to achieve the goal. The effect of the scanning actions on the voxel maps is to increase the amount of free space available for the Puma to move within. A total of 0.07992 $m^3$ (9,990 voxels) of free space was added to the voxel map by the five scans, as illustrated in Fig. 11.

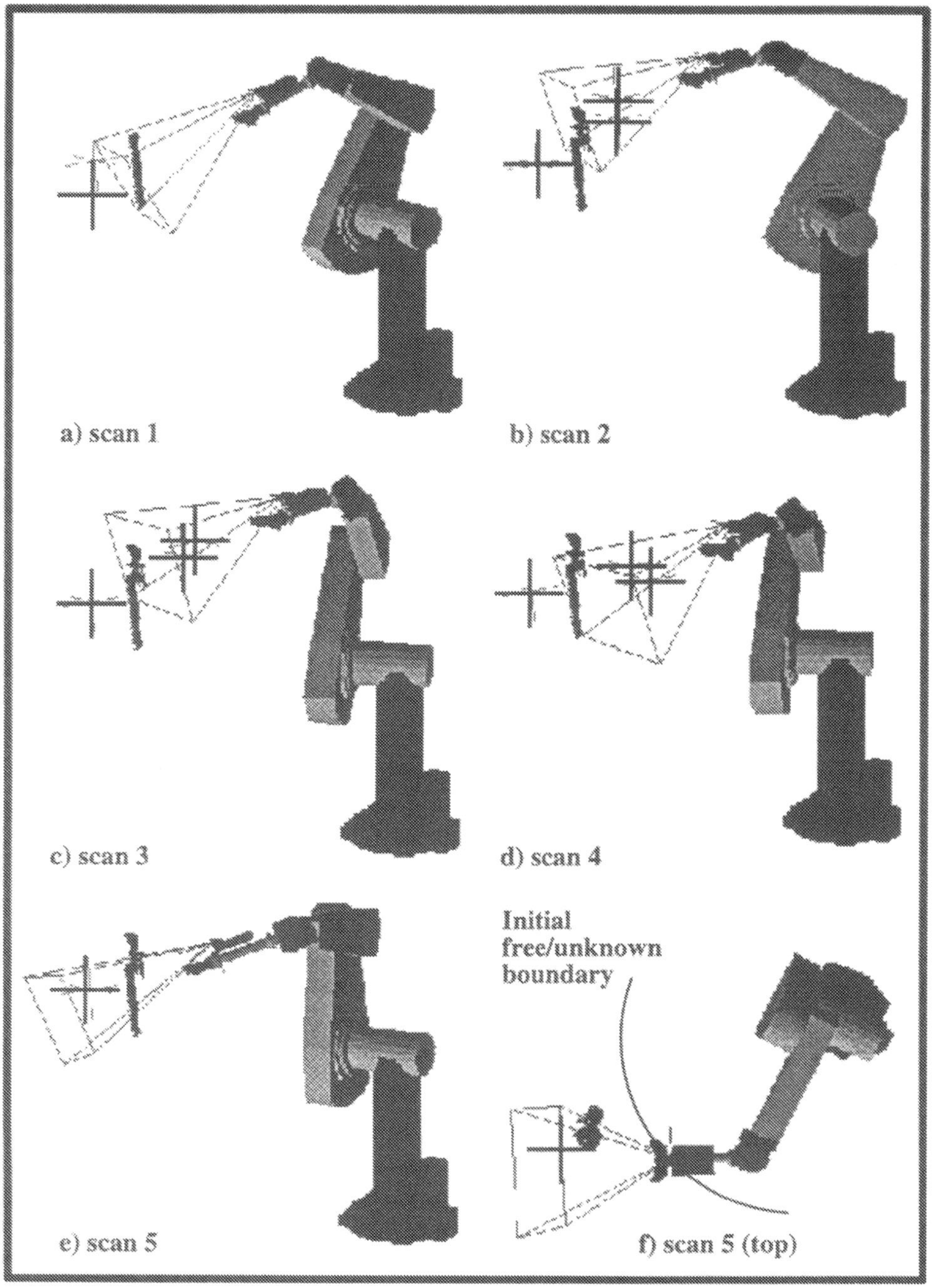


*Fig. 9.* Target scanning results: views of the Puma robot configurations.

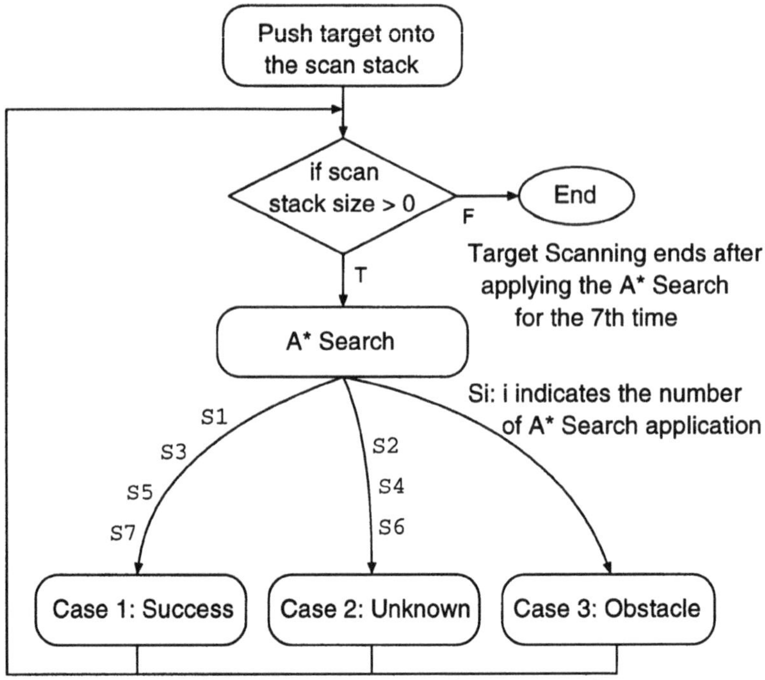


*Fig. 10.* Scanning action sequence for target scanning.

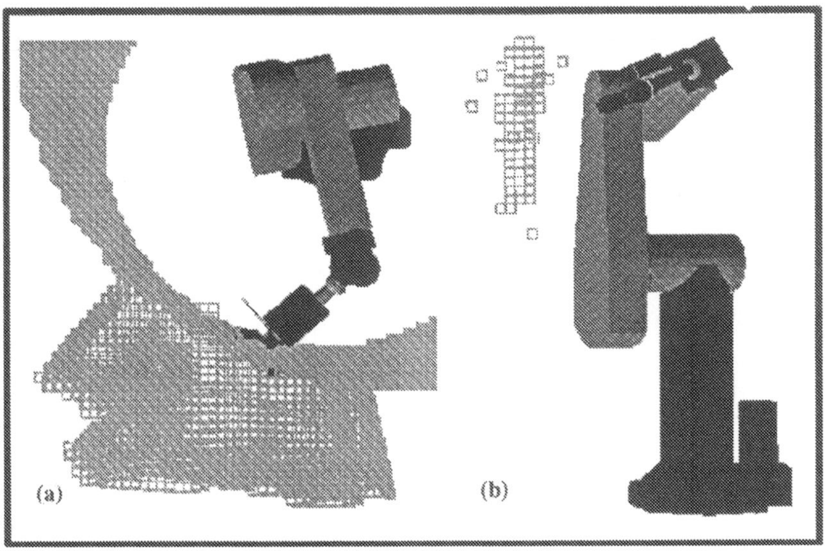


*Fig. 11.* Effects of target scanning on the voxel maps: (a) boundary between expanded free space and unknown space; and (b) obstacle voxels.

6.1.2. Volume Scanning Experiment

The volume scanning experiment was set-up by adding another pole and a number of boxes in the robot workspace. Regions of the workspace to be scanned were selected. The scanning sequence documented here was initiated by selecting a region of space to cover the top portion of the right pole, through to the mid-section of the left pole where the boxes lie just beyond the selected volume.

The volume scanning system was also executed on a MIPS R4000 Risk Chip in an SGI Indigo workstation. It took twenty three (23) scans and 110 minutes to successfully map the selected volume. Fig. 12 shows the amount of CPU time required to plan each scan and the number of *unknown* voxels captured at each scan.

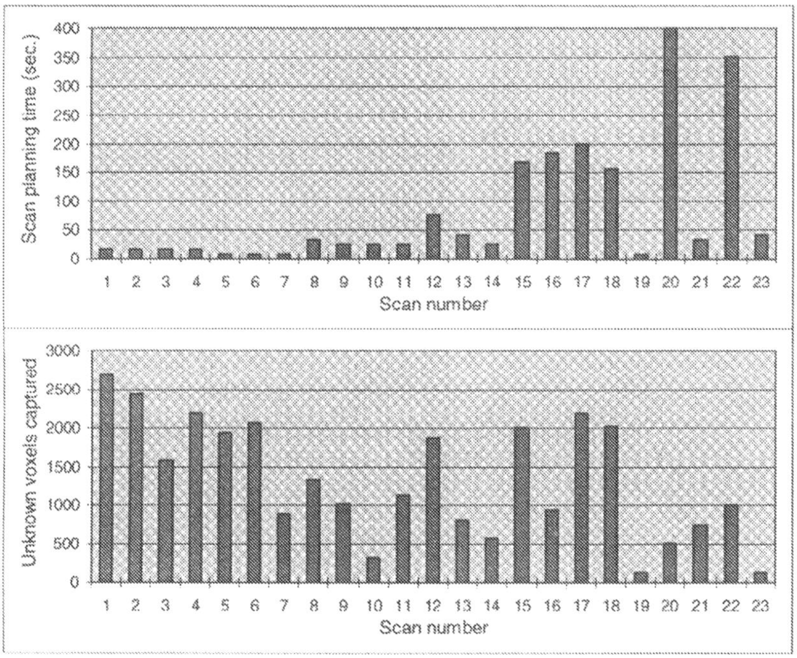


*Fig. 12.* Volume scanning results; CPU times and mapped voxel data.

The planning times did not include the time taken to move the robot, take scans, and update the workspace with the newly acquired scan data. It can be observed that the initial scans were quickly planned, adding free space rapidly and efficiently. However, scanning the backtrack points (scans 12, 15-18) were computationally expensive, as in the target scanning example. Volume scanning required more computing power, as shown for scans 20 and 22, due to the local filter in the target selection process accepting a target for which the *A* Search* failed to plan a path. Alternate targets were selected until one could finally be scanned. The number of *unknown* voxels acquired decreased with each successive scan.

The sequence of planning actions taken by the volume scanning system is shown in Fig. 13. The first eleven workspace targets selected by the target scanning selection process were directly scannable. The 12th target required scanning one intermediate target. The 14th target caused the target scanning system to scan five intermediate targets before giving up and setting it to *unscannable*. The last four targets were immediately scannable. Other voxels were set to *unscannable* since the viewing position selection algorithm failed to find a collision-free position with a clear line of sight to the target.

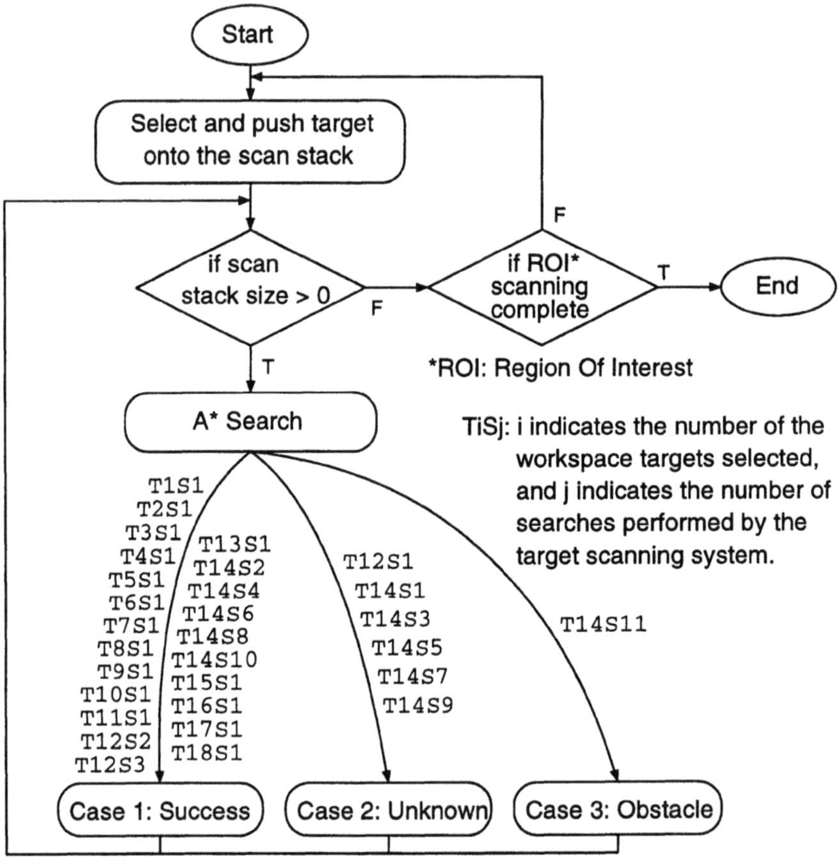


*Fig. 13.* Scanning action sequence for volume scanning.

A total of 0.2375 $m^3$ (about 29,690 voxels) of known free space was added to the workspace by the scanning process. Fig. 14 illustrates the effects of the volume scanning on the voxel maps. Some voxels were set to *obstacle* that should have been *free*, due to the noise in the raw range image. This excluded free space from the robot, hence restricting its motion. This scanning sequence was generated by using a maximum of 70 iterations during the *A* Search*. If the solution was at least case-2, the search would stop if no improvement was achieved for 20 iterations. However, 45 iterations were allowed if the search result was case-3. Only viewing positions that were collision-free and had a clear line of sight were acceptable for invoking the *A* Search*.

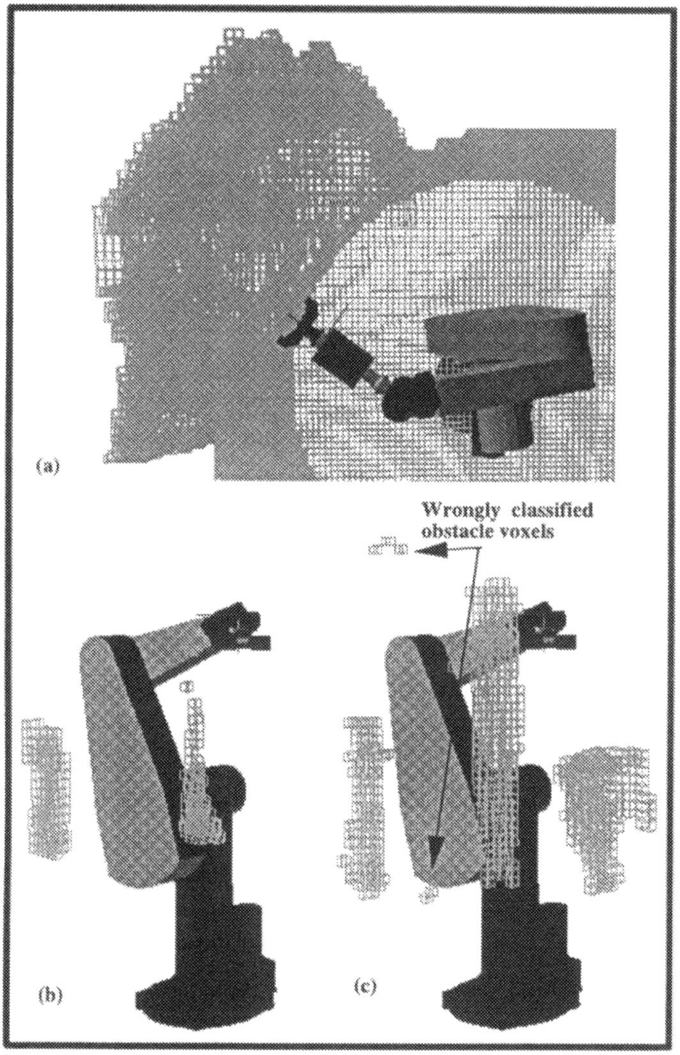


*Fig. 14.* Effects of volume scanning on the voxel maps: (a) boundary between expanded free space and unknown space; (b) *unscannable* voxels; and (c) *obstacle* voxels.

6.1.3. Scanning Performance and Comparison

The region in which the robot used (Puma 560) is effective for scanning is quite limited. As the configuration of the Puma approaches its joint limits, its maneuverability and, therefore, its ability to find acceptable scanning positions decrease. It was found that targets more than 1.0 *m* away from the robot base start to become difficult to scan, especially in a cluttered environment.

From the pictures of the scans taken during the tests, it was observed that the scans were not well registered. This is largely due to the low positioning accuracy of the Puma (*i.e.* ±2 *cm*). Any angular error also increases with the range reading values. It is therefore important to maintain a clearance between the Puma and any workspace obstacles of at least one voxel (which is equal in size to the Puma accuracy).

It was observed from the experiments that both the target and volume scanning examples penetrated to approximately the same depth within the workspace, using very different strategies. Target scanning takes a minimalist approach, by only scanning the minimum volume in order to reach a target, resulting in a *depth-first* like search. However, the need may exist to explore and determine the workspace contents, in which case a volume scan would be necessary. A *breadth-first* search is required for scanning everything in the selected volume from the center of the workspace out.

Although the target scanning method does generally penetrate into the workspace more quickly than a volume scan, it is less efficient in terms of the rate at which it adds known free space to the voxel map compared with the initial scans in volume scanning. This is because, unlike most of targets in volume scanning, most of those selected for target scanning are chosen by the backtrack method, which is computationally expensive.

It was also observed for volume scanning that the initial scans captured significantly more unknown space than the later ones. The final scans required for the volume scanning system actually "fill-in" gaps left by previous scans, and add little information to the voxel maps. Hence, the initial scans are more efficient than the final ones. The need to find the backtrack points also slowed the process because the arm progresses into the workspace and encounters *unknown* voxels. To avoid this case and the need to backtrack, volume scanning was implemented using a *breadth-first* scanning approach. However, unless the entire workspace is selected for scanning, the problem of *unknown* voxels is not completely avoidable.

## 6.2. APPLICATIONS

### 6.2.1. Exploration of Hazardous Waste Storage Tanks

The target application for the Plan-N-Scan project is the autonomous scanning and mapping of hazardous waste storage tanks. A large-degree-of-freedom (snake like) robotic arm is being designed specifically for use in these tanks, will be deployed down a tube and into the tank as shown in Fig. 15. The contents of most tanks are unknown, but they are expected to contain sludge, solid objects (*e.g.* spent fuel and contaminated machinery) and pillars of solidified waste (formed in a similar manner to stalagmites and stalactites in caves). The pillars present the greatest challenge and complication for scanning and maneuvering within the tanks because they are difficult to avoid. Therefore, vertical poles were included in the laboratory robotic test setup to approximate the anticipated environment.

For this application the robot, with a wrist-mounted range camera, has to incrementally scan and plan its safe way among the unknown obstacles. A range camera with a field of view large enough to cover the workspace in approximately 20 to 30 scans would be used. These scans can be rendered on an SGI reality engine with a reasonable screen update performance when changing views. The size and resolution of the voxel map used to represent the waste tank would be determined by the tank's size and the positioning accuracy of the robotic arm.

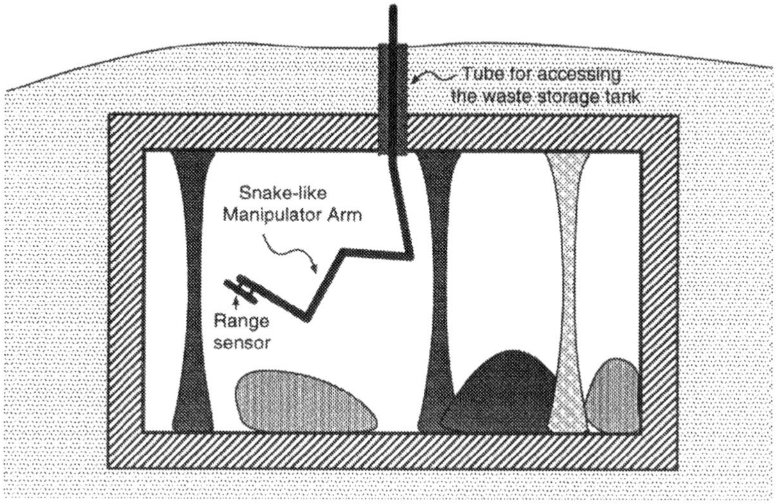


*Fig. 15.* Graphical illustration of the hazardous waste storage tank exploration problem.

### 6.2.2. Remote Excavation

The Plan-N-Scan system, developed in the IRIS project, was also implemented as part of the "Remote Excavator" project that was investigated by SPAR, AECL, and NRC at AECL Chalk River Laboratories. The project involved the teleoperation of an industrial excavator in an unknown environment using the NRC Laser scanners. A "Remote Scout Vehicle" was used for obtaining Laser range scans and transferring them to the Plan-N-Scan system and the excavator controller by means of radio and Ethernet communication tools, as illustrated in Fig. 16.

*Fig. 16.* Illustration of the Remote Excavator Application.

## 7. Conclusions and Future Work

This chapter presented a technique for autonomously scanning and mapping selected targets and volumes in the workspace and planning robot paths to these targets. The novelty of this work stems from the successful combination of collision detection with path planning by performing gaze planning with collision detection in a 3-D world. This technique, implemented into the Plan-N-Scan system, involves a number of innovative and novel algorithms and unique features. First, the generated voxel maps are easily updated using the incrementally acquired sensory data without reconstructing object shapes. In fact, since the voxel map is basically a 3-D grid, any occupancy information can be added to it. The voxel maps are also suitable for real-time collision detection when combined with the spherical manipulator models. Another important feature is the backtracking mechanism built into the target scanning system. It takes advantage of the ability to sequentially acquire scans in order to reach a target that is not immediately viewable. This gives the scanning system the flexibility needed to successfully scan targets, even if they are difficult to view immediately. Another feature is the careful selection of targets and their viewing positions for volume scanning, to direct the *A* Search* to feasible configurations, while weeding out impossible targets. Dividing the scanning system into the selection and scanning of targets gives it a modular design that facilitates both its implementation and maintenance. The performance and reliability of the Plan-N-Scan system was demonstrated through simulation tests and a number of experiments involving a real robot system. This proved the system's ability to incrementally acquire range information and successfully scan targets and workspace volumes as well.

The "Waste Tank Remediation" problem is a typical potential application (among others) which would require a specially designed robotic system. Therefore, Plan-N-Scan was developed to be general and independent of the robotic configuration on which it would be implemented. The concepts behind the scanning system, the representation and search schemes used, and the higher level code are all applicable to other robotic systems and range cameras. Therefore, it would only be necessary to change the code related to the specific hardware components being used, such as the arm kinematics and the range camera characteristics, to adapt it to new systems. The successful completion of the lab implementation paves the way to incorporating the Plan-N-Scan system in many industrial and other applications of the IRIS project.

## Acknowledgment

The scholarship awarded to Miss P. Renton by the Natural Sciences and Engineering Research Council of Canada (NSERC) and the research support of the Institute for Robotics and Intelligent Systems (IRIS) through the IRIS/ISDE-3 project awarded to professor H. ElMaraghy (principal investigator) are gratefully acknowledged. The help and assistance of Mr. N. Burtnyk of NRC-Canada, during the period when Miss Renton conducted this IRIS research at NRC in Ottawa, is also gratefully acknowledged.

## References

1. S. Hedberg: Robots Cleaning Up Hazardous Waste, *AI Expert*, 20-24 (May 1995).
2. K. Joarder and D. Raviv: Autonomous Obstacle Avoidance Using Visual Fixation and Looming, *SPIE Proc.: Intelligent Robots and Computer Vision XI*, Vol. 1825, pp. 733-744 (1992).
3. C. A. Shaffer and G. M. Herb: A Real-Time Robot Arm Collision Avoidance System, *IEEE Trans. on Robotics and Automation* **8,** 149-160 (1992).
4. P. Whaite and F. Ferrie: *Autonomous Exploration: Driven By Uncertainty*, Technical Report TR-CIM-93-17, Centre for Intelligent Machines, McGill University, Montréal, QC, Canada (1993).
5. A. Zelinsky: A Mobile Robot Exploration Algorithm, *IEEE Trans. on Robotics and Automation* **8,** 707-717 (1992).
6. M. Greenspan and N. Burtnyk: *Real Time Collision Detection,* United States Patent Number 5,347,459 (March 1993).
7. J.-A. Beraldin, M. Rioux, F. Blais, and R. A. Couvillon: Digital Three-Dimensional Imaging in the Infrared at the National Research Council of Canada, *SPIE Proc.: Int. Symp. on Optics, Imaging, and Instrumentation: Infrared Technology XX*, Vol. 2269, San Diego, CA, USA, pp. 208-225 (July 1994).
8. M. Greenspan, M. Lipsett, J. Ballantyne, P. Renton, E. Gagnon, and N. Burtnyk: Laser Range Vision for Tele-Excavation, *Proc. 1995 Robotics and Knowledge Based Systems Workshop*, St. Hubert, QC, Canada (Oct. 1995).
9. P. Renton, M. Greenspan, N. Burtnyk, and H. ElMaraghy: Scan-N-Plan: Collision-Free Autonomous Workspace Exploration, *Proc. 1995 Robotics and Knowledge-Based Systems Workshop*, St. Hubert, QC, Canada (Oct. 1995).
10. F. Arman and J. K. Aggarwal, Model-Based Object Recognition in Dense-Range Images - A Review, *ACM Computing Surveys* **25,** 67-108 (1993).
11. S. Motavalli and B. Bidanda, A Part Image Reconstruction System for Reverse Engineering of Design Modifications, *Journal of Manufacturing Systems* **10,** 383-395 (1991).
12. R. A. Jarvis: Range Sensing for Computer Vision, *Three-Dimensional Object Recognition Systems*, A.K. Jain and P.J. Flynn, eds., Elsevier Science Publishers B.V., Amesterdam, The Netherlands, pp. 17-56 (1993).
13. M. Rioux, Digital 3-D Imaging: Theory and Applications, *SPIE Proc., Videometrics III, Int. Symp. on Photonic and Sensors and Controls for Commercial Applications,* Vol. 2350, Boston, MA, USA, pp. 2-15 (Oct./Nov. 1994).
14. J. Maver and R. Bajcsy: Occlusions as a Guide for Planning the Next View, *IEEE Trans. on Pattern Analysis and Machine Intelligence* **15,** 417-432 (1993).
15. J. E. Banta, Y. Zhien, X. Z. Wang, G. Zhang, M. T. Smith, and M. A. Abidi: A 'Best-Next-View' Algorithm for Three-Dimensional Scene Reconstruction Using Range Images, *SPIE Proc., Intelligent Robots and Computer Vision XIV: Algorithms, Techniques, Active Vision and Materials Handling*, Vol. 2588, Philadelphia, PA, USA, pp. 418-429 (Oct. 1995).
16. E. Kruse, R. Gutsche, and F. M. Wahl: Efficient, Iterative, Sensor Based 3-D Map Building Using Rating Functions in Configuration Space, *Proc. IEEE Int. Conf. on Robotics and Automation*, Minneapolis, MN, USA, pp.1067-1072 (April 1996).
17. Z. Chen and C. Huang: Terrain Exploration of a Sensor-Based Robot Moving among Unknown Obstacles of Polygonal Shape, *Robotica* **12,** 33-44 (1994).
18. R. H. T. Chan, P. K. S. Tam, and D. N. K. Leung: Robot Navigation in Unknown Terrains via Multi-Resolution Grid Maps, *IECON'91*, pp. 1138-1143 (1991).

19. E. R. Stuck, A. Manz, D. A. Green, and S. Elgazzar: Map Updating and Path Planning for Real-Time Mobile Robot Navigation, *Proc. IEEE/RSJ/GI Int. Conf. on Intelligent Robots and Systems,* Munich, Germany, pp. 753-760 (Sept. 1994).
20. Y. K. Hwang and N. Ahuja: Gross Motion Planning - A Survey, *ACM Computing Surveys* **24,** 219-291 (1992).
21. E. Cheung and V. Lumelsky: Motion Planning for a Whole-Sensitive Robot Arm Manipulator, *Proc. IEEE Int. Conf. on Robotics and Automation*, Cincinnati, OH, USA, pp. 344-349 (May 1990).
22. J. L. Novak and J. T. Feddema: Capacitive-Based Proximity Sensor for Whole Arm Obstacle Avoidance, *Proc. IEEE Int. Conf. on Robotics and Automation*, Nice, France, pp. 1307-1313 (May 1992).
23. K. K. Gupta and X. M. Zhu: Extracting Polyhedral Models from a Range Image: A Hybrid Approach, *Computer Vision: Systems, Theory and Applications,* A. Basu, ed., World Scientific Press, Singapore (1992).
24. T. O. Binford: Visual Perception by Computer. *Proc. IEEE Conf. on Systems and Control,* Miami, FL, USA (1971).
25. A. A. G. Requicha and H. B. Voelcker: Solid Modeling: A historical Summary and Contemporary Assessment, *IEEE Computer Graphics and Applications* **2,** 9-24 (1982).
26. T. C. Henderson: Efficient 3-D Object Representations for Industrial Vision Systems, *IEEE Trans. on Pattern Analysis and Machine Intelligence* **5,** 609-617 (1983).
27. D. J. Meagher: Geometric Modeling Using Octree Encoding, *Computer Graphics and Image Processing* **19,** 129-147 (1981).
28. M. Greenspan and N. Burtnyk: Obstacle Count Independent Real-Time Collision Avoidance, *Proc. IEEE Int. Conf. on Robotics and Automation*, Minneapolis, MN, USA, pp. 1073-1080 (April 1996).
29. S. K. Singh and M. C. Leu, Manipulator Motion Planning in the Presence of Obstacles and Dynamic Constraints, *The Int. Journal of Robotics Research* **10,** 171-187 (1991).
30. Y. K. Hwang and N. Ahuja: A Potential Field Approach to Path Planning, *IEEE Trans. on Robotics and Automation* **8,** 23-32 (1992).
31. J. Barraquand and J. C. Latombe: A Monte-Carlo Algorithm for Path Planning with Many Degrees of Freedom, *Proc. IEEE Int. Conf. on Robotics and Automation*, Cincinnati, OH, USA, pp. 1712-1717 (May 1990).
32. K. K. Gupta and Z. Guo: Motion Planning for Many Degrees of Freedom: Sequential Search with Backtracking, *IEEE Trans. on Robotics and Automation* **11,** 897-906 (1995).
33. T. Lozano-Pérez: Spatial Planning: A Configuration Space Approach, *IEEE Trans. on Computers* **C-32**, 108-120 (1983).
34. N. J. Nilsson: *Principles of Artificial Intelligence,* Morgan Kaufmann Publishers Inc., Los Altos, CA, USA (1980).
35. K. S. Fu, R. C. Gonzalez, and C. S. G. Lee: *Robotics: Control, Sensing, Vision, and Intelligence*, McGraw-Hill, New York, NY, USA (1987).
36. Y. Roth-Tabak and R. Jain: *Building an Environment Model Using Depth Information,* Technical Report CSE-TR-07-88 U, E.E. and C.S. Dept., University of Michigan, Ann Arbor, MI, USA (1988).
37. P. A. M. Renton: *Plan-N-Scan: Autonomous Workspace Mapping,* Master's Thesis, Department of Mechanical Engineering, McMaster University, Hamilton, ON, Canada (1996).

# 23

# MIXED TRAFFIC CONTROL INVOLVING MANUALLY-CONTROLLED AND AUTOMATICALLY-CONTROLLED VEHICLES IN IVHS

SU-NAN HUANG
STEVEN C. CHAN
WEI REN
*Dept of Electrical Engineering and Computer Sciences*
*University of California at Berkeley*
*Berkeley, CA 94720, U.S.A*

## 1. Introduction

Traffic congestion is a global problem. Intelligent Vehicle Highway Systems (IVHS) have been proposed to help solve this problem (see [1, 2, 3, 4, 5, 6, 7, 8, 9]). The works of [1, 2, 3] propose a fully automated traffic system in which vehicles are automated and organized in platoons through interactive vehicle-to-vehicle communication. This approach can yield superior control quality and has the best potential for throughput improvement but is difficult to implement due to high costs and heavy infrastructure requirements. The reports in [4, 5, 6, 7] suggest an autonomous intelligent cruise control for the IVHS, i.e., to keep a desired velocity when there is no vehicle in front, and to keep a desired spacing (possibly velocity dependent) when there is a vehicle in front. This approach allows a mixture of automated and non-automated vehicles and requires little vehicle-to-vehicle communication. However, it is not a completely self-contained intelligent controller that can handle all possible scenarios such as sudden lane changes, following a vehicle with rapid acceleration or deceleration, emergency braking, switching between human drivers and automatic controllers, etc. In [8, 9], artificial intelligence is used to design fully autonomous vehicle control system which can operate in mixed traffic. The research in this area is mostly concerned with handling complex traffic scenarios and do not attempt to optimize control performance for tight vehicle spacing and passenger comfort. In addition, the previous results do not provide the performance analysis of mixed traffic control, though they claim their control can be used in an environment of automatically controlled and manually controlled vehicles. It is therefore necessary to design a complete intelligent control system that 1) handles all possible scenarios in mixed traffic, 2) optimizes control performance so that tight vehicle spacing can be achieved with minimum passenger discomfort, 3) evaluates the performance of mixed traffic.

In this chapter, we propose a technology which allows for the existence of mixed traffic. We begin by designing an automatic driving controller, called the intelligent vehicle driving system (IVDS). This is a two-layer system: the higher layer analyzes current scenario and infers control objective that associates with a certain index function; the lower layer optimizes the function provided by the upper level. IVDS only uses the measurement of a vehicle's speed and distance relative to the vehicle in front, together with measurements of the vehicle's own state. Consequently, the vehicles equipped with the IVDS can operate together with manually controlled vehicles. Next, a mathematical and rule based model for human drivers is developed. This model attempts to mimic human driver behavior in vehicle following and lane-changing. Finally, we examine the potential benefits of mixed traffic by implementing the human driver model and IVDS on the automated highway simulator SmartPath (see [10, 11]).

## 2. Mixed Traffic System

In a mixed traffic system, a vehicle is either automatically or manually controlled. We require that a manual driver model should be capable of simulating actual human driver behavior, such as vehicle following, lane-change, and driver aggressiveness. We allow the manually-controlled vehicle to enter any lane, with driver aggressiveness determining which particular lane (we discuss this point later).

The behavior of automated vehicles is more complex and so it may helpful to describe how you as a driver would experience a trip within an automated vehicle (see Figure 1). To drive your car over a three lane highway, you enter from an on-ramp and try to join the traffic. You then assign your path and announce your destination by keyboard entry. Thus, your vehicle's on-board computer automatically assigns you the next highway number or the exit number to proceed to. Once the on-board computer has taken over control, it will control the car by use of sensor information so that it co-exists together with manual vehicles on a highway.

With these requirements in our mind, we will design a detailed automated driving system and human driver model.

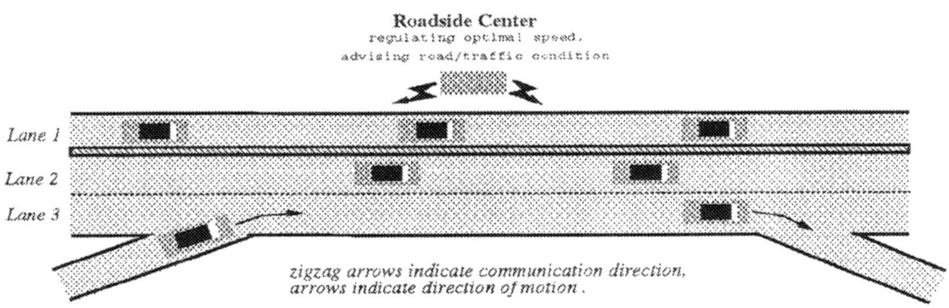


Figure 1: A highway structure

## 3. Vehicle Dynamics

Consider the following simplified nonlinear vehicle model (see [12])

$$\dot{x} = v \tag{1}$$

$$\dot{v} = -\frac{1}{m}(F_w - F_e + F_d) \tag{2}$$

where $x$ is the longitudinal position of the vehicle, $v$ is the velocity of the vehicle, $m$ is the mass of the vehicle, $F_w = K_w v^2$ is the force due to wind resistance, $K_w$ is the aerodynamic drag coefficient, $F_d$ is the force due to mechanical drag, and $F_e$ is the engine traction force, which we assume to evolve under the dynamics

$$\dot{F}_e = -\frac{F_e}{\tau} + \frac{u}{\tau}, \tag{3}$$

where $\tau$ is the engine time constant and $u$ is the throttle/brake input. Differentiating (2) and substituting (3) into (2) yields

$$\dot{a} = -\frac{1}{m\tau}[2\tau K_w v\dot{v} + ma + K_w v^2 + F_d - u] \tag{4}$$

If we assume that we have access to both the velocity and acceleration of the vehicle and the other vehicle parameters, the vehicle model can be linearized by state feedback. Setting

$$u = m\tau(c + 2\tau va K_w + K_w v^2 + F_d)$$

we find that the new linearized vehicle model is given by $\dot{a} = c + pa$, where $c$ is an exogenous control input and $p = -\frac{1}{\tau}$. Now the dynamic of the linearized vehicle model is summarized by three simple state equations

$$\dot{x} = v \tag{5}$$

$$\dot{v} = a \tag{6}$$

$$\dot{a} = pa + c \tag{7}$$

Thus, the equation above can be discretized as

$$x(k+1) = x(k) + v(k)T \tag{8}$$

$$v(k+1) = v(k) + a(k)T \tag{9}$$

$$a(k+1) = (pT + 1)a(k) + c(k)T \tag{10}$$

where $T$ is the sampling time.

## 4. Automated Driving System

Before discussing our automated driving system, we discuss some of our assumptions. The automatically-controlled vehicles in a highway system are equipped with sensors which can provide velocity, acceleration, and distance measurements of the vehicle in front, as well as measurements of the vehicle's own velocity and acceleration. The sensors should also be capable of detecting vehicles in adjacent lanes (see Figure 2). The detection ranges are taken to be constants

- In region 1 and 2, 60 meters.
- In region 3 and 4, ±30 meters from the center of the car.

In addition, we assume that the sensors can detect the signals from highway number and exit markers.

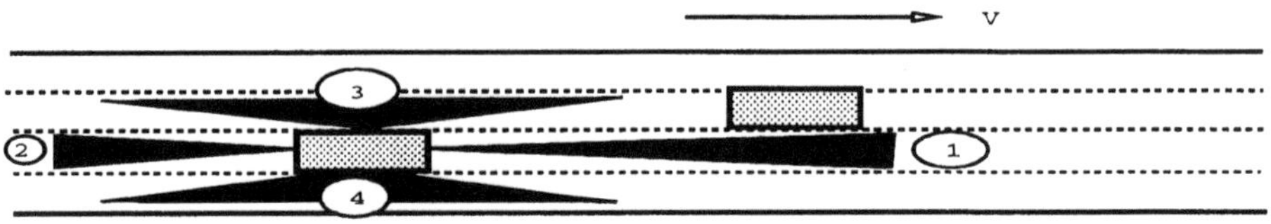


Figure 2: Detection ranges of sensors

With these assumption in our mind, a two-layer hierarchical controller (see Figure 3) is presented. The top layer, called the decision layer, classifies the driving states and the objective of the traffic scenario. The bottom layer, called the control layer, issues control commands to a vehicle. Such a control structure for automatic driving not only is more modular and extensible, but offers superior control performance due to on-line optimizations. The detailed design of each layer will be discussed in the following sections.

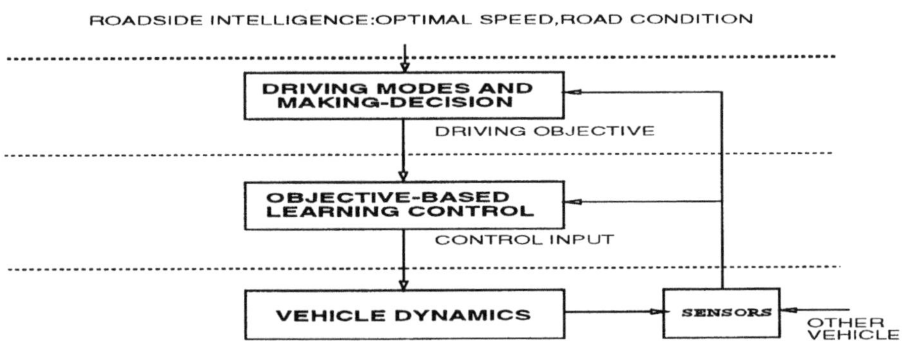


Figure 3: Hierarchical controller

## 4.1. THE DECISION MAKING AT THE TOP LAYER

The flow diagram for the automated driving decision at the top layer is displayed in Figure 4, where $v$ is the velocity of the vehicle controlled, $v_h$ is the velocity of preceding vehicle, $v_{opt}$ is the desired velocity, $s$ is the actual spacing from the preceding vehicle, and $s_d$ is the desired spacing. The decision algorithm makes a comparison between $v_h$ and both $v$ and $v_{opt}$, and a comparison between $s$ and $s_d$. This will help provide a correct driving mode. Several optional conditions in the ellipse block need to be explained

- *check if an exiting signal or flag* is a flag indicating that vehicle will exit from the highway. This flag is set by a human driver.
- *mode keeping* is used to record the mode obtained by the decision layer's last sampling time.
- *check if change−lane flag* is a flag indicating that vehicle is in lane-change mode.
- *check if it is in its assigned lane* is used to determine if the vehicle is in its assigned lane. If *yes*, then the vehicle will search the next branch down decision-tree. If *no*, then the vehicle will try to move to an adjacent lane that is nearest to the assigned lane.
- *lane − change request from another vehicle* is used to determine if the change-lane signal was sent from an adjacent lane. We will discuss this case later.
- *check if it is a closed − most car* is used to determine if the vehicle controlled is at a closed-most position to a WANT-TO-LANE-CHANGE vehicle in an adjacent lane (see the *change lane* part of Section 4.1.2).
- *check space in adjacent lane* is used to check if there is an appropriate space in an adjacent lane to accommodate the vehicle.
- *check if there is a car ahead* is used to determine if the vehicle can "see" a vehicle in front.
- *check if there is a new scenario* is used to check if the vehicle encounters a new scenario in front. It includes three cases:

  case 1 A vehicle does not detect any other vehicle along the longitudinal direction for time $t < 0$. At time $t = 0$, it sees a vehicle ahead.

  case 2 A vehicle finds another vehicle changing lanes from an adjacent lane into its current lane.

  case 3 A vehicle changes lanes.
- *check if preceding vehicle is too slow* is used to determine if the speed $v_h$ of the preceding vehicle is far less than the desired speed $v_d$. We say that if $v_h < v_d - 10m/s$, then the preceding vehicle is too slow.

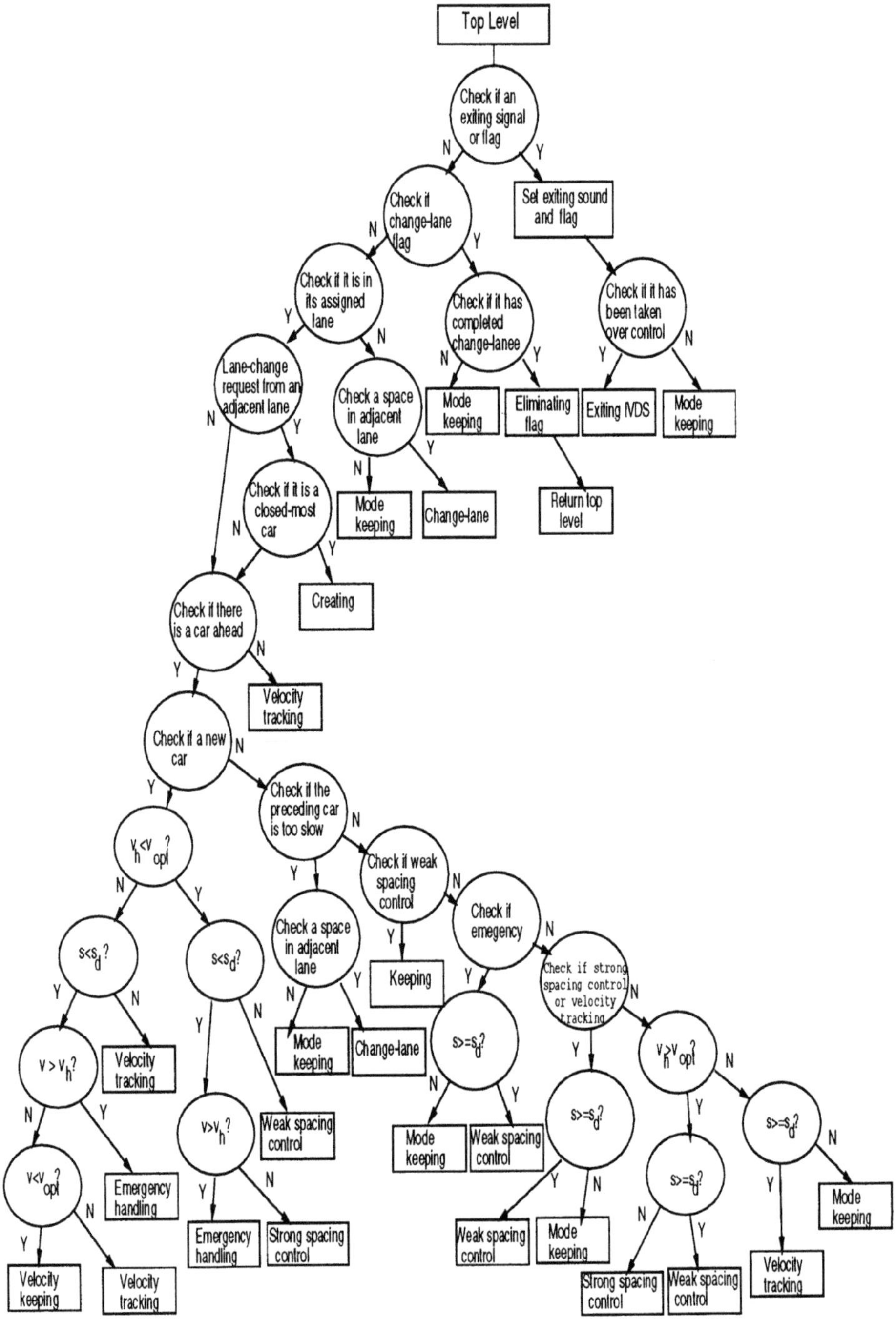


Figure 4: Driving mode decision

Since the other optional conditions are clear, we shall instead discuss the terminal objectives of the decision-tree. There are two types of control behaviors. The first is the longitudinal control behavior, such as velocity keeping mode, velocity tracking mode, weak spacing control mode, and strong spacing control mode. The second type of control behavior is mixed longitudinal/lateral control, such as emergency mode and change-lane mode.

### 4.1.1. LONGITUDINAL MODES AND ASSOCIATED OBJECTIVES

We conduct a classification of basic driving modes and assign appropriate associated control objective functions. The goal of the driving modes is to ensure safety and to achieve a compromise between control performance and passenger comfort.

The control objective function to be optimized determines the control actions at time $kT$ take the following general form

$$J(k) = \gamma_1[v(k) - v_r(k)]^2 + \gamma_2[s(k) - s_r(k)]^2. \tag{11}$$

where $v(k), s(k)$ are the current vehicle velocity and the spacing from the preceding vehicle, $v_r(k), s_r(k)$ are the reference velocity and spacing, $c(k)$ is the control input, and $\gamma_i$ are the weighting coefficients.

Velocity keeping mode. In this mode, the control simply tries to maintain the vehicle velocity, that is

$$v_r(k) = v(k-1) \tag{12}$$

The control objective function is

$$J(k) = \gamma_1[v(k) - v(k-1)]^2. \tag{13}$$

Velocity tracking mode. In this mode, the controller tries to force the vehicle to track a desired velocity, that is

$$v_r(k) = \eta v_{opt} + (1-\eta)v(k-1) \tag{14}$$

The control objective function then takes the form

$$J(k) = \gamma_1[v(k) - \eta v_{opt} - (1-\eta)v(k-1)]^2. \tag{15}$$

where the parameter $\eta$ is used to adjust the aggressiveness of the control. To see the reason for choosing such a control objective, consider the ideal situation

$$v(k) - v_{opt}\eta - (1-\eta)v(k-1) = 0, \quad k \geq 0 \tag{16}$$

Then,

$$v(k) = v_{opt} + [v(0) - v_d](1-\eta)^k \tag{17}$$

Hence, this defines a smooth trajectory from the initial velocity $v(0)$ to the optimal velocity $v_{opt}$. We notice that a smaller value of $\eta$ can cause passenger comfort. Usually, this mode is operated in quite a safe distance. We set $\eta = 0.65$ in this mode.

Weak spacing control mode. In this mode, the control forces the vehicle in a smooth maneuver to follow the preceding vehicle with certain desired spacing, that is

$$s_r(k) = \eta s_d + (1 - \eta)s(k - 1) \tag{18}$$

The control objective function takes the form

$$J(k) = \gamma_2[s(k) - \eta s_d - (1 - \eta)s(k - 1)]^2 \tag{19}$$

where we can again choose a suitable $\eta$ to ensure a smooth trajectory. We set $\eta = 0.65$ in this mode. In fact, the spacing control also requires tracking the speed of the preceding vehicle, that is

$$v_r(k) = \eta v_h(k) + (1 - \eta)v(k - 1) \tag{20}$$

where $v_h(k)$ is the speed of the preceding vehicle. Thus, the modified control objective takes the form

$$\begin{aligned} J(k) &= \gamma_1[v(k) - \eta v_h(k) - (1 - \eta)v(k - 1)]^2 \\ &+ \gamma_2[s(k) - \eta s_d(k) - (1 - \eta)s(k - 1)]^2 \end{aligned} \tag{21}$$

where $v_h(k)$ is the speed of the preceding vehicle.

Strong spacing control mode. In this mode, the control forces the vehicle in a strong maneuver to follow the preceding vehicle with certain desired spacing. The control objective function takes the same form as (21). However, a larger value of $\eta$ is set in the reference trajectory, $\eta = 0.85$ here.

Creating mode. In this mode, the control forces the vehicle in a strong maneuver to follow an imaginary preceding vehicle. Let us illustrate this mode by Figure 5.

Now suppose $A$ is in lane 1 and wishes to change to lane 2 (see Figure 5a). It can do so only if there is an adequate space in lane 2 and no vehicle in lane 2 is planning to move into that space. Car $A$ sets a WANT-TO-LANE-CHANGE light. The vehicles nearby can detect this signal by using their sensors. The detection range has been described in section 3. When car $B$ in lane 2 lies a closed-most position to the WANT-TO-LANE-CHANGE vehicle, it will stop tracking the car ahead. Instead, it will determine how to create a space in lane 2 in order to accommodate $A$'s lane change. $B$'s decision is to track an imaginary car ahead using *creating mode* (see Figure 5b). The imaginary following objective is given as

$$s_r(k) = 40\eta + (1-\eta)(20 - s'_A(k-1) \tag{22}$$

$$v_r(k) = \eta v_B(k) + (1-\eta)v_A(k-1) \tag{23}$$

and objective function is taken as

$$J_{imag}(k) = \gamma_2[s'_A(k) - 40\eta + (1-\eta)(20 - s'_A(k-1))]^2$$
$$+\gamma_1[v_A(k) - \eta v_B(k) + (1-\eta)v_A(k-1)]^2$$

where $\eta = 0.75$, and $s'_A(k)$ represents a longitudinal distance between car $B$ and car $A$. Note that $s'_A(k)$ may take a negative value since it tracks an imaginary vehicle. After the objective $J_{imag}$ is reached (see Figure 5c), the control system eliminates it automatically and then follows actual preceding vehicle.

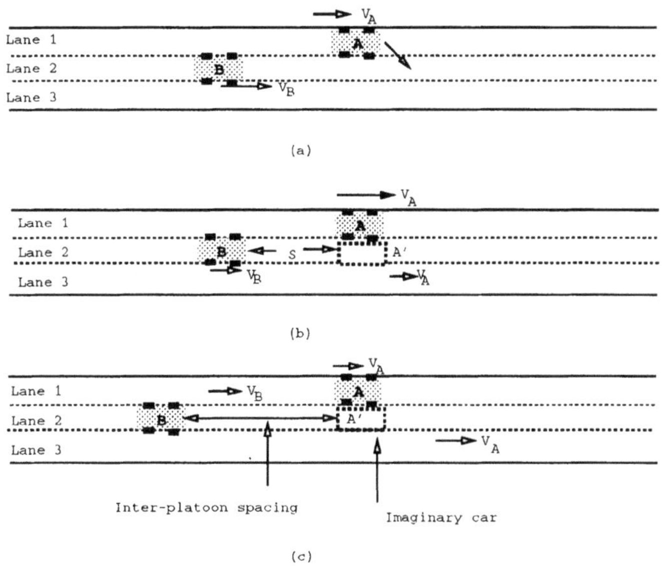


Figure 5: Tracking an imaginary car in right-lane

### 4.1.2. MIXED LONGITUDINAL/LATERAL MODES

So far we have only considered longitudinal behavior. Clearly, there is another important type of mode in vehicle driving, that is mixed longitudinal/lateral behavior.

Emergency handling. In this mode, due to emergency or near-emergency situations, such as sudden lane change, or sudden braking of the preceding vehicle, the control law is required to assess safety by predicting future trajectories, to decide whether to change lanes, to plan a trajectory, and to execute the trajectory.

The logic tree diagram in Figure 6 describes a possible design for this mode in detail. There are three blocks needing explanation. These are the trajectory prediction, trajectory planing, and change-lane. The change-lane will be discussed in another section. Here, we are concerned with the other two cases.

*Trajectory prediction.* This is used to estimate if a collision will occur when

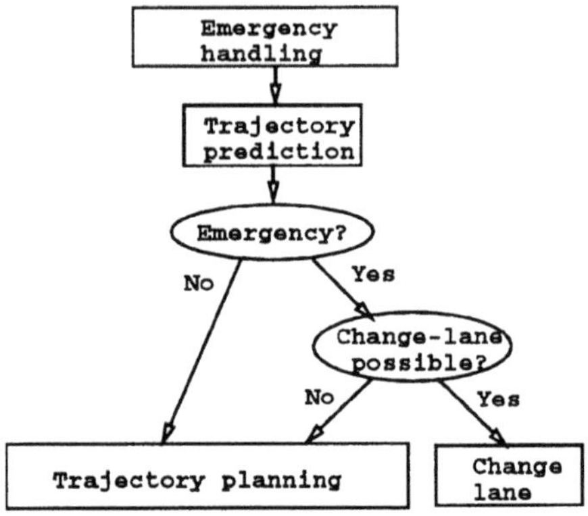


Figure 6: Emergency handling

the vehicle makes use of maximum brake control. We assume that the preceding vehicle in this case will keep the current acceleration on time interval $[t_0, t_0 + t]$, where $t_0$ is the initial time. According to motion equation, we have

$$x_h(t) = s(t_0) + v_h(t_0)t + \frac{1}{2}a_h(t_0)t^2 \tag{24}$$

$$x(t) = v(t_0)t + \frac{1}{2}a_{max,d}t^2 \tag{25}$$

where $s(t_0)$ is the initial distance between the following vehicle and the preceding vehicle, $v_h(t_0)$ is the initial speed of the preceding vehicle, $a_h(t_0)$ is the initial acceleration of the preceding vehicle, $a_{max\_braking}$ is the maximum deceleration of the following vehicle, and $x_h(t)$ and $x(t)$ are the moving position of the preceding vehicle and the following vehicle from time $t_0$ to $t + t_0$, respectively. Thus we have an optimization problem

$$\begin{aligned} min \quad & \{x_h(t) - x(t)\} \\ s.t. \quad & t > 0 \\ & x_h(t) - x(t) \geq d_0 \\ & v_h(t_0) + a_h(t_0)t > 0 \\ & v(t_0) + a_{max,d}t > 0 \end{aligned} \tag{26}$$

where $d_0$ is defined as the minimum separation distance of the vehicle from the preceding vehicle.

If we can not find the solution for $t$, then the situation is dangerous and the vehicle should try to change lanes. Otherwise, IVDS will enter *trajectory planning*.

*Trajectory planning*. This is used to plan a safe trajectory for the vehicle under the emergency situation. The method presented is to solve the following equations

$$x_h(t) = s(t_0) + v_h(t_0)t + \frac{1}{2}a_h(t_0)t^2 \quad (27)$$

$$x(t) = v(t_0)t + \frac{1}{2}a_{des}t^2 \quad (28)$$

where $a_{des}$ is the desired acceleration of the vehicle controlled. Since we wish to track the desired spacing and the speed of the preceding vehicle, we get

$$x_h(t) - x(t) = s_d \quad (29)$$

$$v_h(t_0) + a_h(t_0)t = v(t_0) + a_{des}t \quad (30)$$

Substituting (27) and (28) into (29) and solving combined equations (29)-(30), we get the value of $a_{des}$ and then obtain the control from (10)

$$c(k) = \frac{a_{des} - (pT + 1)a(k)}{T} \quad (31)$$

Change lane. This mode is a lateral behavior mode. We do not use a model to describe the lateral behavior since this chapter focuses on longitudinal control. When a vehicle receives a change-lane command from the decision module, the controller will move thevehicle into an adjacent lane while keeping its speed.

## 4.2. THE ALGORITHM AT THE CONTROL LAYER

We propose an objective-based learning algorithm to complete the task of the control. The algorithm starts with a parameterized control and a control objective function, and then uses a gradient algorithm to adapt the control parameter to optimize the given objective function.

We use the following simple control law parameterized by $\alpha$ for velocity tracking

$$c(k) = Ka(k) + \alpha\Delta s(k) + (1 - \alpha)\Delta v(k), \ (0 < \alpha < 1) \quad (32)$$

with

$$\Delta s(k) = s(k) - s_r(k) \quad (33)$$

$$\Delta v(k) = v(k) - v_r(k) \quad (34)$$

where $K$ is the constant $(K = pT + 1)$.

Consider the objective function

$$J = \frac{1}{2}\Delta s^2(k) + \frac{1}{2}\Delta v^2(k) \quad (35)$$

We can adapt the parameter $\alpha$ using the instantaneous derivation of $J$ with respect $\alpha$ as follows

$$\alpha_{l+1} = \alpha_l - \beta_l \frac{\partial J}{\partial \alpha} |_{\alpha=\alpha_l} \tag{36}$$

where $\beta_l$ is the adaptation step size. To compute the gradient, we note

$$\begin{aligned} \frac{\partial J}{\partial \alpha} &= \Delta s \frac{\partial \Delta s(k)}{\partial \alpha} + \Delta v \frac{\partial \Delta v(k)}{\partial \alpha} \\ &= \Delta s \frac{\partial \Delta s(k)}{\partial s(k)} \frac{\partial s(k)}{\partial c(k)} \frac{\partial c(k)}{\partial \alpha} + \Delta v(k) \frac{\partial \Delta v(k)}{\partial v(k)} \frac{\partial v(k)}{\partial c(k)} \frac{\partial c(k)}{\partial \alpha} \\ &= (\Delta s(k) - \Delta v(k))[\Delta s(k) \frac{\partial s(k)}{\partial c(k)} + \Delta v(k) \frac{\partial v(k)}{\partial c(k)}] \end{aligned} \tag{37}$$

The only unknown in (37) is the value of $\frac{\partial s(k)}{\partial c(k)}$ and $\frac{\partial v(k)}{\partial c(k)}$. In fact, it is not important to estimate the exact value of $\frac{\partial s(k)}{\partial c(k)}$ or $\frac{\partial v(k)}{\partial c(k)}$ because the step size can be adjusted by

$$\epsilon_1 = \beta_l \frac{\partial s(k)}{\partial c(k)} \tag{38}$$

$$\epsilon_2 = \beta_l \frac{\partial v(k)}{\partial c(k)} \tag{39}$$

The key problem is to identify the sign of $\frac{\partial s(k)}{\partial c(k)}$ or $\frac{\partial v(k)}{\partial c(k)}$. This may be achieved by the following approximations

$$\frac{\partial s(k)}{\partial c(k)} \approx \frac{s(k) - s(k-1)}{c(k) - c(k-1)} \tag{40}$$

$$\frac{\partial v(k)}{\partial c(k)} \approx \frac{v(k) - v(k-1)}{c(k) - c(k-1)} \tag{41}$$

Thus, we have

$$\frac{\partial J}{\partial \alpha} \propto (\Delta s(k) - \Delta v(k))[\Delta s(k) sign(\frac{\partial s(k)}{\partial c(k)}) + \epsilon_0 \Delta v(k) sign(\frac{\partial v(k)}{\partial c(k)})],$$
$$(let \ |\frac{\partial v(k)}{\partial c(k)}| = \epsilon_0 |\frac{\partial s(k)}{\partial c(k)}|)$$

## 5. The Human Driver Model

In order to fully examine and evaluate the performance of mixed traffic, we must simulate automated and manual vehicles together. Motivated by this need, we develop a mathematical and rule based model for human driver.

Figure 7 shows one of the driver vehicle systems proposed in [13]. The first block in system represents driver reaction time, followed by central information

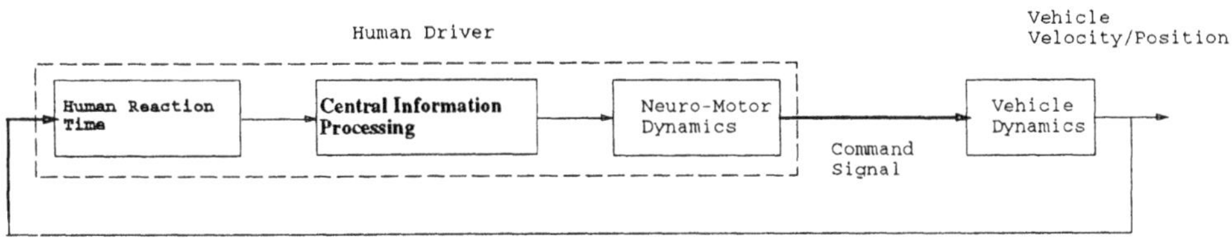


Figure 7: Driver Vehicle System

processing, and then neuro-motor or muscular dynamics. Our model for human drivers is based on this driver vehicle system.

Before engaging in a detailed description of the human driver model, an important issue that should be addressed is the variability in driver behavior — variability among different drivers on a highway and variability among the same driver over time. Clearly, this driver variability must be modeled, and we use random variables to do this.

A particularly important variability among drivers and which affect many aspects of our human driver model is aggressiveness. For example, a highly aggressive driver tends to drive faster, follow another vehicle more closely, change lanes more frequently, and is more likely to perform a dangerous maneuver than a less aggressive driver. To model aggressiveness in driving behavior, each driver is assigned an aggressiveness factor $\alpha$, drawn from a normal distribution; in particular, the $N(50, 144)$ distribution is used. [1] Drivers with $\alpha < 42.0$ are considered low aggressors, drivers with $42.0 \leq \alpha \leq 58.0$ are considered average aggressors, and drivers with $\alpha > 58.0$ are considered high aggressors. The terms low, average, and high aggressiveness will be used in this chapter to refer to these values of the aggressiveness factor $\alpha$. We use random variables based upon driver aggressiveness to model other variability in behavior, such as optimal velocity and lane change frequency.

## 5.1. LONGITUDINAL BEHAVIOR

### 5.1.1 CONTROL LAWS

There are three control laws that model pure longitudinal human driver behavior. The particular law used depends upon traffic conditions. Before presenting the three laws, the following terms are defined:

[1] The notation $N(\mu, \sigma^2)$ will be used in this chapter to refer to a normal distribution with mean $\mu$ and variance $\sigma^2$.

$$\begin{aligned}
x(k) &= \text{vehicle position } (m) \\
x_h(k) &= \text{ahead vehicle position } (m) \\
v(k) &= \text{vehicle velocity } (m/s) \\
v_h(k) &= \text{ahead vehicle velocity } (m/s) \\
a(k) &= \text{vehicle acceleration } (m/s^2) \\
s_{opt}(k,\alpha) &= \text{driver optimal headway } (m) \\
v_{opt}(\alpha) &= \text{driver optimal velocity } (m/s) \\
C(\alpha) &= \text{Calif rule time constant } (s^{-1}) \\
L &= \text{vehicle length } (5\ m)
\end{aligned}$$

Driver optimal headway $s_{opt}$ and driver optimal velocity $v_{opt}$ are determined using random variables that depend upon driver aggressiveness $\alpha$.

$$\begin{aligned}
s_{opt}(k,\alpha) &= C(\alpha)v(k) + L \\
v_{opt}(\alpha) &= \begin{cases} \text{N(27,1)} & \text{if } \alpha = \text{low} \\ \text{N(29,1)} & \text{if } \alpha = \text{med} \\ \text{N(31,1)} & \text{if } \alpha = \text{high} \end{cases}
\end{aligned}$$

where

$$C(\alpha) = \begin{cases} \text{N(1.2, 0.04)} & \text{if } \alpha = \text{low} \\ \text{N(1.0, 0.04)} & \text{if } \alpha = \text{med} \\ \text{N(0.8, 0.04)} & \text{if } \alpha = \text{high} \end{cases}$$

In addition to the above terms, the following variables and constants are used

$$\begin{aligned}
s(k) &= x_h(k) - x(k) - L \\
\Delta v(k) &= v_h(k) - v(k) \\
\Delta s_{opt}(k,\alpha) &= s(k) - s_{opt}(k,\alpha) \\
\Delta v_{opt}(t,\alpha) &= v_{opt}(\alpha) - v(k) \\
s_{buf}(k) &= 0.1v(k) + 1 \\
\Delta v_{buf}(k) &= -0.05v(k) - 1 \\
\lambda_1 &= 12.253\ m/s \\
\lambda_2 &= 0.368\ s^{-1} \\
K &= 0.3\ s^{-2} \\
\tau &= 0.5\ s
\end{aligned}$$

$s(k)$ is the distance between the front bumper of a vehicle and the vehicle ahead of it; $\Delta v(k)$ is the relative velocity of a vehicle with respect to the vehicle ahead of it; $\Delta s_{opt}(k,\alpha)$ is the difference between the optimal headway and the actual headway; $\Delta v_{opt}(k,\alpha)$ is the difference between a driver's optimal velocity and the actual velocity; $s_{buf}(k)$ and $\Delta v_{buf}(k)$ help determine which control law to use (see Section 5.1.2); $\lambda_1$, $\lambda_2$, and $K$ are sensitivity constants; and $\tau$ is a reaction time delay.

Equations (42), (43), and (44) below represent Control Laws 1, 2, and 3 respectively.

$$a(k) = \lambda_2 \Delta v_{opt}(k-\tau,\alpha) \tag{42}$$

$$a(k) = \frac{\lambda_1}{s(k-\tau)} \Delta v(k-\tau) + K \Delta s_{opt}(k-\tau,\alpha) \tag{43}$$

$$a(k) = \frac{\lambda_1}{s(k-\tau)} \Delta v(k-\tau) \tag{44}$$

5.1.2. CONTROL RULES

Figure 8 shows the decision tree used to determine which control law to use at each simulation time step. We outline and explain the decision making here.

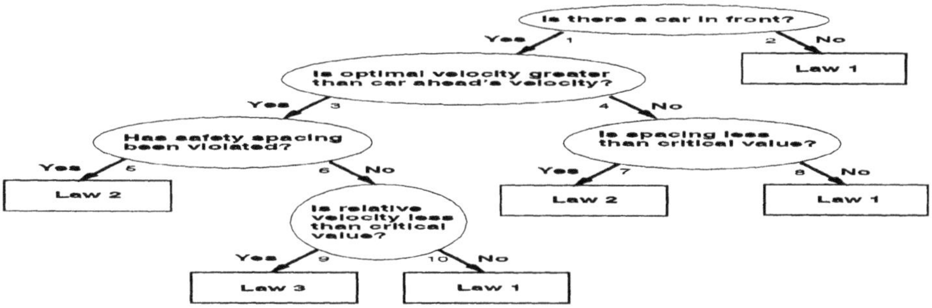


Figure 8: Decision tree for pure longitudinal control.

The first decision to be made is whether there is another vehicle 100 $m$ or less ahead. If there is no such vehicle (branch 2), Control Law 1 is used to track the driver's optimal velocity. An implicit assumption in this rule is that drivers ignore vehicles beyond a range of 100 $m$.

If there is a vehicle within 100 $m$ (branch 1), our model asks whether the driver's optimal velocity is greater than the velocity of the vehicle ahead. If not (branch 4), the driver should track his optimal velocity using Control Law 1; controlling spacing and velocity with respect to the vehicle ahead is not needed since tracking the driver's optimal velocity, which is less than the velocity of the vehicle ahead will guarantee that every spacing and velocity constraint will be meet after a sufficient amount of time. However, before committing to Control Law 1 and disregarding spacing and velocity relative to the vehicle ahead, we ask whether a critical, minimum spacing constraint of 1 $m$ has been violated. If so (branch 7), Control Law 2 must be used instead of Control Law 1 to increase the spacing between the vehicles to a safe value.

On the other hand, if the driver's optimal velocity is greater than the velocity of the vehicle ahead (branch 3), safety spacing must be considered. If safety spacing has been violated (branch 5), Control Law 2 is used to track the velocity of the vehicle ahead, while at the same time maintaining safe headway. Safe spacing is defined as $d_{opt}$ and is considered to have been violated when $d_{opt} < \Delta d_{buf}$. If safety spacing has not been violated (branch 6), the driver should track his optimal velocity using Control Law 1. However, as was done previously, before committing to Control Law 1 and disregarding spacing and velocity relative to the vehicle ahead, we ask whether a critical, minimum velocity constraint has been violated. In particular, the critical minimum relative velocity with respect to the vehicle ahead is defined to have been violated when $\Delta v < \Delta v_{buf}$.

With the pure longitudinal behavior modeled, we now turn our attention to pure lateral behavior. In contrast to longitudinal behavior, lateral behavior is modeled using mostly rules and very little mathematics.

## 5.2. LATERAL BEHAVIOUR

The only type of lateral behavior considered is that for lane changing. We do not model human lane keeping behavior.

The frequency of lane changes is determined by driver aggressiveness. A driver with low aggressiveness changes lane around once every 20 minutes, a driver with average aggressiveness changes lane around once every 10 minutes, and a driver with high aggressiveness changes lane around once every 2 minutes. Using these lane change frequencies, at each time step in the simulation (0.1 seconds) we can model whether a driver attempts to change lanes as a Bernoulli random variable. In particular, we use three Bernoulli random variables with expected values of 0.00008333, 0.0001667, and 0.0008333 to model whether a driver with low, average, and high aggressiveness respectively attempts to change lanes.

In addition to the random lane change behavior, a simple rule for passing is developed. A driver will attempt to change lanes to pass if the vehicle's velocity is less than the its optimal velocity by more than 1 $m/s$. As before, a Bernoulli random variable is used to model passing. The expected value of this Bernoulli random variable is 0.003333.

Changing lancs to cxit is modeled as a Bernoulli random variable with expected value 0.005. Modeling vehicle egress as a Bernoulli random variable has an important implication to our simulation model. We do not assign manual vehicles a path from entrance to egress. We simply force some vehicles to exit using the above random variable.

Figure 9 shows the decision tree used to determine whether a driver should attempt to change lanes. Starting from the top of the tree, we first ask whether the vehicle is on an entrance ramp. If so, the driver should attempt to merge. The likelihood of the driver attempting to merge is a Bernoulli random variable with expected value 1.0 (the E in the boxes in the figure represent the expected value for a Bernoulli random variable representing whether the driver will attempt to change lanes).

We work our way down the tree, asking whether the lane the vehicle is on is an exit, whether the lane the vehicle is next to is an exit, whether the vehicle is traveling less that it's optimal velocity, whether the driver is a high, medium, or low aggressor. If we reach bottom of the tree, the no lane change should be attempted.

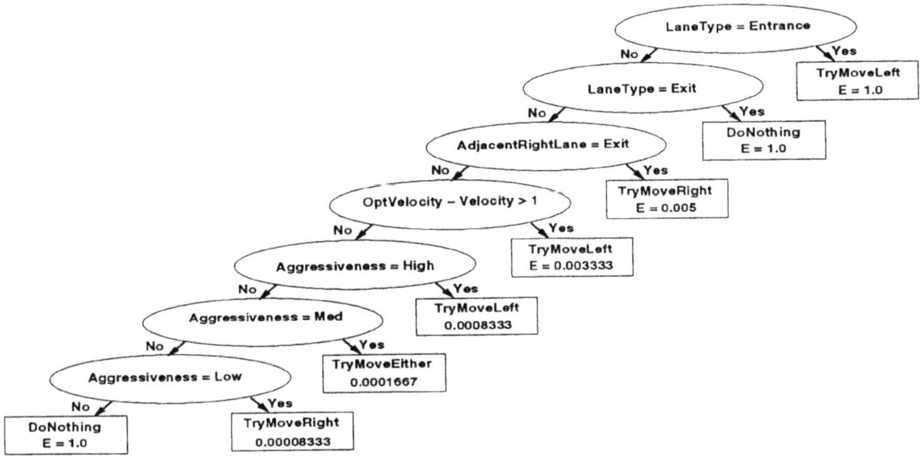


Figure 9: Decision tree to determine whether to change lanes. The E in the boxes represent the expected value for a Bernoulli trial used to determine whether a lane change should occur.

## 6. Simulation Implementation and Performance Analysis of Mixed Traffic Control

Based on the simulation package, SmartPath (see [10, 11]), manual driver and IVDS have been successfully implemented on Silicon Graphics workstations. The program is used to analyze how a mixed highway system would perform in terms of highway capacity, traffic flow, and other performance measures of interest to transportation system planners and drivers.

The simulator uses a circular, that is 1 km in length, has two lanes, and has one entrance and no exits. We attempt to "pump" as many vehicles as possible onto this 1 km stretch of highway. Vehicles appear at the entrance at a rate of 1 vehicle every 4 seconds. The desired inter-vehicle headway spacing of the IVDS-equipped vehicles for the automatic lane is $0.2v + 3$, and for the mixed lane is $v + 10$, where $v$ is the speed of the vehicle controlled. The desired optimal speed is $25m/s$. Inside lane is a mixed lane while outside lane is an automatic lane. We show change-lane simulations running on SmartPath since this is a most complex operation in mixed traffic control.

At time $t = 245.3s$, a vehicle aggressiveness wants it to enter the automatic lane and so it sends a LEFT-LANE-CHANGE light (see Figure 10). Car $B$ driven by the IVDS in the automatic lane finds that it is a closed-most vehicle to Car $A$. It uses *creating mode* to create an appropriate space for $A$. At time $t = 256.9s$, Car $A$ detects the space in automatic lane to be enough for lane change and decides to move over (see Figure 11). Car $B$ uses IVDS technique to follow Car $A$ from the mixed lane (see Figure 12).

To evaluate mixed traffic capacity, the simulator ran $400sec$ with the proportion of the IVDS-equipped vehicles at 30% or 90%. Relationships among observed values of flow, density and speed are shown in Figure 13-18.

These results are based on averages of both lanes. When the proportion of the IVDS-equipped vehicles is less than 30%, the traffic speed is higher. As the density increases it is found that the speed of vehicles decreases. When the proportion of the IVDS-equipped vehicles is more than 30%, the traffic speed is stable from low density to high density. The reason is that the more IVDS-equipped vehicles there are, more stream stability the traffic will have.

Figure 10: *A* attempts to enter automatic lane and sends a want-to-lane-change light and *B* begins to create a space for accommodating Car *A*

Figure 11: *A* tries to enter automatic lane

## 7. Conclusions

In this chapter, we have designed a completely self-contained driving system suitable for mixed traffic. This format allows gradual market penetration of IVDS,

Figure 12: $A$ completes the lane change

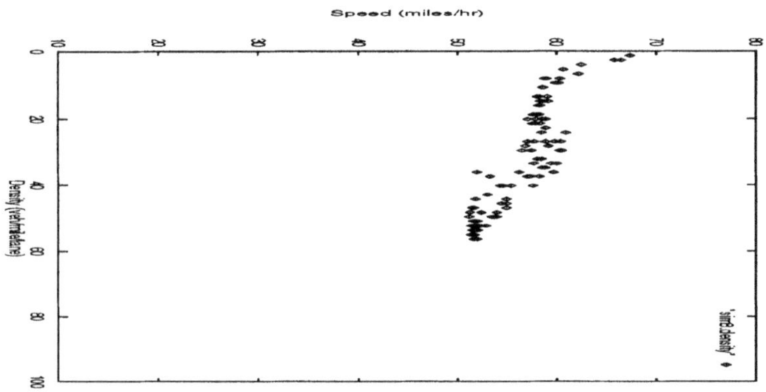


Figure 13: Speed vs. density relationship with the IVDS-equipped vehicles 30%

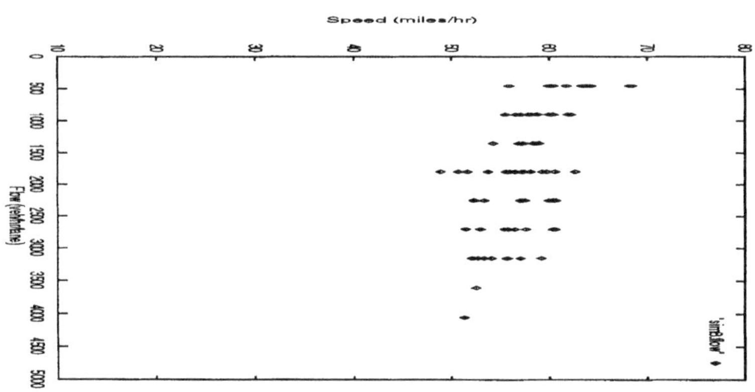


Figure 14: Speed vs. flow relationship with the IVDS-equipped vehicles 30%

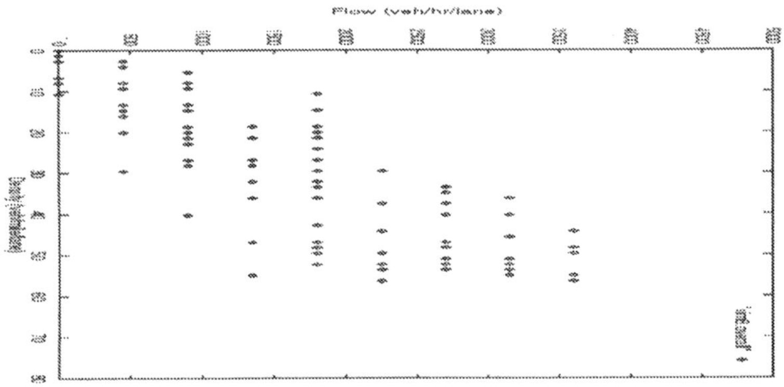

Figure 15: Flow vs. density relationship with the IVDS-equipped vehicles 30%

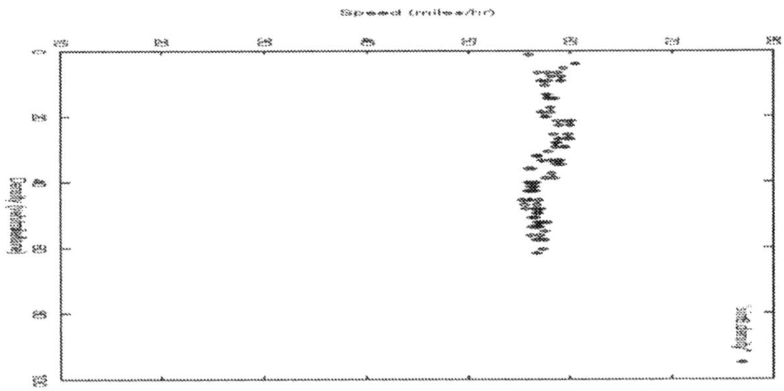

Figure 16: Speed vs. density relationship with the IVDS-equipped vehicles 90%

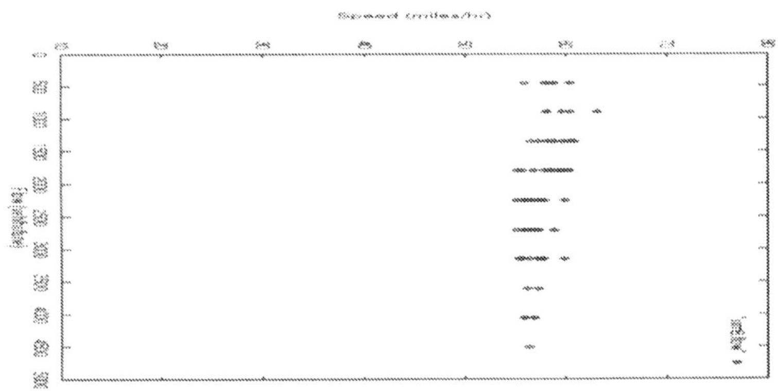

Figure 17: Speed vs. flow relationship with the IVDS-equipped vehicles 90%

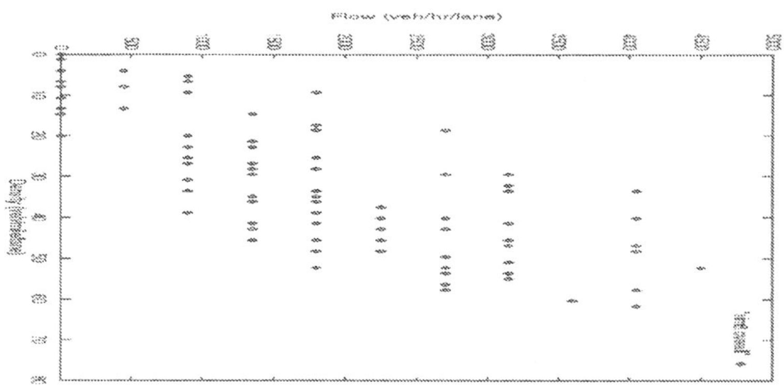

Figure 18: Flow vs. density relationship with the IVDS-equipped vehicles 90%

which makes it attractive from a system implementation standpoint. Mixed traffic simulations with different proportions of automated vehicles have been conducted. It is found that as the number of IVDS-equipped vehicle increase, the traffic capacity and stream stability increases significantly. Taking measured data from current highways, we find that mixed traffic can achieve a peak flow of 4,000 vehicles/hr —two times the current capacity under the condition that the portion of the IVDS-equipped vehicles is more than 30%.

## References


[1] Varaiya,P. and Shladover,S.E.:Sketch of an IVHS systems architecture, PATH Research Rep., UCB-ITS-PRR-91-3

[2] Varaiya, P. Smart cars on smart roads: problems for control, *IEEE Trans.on Automatic Control*, **38**(1993),195-207

[3] Huang, S.N. and Ren,W. :Design of vehicle following control systems with actuator delays, *Int.J.Systems and Science*, **28**(1997),145-151

[4] Ioannou, P.A.; Chien, C.C.: Autonomous intelligent cruise control. *IEEE Transactions on Vehicular Technology*, **42**(4)(1993),657-672.

[5] Chien, C.C.; Ioannou, P.: Automatic vehicle-following, *Proceedings of the 1992 American Control Conference*,Chicago,1992. pp.1748-1752

[6] Ren, W. and Green,D.:Continuous platooning: a new evolutionary and operating concept for automated highway systems, Memorandum No.UCB/ERL M94/24

[7] Swaroop,D.: *String stability of interconnected applications to IVHS*,Ph.D Disseration,1994.

[8] Forbes,J. Huang,T. Kanazawa,K. and Russell,S.: The BATmobile:Towards a bayesian automated Taxi, *Proceedings of Fourteenth International Joint Conference on Artificial Intelligence*, Montreal,Canada,1995.

[9] Niehaus,A.and Stengel,R.E.:Probability-based decision making for automated highway driving, *IEEE Trans.on Vehicular Technology*,**43**(3)(1992),pp.626-6634

[10] Eskafi,F, Khorramabadi,D. and Varaiya,P.: Smartpath: An automated highway system simulator, TECH.REP.PATH Techical Note 94-3

[11] Eskafi,F and Khorramabadi,D.: Smartpath User's Manual. Department of Electrical Engineering and Computer Science and PATH/ITS, UCB-94

[12] Shiekholeslam,S.: *Control of a class of interconnected nonlinear dynamical systems: the platoon problem*, Ph.D.Dissertation,Department of Electrical Engineering and Computer Sciences,University of California at Berkeley, 1991

[13] Bekey,Burnham,G.A G. and Seo,J.:Control theoretic models of human drivers in car following, *Human Factors*, **19**(4)(1977), 399–413.

[14] Gazis, Herman,D.C and Potts,R.B. :Car-following theory of steady state traffic flow, *Operations Research*, **7**(1959),499–505.

[15] Chandler,R.E., Herman,R., and Montroll,E.W.:Traffic dynamics: Studies in car following, *Operations Research*, **6**(1958), 165–184.

[16] Godbole,D. and Lygeros,J.: Longitudinal control of the lead car of a platoon, UCB Rep., TECH MEMO-93-07,1993

[17] Gerdes, J.C.; Hedrick, J.K.: Brake system requirements for platooning on an automated highway. *Proceedings of the 1995 American Control Conference,*1995, pp165-9

# INDEX

International Series on
# MICROPROCESSOR-BASED AND INTELLIGENT SYSTEMS ENGINEERING

*Editor*: Professor S. G. Tzafestas, *National Technical University, Athens, Greece*

1. S.G. Tzafestas (ed.): *Microprocessors in Signal Processing, Measurement and Control.* 1983 ISBN 90-277-1497-5
2. G. Conte and D. Del Corso (eds.): *Multi-Microprocessor Systems for Real-Time Applications.* 1985 ISBN 90-277-2054-1
3. C.J. Georgopoulos: *Interface Fundamentals in Microprocessor-Controlled Systems.* 1985 ISBN 90-277-2127-0
4. N.K. Sinha (ed.): *Microprocessor-Based Control Systems.* 1986 ISBN 90-277-2287-0
5. S.G. Tzafestas and J.K. Pal (eds.): *Real Time Microcomputer Control of Industrial Processes.* 1990 ISBN 0-7923-0779-8
6. S.G. Tzafestas (ed.): *Microprocessors in Robotic and Manufacturing Systems.* 1991 ISBN 0-7923-0780-1
7. N.K. Sinha and G.P. Rao (eds.): *Identification of Continuous-Time Systems.* Methodology and Computer Implementation. 1991 ISBN 0-7923-1336-4
8. G.A. Perdikaris: *Computer Controlled Systems.* Theory and Applications. 1991 ISBN 0-7923-1422-0
9. S.G. Tzafestas (ed.): *Engineering Systems with Intelligence.* Concepts, Tools and Applications. 1991 ISBN 0-7923-1500-6
10. S.G. Tzafestas (ed.): *Robotic Systems.* Advanced Techniques and Applications. 1992 ISBN 0-7923-1749-1
11. S.G. Tzafestas and A.N. Venetsanopoulos (eds.): *Fuzzy Reasoning in Information, Decision and Control Systems.* 1994 ISBN 0-7923-2643-1
12. A.D. Pouliezos and G.S. Stavrakakis: *Real Time Fault Monitoring of Industrial Processes.* 1994 ISBN 0-7923-2737-3
13. S.H. Kim: *Learning and Coordination.* Enhancing Agent Performance through Distributed Decision Making. 1994 ISBN 0-7923-3046-3
14. S.G. Tzafestas and H.B. Verbruggen (eds.): *Artificial Intelligence in Industrial Decision Making, Control and Automation.* 1995 ISBN 0-7923-3320-9
15. Y.-H. Song, A. Johns and R. Aggarwal: *Computational Intelligence Applications to Power Systems.* 1996 ISBM 0-7923-4075-2
16. S.G. Tzafestas (ed.): *Methods and Applications of Intelligent Control.* 1997 ISBN 0-7923-4624-6
17. L.I. Slutski: *Remote Manipulation Systems.* Quality Evaluation and Improvement. 1998 ISBN 0-7932-4822-2
18. S.G. Tzafestas (ed.): *Advances in Intelligent Autonomous Systems.* 1999 ISBN 0-7923-5580-6

KLUWER ACADEMIC PUBLISHERS – DORDRECHT / BOSTON / LONDON

GPSR Compliance
The European Union's (EU) General Product Safety Regulation (GPSR) is a set of rules that requires consumer products to be safe and our obligations to ensure this.

If you have any concerns about our products, you can contact us on

ProductSafety@springernature.com

In case Publisher is established outside the EU, the EU authorized representative is:

Springer Nature Customer Service Center GmbH
Europaplatz 3
69115 Heidelberg, Germany

www.ingramcontent.com/pod-product-compliance
Ingram Content Group UK Ltd.
Pitfield, Milton Keynes, MK11 3LW, UK
UKHW020225290926
15470UKWH00001B/379

* 9 7 8 0 7 9 2 3 5 5 8 0 9 *